European Large Lakes – Ecosystem changes and their ecological and socioeconomic impacts

Developments in Hydrobiology 199

Series editor

K. Martens

European Large Lakes
Ecosystem changes and their ecological and socioeconomic impacts

Edited by

Tiina Nõges[1], Reiner Eckmann[2], Külli Kangur[1], Peeter Nõges[1,3], Anu Reinart[4], Gulnara Roll[5], Heikki Simola[6] & Markku Viljanen[6]

[1] *Centre for Limnology, Institute of Agricultural and Environmental Sciences, Estonian University of Life Sciences, 61101 Rannu, Tartumaa, Estonia*

[2] *University of Konstanz, Limnological Institute, D-78457 Konstanz, Germany*

[3] *European Commission – Joint Research Centre, Institute for Environment and Sustainability Via Enrico Fermi 1, 21020 Ispra (VA), Italy*

[4] *Tartu Observatory, Tõravere, 61602 Tartu County, Estonia*

[5] *Peipsi Centre for Transboundary Cooperation, Aleksandri 9, Tartu, Estonia*

[6] *University of Joensuu, Faculty of Biosciences, Ecological Research Institute, P.O. Box 111, FI-80101 Joensuu, Finland*

Reprinted from Hydrobiologia, Volume 599 (2008)

Springer

Library of Congress Cataloging-in-Publication Data

A C.I.P. Catalogue record for this book is available from the Library of Congress.

ISBN-13: 978-1-4020-8378-5

Published by Springer,
P.O. Box 17, 3300 AA Dordrecht, The Netherlands

Cite this publication as Hydrobiologia vol. 599 (2008).

Cover illustration: Lake Võrtsjärv, Estonia. Photo: Priit Zingel.

Printed on acid-free paper

Printed in the Netherlands

TABLE OF CONTENTS

WATER POLICY AND SOCIOECONOMIC ASPECTS OF LARGE LAKE MANAGEMENT

Hydrobiologia (2008) 599:1–2
DOI 10.1007/s10750-008-9304-5

ELLS 2007

Preface

Large lakes are important because of their size and ecological distinctiveness, as well as their economic and cultural value. Optimal management of them requires a proper understanding of anthropogenic impacts, both on the lake ecosystems, as such and on the services they provide for society. The specific structural and functional properties of large lakes, e.g. morphology, hydrography, biogeochemical cycles, and food-web structure, are all directly related to their size. These vulnerable ecosystems often suffer from accelerated eutrophication, over-fishing, toxic contamination, and invasive species. Large lakes offer socio-economic benefits and could be used in many ways, and are often areas in which economic, cultural and political interests overlap. These multiple uses create potential risks for the health and functioning of the ecosystem. Dissemination of information about the risks caused by human activities is the first step toward encouraging and enabling the community to participate in decision-making about the use and protection of large lakes. Several large lakes in Europe (Lakes Geneva, Constance, Peipsi and Maggiore, for instance) or their catchment areas (those of Lakes Ladoga, Vänern and Saimaa) are shared between two or more countries, which makes international cooperation a prerequisite for their sustainable management.

Guest Editors: T. Nõges, R. Eckmann, K. Kangur, P. Nõges, A. Reinart, G. Roll, H. Simola and M. Viljanen
European Large Lakes—Ecosystem changes and their ecological and socioeconomic impacts

The European Large Lakes Symposium (ELLS) 2006, which took place in Tartu, Estonia, 11–15 September, 2006, focused especially on the ecosystems of European large lakes and their ecological and socio-economic impacts. The ELLS grew out of the International Lake Ladoga Symposia organized in 1993, 1996, 1999, and 2002, which improved our understanding of the structure and functioning not only of Lake Ladoga, but also of other large northern lake ecosystems. The group of problems regarding the present status of large lakes and the directions of change are much the same in all these cases: threats caused by direct human impact and by climate change, protection needs and restoration measures. It has therefore become evident that international exchange of opinions and scientific information from large lake research in Europe is necessary. The ELLS provided a platform for (i) discussing new scientific findings regarding the functioning of large lake ecosystems under the influence of anthropogenic and climatic stressors, (ii) enhancing the communication and exchange of ideas among scientists, water managers and politicians, and (iii) fostering international cooperation in all aspects of investigation and management of both national and transnational European water bodies.

The ELLS was organized by the International Organizing Committee including the following members: Dr. Tiina Nõges (Chair; Estonia), Dr. Markku Viljanen (Vice chair; Finland), M.A. Tuula Toivanen (Secretary; Finland), M.Sc. Ain Järvalt (Estonia),

M.A. Kati Kangur (Estonia), Dr. Külli Kangur (Estonia), Dr. Veljo Kisand (Estonia), Dr. Anu Reinart (Estonia), Dr. Gulnara Roll (Estonia), M.Sc. Lea Tuvikene (Estonia), Dr. Peeter Nõges (European Commission), Prof. Reiner Eckmann (Germany), Prof. Valentina G. Drabkova (Russia), Prof. Vladislav A. Rumyantsev (Russia) and Dr. Niklas Strömbeck (Sweden). The practical arrangements were made by the Centre for Limnology, Estonian University of Life Sciences; University of Joensuu, Finland; and The Peipsi Centre for Transboundary Cooperation, Estonia.

The International Advisory Committee comprised Prof. Dr. Martin Dokulil (Austria), Dr. Glen George (UK), Prof. Erik Jeppesen and Prof. Dr. Sven Erik Jørgensen (Denmark), Prof. Roger Jones and Prof. Jouko Sarvala (Finland), Prof. Ulrich Lemmin and Prof. Dr. Alfred Johny Wüest (Switzerland), Dr. Mohiuddin Munawar (Canada), Prof. Judit Padisák, (Hungary), Dr. Anne Lyche Solheim (Norway), Prof. Dr. Ulrich Sommer (Germany), Dr. Oleg A. Timoshkin (Russia) and Dr. Gesa Weyhenmeyer (Sweden).

ELLS had 170 participants from 20 countries: Austria (3), Belarus (1), Canada (1), China (2), Czech Republic (3), Estonia (56), Finland (33), France (2), Germany (11), Hungary (6), Italy (2), Latvia (3), Lithuania (3), Poland (1), Russian Federation (28), Sweden (4), Switzerland (2), The Netherlands (6), United Kingdom (2), USA (1).

The themes at ELLS were as follows:

1. Climate change and anthropogenic impacts on large lake ecosystems (keynote speaker Dr. Glen D. George).
2. Food web interactions and dynamics (keynote speaker Dr. Alois Herzig).
3. Modeling tools in large lakes research (keynote speaker Dr. Thorsten Blenckner).
4. The Water Framework Directive and large lakes (keynote speaker Dr. Anne Lyche Solheim).
5. Socio-economic aspects of water resource and catchment management (keynote speaker Dr. Ragnar Löfstedt).
6. Policy support systems for the sustainable management of large lakes affected by global changes (keynote speaker Dr. Ramesh Gulati).

The Guest Editors of this Special Issue sincerely thank all organizers and presenters of ELLS, all contributors to the Special Issue and all anonymous reviewers for their invaluable help in improving the scientific quality of the manuscripts. Our special gratitude goes to our secretary Tuula Toivanen for her immense technical assistance and her always patient and friendly attitude, to the staff of Theoretical and Cell Biology Consultancy led by Paul S. Agutter for their highly professional language editing, and to Judit Padisák, Gesa Weyhenmeyer, Jouko Sarvala and Thorsten Blenckner for their help and support during the critical stages of the editing process.

The ELLS organizers and the Guest Editors are greatly indepted to the Estonian Environmental Investment Centre for providing financial support for ELLS organizing and editing of this Special Issue.

Hydrobiologia (2008) 599:3–11
DOI 10.1007/s10750-007-9208-9

ELLS 2007

Nutrients and phytoplankton in Lake Peipsi during two periods that differed in water level and temperature

Marina Haldna · Anu Milius · Reet Laugaste · Külli Kangur

Abstract Data for the vegetation periods (May–November) of 1985–2003 were used to collate the nutrient content and biomass of the most important phytoplankton groups in Lake Peipsi (Estonia). Two periods differing in external nutrient load and water level were compared by analysis of variance. The years 1985–1988 were characterized by the highest loads of nitrogen and phosphorus, high water level and cool summers. The years 2000–2003 were distinguished by low or medium water levels and warm summers. The first period showed statistically significantly higher values of total nitrogen (N_{tot}) and a higher N_{tot}:P_{tot} mass ratio. The second period showed a higher content of total phosphorus (P_{tot}), a higher ratio of dissolved inorganic compounds N to P and higher phytoplankton and cyanobacterial biomasses. Comparison between parts of the lake demonstrated that the differences between the two periods were more evident in the shallower and strongly eutrophic parts, Lake Pihkva and Lake Lämmijärv, than in the largest and deepest part, the moderately eutrophic Lake Peipsi *s.s.* Temperature and water level acted synergistically and evidently influenced phytoplankton via nutrients, promoting internal loading when the water level was low and the temperature high. The effect of water level was stronger in the shallowest part, Lake Pihkva. The difference in P_{tot} content between the southern and northern parts was twofold; the N_{tot}:P_{tot} mass ratio was significantly lower in the southern parts, and phytoplankton biomass (particularly the biomass of cyanobacteria) was significantly higher for Lake Pihkva and Lake Lämmijärv than for Lake Peipsi *s.s.*

Keywords Large and shallow lake · Water level · Nutrients · Phytoplankton · Cyanobacteria

Guest editors: T. Nõges, R. Eckmann, K. Kangur, P. Nõges, A. Reinart, G. Roll, H. Simola & M. Viljanen
European Large Lakes – Ecosystem changes and their ecological and socioeconomic impacts

M. Haldna · A. Milius · R. Laugaste · K. Kangur (✉)
Institute of Agricultural and Environmental Sciences, Estonian University of Life Sciences, Kreutzwaldi 64, 51014 Tartu, Estonia
e-mail: kylli.kangur@emu.ee

M. Haldna
e-mail: marina.haldna@emu.ee

A. Milius
e-mail: anu.milius@emu.ee

R. Laugaste
e-mail: reet.laugaste@emu.ee

Introduction

The water temperature and water level in a lake affect water chemistry (nutrients) and biota (plankton, fish) both directly and indirectly. High temperature promotes resuspension of phosphorus from sediments

(Pettersson et al., 2003), decrease in the N:P ratio and cyanobacterial blooms. Weather-driven changes can exceed or prevent eutrophication process in the lake (Padisak & Koncsos, 2002). Nagid et al. (2001) and Nõges et al. (2003) stressed the increase in internal loading during periods of low water level. In the large, shallow, moderately eutrophic Rybinsk Reservoir, water content is the most significant factor in the control of phytoplankton, and the highest values of chlorophyll *a* (Chl *a*) occur in periods of extremely low water level (Mineeva & Litvinov, 1998). The effect of warm weather on shallow lakes is particularly strong when it coincides with low water level. It is evident that the effect of water level is greater in shallow lakes: the reloading of phosphorus from the sediment when the water is shallow is more intensive because of wind action, as observed in stormy days in Lake Pihkva in August 2003 (personal data). In Reynolds and Petersen (2000), the direct relationship between nutrients and phytoplankton, primarily cyanobacteria, is weak and nutrients are not an issue when the physical requirements of algae (water temperature, insolation, water immobility) are satisfied. Nevertheless, the connection between water-blooming cyanobacteria and phosphorus is well documented.

Long-term investigations of Lake Peipsi (since 1962) have shown that high phytoplankton biomass converges with periods of low water level (Laugaste et al., 2001). During years with maximal nutrient loading, as in the 1980s, the lake was relatively poor in phytoplankton when there were high water levels. There was a more than sevenfold decrease in the application of mineral fertilizers in Estonia between the middle of the 1980s and the end of the 1990s (Leisk & Loigu, 2001). The external load entering Lake Peipsi from Estonian rivers decreased 2.4 times for N (from 14.62 to 6.19 t yr^{-1}) but remained at almost the same level for P (from 199 to 196 t yr^{-1}); total point source loads were reduced by 42% for nitrogen and 21% for phosphorus (Mourad et al., 2006). The concentrations of nitrogen and phosphorus mineral compounds in the River Velikaya on the Russian side decreased from 0.8 mg l^{-1} (averaged over 1983–1991) to 0.37 (averaged over 1997–2001) for nitrogen and from 0.032 to 0.020 for phosphorus (Nõges et al., 2004). Long-term datasets of nutrients and phytoplankton populations in Lake Peipsi are a valuable resource for studying the responses of phytoplankton to changes in water level and water temperature and, in particular, the impact of climate-driven changes via nutrients on phytoplankton biomass and component groups. The aim of this work was to follow the effects of water level and temperature on nutrients and on phytoplankton and its major groups (diatoms, cyanobacteria and cryptophytes) during two periods that differed in water level and temperature.

Study site

Lake Peipsi *s.l.* (3,558 km^2, mean depth 7.1 m), located on the border of Estonia and Russia, is the largest transboundary lake in Europe. It consists of three parts that differ in trophic state: the moderately eutrophic clear-water Lake Peipsi *s.s.* (2,611 km^2, mean depth 8.4 m, maximum depth 12.9 m), the highly eutrophic Lake Pihkva (708 km^2, mean depth 3.8 m, maximum depth 5.3 m), and the narrow Lake Lämmijärv (236 km^2, mean depth 2.5 m, maximum depth 15.3 m) connecting the former two. Lake Pihkva is situated on the Russian border, and material from this lake was only sporadically available. Lake Peipsi is well mixed by the wind; no stratification of temperature, O_2 or hydrochemical parameters occur during the ice-free period. Water level is not regulated; the reference water level is 30 m a.s.l. (200 cm according to the Mustvee hydrometric station). Diatoms dominate in spring and autumn, and also in summer in some years. Besides the large filamentous forms of *Aulacoseira islandica* (O. Müller) Sim., *A. granulata* (Ehr.) Sim. and *Stephanodiscus binderanus* (Kütz.) Krieger, unicellular centric species (genera *Cyclotella, Puncticulata, Stephanodiscus, Cyclostephanos*) and, to a lesser extent, pennates such as *Asterionella formosa* Hass., are also abundant. Cyanobacteria yield maximum biomass in summer and autumn. A summer cyanobacterial bloom occurs every year, even if the weather is cool. The most conspicuous forms are *Gloeotrichia echinulata* (J. S. Smith) P. Richter in summer in Lake Peipsi *s.s.* and *Aphanizomenon flos-aquae* (L.) Ralfs in autumn in the southern parts. The genera *Microcystis* and *Anabaena* are also important, particularly since 2002.

Materials and methods

Water samples for nutrient analysis were collected from April–May to October–November 1985–2005. Since 1992, all stations have been located in the Estonian area of the lake, except for nine on the Russian side, sampled in May 1992, October 2001 and 2002 and August 2003–2005. Depending on the year, the number of sampling sites varied from 5 to 32. The chemical composition of the water (total phosphorus, P_{tot}; orthophosphate ion, PO_4-P; total nitrogen, N_{tot}; ammonium ion, NH_4-N; nitrate ion, NO_3-N; nitrite ion, NO_2-N (the last three joined as DIN); and silicon, Si) was analysed at the Institute of Zoology and Botany during 1985–1992, and at Tartu Environmental Researchers Ltd, Estonia, during 1992–2005. The two laboratories mostly employed identical methods.

Summer (July or August) phytoplankton material covers the period 1985–1991. Monthly samples were collected from May (April) to November in 1997–2005. The methods for collecting samples and treating hydrochemical analyses are described in detail in Möls et al. (1996). Phytoplankton samples were collected and treated by the same person during the two periods; the methods are described in Laugaste et al. (2001). In the present study, the following summer phytoplankton parameters were analysed: Chl *a*, total biomass (FBM) and the biomasses of cyanobacteria (CY), diatoms (BAC), cryptophytes (CRYP), chlorophytes (CHL), dinophytes (DINO) and chrysophytes (CHR).

Water temperature and water level data were obtained from periodical issues of the Hydrometeorological Service and from the Institute of Meteorology and Hydrology of the Estonian Ministry of Environment. To examine the influence of water level and water temperature on nutrients, total phytoplankton and phytoplankton groups, we distinguished two time periods: 1985–1987 as the high water level period and 2001–2003 as the low water level period. The latter period was warmer, especially the summers. Table 1 illustrates the seasonal variation of water temperature and water level in Lake Peipsi for the two study periods. Figure 1 presents the monthly average water levels for the study periods.

Statistical methods

All chemical, physical and plankton variables were log-transformed to improve their statistical properties. We used general linear modelling techniques provided by SAS, Release 8.1 (SAS Institute Inc., 1999), especially the MIXED and GLM procedures. In calculating geometrical means, 95% tolerance limits and differences between the periods or between the parts of Lake Peipsi, we used ANOVA, the factors being period, lake part and the effect of their interaction. To analyse seasonality, we used a large regression model developed by Möls et al. (2004) and Möls (2005). This mathematical model was developed especially for Lake Peipsi. It has 70 parameters including square root of depth (m); latitude; longitude; a six-component beta-presentation for yearly (long-period) dependence transformations of the year number a1–a6, where

Table 1 Mean, minimum (min) and maximum (max) water level and water temperature at Mustvee hydrometric station in Lake Peipsi s.s. during two periods, 1985–1987 and 2001–2003

	1985–1987			2001–2003		
	Mean	Min	Max	Mean	Min	Max
Water level (cm)						
May	246	208	277	208	138	241
July–August	230	195	289	184	151	221
October–November	232	189	297	152	113	181
All months	220	146	297	173	111	247
Water temperature (°C)						
May	9.0	0.2	19.8	12.2	2.3	16.2
July–August	18.2	12.7	23.1	21.5	14.0	27.4
October–November	4.6	0.1	12.1	4.0	0.1	10.9
All months	11.3	0.1	23.1	13.5	0.1	27.4

Reference water level is 30 m a.s.l. (200 cm according to the Mustvee hydrometric station)

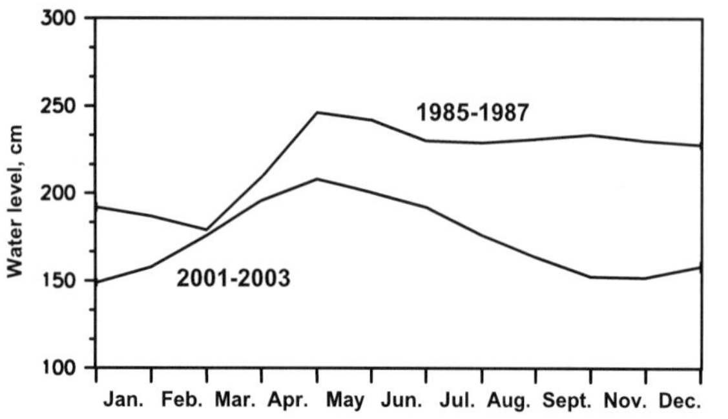

Fig. 1 Mean water level in Lake Peipsi *s.s.* during periods of high (1985–1987) and low (2001–2003) water levels

$a_i = N((\text{year-1920})/10;\ \mu_i,\ 1.4)\ i = 1,...6)$ (density function of the normal distribution), $\mu_i = \{3, 4.5, 6, 7, 8, 8.4\}$; a three-component beta-presentation for seasonal dependence, the ß-functions: $t_{35} = t^3(1-t)^5$, $t_{44} = t^4(1-t)^4$, $t_{53} = t^5(1-t)^3$, where t = number of days in year/365; and a set-of-interactions term. Water variables were predicted for every 20 days in a year for each period. These predictions were used to construct graphs of seasonal changes with the corresponding confidence limits.

Results

Nutrients

Comparison of nutrient concentrations in the two periods revealed that the mean content of P_{tot} was significantly higher during the low water level period (2001–2003), and the increase was markedly higher in the shallower lakes Lämmijärv and Pihkva (Table 2). Unlike P_{tot}, the concentration of N_{tot} was lower in the low water level period, but the differences in N_{tot} in the shallower parts of the lake were not statistically significant. The mass ratio N_{tot}:P_{tot} was higher (21) for the high water level period than for the low water level period (14). The concentrations of the mineral forms of P and N showed an inverse trend. Although the mean value of PO_4-P for the whole lake was slightly higher in the high water level period, this difference was not statistically significant. The mean concentration of DIN was significantly higher in the low water level period, mainly because of the higher content of NO_3-N; for this reason, the mass ratio DIN:PO_4-P was higher as well.

The seasonal trends of P compounds in Lake Peipsi *s.s.* and Lake Lämmijärv in the two periods were similar: the nutrient content was minimum from early spring to June, and thereafter started to increase towards autumn (Fig. 2). The increase in P_{tot} and PO_4-P started earlier when the water level was low and the summer warmer, and higher values of phosphates occurred during late summer and autumn. Seasonal variations in N_{tot} concentration were more pronounced in the high water level period and were synchronous in Lake Peipsi *s.s.* and Lake Lämmijärv (Fig. 2). The maximum N_{tot} content was established in early spring, while its minimum content was recorded in late June and July. In contrast to N_{tot}, no trend was apparent in the seasonality of DIN during the high water level period (Fig. 2), whereas the DIN content during the low water level period was very high in early spring before the onset of the spring phytoplankton bloom.

Phytoplankton

Significant differences between the two periods were noted in total summer biomass and Chl *a* as well as in some phytoplankton groups. The biomass and Chl *a* in the low water level period exceeded the corresponding values in the high water level period two to threefold (Table 2, Fig. 2). The most significant (three to sevenfold) increase occurred in CY biomass. The biomass of BAC increased up to twofold; however, the upper limit of BAC biomass was threefold lower during the low water level period. Among the minor groups, the growth of DINO (3–4 times) was quite marked. A significant decrease (3–4 times) was observed in CRYP biomass; the decrease in CHL in some parts of the lake was less significant. The parts of the lake that differed in trophic state also differed in the changes in phytoplankton groups. The increases in total biomass, Chl *a*, CY and DINO in the low water level period were most pronounced in the southern parts. Water transparency diminished about 1.5 times in Lake Lämmijärv and Lake Pihkva in the second period. A growth in diatom biomass was obvious in the northern part, Peipsi *s.s.*, while there was even some decrease in the southern parts. The biomass of cryptophytes diminished most in Lake Pihkva, about eight times (in Lake Peipsi *s.s.* and Lake Lämmijärv about three times).

Table 2 Nutrients (total phosphorus: P_{tot}; orthophosphate ion: PO_4-P; total nitrogen: N_{tot}; ammonium ion: NH_4-N; nitrate ion: NO_3-N; nitrite ion: NO_2-N (the last three combined as DIN); and silicon: Si), transparency (Secchi), chlorophyll *a* (Chl *a*) and phytoplankton (total biomass (FBM), biomasses of cyanobacteria (CY), diatoms (BAC), cryptophytes (CRYP), chlorophytes (CHL), dinophytes (DINO) and chrysophytes (CHR) in water in Lake Peipsi and its three parts during periods of high (1985–1987; period 1) and low (2001–2003; period 2) water levels

Variable	Unit	Period	N	Mean	95% tolerance limits		Lake Peipsi *s.s.*	Lake Lämmijärv	Lake Pihkva
P_{tot}	mg P m^{-3}	1	471	40*	14	118	32*	59*	64*
		2	129	53	17	161	41	91	138
PO_4P	mg P m^{-3}	1	222	13	2	92	10	19	18*
		2	129	10	2	62	7	18	46
N_{tot}	mg N m^{-3}	1	466	841*	336	2107	756*	1080	1042
		2	129	712	375	1351	642	933	963
DIN	mg N m^{-3}	1	222	45*	8	254	44*	59*	45*
		2	129	110	29	412	106	126	108
N_{tot}:P_{tot}		1	461	21*	6	72	24*	18*	16*
		2	129	14	6	33	15	10	7
DIN:PO_4		1	243	3.3	0.2	45.8	3.5	3.1	2.9
		2	129	11.2	1	104	14.7	7	2.3
Si	mg l^{-1}	1	233	0.8	0.2	3.8	0.7	1	0.9*
		2	129	0.8	0.1	8.1	0.7	0.7	1.9
Secchi	m	1	718	1.6*	0.7	3.7	1.8	1.2*	1.1*
		2	122	1.4	0.5	3.6	1.7	0.9	0.7
Chl *a*	mg m^{-3}	1	488	12.6*	2.5	64.1	11.4*	12.9*	17.2*
		2	132	23.7	7.2	78.1	20.1	35	44.8
FBM	g m^{-3}	1	101	3.4*	0.3	34.5	2.5*	8.5	9.8
		2	77	9.2	2.6	33	7.5	14.3	22.1
CY	g m^{-3}	1	84	0.40*	0.02	10.04	0.24*	1.52*	1.52*
		2	77	3.86	0.25	59.09	2.89	7.55	13.19
BAC	g m^{-3}	1	90	1.07	0.02	60.71	0.60*	4.08	6.37
		2	77	2.06	0.14	30.37	1.58	3.83	5.68
CHL	g m^{-3}	1	90	0.18*	0.02	1.8	0.13	0.54*	0.52
		2	77	0.15	0.03	0.86	0.14	0.17	0.21
CHR	g m^{-3}	1	55	0.004	0	0.076	0.003	0.011	0.011
		2	54	0.025	0.003	0.209	0.025	0.022	0.029
CRYP	g m^{-3}	1	81	0.44*	0.03	5.75	0.45*	0.33*	0.42*
		2	77	0.14	0.02	1.15	0.15	0.12	0.05
DINO	g m^{-3}	1	45	0.05	0	0.76	0.06	0.03	0.05*
		2	46	0.18	0.01	3.56	0.16	0.12	1.91

* The geometrical means of the first and second periods are significantly ($P < 0.05$) different

Chlorophytes maintained their quantity in Lake Peipsi *s.s.* but declined 2.5–3 times in the southern parts of the lake.

Some phytoplankton groups revealed significant Pearson correlations (logarithmic values, $P < 0.0001$) with nutrients and water temperature: CY with total P ($r = 0.52$), DIN:PO_4-P ($r = -0.35$), N_{tot}:P_{tot} ($r = -0.33$), NO_3-N ($r = -0.31$) and water temperature at the time of sampling ($r = 0.47$); BAC with P_{tot} ($r = 0.36$), PO_4-P ($r = 0.33$) and water

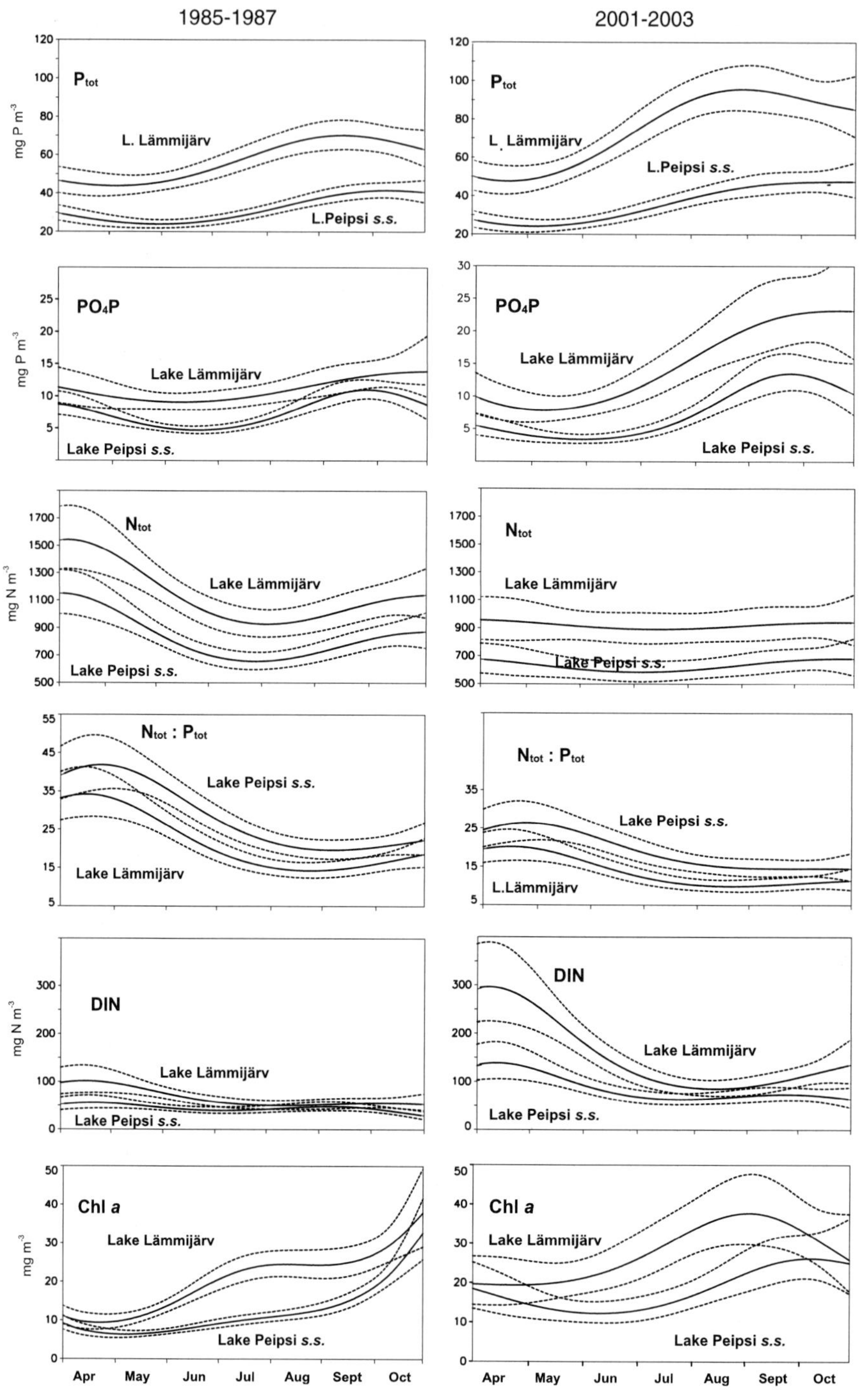

Fig. 2 Seasonal patterns of nutrients (total phosphorus: P_{tot}; orthophosphate ion: PO_4P; total nitrogen: N_{tot}; ammonium ion; NH_4N; nitrate ion: NO_3N; nitrite ion: NO_2-N (the last three combined as DIN); and N_{tot}:P_{tot} mass ratio) and chlorophyll *a* in Lake Peipsi *s.s.* and Lake Lämmijärv during periods of high (1985–1987) and low (2001–2003) water levels. Predicted mean and 95% confidence limits for the true value are estimated by mathematical model described in the part of statistical methods

temperature at the time of sampling ($r = -0.35$). Cryptophytes showed no correlation with nutrients. Weak positive correlations appeared between CRYP and water level, and between CRYP and N_{tot}:P_{tot}, while negative correlations occurred between BAC and N_{tot}:P_{tot}.

Discussion

A substantial difference in both physicochemical parameters and phytoplankton between the high and low water level periods was recorded in our study. On the basis of long-term data for Lake Peipsi, a water condition factor (WCF) combining water temperature and water level was developed by Tõnu Möls (Milius et al., 2005). In the WCF, water level had a more marked effect on nutrients than water temperature. The P_{tot} concentration showed an inverse relationship with water level; it was higher when the water level was lower. The influence of water level on N_{tot} content was positive and somewhat less marked than on P_{tot}. The effects of water level and water temperature on PO_4-P and DIN were also opposite. The results of the present study were entirely consistent with those data.

In an earlier article (Milius et al., 2005), we applied a classical canonical model without considering seasonal effects; in this study, we examined the seasonality of nutrients. The regular increase in P content during summer is characteristic of shallow productive water bodies with intermittent mixing in the summer season (Prepas & Trew, 1983; Carvalho & Kirika, 2003; Søndergaard et al., 2003). The seasonal patterns of N_{tot} and DIN were different. According to Prepas & Trew (1983), inorganic N does not follow the same pattern as N_{tot}. Pettersson et al. (2003) compared the seasonality of nutrients and chlorophyll in Lake Erken in cold and warm periods, and found elevated phosphate, ammonium and Chl *a* levels in the warm period during August and in autumn. The authors explain these results by the increasing activity of bacteria, which increases the uptake of oxygen and the release of phosphate and ammonium from sediment into the water. Comparison of our data with their graphs shows that the dynamics of PO_4-P and Chl *a* fit the data presented by Pettersson et al. (2003) closely, while the earlier increase in PO_4-P is related to the absence of permanent stratification in the lake. As a result of water mixing, the PO_4-P ions released from fresh sediments are transported to the photic zone, which renders the continuation of photosynthesis possible. In Lake Peipsi, the higher values of DIN during the low water level period were due to the higher content of nitrates, not ammonium, evidently resulting from more intensive bacterial nitrification. An increase in NO_3-N is characteristic of all parts of the lake.

In terms of mean values, P_{tot} has increased in recent years, especially in Lake Lämmijärv and Lake Pihkva (Kangur et al., 2003). The polarity of Lake Peipsi has increased more with respect of P_{tot} than to N_{tot} (Kangur & Möls, this issue). In shallow lakes, summer P concentrations are largely controlled by internal processes, and P release from sediment into lake water depends on the sediment surface:water column ratio, which means that it is more intensive in large and shallow lakes (Søndergaard et al., 2003). This is confirmed by the increase in the total and mineral phosphorus values; also, the N:P ratio decreased much more in the shallower Lake Pihkva than in Lake Peipsi *s.s.* (Table 2). In the shallower part, phosphorus reloading from sediment when the water level is low is also more intensive owing to wind action, as observed in Lake Pihkva on stormy days in August 2003.

The concentrations of the mineral forms of nitrogen and phosphorus exceeded the values that limit the growth of phytoplankton groups. According to Wilander & Persson (2001), N-deficiency appears at a DIN concentration of 30 mg N m^{-3} or lower. Only in Lake Peipsi *s.s.* did the lowest values temporarily approach this limit (described by Gammeter & Zimmermann, 2000; Dokulil & Treubner, 2000). The ratio DIN:PO_4-P should be more directly related to phytoplankton than the ratio of total N to total P. Although ammonia and nitrate-N are associated in opposite ways with water level and temperature, the correlations between DIN and phytoplankton were stronger than those between N_{tot} and phytoplankton. The ratio of the mineral forms of N to P was higher during the low water level period, and there was an increasing dominance of cyanobacteria that are unable to fix N_2 (species of *Microcystis*). In most articles, phosphorus and the N:P ratio are regarded as crucial, while weaker correlations have been found with total nitrogen and its mineral forms. Similarly, in our study, correlations with nitrogen were weak or absent. In general, CRYP and CHL prefer a higher nitrogen content and N:P ratio (Planas, 1991; Wilk-Woźniak & Ligęza, 2003); this may explain the decrease in biomass of these groups during the low water level period. The low abundance of CY in the 1980s was predictably related to the high N:P ratio in this period (Nõges et al., 2004). Our earlier results showed that the high ratio was clearly caused by the high water level, besides the high external load (Nõges et al., 2003). On the other hand, the low N:P ratio values in the second period were not only due to the reduced external load but also to the lower water level in combination with warm summers.

As a rule, water quality deteriorates during the warm period (Pettersson et al., 2003; Søndergaard et al, 2003). One should keep in mind the different seasonal dynamics of water temperature in different years: years with similar mean water temperatures may have very different seasonal temperatures, which cause, e.g. the domination of diatoms in cool summers and cyanobacteria in warm autumns. Carvalho and Kirika (2003) found no relationship between phytoplankton Chl *a* content and annual mean water temperature. On the other hand, spring water temperature has a strong effect on summer chemical conditions (George et al., 2000), and phytoplankton biomass in summer depends on the total P content in spring (Krzywosz, 1999). However, correlations between water temperature in spring and nutrients in summer were very weak in Lake Peipsi ($r = 0.2$); as for the phytoplankton groups, only a weak positive effect on CY was revealed ($r = 0.26$, $P < 0.0001$). Water temperature at the sampling time showed significant positive correlations with the biomass of CY and CRYP, and a negative correlation with diatoms.

Dominance of diatoms is more associated with silica. However, we found no correlation with silicon in Lake Peipsi, where the mean silicon content exceeded the limiting value of 0.5 mg l^{-1} reported in the literature (Wetzel, 2001). The lower silicon:nitrogen mass ratio for the southern parts was obviously integral to the decreasing diatom biomass (Table 2). The increase in dinoflagellate biomass in the low water level period is possibly also related to the higher P content. A parallel dynamics of DINO and CY in lakes in Great Britain was stressed by Fogg (1965).

On the basis of the water condition factor (WCF), CY were most strongly affected by weather, while FBM, BAC and CRYP were less affected (Milius et al., 2005). Water temperature at the sampling time was the most important factor for cyanobacteria and diatoms (opposite for CY and BAC), and water level for CRYP. It was obvious that cyanobacteria and the N:P ratio followed opposite courses. Water level in the previous year displayed strong negative correlations with the mineral forms of N and P. When the parts of the lake were compared, the correlations between the WCF and corresponding phytoplankton values were evidently stronger in Lake Pihkva and Lake Lämmijärv than in Lake Peipsi *s.s.*

Conclusions

Data from two periods with different external nutrient load and water level in Lake Peipsi demonstrated the synergistic effect of water level and water temperature on nutrients and, via nutrients, on phytoplankton. In the period of high external N and P load, high water levels and cool summers (1985–1987), there were higher values of N_{tot} and a higher $N_{tot}:P_{tot}$ mass ratio. The period with lower water level and warmer summers (2001–2003) was characterized by higher P_{tot} content, lower $N_{tot}:P_{tot}$ mass ratio and higher DIN:PO_4-P mass ratio, and by higher Chl *a* content, phytoplankton and cyanobacterial biomass and lower water transparency. An increasing dominance of the cyanobacteria that cannot fix N_2 (species of *Microcystis*) in the low water period may be explained by the higher ratio of the mineral forms of N to P in this period. The effect of water level and of the mechanical influence of the wind and waves was stronger in the shallowest part, Lake Pihkva. Our studies confirmed that lower water levels and warmer summers resulted in a deterioration of water quality in spite of the decreasing external load.

Acknowledgements The research was supported by the Estonian target financed project SF 0362483s03 and the Estonian Science Foundation (grants 6008, 6820). Data from the Estonian State monitoring programme were used in this study. We are indebted to Mrs. Ester Jaigma for revising the English text of this article. The contribution of the anonymous referees is greatly appreciated.

References

Carvalho, L. & A. Kirika, 2003. Changes in shallow lake functioning: response to climate change and nutrient reduction. Hydrobiologia 506–509: 789–796.

Dokulil, M. T. & K. Treubner, 2000. Cyanobacterial dominance in lakes. Hydrobiologia 438: 1–12.

Fogg, G. E., 1965. Algal cultures and phytoplankton ecology. The University of Wisconsin Press, Madison and Milwaukee.

Gammeter, S. & U. Zimmermann, 2000. Changes in phytoplankton productivity and composition during reoligotrophication in two Swiss lakes. Verhandlungen der internationale Vereinigung für Limnologie 27: 2190–2193.

George, D. G., J. F. Talling & E. Rigg, 2000. Factors influencing the temporal coherence of five lakes in the English Lake District. Freshwater Biology 43: 449–461.

Kangur, K., T. Möls, A. Milius & R. Laugaste, 2003. Phytoplankton response to changed nutrient level in Lake Peipsi (Estonia) in 1992–2001. Hydrobiologia 506–509: 265–272.

Kangur, K. & T. Möls. Changes in spatial distribution of phosphorus and nitrogen in large north-temperate lowland Lake Peipsi (Estonia/Russia). Hydrobiologia (this issue). doi:10.1007/s10750-007-9204-0

Krzywosz, W., 1999. Relationships between trophic indicators and phytoplankton composition, biomass and photosynthetic activity in small eutrophic lakes. Polish Journal of Ecology 47: 15–27.

Laugaste, R., P. Nõges, T. Nõges, V. V. Yastremskij, A. Milius & I. Ott, 2001. Algae. In Pihu, E. & J. Haberman (eds), Lake Peipsi. Flora and fauna. Sulemees Publishers, Tartu: 31–49.

Leisk, Ü. & E. Loigu, 2001. Nutrients. In Nõges, T. (ed.), Lake Peipsi. Meteorology. Hydrology. Hydrochemistry. Sulemees Publishers, Tartu: 79–82.

Milius, A., R. Laugaste, T. Möls, M. Haldna & K. Kangur, 2005. Weather conditions and water level as factors determining phytoplankton biomass and nutrient content in Lake Peipsi. Proceedings of the Estonian Academy of Sciences. Biology Ecology 54: 5–17.

Mineeva, N. M. & A. S. Litvinov, 1998. Long-term variation of chlorophyll content in Rybinsk reservoir (Russia) in relation to its hydrological regime. NATO ASI Series 42: 159–183.

Möls, T., 2005. Linear statistical methods for Estonian freshwater waterbodies. Estonian Naturalists' Society, Tartu.

Möls, T., H. Starast, A. Milius & A. Lindpere, 1996. The hydrochemical state of Lake Peipsi–Pihkva. Hydrobiologia 338: 37–47.

Möls, T., K. Kangur, M. Haldna, A. Milius, J. Haberman, R. Laugaste & M. Möls, 2004. The synthetic hydrochemical and hydrobiological database (SD) for Lake Peipsi release 1.0). Tartu (CD).

Mourad, D. S., M. van der Perk & K. Piirimäe, 2006. Changes in nutrient emissions, fluxes and retention in north-eastern European lowland drainage basin. Environment Monitoring and Assessment 120: 415–448.

Nagid, E. J., D. E. Canfield & M. V. Hoyer, 2001. Wind-induced increases in trophic state characteristics of a large (27 km^2), shallow (1.5 m mean depth) Florida lake. Hydrobiologia 455: 97–110.

Nõges, T., P. Nõges & R. Laugaste, 2003. Water level as the mediator between climate change and phytoplankton composition in a large shallow temperate lake. Hydrobiologia, 506–509: 257–263.

Nõges, T., I. Tõnno, R. Laugaste, E. Loigu & B. Skakalski, 2004. The impact of changes in nutrient loading on cyanobacterial dominance in Lake Peipsi (Estonia/Russia). Archiv Für Hydrobiologie 160: 261–279.

Padisak, J. & I. Koncsos, 2002. Trend and noise: long-term changes of phytoplankton in the Keszthely Basin of Lake Balaton, Hungary. Verhandlungen der internationalen Vereinigung für Limnologie 28: 194–203.

Pettersson, K., K. Grust, G. Weyhenmeyer & T. Blenckner, 2003. Seasonality of chlorophyll and nutrients in Lake Erken – effects of weather conditions. Hydrobiologia: 506–509, 75–81.

Planas, D., 1991. Factors controlling phytoplankton community structure in an alkaline versus softwater lake. Oecologia aquatica 10: 95–111.

Prepas, E. E. & D. O. Trew, 1983. Evaluation of the phosphorus–chloropyll relationship for lakes of the Precambrian Shield in Western Canada. Canadian Journal of Fisheries and Aquatic Sciences 40: 27–35.

Reynolds, C. S. & A. C. Petersen, 2000. The distribution of planktonic Cyanobacteria in Irish lakes in relation to their trophic states. Hydrobiologia 424: 91–99.

SAS Institute Inc., 1999. SAS OnlineDoc, version 8. SAS Institute Inc., Cary.

Søndergaard, M., J. P. Jensen & E. Jeppesen, 2003. Role of sediment and internal loading of phosphorus in shallow lakes. Hydrobiologia 506–509: 135–145.

Wetzel, R. G., 2001. Limnology. lake and river ecosystems, 3rd edn. Academic Press, USA.

Wilander, A. & G. Persson, 2001. Recovery from eutrophication: experiences of reduced phosphorus input to the four largest lakes of Sweden. Ambio 30: 475–485.

Wilk-Woźniak; E. & S. Ligęza, 2003. Phytoplankton–nutrient relationships during the early spring and the late autumn in shallow and polluted reservoir. Oceanological and Hydrobiological Studies 32: 75–87.

Hydrobiologia (2008) 599:13–21
DOI 10.1007/s10750-007-9205-z

ELLS 2007

Pollen, diatom and plant macrofossil assemblages indicate a low water level phase of Lake Peipsi at the beginning of the Holocene

Tiit Hang · Volli Kalm · Kersti Kihno · Martynas Milkevičius

Abstract During the Fennoscandian ice recession from the eastern Baltic area, the water level in the Lake Peipsi basin was decreasing and reached a minimum at the end of the Younger Dryas Chronozone. The low lake level episode is represented in the basal deposits by a ca. 0.8 m thick bed of coarse detritus gyttja dated to 9.6–9.1 ka ^{14}C BP. The gyttja lies at an elevation of 20–21 m a.s.l., i.e. about 9 m below the present lake level (30 m a.s.l.), and is buried under a 1.5–2 m bed of fine to medium-grained sand. The overall pollen data indicate a shallow-water littoral environment during the gyttja accumulation in the area studied. The diatom flora in the gyttja is dominated by shallow-water benthic and epiphytic taxa, indicating eutrophic littoral conditions at the time of gyttja accumulation. In the uppermost portion of the gyttja sequence the pollen and diatom successions indicate a short period of land emergence when swampy conditions prevailed in the surroundings.

Keywords Diatoms · Pollen · Plant macrofossils · Lake Peipsi · Buried organic beds · Lake level change · Early Holocene

Introduction

On the basis of palynological (Sarv & Ilves, 1975; Hang et al., 1995, 2001) and geomorphological (Hang et al., 1964; Raukas & Rähni, 1969; Hang & Miidel, 1999) data, a low water phase in the early Holocene development of Lake Peipsi (Fig. 1) has been inferred. The early palynological evidence concerned the surrounding wetlands (Sarv & Ilves, 1975) and suggested that the lowest water level occurred at the beginning of the Preboreal Chronozone (*sensu stricto*, Mangerud et al., 1974). More recent research (Hang et al., 2001) has shown that the lake level may already have been 10 m below the current water table (30 m a.s.l.) at the end of the Younger Dryas Chronozone. Recently we discovered a layer of coarse detritus gyttja in the south-western part of the basin of Lake Peipsi proper (Fig. 1), located 9.2–8.2 m below the current mean water level and buried under a sand layer. Pollen, diatom and plant macrofossil analyses of the gyttja layer reveal an extremely low water level period at the time of the gyttja accumulation. The new biostratigraphic data

Guest editors: T. Nõges, R. Eckmann, K. Kangur, P. Nõges, A. Reinart, G. Roll, H. Simola & M. Viljanen
European Large Lakes—Ecosystem changes and their ecological and socioeconomic impacts

T. Hang (✉) · V. Kalm · M. Milkevičius
Institute of Geology, University of Tartu, Vanemuise 46, Tartu 51014, Estonia
e-mail: Tiit.Hang@ut.ee

K. Kihno
Department of History, University of Tartu, Lossi 3, Tartu 51003, Estonia

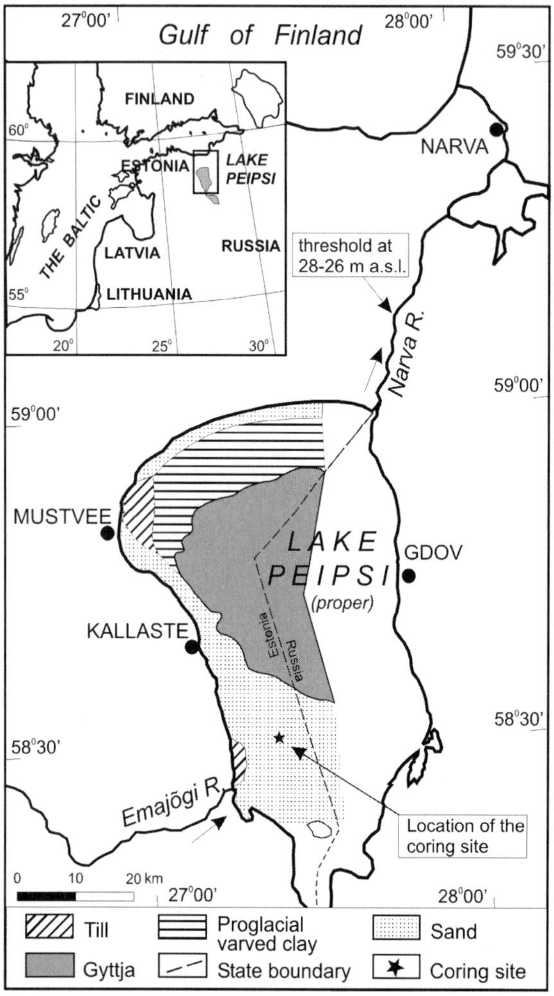

Fig. 1 Location of the sediment sequences investigated in Lake Peipsi, eastern Estonia. Basal deposits according to Hang et al. (2001)

and a synthesis of earlier research enable us to revise the previous environmental reconstructions of Lake Peipsi at its lowest level at the beginning of the Holocene.

Lake Peipsi (30 m a.s.l.) is a rather shallow lake with a mean water depth of about 8 m (max 15 m). It occupies a 50–60 m deep glacially eroded bedrock depression (Miidel et al., 2001). The sequence of lacustrine sediments in the lake depression begins with annually laminated or homogenous glaciolacustrine clay up to 10 m thick. The clay is covered by a calcareous gyttja layer, the thickness of which varies from about 2 cm to 1.9 m, followed by 5 m thick pure gyttja layer. In the southern part of Lake Peipsi proper (Fig. 1) the lake bottom is at present covered by a ca. 2 m thick fine to medium-grained lacustrine sand (Hang et al., 2001). This sand covers the coarse detritus gyttja layer discussed in the current article, which according to AMS ^{14}C dates accumulated around 9.6–9.1 ka ^{14}C BP.

In the central part of Lake Peipsi proper the transition from glaciolacustrine clays to Holocene lake sediments is distinct and in places marked by sand and silt rich in organic remains: the freshwater bivalves *Pisidium amnicum* (O.F. Müller 1774) and two species of snails, *Valvata profunda* (Clessin 1887) and *Valvata depressa* (Pfeiffer 1828), which point to a rather shallow water upper littoral environment (Hang et al., 2001) at the time of transition from glaciolacustrine to Holocene lake sedimentation. The ostracod fauna in the overlaying lake marl sequence, particularly *Ilyocypris bradyi* and *Herpetocypris reptans* (Niinemets, 1999), indicate a water depth of around 4 m during the period (ca. 1,000 yrs) of lake marl accumulation. At the beginning of the succeeding gyttja deposition the water level of Lake Peipsi proper was still so low that it was isolated from the body of water in the southern part of the lake depression (Davydova & Kimmel, 1991; Hang & Miidel, 1999).

Materials and methods

The sites of buried coarse detritus gyttja (58°30′06″ N; 27°23′21″ E; Fig. 1) reported in Hang et al. (2001) were revisited and five parallel sediment sequences were obtained by coring through the lake ice at a water depth of 7.30 m. A Russian type peat corer with a 1 m long and 5 cm diameter chamber was used. The samples were wrapped in plastic film and placed in suitable lengths of a U-shaped PVC trough for transport and storage.

Colour determination of the sediment followed the Munsell soil colour chart (Munsell Color Company, 1998) and the pH of fresh water-saturated sediments was measured with an Evikon pH meter E6121.

The organic content as loss-on-ignition (LOI) was estimated in 50 continuous 2 cm sub-samples from the sediment sequence 8.20–9.20 m. The LOI was estimated from dried samples by incineration at 500°C for 2 h.

Pollen samples were prepared following the standard technique (Berglund & Ralska-Jasiewiczowa, 1986) with additional flotation treatment of highly minerogenic samples with a heavy liquid (CdJ_2 and KJ solution with specific gravity 2.2 g cm^{-3}). Normally, 500 land pollen grains were counted per sample and aquatics, spores and coenobia of the green alga *Pediastrum* were identified. The percentage pollen diagram was compiled using *Tilia*-*Tilia**Graph software (Grimm, 1992).

Samples for macrofossil analyses were enriched on a 0.25 mm mesh sieve by washing with tap water. Identification was performed by Ms. Sirje Hiie using a Nikon SMZ800 stereomicroscope at 10–63× magnification, employing the keys of Katz et al. (1965, 1977) and Schoch et al. (1988) and the reference collection of seeds and fruits in the Laboratory of Geoarchaeology and Ancient Technology at the Institute of History of the Tallinn University. A Jenaval microscope with 400× magnification was used to identify plant tissues.

Sediment sub-sampling and slide preparation for diatom analyses followed the standard technique outlined by Battarbee (1986). Diatoms were counted using a Zeiss III RS microscope with a 100× Plan-apo phase contrast objective and 10× eye-pieces. At least 500 diatom valves were counted per sample. Broken valves were counted as a unit if at least two-thirds had remained. Half valves were counted as a half but smaller pieces were not counted (Miettinen, 2002). Taxonomy and grouping of diatoms by biotype and pH and salinity preferences followed Krammer & Lange-Bertalot (1986, 1988, 1991a, b).

Results

Sediment lithostratigraphy

The late glacial/Holocene sediment sequence analysed consists of four lithostratigraphic layers. Late Weichselian till at the bottom of the lake depression is overlaid by medium-grained sand with a low organic matter content (Fig. 2, depth 9.20–9.06 m). The sand is followed by a coarse detritus gyttja layer (9.06–8.33 m) in which the organic content increases from 5–10% in the lower part (9.06–8.90 m) to a maximum of 25–40% in the upper part (8.65–8.40 m). The topmost 7 cm (8.40–8.33 m) of the gyttja interval is characterised by a rapid decrease in organic content from 35% to 5%. This change marks the transition from gyttja to the uppermost sediment layer, the fine-grained lacustrine sand at a depth of 8.33–7.30 m (Fig. 2).

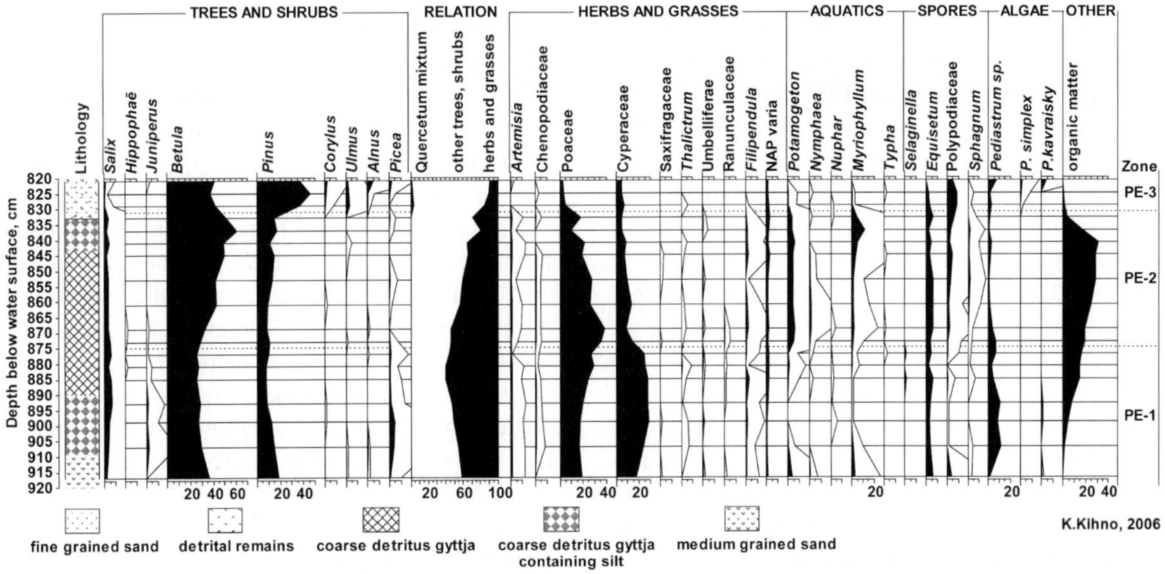

Fig. 2 Pollen percentage diagram with indication of the local pollen assemblage zones (LPAZ: PE-1–PE-3), relative proportions of the main terrestrial groups, lithostratigraphy and LOI of sediments. The black areas on the diagram show the actual pollen in percentages, while the white areas show the percentages multiplied 10-fold

The colour of the gyttja layer changes from dark reddish brown (Munsell colour designation: 5YR/3-2) in the lower part (9.0–8.6 m) into greenish black (Gley1/2.5–1) in the upper portion (8.6–8.43 m), and back to reddish brown (5YR/2.5–1) in the topmost 10 cm. The colour change is apparently due to variable oxidation of Fe-compounds in the sediments, reflecting predominantly ferrous (Fe^{2+}), i.e. more reduced, compounds in the middle part of the gyttja layer. The pH of the gyttja interval varies between 7.18 and 7.37 with the lowest value in the upper part (at 8.5 m), where the organic content reaches its maximum. The gyttja layer contains a visible number of plant macrofossils. The first AMS radiocarbon dates of these macrofossils show that the gyttja deposition took place between 9.6 and 9.1 ka ^{14}C BP.

Pollen data

The percentage pollen diagram was plotted, taking the sum of arboreal (AP) and non-arboreal (NAP) pollen as 100% (Fig. 2). Three local pollen assemblage zones (LPAZ: PE-1, PE-2, PE-3) were established from the most characteristic changes in pollen composition (Fig. 2).

PE-1 (9.20–8.75 m): in this LPAZ the dominant (26–28%) arboreal pollen taxon on average is *Betula*, while *Pinus* has low relative values (8–11%). The only exception to that rule is the lowermost sample from the massive sand overlying the till, where these taxa have values of 36% and 19%, respectively. The *Picea, Ulmus, Corylus* and *Alnus* pollen grains in the samples are believed to represent redeposited material. *Salix* pollen is present regularly throughout the whole diagram at values of 1–7%. The light-demanding shrubs *Juniperus* and *Hippophae* also occur in this LPAZ. The sum of grass and herb pollen, mainly attributable to Cyperaceae (24–28%) and Poaceae (17–27%), accounts for up to 50–60% of the total pollen. In addition, *Artemisia*, Chenopodiaceae, *Filipendula* and *Thalictrum* are constantly present in PE-1 and the succeeding LPAZ, PE-2 (Fig. 2). Aquatics (*Potamogeton, Nymphaea, Nuphar, Myriophyllum*) are present in this LPAZ and *Pediastrum* coenobia are abundant here compared to zones PE-2 and PE-3. Also, spores of *Lycopodium* (not shown in Fig. 2), *Equisetum*, Polypodiaceae, *Sphagnum* and *Selaginella* were recorded in PE-1.

PE-2 (8.75–8.30 m): in this LPAZ, *Betula* pollen dominates over *Pinus*, gaining its maximum values (60%) near the upper boundary of the zone. *Ulmus, Alnus, Picea* and *Corylus* are present as scattered finds. Pollen of *Juniperus* is present (0.2%) only in the lower part of the zone. At the lower boundary of the zone, Poaceae pollen accounts for up to 38% of the total pollen and decreases upwards to 15–20%. A similar upwards decrease (from 25% to 8%) was recorded in the abundance of Cyperaceae pollen. Comparatively high values (10–15%) of aquatic pollen correspond to the level of the lowest numbers of *Pediastrum*. In the upper part of the zone, at 8.37 m, there is a significant increase (12%) in *Myriophyllum*.

PE-3 (8.30–8.20 m): the uppermost pollen zone, PE-3, in the sequence is established on the basis of the samples in which the concentration of pollen from broad-leaved trees (Quercetum mixtum) and *Corylus* is considerably higher than in PE-1 and PE-2 (3.3% and 2.2%, respectively). In this LPAZ the amount of *Pinus* pollen rises rapidly to 46% and the proportion of *Alnus* reaches 5%. After its peak in LPAZ PE-2, the amount of *Betula* pollen decreases to 37%. The proportion of herb pollen is at its lowest value (9.5%) in the whole diagram here, while spores of Polypodiaceae are more frequent (8%) than in PE-2. Coenobia of *Pediastrum simplex* and *P. kavraisky* were identified in this zone. At the boundary between PE-2 and PE-3 there is a marked decrease in the quantity of aquatic pollen.

Plant macrofossils

Macroscopic remains of 15 plant taxa were recorded from the coarse detritus gyttja sequence. The highest concentration of macrofossils was in the lower part of the section studied (8.48–8.74 m). Submerged aquatics (eight taxa) and littoral helophytes and wetland plants (five taxa) are characteristic of the assemblage. Table 1 gives a detailed overview of the distribution and grouping of the plant macrofossils identified.

Diatom assemblages

About 67 diatom taxa were identified from the sediment sequence analysed. Genera with the greatest

Table 1 Plant macrofossils recorded from the coarse detritus gyttja interval of the sediment core investigated from Lake Peipsi

Groups	Taxa	Remains	Depth (cm)												Sum
			817–819	822–824	826–829	829–831	833–835	848–851	851–853	860–863	863–865	871–874	880–882	888–890	
Aquatic	*Nuphar lutea* (L.) Sm.	Seed				1					1				2
	Potamogeton natans L.	Fruit								5					5
	Potamogeton sp.	Fruit									2				2
	Potamogeton sp.	Stem f				1	1	1				1			4
	Chara sp.	Oogone						3		5		2			10
	Myriophyllum sp.	Seed			3										3
	Myriophyllum sp.	Leaf f						1		1		1	1		4
	Ranunculus subg Batrachium	Fruit						2		1					3
	Alisma plantago-aquatica L.	Seed											1		1
	Nymphaea alba L.	Seed						1							1
	Sum of remains				3	2	1	8		12	3	4	2		35
	Sum of taxa				1	2	1	5		4	2	3	2		
Plants of wetlands and shores	*Eleocharis* sp.	Fruit								1					1
	Phragmites australis (Cav) Trin.ex Steud.	Stem f							*						
	Menyanthes trifoliata L.	Seed		1											1
	Schoenoplectus lacustris (L.) Palla	Fruit						5		14		6			25
	Carex spp.	Fruit								2					2
	Sum of remains			1				5		17		6			29
	Sum of taxa			1				1	1	3		1			
Woodland and scrubland	*Alnus* sp.	Fruit								1					1
	Betula pubescens Ehrh.	cs/fruit								1		1			2
	Sum of remains									2		1			3
	Sum of taxa									2		1			
	Mosses and other plant remains					*				*		*	*	*	
	Fragments of tree bark		*	*										*	
	Stems and leaves						*		*						
Total sum of remains				1	3	2	1	13		31	3	11	2		67
Total sum of taxa				1	1	2	1	6	1	9	2	5	2		

cs, catkin scale; f, fragment; *, species not identified

Location of the site indicated in Fig. 1

number of taxa were: *Navicula* (13), *Epithemia* (6), *Cymbella* (5) and *Nitzschia* (5). The most frequent diatoms were *Navicula scutelloides* W. Smith (reaching a share of 71% at 8.27 m), *Ellerbeckia arenaria* (Moore) Crawford (32% at 8.20 m), *Gyrosigma attenuatum* (Kützing) Rabenhorst (24% at 8.53 m), *Epithemia adnata* (Kützing) Brébisson (15% at 8.39 m) and *Synedra ulna* (Nitzsch) Ehrenberg (11% at 8.60 m). The dominant diatom species are presented in Fig. 3. Benthic and epiphytic species dominate throughout the sediment sequence. Planktonic species were identified in low quantities in the lower part of sequence at 9.00 m and 8.85 m (reaching a maximum share of 5%). The diatom flora is dominated by freshwater forms. Alkaliphilous diatoms that thrive in high pH waters dominate, reaching up to 82% at 8.25 m. On the basis of the diatom composition and succession in the sediment sequence, three local diatom zones (LDZ) were distinguished (Fig. 3).

LDZ-1 (9.00–8.37 m): in this zone, shallow water benthic and epiphytic species dominate, notably the benthic *Ellerbeckia arenaria* (up to 30%) in the lower part (9.00–8.65 m), while the benthic *Gyrosigma attenuatum* and epiphytic *Epithemia adnata* are more abundant (24% and 15%, respectively) in the upper part (8.65–8.37 m). The epiphytic *Synedra ulna* and the benthic *Navicula oblonga* Kützing are present throughout LDZ-1 in amounts between 5% and 12%. Greater quantities of the benthic/pseudoplanktonic *Campylodiscus hibernicus* Ehrenberg (up to 10%) were observed only in the lowermost (9.00–8.90 m) part of LDZ-1.

LDZ-2 (8.37–8.28 m): relatively few diatom frustules (e.g. *Cocconeis placentula* Ehrenberg, *Epithemia adnata, Fragilaria pinnata* Ehrenberg) were found in this interval, probably indicating the emergence of the sediments above the water level.

LDZ-3 (8.28–8.20 m): the lower portion of this zone (8.28–8.23 m) comprises mainly benthic *Navicula* taxa (especially *N. scutelloides*—up to 71%), as well as epiphytic diatom species from the genera *Epithemia* and *Cocconeis* (Fig. 3). In the upper part of LDZ-3 the benthic *Ellerbeckia arenaria* and the epiphytic *Epithemia adnata* account for up to 32% and 15%, respectively. The amount of *N. scutelloides* decreases in the upper layers of the zone to 1–2%.

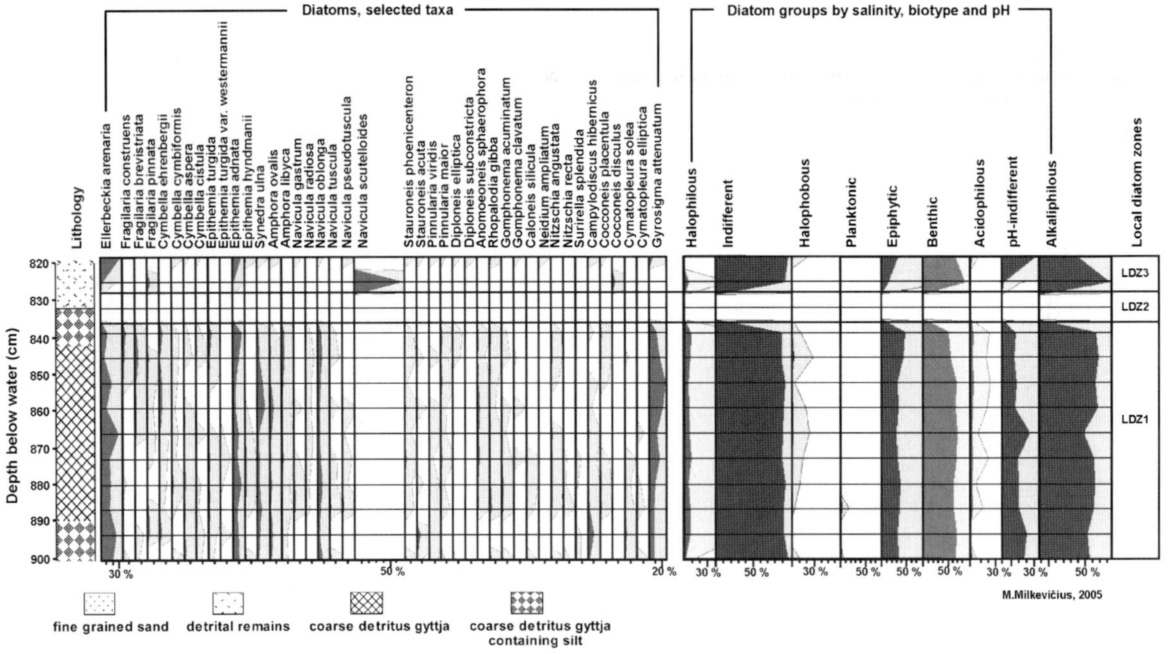

Fig. 3 Diagram displaying succession of selected diatom taxa and diatom assemblage zones (DAZ). Grouping of taxa according to salinity, habitat and pH preference. Solid curve represents the actual percentage, while empty curves show the percentage values multiplied by 10

Discussion

This study focuses on the 0.75–0.80 m thick continuous coarse detritus gyttja layer resting on top of almost purely minerogenic glaciolacustrine sand. The presence of aquatic pollen and benthic diatoms in the glaciolacustrine sand and in the lowermost portion of the organic-rich bed demonstrates that the area had not entirely emerged from the lake waters either before or at the beginning of the gyttja deposition. However, the dominance of shallow water benthic diatom species (e.g. *Ellerbeckia arenaria*) and wood fragments in the sediments imply comparatively shallow water conditions at that time. The LPAZ PE-1 (9.20–8.75 m), which includes both the underlying lacustrine sand and the lower part of gyttja, reflects a typical moist lake shore vegetation in which Cyperaceae and Poaceae dominate, while the presence of the green alga *Pediastrum* indicates a lake environment at the site. As the sediment was accumulating in this shallow-lake environment, there was an increase in organic content (Fig. 2) and a corresponding decrease in terrigenous and possible authigenic components. The amount of *Pediastrum* also decreased in the upper portion of the gyttja layer. The pollen composition and macrofossils *Myriophyllum, Potamogeton*, *Nymphaea alba*, *Nuphar lutea*, *Schoenoplectus lacustris*, *Eleocharis* and *Ranunculus* as well as the presence of oogonia of the green alga *Chara* sp. in LPAZ zone PE-2 indicate that the site was continuously inundated until the end of the gyttja accumulation. This inference is supported by the overall diatom composition in LDZ-1 (9.00–8.37 m), which indicates a nutrient-rich, littoral, very shallow water sedimentary environment. From this we conclude that the accumulation of the coarse detritus gyttja took place under conditions of decreasing water depth in a nearshore freshwater environment.

A biostratigraphically distinct level was determined in the topmost part of the buried gyttja layer at 8.37–8.32 m, where a peak of *Myriophyllum* is accompanied by a maximum of *Betula* pollen (Fig. 2) and the presence of moss remains in the pollen slides. Most finds of *Myriophyllum* are *M. verticillatum* L. According to Mäemets (2002), this species occurs in different water bodies, including shallow muddy ones, and tolerates organic sediments better than *M. spicatum* L. The same sediment interval (LDZ-2) was almost devoid of diatoms (Fig. 3), which may indicate temporary terrestrial conditions at the site. This conclusion is supported by the dark reddish colour (5YR/2.5–1) in the topmost (8.40–8.33 m) layer of gyttja, indicating more oxidizing conditions, compared to the greenish black (Gley 1/2.5–1) gyttja below that level. In the sediments, the 8.37–8.32 m level is characterised by a very rapid decrease in organic matter (Fig. 2) and a corresponding increase in mineral matter. According to our interpretation, the onset of the phase of massive sand accumulation reflects a change in sedimentary environment caused by an accelerated water level rise, which followed a temporary very low stand of the lake.

The fine sand that covers the gyttja has variable (1–2 m) thickness (Hang et al., 2001) and its pollen composition shows rapid changes in tree pollen, a decrease in herb and aquatics pollen and an increase in the alga *Pediastrum* (Fig. 2). The composition of the diatom assemblage in the sand (LDZ-3) indicates a shallow water littoral or swampy environment. According to our interpretation, the sand overlying the gyttja was redeposited because of wave erosion in the surrounding foreshore area under conditions of a continuously rising water level.

As shown by the first radiocarbon dating results, the deposition of the discussed course detritus gyttja may have lasted ca. 500 years and the onset of the water level rise and deposition of sand on to the gyttja took place after 9.1 ka ^{14}C BP. Our data show that the Lake Peipsi water level reached its early Holocene minimum (ca. 22 m a.s.l.) at the end of gyttja accumulation before 9.1 ka ^{14}C BP.

The isolation of Lake Peipsi from the Baltic Ice Lake (BIL) and following drainage down to the currently described lowest lake level need further investigation. Shore displacement curves for the southern part of Lake Peipsi proper (Hang et al., 1995; Rosentau, 2006) display a rapid regression between 12.5 and 10.2 ka ^{14}C yrs BP with the minimum water level about 10.2 ka ^{14}C yrs BP, slightly after the last drainage event of the BIL (ca. 10.3 ka ^{14}C yrs BP). Our data show that the low stand of Lake Peipsi reached its minimum during the early Holocene ca. 9.1 ka ^{14}C yrs BP and was thereafter followed by a continuous rise in water level up to the present. Recent GIS-based simulations of the development of proglacial lakes in eastern Estonia (Rosentau, 2006) demonstrate that the final strait-like connection between the BIL and the Glacial Lake Peipsi ceased at the Narva River valley (Fig. 1) if the proglacial lake level was lowered to 35–32 m. a.s.l. Simulation of the lake level to the present threshold altitude in the Narva River valley (28–26 m a.s.l.; Hang & Miidel, 1999) shows a rather small and shallow body of water within the depression of modern Lake Peipsi proper. This simulation displays a water level of ca. 23 m a.s.l. in our study area, which conforms closely to our estimates of the minimum water level (22 m a.s.l.). Thus, the decrease in Lake Peipsi water level to its minimum can be explained by lake drainage through the Narva River valley after the final drainage of the Baltic Ice Lake. The onset of the continuous Holocene water level rise is most probably related to the more intensive tectonic uplift in the northern part of the lake depression compared to the southern regions, which hampered the only northern outflow. Currently, the impact of a different tectonic uplift is apparent in the emergence of northern shores and the inundation of low coastal areas in the southern part of the lake depression.

Conclusions

We discovered a 0.75–0.80 m thick sand-covered continuous coarse detritus gyttja layer at the bottom of Lake Peipsi. The pollen and diatom compositions of the gyttja indicate accumulation under conditions of decreasing water depth in a nearshore environment.

The pollen composition of the lower part of the gyttja is typical of moist lake shore vegetation. Diatoms from the same interval indicate a nutrient-rich, very shallow-littoral environment.

An increase in *Myriophyllum* pollen, accompanied by the presence of *Betula* pollen and moss remains and the absence of diatoms in the uppermost 5–7 cm of the gyttja interval, imply a temporary emergence of the sediment upper layer.

The Lake Peipsi water level during the early Holocene low stand reached its minimum (ca. 22 m a.s.l.) at the end of the gyttja accumulation.

The onset of the water level rise and deposition of sand on to the gyttja took place after 9.1 ka ^{14}C PB.

Acknowledgements This article derives from the project supported by Estonian Science Foundation Grant 5370, Estonian target funding project 0182530s03 and University of Tartu project PFLAJ 05909. We thank S. Hiie for identification of plant macrofossils and C.J. Caseldine and M. Kuura for improving the language. The journal reviewers and the guest editor of the current issue, H. Simola, are acknowledged for valuable criticisms.

References

Battarbee, R. W., 1986. Diatom analysis. In Berglund, B. E. (ed.), Handbook of Holocene Palaeoecology and Palaeohydrology. Wiley & Sons Ltd., Chichester: 527–570.

Berglund, B. E. & M. Ralska-Jasiewiczowa, 1986. Pollen analysis and pollen diagrams. In Berglund, B. E. (ed.), Handbook of Holocene Palaeoecology and Palaeohydrology. Wiley & Sons Ltd., Chichester: 455–484.

Davydova, N. & K. Kimmel, 1991. Palaeogeography of the Lake Peipsi on the basis of biostratigraphical studies of bottom sediments. Proceedings of Estonian Academy of Sciences, Geology 40: 16–23.

Grimm, E., 1992. *TILIA-TILIA**GRAPH Computer Program. Illinois State Museum.

Hang, E., T. Liblik & E. Linkrus, 1964. On the relations between Estonian valley terraces and lake and sea levels in the late glacial and Holocene periods. Transactions of the Tartu State University 156, Publications on Geography IV: 29–42.

Hang, T., A. Miidel & R. Pirrus, 1995. Late Weichselian and Holocene water-level changes of Lake Peipsi, eastern Estonia. PACT 50: 121–131.

Hang, T. & A. Miidel, 1999. Holocene history of the lake. In Miidel, A. & A. Raukas (eds.), Lake Peipsi Geology. Sulemees Publishers, Tallinn: 131–135.

Hang, T., A. Miidel, V. Kalm & K. Kimmel, 2001. New data on the distribution and stratigraphy of the bottom deposits of Lake Peipsi, eastern Estonia. Proceedings of the Estonian Academy of Sciences, Geology 50: 233–253.

Katz, N. Y., S. V. Katz & M. G. Kipiani, 1965. Atlas i opredelitel plodov i semyan vstrechayushchikhsya v chetvertichnykh otlozheniyakh SSSR. Nauka, Moscow: 366 pp (Atlas and keys of fruits and seeds occurring in the Quaternary deposits of the U.S.S.R (in Russian)).

Katz, N. Y., S. V. Katz & E. I. Skobejeva, 1977. Atlas rastitelnykh ostatkov v torfakh. Nedra, Moscow: 372 pp (Atlas of plant remains in peat (in Russian)).

Krammer, K. & H. Lange-Bertalot, 1986. Bacillariophyceae: Naviculaceae. Süsswasserflora von Mitteleuropa 2/1. Gustav Fisher Verlag, Stuttgart-New York.

Krammer, K. & H. Lange-Bertalot, 1988. Bacillariophyceae: Bacillariaceae, Epithemiaceae, Surirellaceae. Süsswasserflora von Mitteleuropa 2/2. Gustav Fisher Verlag, Stuttgart-New York.

Krammer, K. & H. Lange-Bertalot, 1991a. Bacillariophyceae: Centrales, Fragilariaceae, Eunotiaceae. Süsswasserflora von Mitteleuropa 2/3. Gustav Fisher Verlag, Stuttgart-New York.

Krammer, K. & H. Lange-Bertalot, 1991b. Bacillariophyceae: Achnanthaceae & Kritsche Ergänzerungen zu Navicula (Lineolatae) und Gomphonema. Süsswasserflora von Mitteleuropa 2/4. Gustav Fisher Verlag, Stuttgart-New York.

Mäemets, H., 2002. Commented list of macrophyte taxa of Lake Võrtsjärv. Proceedings of Estonian Academy of Sciences, Biology, Ecology 51: 5–25.

Mangerud, J., S. Andersen, B. Berglund & J. Donner, 1974. Quaternary stratigraphy of Norden, a proposal for terminology and classification. Boreas 3: 109–128.

Miettinen, A., 2002. Relative sea level changes in the eastern part of the Gulf of Finland during the last 8000 years. Annales Academiae Scientiarium Fennicae, Geologica-Geographica 162: 12–14.

Miidel, A., R. Noormets, T. Hang, T. Flodén & M. Bjerkéus, 2001. Bedrock geology and topography of the Lake Peipsi depression, eastern Estonia. GFF 123: 15–22.

Munsell Color Company, 1998. Munsell Soil Color Charts. Munsell Color, New York.

Niinemets, E., 1999. Ostracods. In Miidel, A. & A. Raukas (eds.), Lake Peipsi Geology. Sulemees Publishers, Tallinn: 90–97.

Raukas, A. & E. Rähni, 1969. On the geological development of the Peipsi-Pihkva depression and the basins distributed in that region. Proceedings of Estonian Academy of Sciences, Chemistry and Geology 18, 113–127 (in Russian).

Rosentau, A., 2006. Development of Proglacial Lakes in Estonia. Dissertationes Geologicae Universitatis Tartuensis 18, Tartu University Press, Tartu: 48 pp.

Sarv, A. & E. Ilves, 1975. Über das Alter der holozänen Ablagerungen im Mündungsgebiet des Flusses Emajõgi (Saviku). Proceedings of Estonian Academy of Sciences, Chemistry and Geology 24: 64–69.

Schoch, W. H., B. Pawlik & F. H. Schweingruber, 1988. Botanische makroreste. P. Haupt, Bern & Stuttgart: 205 pp.

Hydrobiologia (2008) 599:23–30
DOI 10.1007/s10750-007-9206-y

ELLS 2007

Water level changes in a large shallow lake as reflected by the plankton:periphyton-ratio of sedimentary diatoms

Atko Heinsalu · Helen Luup · Tiiu Alliksaar · Peeter Nõges · Tiina Nõges

Abstract Biostratigraphic diatom analyses were carried out on a short sediment core from the large shallow-water Lake Võrtsjärv, Estonia, in order to relate the diatom composition to the instrumental water level record. We dated the sediment core by radiometric methods (^{210}Pb, ^{137}Cs, ^{241}Am) and spheroidal fly-ash particle abundance chronology and evaluated the statistical significance of the relationships between the percentage of planktonic diatoms and the water level continuously monitored since 1871. Before the 1960s, the percentage of planktonic diatoms in the sediment showed quite strong positive relationship to water level. The impact of eutrophication after the 1960s presumably masked the influence of water level changes on the diatom community. In addition, statistical analysis of the upper part of the sediment core (1970—present day) together with measured limnological parameters of the lake showed that water transparency had the strongest influence on diatoms, while temperature, pH and alkalinity had lesser impacts. Our study shows that the planktonic:periphytic diatom ratio in the sediment can be used to track overall trends of the lake-level changes in Lake Võrtsjärv before the onset of cultural eutrophication; however, the results have to be interpreted carefully, taking into consideration other possible limnological factors such as water transparency, nutrients and wind.

Keywords Paleolimnology · Sediment diatoms · Water level changes · Lake Võrtsjärv · Estonia

Guest editors: T. Nõges, R. Eckmann, K. Kangur, P. Nõges, A. Reinart, G. Roll, H. Simola & M. Viljanen.
European Large Lakes—Ecosystem changes and their ecological and socioeconomic impacts.

A. Heinsalu (✉) · T. Alliksaar
Institute of Geology, Tallinn University of Technology, Ehitajate tee 5, 19086 Tallinn, Estonia
e-mail: heinsalu@gi.ee

H. Luup · P. Nõges · T. Nõges
Centre for Limnology, Institute of Agricultural and Environmental Sciences, Estonian University of Life Sciences, 61101 Rannu, Tartumaa, Estonia

P. Nõges
Institute for Environment and Sustainability, European Commission – Joint Research Centre, 21020 Ispra (VA), Italy

Introduction

Many lakes in temperate regions show fluctuations in the water level on seasonal, annual and long-term time scales in response to variations in their water balance (Vassiljev, 1997). Water level changes, presumably controlled by climatic variability and climate change, affect biological productivity, which has implications for water quality and the ecological state of these lakes. There is an increasing need to understand the long-term variability of climate and its impact on aquatic ecosystems. However, the available instrumental records are too short to capture the whole range of post-glacial variability and to predict future changes. Lake sediments provide natural archives that record the response of lake ecosystems to environmental changes, and different sediment proxies have great potential for reconstructing long-term variations in climatic and climate-driven parameters beyond the range of instrumental records. Diatoms are abundant and well preserved in sediments, and their ecology and species sensitivity to changing hydrological conditions such as water depth, nutrient concentration etc. make them applicable for reconstructing past environmental changes (Stoermer & Smol, 1999), including water levels. Many lake-level reconstructions based on sub-fossil diatoms have used changes in diatom habitat groups (i.e. the ratio of planktonic to periphytic diatoms) in sediments (e.g. Barker et al., 1994; Hyvärinen & Alhonen, 1994; Wolin, 1996; Stone & Fritz, 2004). Planktonic diatoms contribute frustules to the sediment in pelagic deep-water areas, while benthic and epiphytic diatoms are primarily associated with shallower littoral habitats closer to shores (Wolin & Duthie, 1999); thus, a rise in the lake level is commonly recorded in sediments as an increase in the share of planktonic forms.

Lake Võrtsjärv is one of the most thoroughly investigated lakes in Estonia: instrumental records of its water level date back to 1871 and water physical and chemical parameters and biota have been continuously monitored over the past 40 years. Long-term records have shown that the water level in Võrtsjärv is positively correlated with the North Atlantic Oscillation (NAO) winter index (Järvet, 2004) and that climatic variability is the most important factor for the hydrology and the ecosystem of the lake. Phytoplankton records from 1964 to 2000 show a positive correlation between the water level and the planktonic diatom biomass in spring (Nõges et al., 2003; Nõges, 2004), so the sediment diatom record from Lake Võrtsjärv has great potential for reconstructing the relationship between lake-level oscillations and climate variability. The objective of this study was to examine whether the trends and directions of past lake-level changes in Lake Võrtsjärv can be inferred from sub-fossil sediment diatom communities.

Study site

Lake Võrtsjärv is situated in central Estonia and is the second largest lake in the country with a surface area of 270 km^2 and a catchment area of 3,374 km^2. This very shallow (maximum depth 6 m, mean depth 2.8 m) non-stratified lake has six main inflows and one outflow that carries the water into Lake Peipsi. Owing to its shallowness and large wind-exposed area, the water-body is turbid (Secchi depth ranges from 0.5 to 1.0 m during the ice-free period). The lake is eutrophic, characterised by mean concentrations of about 2 mg l^{-1} total nitrogen and 50 μg l^{-1} total phosphorus.

The ecosystem of Lake Võrtsjärv is strongly influenced by the large amplitude of water level oscillations. The average annual water level fluctuation is about 1.4 m and the difference between the extreme recorded water levels is more than 3 m (32.2–35.3 m a.s.l.).

Materials and methods

A sediment core was taken in March 2003 from the ice in the southern part of the lake at 58°09′42″N and 26°04′10″E (water depth 1.40 m) using a freeze corer (Glew et al., 2001). The 90-cm core was sub-sampled into continuous 1-cm thick slices, which were used for different analyses.

The chronology of the core was based on ^{210}Pb activity measurements and the constant rate of supply (CRS) model, which was applied to calculate the ^{210}Pb age-scale for the core (Appleby & Oldfield, 1978). This was then corrected by reference dates of artificial radionuclides (^{137}Cs and ^{241}Am). The ^{210}Pb chronology was independently validated by the analysis of spheroidal fly-ash particles (SFAP), the stratigraphy of which is related to the history of fossil-fuel consumption in Estonia (Alliksaar, 2000). The SFAP were enumerated microscopically at 250× magnification in chemically digested samples (Rose, 1990).

Basic properties of the sediment were determined by standard methods: the water content was determined by drying the samples to constant weight at 105°C; the organic matter and the carbonate contents were measured as loss on ignition by heating the samples at 550°C for 4 h and at 950°C for 2 h, respectively (Heiri et al., 2001).

We prepared diatom slides for microscopic analysis using the standard H_2O_2 method (Battarbee et al., 2001) and mounted the diatom suspensions with Naphrax. The diatoms were identified and counted under a Zeiss microscope with an oil immersion objective and phase contrast (1,000× magnification). At least 400 non-*Fragilaria* diatom valves per sample were counted. Identification was based primarily on Krammer & Lange-Bertalot (1986–1991). Diatoms were grouped according to their habitats into planktonic and periphytic forms, the latter including benthic, epilithic and epiphytic forms; for the ecological grouping of Lake Võrtsjärv diatoms see Pork & Kõvask (1973). Because of the high abundance of small-sized periphytic fragilarioid species in the sediment assemblages, these diatoms were excluded from the diatom frequency calculations and from the planktonic:periphytic ratio calculations to avoid the over-representation of these taxa. A similar strategy has been adopted by several other authors (e.g. Battarbee, 1986; Barker et al., 1994).

For the most recent 40-year period, in which water quality data from Lake Võrtsjärv are also available, we conducted principal components analysis (PCA) using Canoco for Windows 4.5 (ter Braak & Šmilauer, 2002) to identify the environmental variables that appear to relate changes in diatom assemblages in the upper part of the sediment core.

Correlating time series afflicted with a trend, or other serial dependencies, may lead to spurious correlations that do not represent actual mechanistic relationships, but occur only due to the presence of these instationarities (Chatfield, 1996). Therefore the time series of the percentage of planktonic diatoms in the sediment of Lake Võrtsjärv and the 7-year moving average of the lake water level were first log-transformed (to remove the influence of non-normal distribution of the data) and then detrended using the Time Series analysis module of Statistica for Windows 6.0 (StatSoft, Inc., 2001). Detrending was done to avoid coincidental correlation of the analysed data, which may occur purely because of a common long-term trend even if the series are not related to each other.

Results

The total ^{210}Pb activity in Lake Võrtsjärv sediments (Fig. 1a) showed an irregular profile: a distinctive down-core decrease within the uppermost 22 cm, then an increase, then another downward decline below the 27 cm level. At 94 cm core depth, the total ^{210}Pb and the supporting ^{226}Ra reached an equilibrium that corresponds to ca. 150 years of sediment accumulation. A sharp peak of the artificial radionuclides ^{137}Cs and ^{241}Am (Fig. 1b) at core-depth 35–40 cm clearly marked the fallout from atmospheric nuclear tests of the early 1960s. The datings indicated that dry-mass sediment accumulation rates have been variable (Fig. 1c), with higher values during the 1950s and 1980s.

The stratigraphy of SFAP, representing products of high-temperature fossil-fuel combustion, confirmed the accuracy of the ^{210}Pb chronology. The SFAP abundance profile in lake sediments plotted on the ^{210}Pb age-scale (Fig. 1d) showed the features characteristic of SFAP history over a wider geographical area. A small but steady rise in the particle concentration changed to a sharp increase after the Second World War, when there was a considerable rise in energy demand and several power plants were established (Heinsalu et al., 2007). The peak in particle concentration was followed by a recent decline, a feature that is also common to most sediment cores in Europe and is caused by a decrease in fossil energy production or an increased efficiency of particle removal from the air emissions, or both.

The uppermost 18 cm of the sediment core (1992–2003, Fig. 2) was poorly compacted with a water content of 91–99%. The carbonate content started to increase in 1950s and peaked sharply during the 1980s (Fig. 2). The organic matter content has also risen during the past 15 years.

For diatom analysis, 32 sub-samples in the 0–90 cm sequence of the sediment core were studied. Altogether, 115 diatom taxa were identified (Fig. 3). Small-sized *Fragilaria* taxa, namely *Fragilaria brevistriata* Grunow, *F. construens* (Ehrenberg) Grunow and *F. pinnata* Ehrenberg dominated, making up 40–65% of the assemblages. The diatom species composition showed no major stratigraphic changes; however, the share of planktonic and periphytic species fluctuated within the sequence. From the 1960s until the present, the relative abundance of the eutrophic planktonic diatom *Aulacoseira ambigua* (Grunow) Simonsen has increased significantly, exceeding 70% in the uppermost samples. Another eutrophic planktonic species, *Aulacoseira granulata*

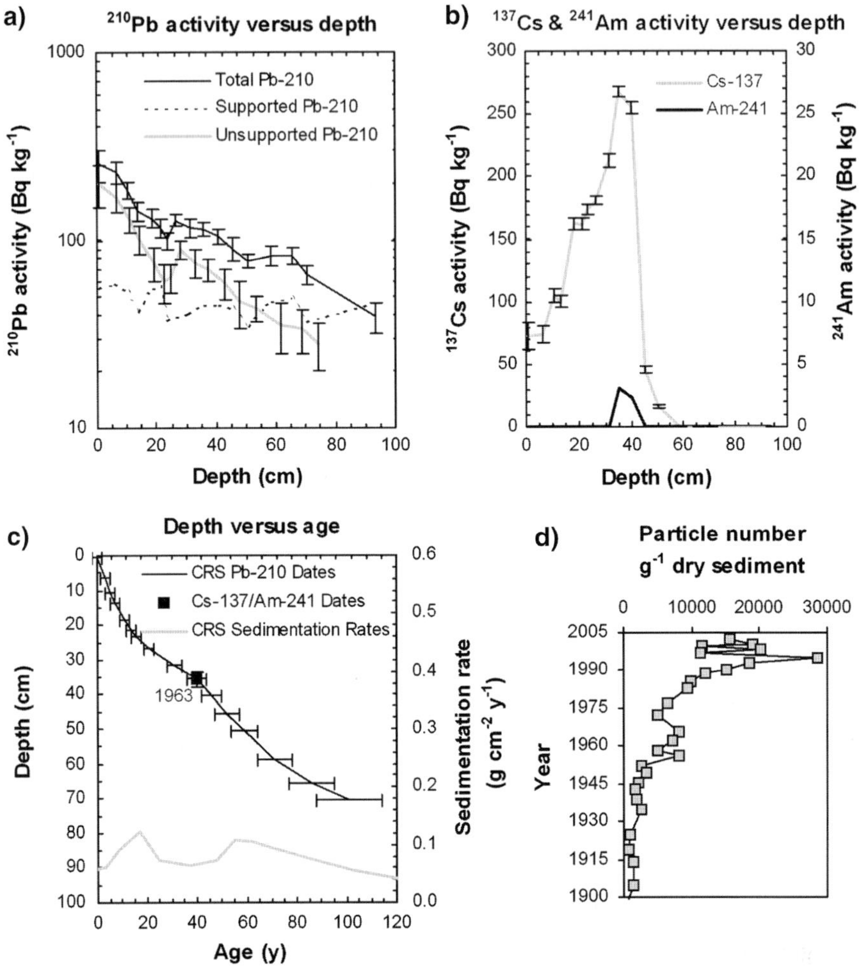

Fig. 1 Chronology of the Lake Võrtsjärv sediment core: (**a**) total, supported and unsupported ^{210}Pb activities vs. core-depth, cm; (**b**) ^{137}Cs and ^{241}Am activities vs. core-depth, cm; (**c**) modelled age of sediments using ^{210}Pb CRS model and ^{137}Cs and ^{241}Am stratigraphy and calculated sedimentation rate (g cm^{-2} y^{-1}); (**d**) sediment concentration profile of spheroidal fly-ash particles (number g^{-1} dry sediment) plotted against the ^{210}Pb age-scale

(Ehrenberg) Simonsen, also increased in the most recent sediment, in which *Stephanodiscus hantzschii* Grunow also appeared.

The PCA for the uppermost part of the sediment core deposited since 1970 was based on percentages of diatom taxa and five environmental variables. The planktonic and periphytic groups of diatoms were used as supplementary variables (Fig. 4). The first principal axis explained 78.2% and the second a further 14.6% of the variation in the data set. The Secchi transparency, which was strongly and negatively correlated with the first axis, turned out to be an important factor in explaining the abundance of periphytic diatoms, while water temperature, pH and alkalinity, positively correlated with the first axis, were related to the abundance of planktonic taxa. The shorter arrows for the latter variables indicate weaker correlations with the changes in diatom assemblages. The water level showed a very low correlation with the first axis but the variable best correlated with the second principal axis.

Correlation analyses indicated that a higher water level supported a generally higher percentage of planktonic forms in the diatom assemblage; however, the relationship had low significance ($r = 0.32$, $P = 0.096$) if the whole studied data set was considered. Splitting the time-series data into two parts, at the 1960 level, which marks the onset of pronounced eutrophication, revealed that in the pre-1960s strata, the percentage of planktonic diatoms was strongly and significantly related with higher water levels in the lake ($r = 0.60$, $P = 0.014$), while after the 1960s the correlation between these indices was lacking ($r = 0.15$, $P = 0.62$; Fig. 5).

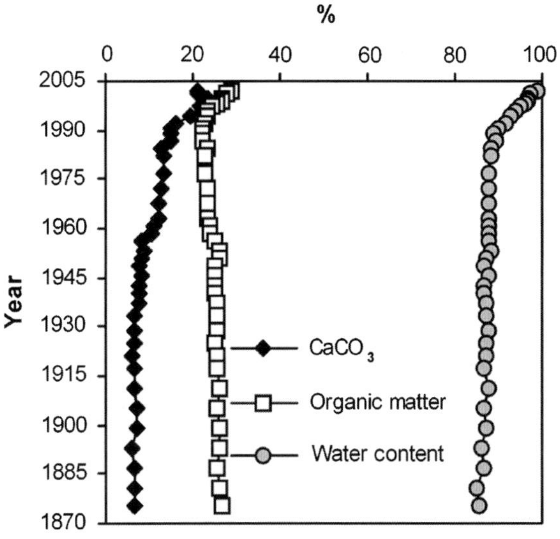

Fig. 2 Depth profiles of sediment physical and chemical properties of Lake Võrtsjärv ($CaCO_3$ and organic matter are given as percentages of dried sediment weight; water content is expressed as percentage of fresh sediment)

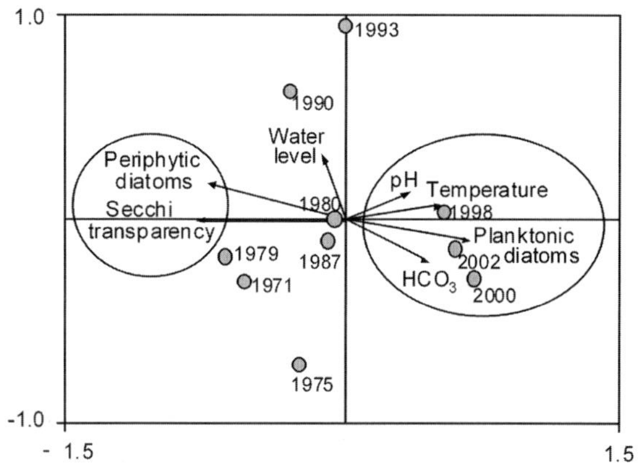

Fig. 4 Principal components analysis biplot showing Lake Võrtsjärv environmental variables (*arrows*) combined with the sediment diatom samples studied (*filled circles*) over the period 1971–2002. The first PCA axis explains 78.2% and the second a further 14.6% of the variation in the data set

Discussion

Within the period investigated (ca. 1840 to the present), the sedimentary diatom flora in Lake Võrtsjärv shows no prominent changes in species composition (Fig. 3) and is indicative of moderately eutrophic shallow-water conditions. This paleolimnological information on sediment diatoms confirms the conclusion derived by Nõges & Nõges (2006) that the ecological status of Lake Võrtsjärv has not deviated substantially from the reference conditions, and that the overall ecological quality of the lake could be estimated as 'good' in terms of the EU Water Framework Directive.

In the majority of lake sediment records studied elsewhere (e.g. Clarke et al., 2005), the share of planktonic diatoms has increased during the past century in consequence of increased anthropogenic nutrient loadings that have stimulated aquatic primary production and led to undesirable eutrophication effects. This is also true for Lake Võrtsjärv, where the oscillations of the ratio of planktonic to periphytic diatoms have become stronger and more frequent during the last 40 years, simultaneously with a progressive increase in the proportion of planktonic diatoms, mostly accounted for by the eutrophic *Aulacoseira ambigua* (Fig. 3).

Fig. 3 Diatom stratigraphy of Lake Võrtsjärv with relative frequencies (%) of the most abundant taxa

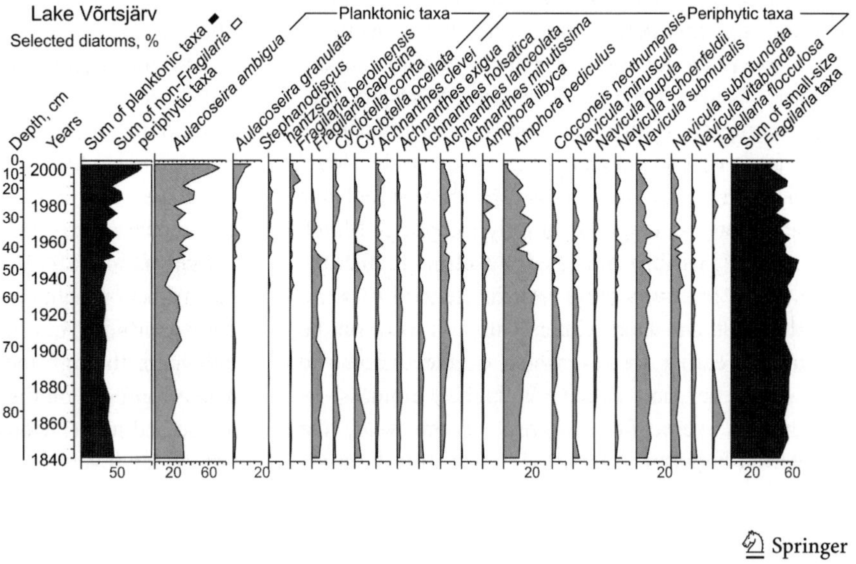

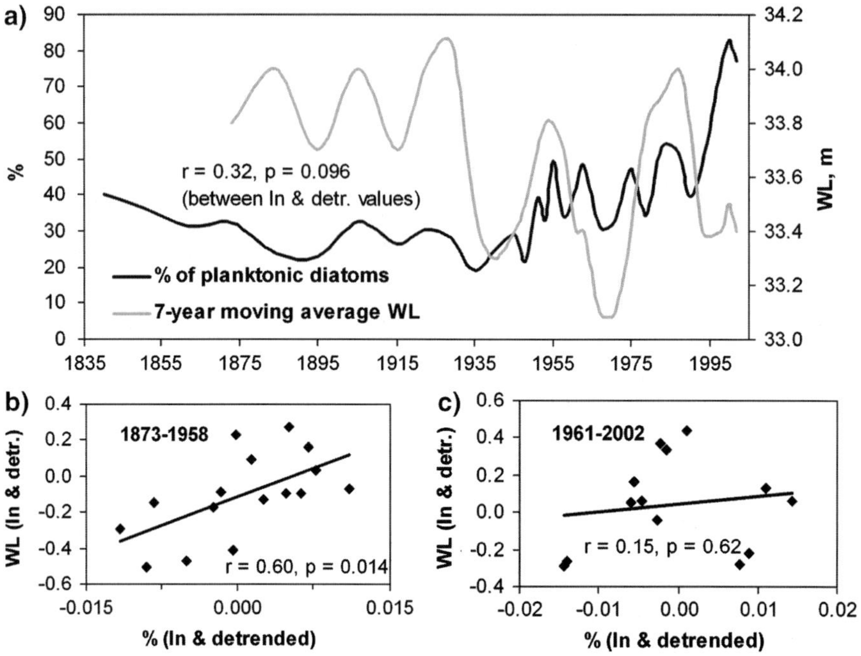

Fig. 5 Relationship between measured water level (WL; 7-year moving average) in Lake Võrtsjärv and the percentage of planktonic forms among diatoms in the sediment core layers over the years (**a**) 1873–2002; (**b**) 1873–1958; (**c**) 1961–2002. All correlations are calculated between log-transformed and detrended values; for more details see the "Materials and methods" section

These changes in sediment diatom assemblages evidently indicate increased human disturbance and accelerated eutrophication during the past 40 years. Our conclusion about the progressive eutrophication since the 1960s is also supported by the increased carbonate content in the sediments over this period. Carbonate precipitation is enhanced at higher pH resulting from intensive photosynthesis (Wetzel, 1983) and, in this respect, increased carbonate precipitation is a sign of eutrophication.

Changes in the planktonic:periphytic diatom ratio in the sediment record have been used as lake-level fluctuation signals in many lakes (Wolin & Duthie, 1999). Higher percentages of open-water planktonic diatoms are expected at high water levels, while at lower levels the share of periphytic diatoms should increase. Studies conducted on Lake Võrtsjärv diatom communities since the 1960s have revealed a weak positive correspondence between water level and planktonic diatom biomass. This correspondence appears strong and significant only in spring (March and April), possibly mediated by climatic factors. High water levels in these months generally correspond to mild and short winters with less snow on the ice and an earlier ice-off, which enable diatoms to develop earlier and achieve a higher biomass by March and April. For the rest of the year, the correlation between water level and diatom biomass in the lake turned out to be non-significant or even negative (Nõges et al., 2003; Nõges, 2004). In our sediment study, the relationship between the frequency of planktonic diatoms and the water level had low significance if the whole available data set was considered (Fig. 5). After the 1960s, when eutrophication was presumably accelerated, the water level ceased to be the main driver determining the planktonic:periphytic diatom ratio, so the correlation between these indices became non-significant for this period (Fig. 5). Thus, our study suggests that since the 1960s, eutrophication has affected the sediment diatom composition in Lake Võrtsjärv more strongly than climatically induced water level fluctuations.

In large shallow lakes like Võrtsjärv, reconstruction of water level changes using the planktonic:periphytic ratio of sediment diatoms may be complicated, as several environmental factors may cause the same signals. If water level falls, sediments become more exposed to wave-induced re-suspension (Nõges et al., 1999). Shallower conditions have a dual effect on light conditions, leading on one hand to higher turbidity but on the other to shortening of the light path through the water column. The net effect depends mostly on the compactness and the grain size of the sediment. Moreover, stronger resuspension at low water levels leads to more intensive release of phosphorus from the sediment and, correspondingly, to enhanced internal loading, which has a positive effect on plankton productivity (Nõges & Nõges, 1999). Statistical analysis of the diatom assemblages in the upper part of the sediment core together with the measured water indices showed that water transparency affecting periphytic diatoms was the main determinant in the system, while water temperature, pH and alkalinity were more related to planktonic species and had a weaker influence (Fig. 4). Nevertheless, the strong influence of water transparency on the diatom community may be indirectly related to water level changes.

Considering only the earlier sediment diatom data, pertaining to the pre-eutrophication period before the 1960s, the percentage of planktonic diatoms showed a strong positive correlation with the instrumentally registered water levels in Lake Võrtsjärv (Fig. 5). We suppose that there was no systematic error in the chronology despite the steady increase in the standard error of the ^{210}Pb dates with time, especially before the 1920s, when it might to some extent hamper the comparison of sediment diatom data with the instrumental water level record. Our study suggests that the composition of sediment diatoms can still be considered an applicable indicator for reconstructing the pre-eutrophication water levels in Lake Võrtsjärv, providing information about changing hydrological conditions and thus about other climate change indicators in the long-term perspective.

Acknowledgements Funding for this research was provided by the Estonian Ministry of Education (SF0362480s03 and SF0332710s06), by Estonian Science Foundation grants (5738 and 5923), and by the European Union project CLIME (EVK1-CT-2002-00121).

References

Alliksaar, T., 2000. Spatial and temporal variability of the distribution of spherical fly-ash particles in sediments in Estonia. Tallinn Pedagogical University, Dissertations on Natural Sciences 4: 1–44.

Appleby, P. G. & F. Oldfield, 1978. The calculation of ^{210}Pb dates assuming a constant rate of supply of unsupported ^{210}Pb to the sediment. Catena 5: 1–8.

Barker, P. A., N. Roberts, H. F. Lamb, S. van den Kaars & A. Benkaddour, 1994. Interpretation of Holocene lake-level change from diatom assemblages in Lake Sidi Ali, Middle Atlas, Morocco. Journal of Paleolimnology 12: 223–234.

Battarbee, R. W., 1986. Diatom analysis. In Berglund, B. E. (ed.), Handbook of Holocene Palaeoecology and Palaeohydrology. John Wiley & Sons, Chichester: 527–570.

Battarbee, R., V. J. Jones, R. J. Flower, N. G. Cameron, H. Bennion, L. Carvalho & S. Juggins, 2001. Diatoms. In Smol, J. P., H. J. B. Birks & W. M. Last (eds), Tracking Environmental Change Using Lake Sediments, Vol. 3: Terrestrial, Algal, and Siliceous Indicators. Kluwer Academic Publishers, Dordrecht: 155–202.

Chatfield, C., 1996. The Analysis of Time Series, 5th edn. Chapman & Hall, New York.

Clarke, G., M. Kernan, A. Marchetto, S. Sorvari & J. Catalan, 2005. Using diatoms to assess geographical patterns of change in high-altitude European lakes from pre-industrial times to the present day. Aquatic Sciences 67: 224–236.

Glew, J. R., J. P. Smol & W. M. Last, 2001. Sediment core collection and extrusion. In Last, W. M. & J. P. Smol (eds), Tracking Environmental Change Using Lake Sediments, Vol. 1: Basin Analysis, Coring, and Chronological Techniques. Kluwer Academic Publishers, Dordrecht: 73–105.

Heinsalu, A., T. Alliksaar, A. Leeben & T. Nõges, 2007. Sediment diatom assemblages and composition of pore-water dissolved organic matter reflect recent eutrophication history of Lake Peipsi (Estonia/Russia). Hydrobiologia 584: 133–143.

Heiri, O., A. F. Lotter & G. Lemcke, 2001. Loss on ignition as a method for estimating organic and carbonate content in sediments: reproducibility and comparability of results. Journal of Paleolimnology 25: 101–110.

Hyvärinen, H. & P. Alhonen, 1994. Holocene lake-level changes in the Fennoscandian tree-line region, western Finnish Lapland: diatom and cladoceran evidence. The Holocene 4: 251–258.

Järvet, A., 2004. Hydrology of Lake Võrtsjärv. In Haberman, J., E. Pihu & A. Raukas (eds), Lake Võrtsjärv. Estonian Encyclopaedia Publishers, Tallinn: 105–139.

Krammer, K. & H. Lange-Bertalot, 1986–1991. Bacillariophyceae. In Ettl, H., G. Gärtner, J. Gerloff, H. Heynig & D. Mollenhauer (eds), Süßwasserflora von Mitteleuropa, Vol. 2(1–4). Gustav Fischer Verlag, Stuttgart.

Nõges, T., 2004. Reflection of the changes of the North Atlantic Oscillation Index and the Gulf Stream Position Index in the hydrology and phytoplankton of Võrtsjärv, a large, shallow lake in Estonia. Boreal Environment Research 9: 401–407.

Nõges, T. & P. Nõges, 1999. The effect of extreme water level decrease on hydrochemistry and phytoplankton in a shallow eutrophic lake. Hydrobiologia 408–409: 277–283.

Nõges, P. & T. Nõges, 2006. Indicators and criteria to assess ecological status of the large shallow temperate polymictic lakes Peipsi (Estonia/Russia) and Võrtsjärv (Estonia). Boreal Environment Research 11: 67–80.

Nõges, T., P. Nõges & R. Laugaste, 2003. Water level as the mediator between climate change and phytoplankton composition in a large shallow temperate lake. Hydrobiologia 506–509: 257–263.

Nõges, P., L. Tuvikene, T. Nõges & A. Kisand, 1999. Primary production, sedimentation and resuspension in large shallow Lake Võrtsjärv. Aquatic Sciences 61: 168–182.

Pork, M. & V. Kõvask, 1973. Vetikad. In Timm, T. (ed.), Võrtsjärv. Valgus, Tallinn: 95–99 (In Estonian with English summary).
Rose, N. L., 1990. A method for the selective removal of inorganic ash particles from lake sediments. Journal of Paleolimnology 4: 61–67.
StatSoft, Inc., 2001. STATISTICA (data analysis software system), version 6. http://www.statsoft.com.
Stoermer, E. F. & J. P. Smol (eds), 1999. The Diatoms: Application for the Environmental and Earth Sciences. University Press, Cambridge.
Stone, J. R. & S. C. Fritz, 2004. Three-dimensional modeling of lacustrine diatom habitat areas: Improving paleolimnological interpretation of planktic:benthic ratios. Limnology and Oceanography 49: 1540–1548.
ter Braak, C. J. F. & P. Šmilauer, 2002. CANOCO Reference manual and CanoDraw for Windows Users guide: Software for Canonical Community Ordination (version 4.5). Microcomputer Power, Ithaca, New York.
Vassiljev, J., 1997. Simulating the paleorecord of northern European lakes using a coupled lake catchment model. Lundqua Thesis 41: 1–19.
Wetzel, R. G., 1983. Limnology. Saunders College Publishing, Philadelphia.
Wolin, J. A., 1996. Late Holocene lake-level and lake development signals in Lower Herring Lake, Michigan. Journal of Paleolimnology 15: 19–45.
Wolin, J. A. & H. C. Duthie, 1999. Diatoms as indicators of water level changes in freshwater lakes. In Stoermer, E. F. & J. P. Smol (eds), The Diatoms: Application for the Environmental and Earth Sciences. University Press, Cambridge: 183–202.

Hydrobiologia (2008) 599:31–39
DOI 10.1007/s10750-007-9204-0

ELLS 2007

Changes in spatial distribution of phosphorus and nitrogen in the large north-temperate lowland Lake Peipsi (Estonia/Russia)

Külli Kangur · Tõnu Möls

Abstract We investigated changes in the spatial distribution of nitrogen (N) and phosphorus (P) in Lake Peipsi using limnological data from 1970 to 2005. The results show differences in nutrient content between the northern and southern parts of the lake (polarity) and indicate possible causes of eutrophication of this large international lake. The results show a steady gradient in total P (TP) and total N (TN) content along the lake: the northern and deepest part, Lake Peipsi s.s., is significantly less loaded with nutrients than the southern and very shallow part, Lake Pihkva, into which the main inflow, the Velikaya River, discharges. However, the long-term temporal patterns of N and P polarity are different. Statistical analysis, using a parametric functions technique in the framework of general linear analysis provided by the SAS procedures GLM and MIXED, revealed that the polarity of N compounds has been relatively stable over the years and can be related to differences in natural conditions between different parts of the lake. Our study indicates that Lake Peipsi is quite resistant to year-to-year changes in N load, and the in-lake N concentrations are quite stable on a long-term scale. In contrast, the increasing difference in P concentrations between the northern and southern parts of the lake clearly shows that the input of P from the south is increasing. Our results confirm that the anthropogenic input of P is the main reason for the deterioration of the Lake Peipsi ecosystem.

Keywords Phosphorus · Nitrogen · In-lake gradients · Pollution · Eutrophication

Guest editors: T. Nõges, R. Eckmann, K. Kangur, P. Nõges, A. Reinart, G. Roll, H. Simola & M. Viljanen
European Large Lakes—Ecosystem changes and their ecological and socioeconomic impacts

K. Kangur (✉) · T. Möls
Centre for Limnology, Institute of Agricultural and Environmental Sciences, Estonian University of Life Sciences, 61101 Rannu, Estonia
e-mail: kylli.kangur@emu.ee

Introduction

Large shallow lakes are unique and vulnerable ecosystems. Many of the structural and functional properties of large lakes, e.g. their morphology, hydrography, biogeochemical cycles and food-web structure, are directly related to lake size (Tilzer, 1990). Differences in natural conditions between parts of a lake can lead to prominent lake-wide gradients of water quality and heterogeneity in biota. Human impact may also lead to increased differences in nutrient concentrations between different parts of a large lake.

Lake Peipsi is the largest transboundary lake in Europe, shared between Estonia and Russia. The lake is elongated in the north-south direction, and the major part of the catchment area is situated to the south (Jaani, 2001a). Differences in natural conditions (topography, water depth, relative size of catchment area) among parts of the lake may determine the varying sensitivity and response of the ecosystem to eutrophication in addition to natural processes (e.g. fluctuations of water level and temperature). Interactions between natural factors and human impact are complicated and long-term studies are needed to elucidate them. In addition, the transboundary conditions complicate the implementation of policies that might prevent or mitigate environmental damage in the Lake Peipsi region. Unfortunately, long-term data on the nutrient emissions from the Russian part of the catchment area are not currently available (Mourad et al., 2006).

According to Battarbee et al. (2005), the natural evolution of a lake is defined as ontogenesis, whereas eutrophication is a result of external nutrient loading from human sources. Both these processes may induce increase in bioproductivity and changes in the biological structure of a lake. It is not always easy to distinguish between natural and human-induced processes, and they may amplify each other (Padisák & Koncsos, 2002).

As in many shallow lakes in Europe, eutrophication is the most serious environmental problem for Lake Peipsi. Owing to its shallowness and relatively long water residence time (about two years), the ecological state of the lake is also strongly influenced by natural processes, among which periodic fluctuations of water level and temperature are the most important (Kangur et al., 2003; Milius et al., 2005).

Previous long-term investigations have demonstrated that the water characteristics and biological communities in Lake Peipsi change from north to south (e.g. Laugaste et al., 2001; Starast et al., 2001; Kangur et al., 2002; Milius et al., 2005). A difference in conditions between the opposite ends of a lake basin is termed polarity. This study focuses on the polarity of N and P compounds as primary indicators of variable human impact on Lake Peipsi. We examine the spatial distributions of N and P and compare the temporal patterns of N and P polarity in Lake Peipsi. The aim of the study is to clarify whether the differences in nutrient content between the northern and southern parts of the lake have remained relatively stable over the years, showing that they are mainly due to natural environmental conditions, or whether they are increasing and showing greater human impact. We try to identify the primary effects of nutrient enrichment.

Materials and methods

Study site

Lake Peipsi is a large shallow lowland lake (surface area: 3,555 km^2), consisting of three limnologically different parts (Fig. 1). The northern part, Lake Peipsi s.s. (sensu stricto) is the largest and has the greatest mean depth (Table 1). The southern part, Lake Pihkva (Pskov), has a mean depth only half that of Lake Peipsi s.s. The strait between them is known as Lake Lämmijärv. According to Estonian lake typology, Lake Peipsi is a unique water body and belongs to a type of its own as a large unstratified eutrophic lake with a light (oligohumic) water of medium hardness (average 2.29 mEq l^{-1} in 1985–2005).

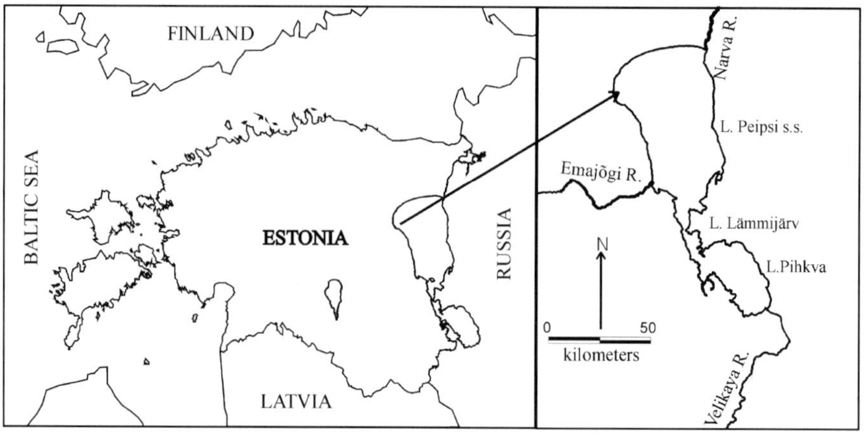

Fig. 1 Location of Lake Peipsi and its three parts

Table 1 Selected morphometric, chemical and phytoplankton characteristics of Lake Peipsi and its three parts (Lake Peipsi s.s., Lake Lämmijärv and Lake Pihkva)

Characteristic	Lake Peipsi s.s.	Lake Lämmijärv	Lake Pihkva	Whole Lake Peipsi
Surface area, km^2	2611	236	708	3555
Maximum depth, m	12.9	15.3	5.3	15.3
Mean depth, m	8.3	2.5	3.8	7.1
Water volume, km^3	21.79	0.60	2.68	25.07
TP, mg P m^{-3}	39.8 (16.4–96.4)	77.4 (38.4–156.0)	143.2 (82.2–249.2)	50.2 (16.3–154.2)
PO_4-P, mg P m^{-3}	6.6 (1.3–34.5)	12.1 (2.0–71.5)	31.4 (5.9–168.3)	8.3 (1.3–54.1)
TN, mg N m^{-3}	658 (375–1155)	856 (525–1397)	1006 (527–1919)	718 (389–1327)
DIN, mg N m^{-3}	121 (29–494)	130 (33–520)	108 (35–339)	122 (31–489)
TN:TP mass ratio	16.5 (6.4–42.7)	11.1 (6.1–20.3)	7.0 (5.6–8.8)	14.3 (5.4–38.1)
DIN:PO_4-P mass ratio	18.4 (1.9–179.5)	10.8 (1.0–119.6)	3.4 (0.6–18.8)	14.7 (1.3–166.1)
Chlorophyll *a* content, mg m^{-3}	17.5 (5.4–56.2)	28.5 (9.1–89.5)	53.5 (23.9–119.6)	20.9 (5.7–76.5)
Secchi depth, m	1.7 (1.6–1.8)	0.9 (0.8–1.0)	0.7 (0.5–0.8)	1.4 (1.3–1.5)
OECD (1982) classification	Eutrophic	Eutrophic/hypertrophic	Hypertrophic	Eutrophic

Water quality parameters are presented as geometric means and 90% tolerance limits are given in brackets. These estimates correspond to the open water periods (Julian days 100–310 within each year) in 2001–2005

The catchment area of Lake Peipsi (47,800 km^2, including lake surface) extends from 56°08′ to 59°13′ N and from 25°36′ to 30°16′ E (Jaani, 2001a). The catchment is shared between Russia (27,917 km^2), Estonia (16,323 km^2) and Latvia (3,560 km^2). The drainage basin is flat, with the highest point 318 m above sea level. The main inflows are the Velikaya River in Russia and the Emajõgi River in Estonia with catchment areas of 25,200 km^2 and 9,745 km^2, respectively. The outflowing Narva River (mean annual discharge 399 m^3 s^{-1}) runs into the Gulf of Finland in the Baltic Sea. The residence time of water in the whole lake is about 2 years. The water level is not regulated. Natural water level fluctuations have shown an overall range of 3.04 m over the last 80 years, with an average annual range of 1.15 m (Jaani, 2001b).

Due to the large surface area and relative shallowness of the lake, temperature stratification is unstable and can even be disturbed by a moderate wind or undulation. Therefore, the lake water is usually rich in oxygen during the open water period. The lake is typically ice-covered from December to April, and during that period the near-bottom water frequently suffers from oxygen deficiency. According to the OECD (1982) classification, the present-day conditions characterize Lake Peipsi s.s. as an eutrophic waterbody, while the trophic status of Lake Lämmijärv is close to hypertrophic and Lake Pihkva is a hypertrophic basin (Table 1).

Sampling and analyses of water

This study is based on a large dataset for Lake Peipsi, which contains more than 120,000 measurements of different hydrochemical and hydrobiological variables from 1950 to 2005 (Möls, 2005). Data for dissolved inorganic P (PO_4-P) are available since 1970 and for dissolved inorganic nitrogen (DIN = NH_4-N + NO_3-N + NO_2-N) since 1975. Total phosphorus (TP) and total nitrogen (TN) were analysed from 1985 to 2005, but earlier data are absent. The data were averaged over 5-year periods (geometric mean values for the open water period, days 100–310 within the year) to reduce the effect of inter-annual variation.

Most studies since 1992 have been on the Estonian part, but several joint Estonian–Russian expeditions to the whole lake have also been conducted. Depending on the year, the number of sampling sites monitored has varied between 3 and 41 (Kangur et al., 2002). Seasonal (or monthly) water samples for routine hydrochemical analysis (Starast et al., 2001) were obtained from the surface layer of 0.1–1.0 m and from the near-bottom layer of water (0.5 m above bottom), both with a Ruttner sampler.

TP and PO_4-P were determined in the water samples by the ammonium molybdate spectrometric method (EVS-EN 1189). TN, ammonium nitrogen (NH_4-N), nitrite nitrogen (NO_2-N) and nitrate nitrogen (NO_3-N) were measured by flow analyses and spectrometric methods (EVS-EN ISO 10304-1, 11732, 13395). Chemical analyses were performed at the Institute of Zoology and Botany of the Estonian University of Life Sciences, and since 1992 at Tartu Environmental Researches Ltd, Estonia.

Data analysis

Since the dataset used is highly unbalanced, we have estimated the content of nutrients in Lake Peipsi by covariance analysis, using a large generalized linear model that depends on 70 parameters. The model parameters (terms) include various functional transformations of the observation year, day within year, geographical coordinates and sampling depth (Möls et al., 2004; Möls, 2005). For the analyses we used procedures provided by the SAS System, Release 8.2 (SAS Institute Inc., 1999), especially the GLM procedure. For each dependent variable, the parameters of the model were fitted to the hydrochemical dataset so that the estimated model presents all the essential information contained in the data.

Prior to analysis, the chemical variables (TP, PO_4-P, TN, DIN) were log_2-transformed to make their residual distribution closer to the normal distribution needed for correct statistical inferences. To enable logarithms to be calculated, we replaced zeroes with small positive numbers approximately equal to half the detection limit of the relevant parameter. The whole data array was filtered twice iteratively by inspecting studentized residuals and excluding observations with absolute studentized residuals >3.0. The means and predicted values of the logarithmically transformed variables, after they were reconverted to the natural scale, are referred to throughout the article as geometric means.

Using the basic linear model, the logarithmic chemical variables were predicted for all the factor combinations 'year × day of the year × geographical point' from a formal regular grid and the predicted values were thereafter averaged separately for Lake Pihkva and Lake Peipsi s.s. (Möls, 2005). The concentrations were averaged over the Julian days 100 to 310 within each year and for the 1 m surface samples only. The difference between the two within-basin averages is called Logarithmic Polarity, and mathematically presented as a parametric function for the basic linear model; it was estimated, together with the corresponding 95% confidence limits, using the SAS GLM procedure. The resulting estimates were used to construct polarity graphs. The Y-axis on these graphs represents a binary logarithm of the ratio between the geometric means of water variables in the two basins. For example, if in a given year the logarithmic TP polarity was 2, then the geometric mean of TP content in Lake Pihkva in that year was four times higher than the corresponding mean in Lake Peipsi s.s. The SAS GLM procedure was also used to test hypotheses about trends and differences in 5-year means of water variables.

Results

Changes in the spatial distribution of nutrients

A comparison between the three parts of Lake Peipsi reveals steadily differing concentrations of TP and TN: throughout the period studied, the northern Lake Peipsi s.s. was significantly ($P < 0.0002$) poorer in nutrients than the southern Lake Pihkva (Fig. 2). However, the spatial distribution patterns of TP and TN are not similar on a long-term scale. In Lake Peipsi s.s., the TP content has not changed significantly ($P > 0.6$). In contrast, a continuous increase in TP concentration can be observed in Lake Pihkva ($P = 0.0006$); it has doubled during the past two decades (Fig. 2), which clearly indicates eutrophication of the lake. Compared to the TP concentration, the TN concentration in the lake water has remained relatively stable on the long-term scale: the small changes visible in Fig. 2 are not statistically significant ($P > 0.1$).

The long-term patterns of the TN:TP mass ratio are different in the southern and northern parts of the lake (Fig. 2). In Lake Peipsi s.s., this ratio has not changed significantly ($P > 0.1$), but it has decreased steadily in Lake Pihkva ($P = 0.0142$).

Long-term patterns in the spatial distribution of the mineral forms of nutrients (PO_4-P and DIN) are complicated (Fig. 3). We could detect no statistically significant large-scale linear or quadratic tendencies

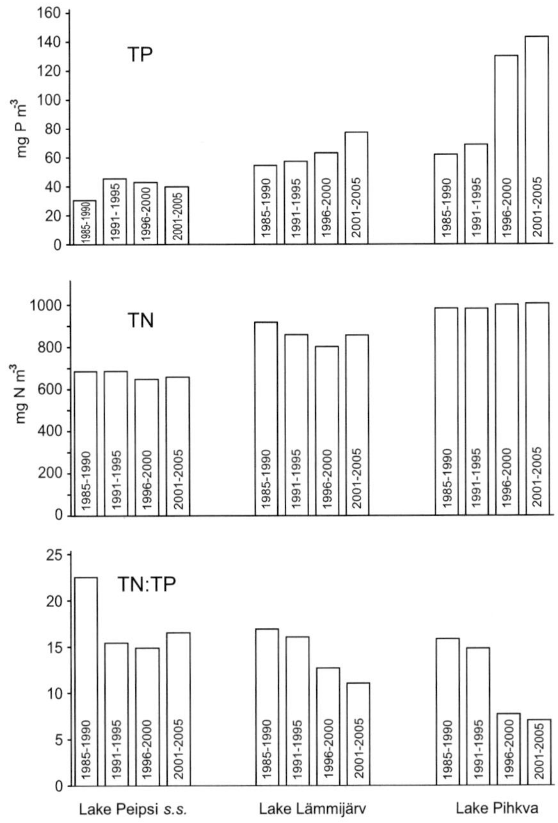

Fig. 2 Concentrations of total phosphorus (TP), total nitrogen (TN) and the TN:TP mass ratio averaged for 5-year periods in the three parts of Lake Peipsi in 1985–2005

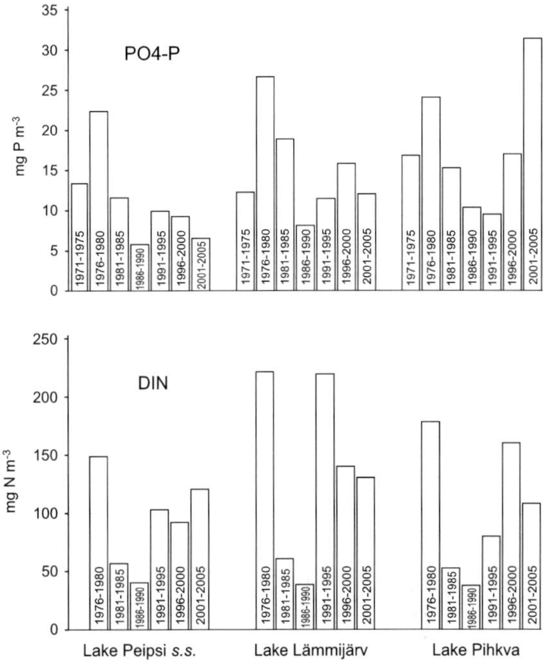

Fig. 3 Concentrations of dissolved inorganic phosphorus (PO_4-P) and dissolved inorganic nitrogen (DIN) averaged for 5-year periods in the three parts of Lake Peipsi

($P > 0.2$). The highest values of PO_4-P content were measured in the late 1970s (Fig. 3). However, the PO_4-P content of Lake Pihkva has been increasing during the past few years. The levels of PO_4-P and DIN were relatively low in the late 1980s (Fig. 3), but since then higher values of DIN have again been measured in the whole lake.

Long-term changes in the polarity of nutrients

The key nutrients (P and N) show different patterns of polarity (Figs. 4, 5). The polarity of P compounds has increased during recent years. In 1995, the TP content was 1.3 times higher in Lake Pihkva than in Lake Peipsi s.s.; in 2005, the ratio was about 3 (Fig. 4). The changes of PO_4-P polarity have been similar: the ratio increased from 1 to 4 over the period 1995–2005 (Fig. 6). However, the spatial gradient of PO_4-P in 1970–1995 was relatively small, the logarithmic polarity ranging between 0 and 0.6 (Fig. 6).

The polarity of TN has been rather stable on the long-term scale, the concentration in Lake Pihkva

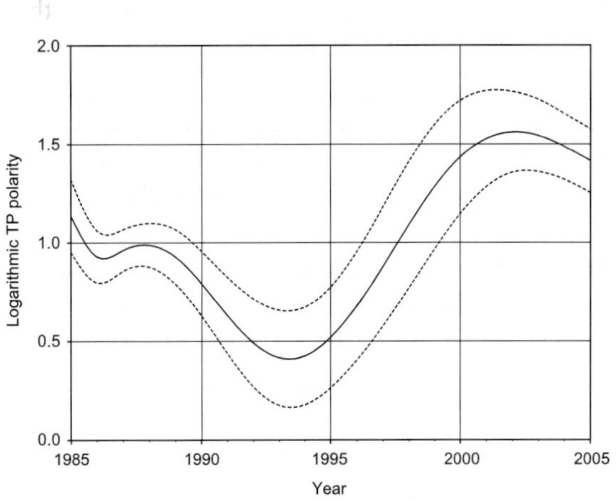

Fig. 4 Logarithmic polarity of total phosphorus (TP) in Lake Peipsi in 1985–2005. Polarity is defined as the binary logarithm of the ratio (Mean TP in Lake Pihkva):(Mean TP in Lake Peipsi s.s.). In this and the following Figs. 5–7, the surface water means for each variable were estimated by the model for days 100–310 within the year. Dashed lines show 95% confidence limits for the polarity estimate

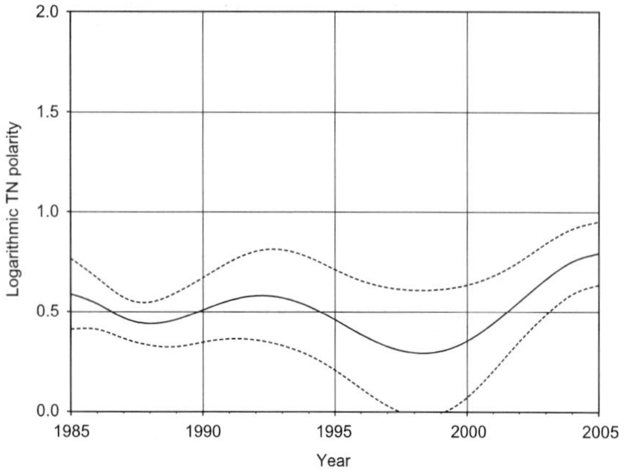

Fig. 5 Logarithmic polarity of total nitrogen (TN) in Lake Peipsi in 1985–2005. Polarity as indicated by the ratio (Mean TN in Lake Pihkva) (Mean TN in Lake Peipsi s.s.); cf. Fig. 4 caption for details

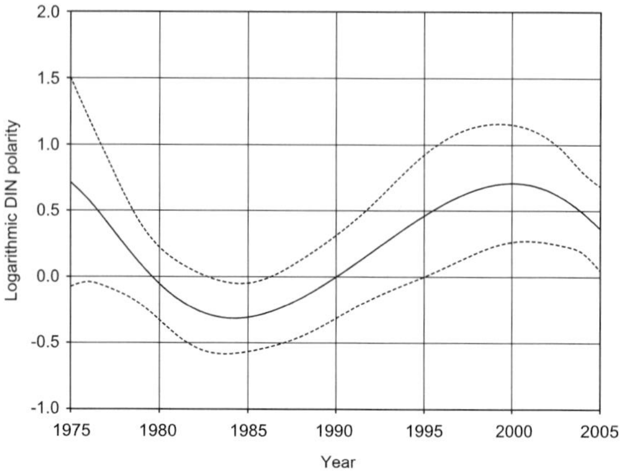

Fig. 7 Logarithmic polarity of dissolved inorganic N (DIN) in Lake Peipsi in 1975–2005. Polarity as indicated by the ratio (Mean DIN in Lake Pihkva):(Mean DIN in Lake Peipsi s.s.). For calculation details see Fig. 4 caption

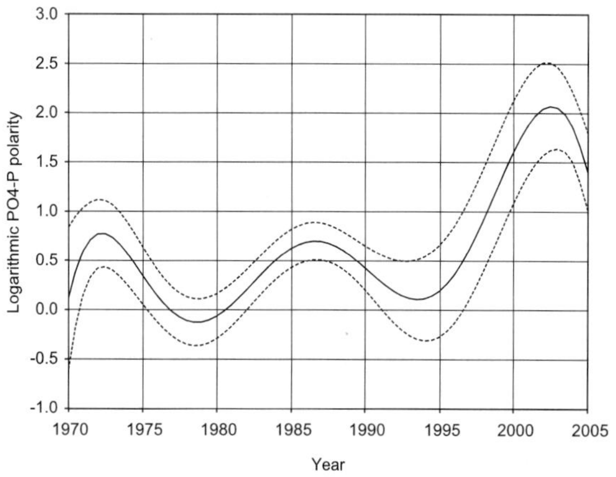

Fig. 6 Logarithmic polarity of dissolved inorganic P (PO_4-P) in Lake Peipsi in 1970–2005. Polarity as indicated by the ratio (Mean PO_4-P in Lake Pihkva):(Mean PO_4-P in Lake Peipsi s.s.). For calculation details see Fig. 4 caption

being about 1.4 times higher than in Lake Peipsi s.s. (Fig. 5). However, the difference in TN concentration between the northern and southern parts of the lake has slightly increased during recent years (Fig. 5). The temporal pattern of DIN polarity has also been fluctuating, with almost no gradient observed during the 1980s (Fig. 7).

Discussion

Eutrophication is well known to be the main reason for ecological deterioration of almost all lowland lakes worldwide (Gulati & van Donk, 2002; Moss et al., 2005). Carpenter (2003) defines eutrophication as the degradation of lake water quality by excessive inputs of P. In most lakes, P concentration limits productivity (Håkanson et al., 2003). Even a small excess of phosphorus runoff can cause eutrophication, while reversing this process is difficult and slow (Carpenter, 2005). These statements are supported by the results of the present study on Lake Peipsi. Our results also support the notion of Scheffer (2005) that the critical nutrient loading may be much lower for large lakes than for small ones.

Our results demonstrate that the patterns of spatial distribution of N and P in the lake are dissimilar on the long-term scale, which indicates different causative factors. The polarity of TN has remained relatively stable over the years, while the difference in P content between the northern and southern parts of the lake has been increasing. These results suggest that the increasing polarity of TP content is the primary eutrophication phenomenon in the lake and indicates increased P input from the south (e.g. the Russian part of the lake's catchment). An increasing concentration of P has been observed in Lake Pihkva, where the inflow of the Velikaya River contributes more than half the entire inflow to the whole of Lake Peipsi. Although long-term data on the external P and N loads from the Russian part of the catchment area are not available, previous assessments have indicated that most (about 65%) of the nutrients are carried into the lake via the Velikaya River

(Stålnacke et al., 2002). Despite dramatic reductions of N and P emissions in the drainage basin in the 1990s (Mourad et al., 2006), the overall mean TP content in the lake was higher in the 1990s than in the 1980s (Fig. 2).

Temporal stability in the polarity of TN content is probably caused mainly by differences in natural conditions between parts of the lake. The relative catchment size of Lake Pihkva is 2.5 times larger than that of Lake Peipsi s.s. in terms of lake surface area and 5 times larger in terms of volume. Therefore, the relative natural background loading level of Lake Pihkva is much bigger than that of Lake Peipsi s.s. Our results indicate that Lake Peipsi is highly resilient to year-to-year changes in N load, and the in-lake N concentrations are rather stable on the long-term scale. This might be due to in-lake processes: N_2 fixation by heterotrophic bacteria and cyanobacteria, and reduction of nitrate to N_2 by denitrification, might smooth out variations in N load. Within a certain trophic level, the internal mechanisms of matter circulation in a lake can compensate for variations in the external load (Punning et al., 1999). Similarly, Stålnacke et al. (2004) suggest that the dramatic decline in agricultural intensity in Eastern Europe after the late 1980s has led only to a slow and limited reduction in nutrient levels in many rivers. However, a thorough investigation of streams and rivers of the Estonian part of the Lake Peipsi drainage area detected statistically significantly declining trends in TN from 1984 to 2001 (Iital et al., 2005).

According to Carpenter (2003), lake P dynamics can be explained by a critical threshold value. Once the P level in the lake exceeds this threshold, it may take many years to return to low P levels; in some cases, the lake will never recover. In the case of Lake Peipsi, this threshold level of P seems to have been exceeded and the lake is "full" of P. Even a small emission of P from external sources, plus release of P from sediments into the water (Kangur & Kangur, 2008), have pushed the ecosystem of Lake Peipsi over a threshold to a degraded state. The limnological time-series data from 1950 to 2005 indicate deterioration of lake water quality and adverse changes in the whole ecosystem of Lake Peipsi. Eutrophication has led to an undesirable growth of algae, massive blooms of cyanobacteria accompanied by oxygen depletion during the night and fish kills, low water transparency and siltation of the lake bottom (Laugaste et al., 2001; Kangur et al., 2003, 2005). The eutrophication phenomena are most intensive in the southern part of the lake, Lake Pihkva. Doubled TP, PO_4-P and chlorophyll *a* contents, increased total alkalinity (HCO_3^-) and pH, as well as decline in water transparency and winter oxygen conditions, indicate a clear rise in the trophic level of Lake Pihkva (Kangur et al., 2007).

Many environmental changes can be reversed only slowly, and some changes are irreversible (Ludwig et al., 2005; Millennium Ecosystem Assessment, 2005). According to Carpenter (2005), it takes more than a century to reduce the phosphorus concentrations of agricultural soils to levels that do not degrade freshwater quality. Therefore, long-term horizons are necessary for making many environmental decisions, because they may have consequences that span many generations. Although the nutrient load from the Estonian part of the Lake Peipsi catchment has decreased since the early 1990s (Leisk & Loigu, 2001; Iital et al., 2005), we have been unable to observe any significant improvement in lake water quality. Judging by the state of the biota, the trophic situation in the lake has not improved.

Conclusions

The results of this study confirm previous assessments that high P content is the main reason for degradation of shallow lowland lakes. Our results indicate that the polarity of P content in Peipsi is increasing, which indicates pollution from the Velikaya River catchment. The relatively stable TN polarity pattern demonstrates that internal processes (e.g. N_2 fixation by cyanobacteria and heterotrophic bacteria, and bacterial denitrification) can largely compensate for year-to-year changes in the external N load. Our results demonstrate that eutrophication phenomena in this large shallow well-mixed lake with a relatively long residence time (about 2 years) are quite stable on the long-term scale and the observed ecosystem changes are probably of long duration or irreversible.

Acknowledgements We thank Marina Haldna for the mathematical calculations. Funding for this research was provided by the target-financed project SF 0362483s03 of the Estonian Ministry of Education and by Estonian Science Foundation grants 6820 and 6008. We used data obtained within the framework of the Estonian State Monitoring Programme on Lake Peipsi. We are grateful to the reviewers for their constructive comments on the manuscript.

References

Battarbee, R. W., N. J. Anderson, E. Jeppesen & P. R. Leavitt, 2005. Combining palaeolimnological and limnological approaches in assessing lake ecosystem response to nutrient reduction. Freshwater Biology 50: 1772–1780.

Carpenter, S. R., 2003. Regime shifts in lake ecosystems: pattern and variation. Vol. 15, Excellence in Ecology Series, Ecology Institute, Oldendorf/Luhe, Germany.

Carpenter, S. R., 2005. Eutrophication of aquatic ecosystems: bistability and soil phosphorus. Proceedings of the National Academy of Sciences 102: 10002–10005.

EVS-EN 1189, Determination of phosphorus – Ammonium molybdate spectrometric method.

EVS-EN ISO 10304-1, Determination of dissolved fluoride, chloride, nitrate and sulphate ions using liquid chromatography of ions.

EVS-EN ISO 11732, Determination of ammonium nitrogen by flow analysis and spectrometric detection.

EVS-EN ISO 13395, Determination of nitrite nitrogen and nitrate nitrogen and the sum of both by flow analysis and spectrometric detection.

Gulati, R. D. & E. van Donk, 2002. Lakes in the Netherlands, their origin, eutrophication and restoration: state-of-the-art review. Hydrobiologia 478: 73–106.

Håkanson, L., A. P. Ostapenia & V. V. Boulion, 2003. A mass-balance model for phosphorus in lakes accounting for biouptake and retention in biota. Freshwater Biology 48: 928–950.

Iital, A., P. Stålnacke, J. Deelstra, E. Loigu & M. Pihlak, 2005. Effects of large-scale changes in emissions on nutrient concentrations in Estonian rivers in the Lake Peipsi drainage basin. Journal of Hydrology 304: 261–273.

Jaani, A., 2001a. The location, size and general characterization of Lake Peipsi and its catchment area. In Nõges, T. (ed.), Lake Peipsi. Meteorology, Hydrology, Hydrochemistry. Sulemees Publishers, Tartu, 10–17.

Jaani, A., 2001b. Water regime. In Nõges, T. (ed.), Lake Peipsi. Meteorology, Hydrology, Hydrochemistry. Sulemees Publishers, Tartu, 65–72.

Kangur, K., A. Milius, T. Möls, R. Laugaste & J. Haberman, 2002. Lake Peipsi: changes in nutrient elements and plankton communities in the last decade. Journal of Aquatic Ecosystem Health and Management 5: 363–377.

Kangur, K., T. Möls, A. Milius & R. Laugaste, 2003. Phytoplankton response to changed nutrient level in Lake Peipsi (Estonia) in 1992–2001. Hydrobiologia 506–509: 265–272.

Kangur, K., A. Kangur, P. Kangur & R. Laugaste, 2005. Fish kill in Lake Peipsi in summer 2002 as a synergistic effect of cyanobacterial bloom, high temperature and low water level. Proceedings of the Estonian Academy of Sciences. Biology Ecology 54: 67–80.

Kangur, M., K. Kangur, R. Laugaste, J.-M. Punning & T. Möls, 2007. Combining limnological and palaeolimnological approaches in assessing degradation of Lake Pskov. Hydrobiologia 584: 121–132.

Kangur, M. & K. Kangur, 2008. Variations of surface sediment phosphorus fractions and water phosphorus content in the large shallow Lake Peipsi. Boreal Environment Research (submitted).

Laugaste, R., T. Nõges, P. Nõges, V. V. Jastremskij, A. Milius & I. Ott, 2001. Algae. In Pihu, E. & J. Haberman (eds), Lake Peipsi. Flora and Fauna. Sulemees Publishers, Tartu, 31–49.

Leisk, Ü. & E. Loigu, 2001. Nutrients. In Nõges, T. (ed.), Lake Peipsi. Meteorology, Hydrology, Hydrochemistry. Sulemees Publishers, Tartu, 79–82.

Ludwig, D., W. A. Brock & S.R. Carpenter, 2005. Uncertainty in Discount Models and Environmental Accounting. Ecology and Society 10: 13. (on line) http://ecologyandsociety.org/vol10/iss2/art13.

Milius, A., R. Laugaste, T. Möls, M. Haldna & K. Kangur, 2005. Weather conditions and water level as factors determining phytoplankton biomass and nutrient content in Lake Peipsi. Proceedings of the Estonian Academy of Sciences. Biology Ecology 54: 5–17.

Millennium Ecosystem Assessment, 2005. Ecosystems and human well-being: general synthesis. Island Press, Washington, D.C., USA. (online) http://maweb.org/en/Products.Synthesis.aspx.

Möls, T., 2005. Linear statistical methods for Estonian freshwater waterbodies. Estonian Naturalists' Society, Tartu, 256 pp (in Estonian with English summary).

Möls, T., K. Kangur, M. Haldna, A. Milius, J. Haberman, R. Laugaste & M. Möls, 2004. The Synthetic Hydrochemical and Hydrobiological Database (SD) for Lake Peipsi (Release 1.0). Tartu, CD.

Moss, B., T. Barker, D. Stephen, A. E. Williams, D. Balayla, M. Beklioglu & L. Carvalho, 2005. Consequences of reduced nutrient loading on a lake system in a lowland catchment: deviations from the norm? Freshwater Biology 50: 1687–1705.

Mourad, D. S. J., M. Van der Perk & K. Piirimäe, 2006. Changes in nutrient emissions, fluxes and retention in a North-Eastern European lowland drainage basin. Environmental Monitoring and Assessment 120: 415–448.

OECD, 1982. Eutrophication of water, monitoring, assessment and control. Organization for Economic Cooperation and Development (O.E.C.D.), Paris, 150 pp.

Padisák, J. & I. Koncsos, 2002. Trend and noise: long-term changes of phytoplankton in the Keszthely Basin of Lake Balaton, Hungary. Verhandlungen der Internationalen Vereinigung für Theoretische und Angewandte Limnologie 28: 194–203.

Punning, J.-M., J. Terasmaa & K. Tõugu, 1999. Changes in the human impact on Lake Ruusmäe in recent decades and their reflection in the sedimentary organic matter. Proceedings of the Estonian Academy of Sciences. Biology Ecology 48: 130–142.

SAS Institute Inc., 1999. SAS Online Doc, Version 8. Cary, NC, SAS Institute Inc.

Scheffer, M., 2005. Shallow lakes and deep insights. Shallow Lakes 2005. Dalfsen, The Netherlands, Book of abstracts: 7.

Stålnacke, P., Ü. Sults, A. Vasiliev, B. Skakalsky, A. Botina, G. Roll, K. Pachel & T. Maltsman, 2002. An assessment of riverine loads of nutrients to Lake Peipsi, 1995–1998. Archiv für Hydrobiologie 141: 437–457.

Stålnacke, P., S. M. Vandsemb, A. Vassiljev, A. Grimvall & G. Jolankai, 2004. Changes in nutrient levels in some Eastern European rivers in response to large-scale changes in agriculture. Water Science and Technology 49: 28–36.

Starast, H., A. Milius, T. Möls & A. Lindpere, 2001. Hydrochemistry of Lake Peipsi. In Nõges, T. (ed.), Lake Peipsi: Meteorology, Hydrology, Hydrochemistry. Sulemees Publishers, Tartu, 97–131.

Tilzer, M. M., 1990. Specific properties of large lakes. In Tilzer, M. M. & C. Serruya (eds), Large Lakes: Ecological Structure and Function. Springer-Verlag Berlin Heidelberg, 39–43.

Hydrobiologia (2008) 599:41–48
DOI 10.1007/s10750-007-9211-1

ELLS 2007

Recent trends in Lake Ladoga ice cover

Sergey G. Karetnikov · Mikhail A. Naumenko

Abstract The condition of the ice on a large lake is an important sensitive indicator of regional climate change. We have investigated the ice regime on Lake Ladoga using aircraft surveys and satellite images. Nearly 1,000 surveys of the lake's ice cover covering the years from 1943 to the present were analyzed to determine the first basic statistical characteristics of Ladoga ice cover based on long-term observation. The aim was to estimate long-term trends in the ice cover. Analysis showed that the average period of ice-cover formation is up to 1.5 times longer than the period of ice-cover destruction. The inter-annual trends in first and last ice dates, and in the duration of ice cover, are a few days per century, i.e. significant but rather small. We created the Relative Ice Cover Index (RICI), a dimensionless single-number variable that allowed the characteristics of winter seasons to be compared. A statistically significant downward trend of no more than 12% in RICI was observed over the 63-winter periods. The distribution of RICI was compared with the North Atlantic Oscillation Index (NAO) in an effort to explain the ice cover trend.

Keywords Lake Ladoga · Ice cover · Climatic trend

Introduction

Ice cover on a lake affects mass and energy exchanges between the lake surface and the atmosphere. Its spatial and temporal characteristics are of considerable interest because of their great importance for the fishing industry and the winter aquatic system. Ice cover is a major component of the annual thermal cycles of lakes; it impacts on weather, navigation, and shoreline processes. In addition, it influences thermal/density structures, with consequences for convective overturn, oxygenation, currents and lake level. Ice cover reduces the penetration of sunlight available for photosynthesis within the water body. The potential economic impacts of climate warming include prolongation of the navigation season on the one hand, and on the another a reduced possibility of transit over stable ice to monasteries on the islands of Valaam and Konevits (Fig. 1).

The year-to-year variation in lake ice conditions is determined not only by the heat flux on the water surface in winter but also by the distribution of depths, and depends on the severity of winter in the

Guest editors: T. Nõges, R. Eckmann, K. Kangur, P. Nõges, A. Reinart, G. Roll, H. Simola & M. Viljanen
European Large Lakes—Ecosystem changes and their ecological and socioeconomic impacts

S. G. Karetnikov (✉) · M. A. Naumenko
Institute of Limnology of Russian Academy of Sciences, Sevastyanov St., 9, 196105 St. Petersburg, Russia
e-mail: sergeyka55@mail.ru

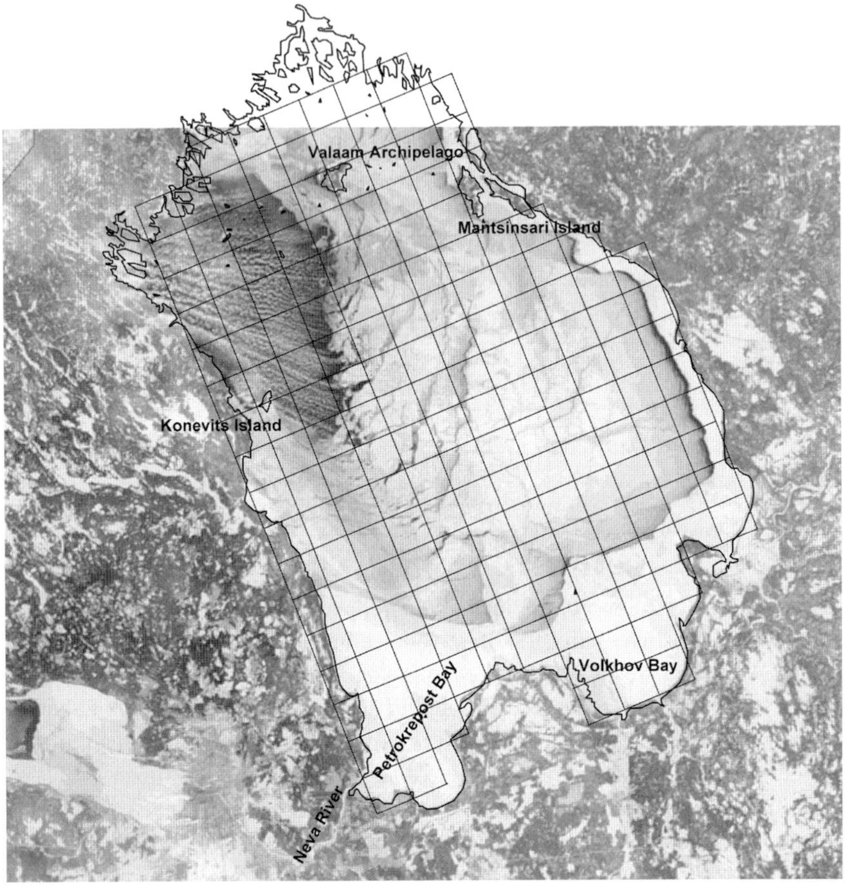

Fig. 1 MODIS Lake Ladoga ice cover image for February 2nd 2006, with a grid to calculate mean lake ice concentration

region. Partial ice cover reduces heat and momentum exchange between the atmosphere and the water mass, while full ice cover stops it completely. Even thin ice cover markedly reduces heat loss from the water surface. Thus, ice-free water depletes the heat storage of the water mass; low heat storage influences the vernal thermal regime.

The thermal and ice regimes of the large lakes in North-West Russia were investigated by Tikhomirov (1966). Medres (1957) plotted the freezing and ice-breakup patterns in Lake Ladoga under average conditions. Marszelewski & Skowron (2006) studied ice cover in Polish lakes. Assel et al. (2003) investigated recent trends in ice cover on the Laurentian Great Lakes. Wang et al. (2006) applied a sea ice dynamics model to simulate ice events in a large lake. Palecki & Barry (1986) derived statistical relationships between the freeze-up and ice-breakup dates and the air temperature for 63 lakes in Finland.

The purpose of our study was to calculate the basic statistical characteristics for Lake Ladoga, which include the dates and durations of both ice phenomena. A further goal was to plot, and derives a formula for, the typical seasonal course of percentage ice cover. We then investigated the inter-annual variations of all the aforementioned parameters.

Continuous observations of ice conditions can be used as a basis for local climate analysis. Climate changes in north-western Europe are reflected in the increasing trend of the North Atlantic Oscillation (NAO) Index. Climate change strongly affects the hydrology and biology of a shallow lake (Nõges, 2004). In an effort to explain our observations we have developed a Relative Ice Cover Index (RICI) (see below), which represents each winter as a single dimensionless number, for comparison with the NAO Index.

Material and methods

Lake Ladoga is the largest dimictic lake in Europe with a total area of 17,700 km^2, average depth 47 m and maximum depth 230 m. Ice phenomena on the surface of Lake Ladoga can be seen for more than half the year (from the beginning of November until the end of May).

The spatial distribution of ice over Ladoga has been observed regularly since 1943. Aircraft surveys were made by the Hydrometeorological Service until 1992 and charts of the ice concentration on Ladoga were published approximately twice a month. In addition to the aircraft surveys, satellite images on the visual, infrared and microwave channels have been made since 1971. Information on ice cover before 1982 was obtained from published tables (Medres, 1957; Lebedev & Medres, 1966; Usachev et al., 1985). We have continued to calculate ice cover concentration (percentage area covered by ice) by aircraft surveys and satellite images using the same methods. Recently, NOAA and MODIS satellite images have became the main sources accept for studying ice cover of the lake for in situ information and scientific purposes.

The increase in the number of surveys per ice-cover period from early 1940s to the present is shown in Fig. 2. On an average, 15 surveys were carried out each winter; the maximum was 37 surveys in 2006. Generally, ice characteristics were mostly observed in Ladoga during April. This is related to the growing interest in ice conditions before and throughout the time of breakup. Also, the weather becomes clearer at the beginning of spring.

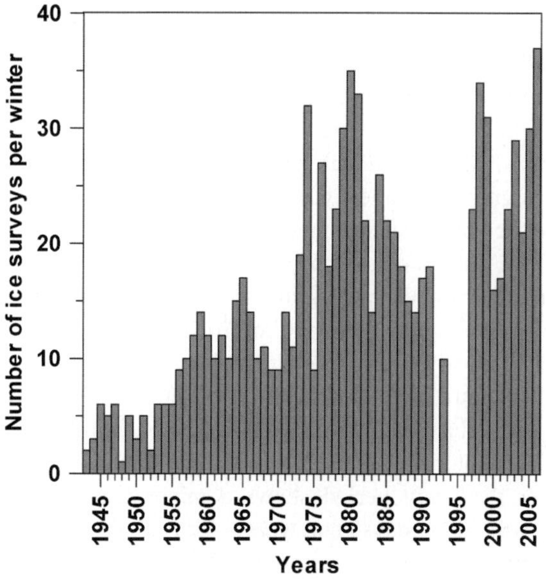

Fig. 2 The number of ice surveys per winter from 1943 until 2006

Aircraft and satellite information on Ladoga ice cover conditions has been collected from 1943 through 2006. However, ice cover data for the period 1992–1996 are missing.

In order to calculate the extent of lake ice cover, taking its concentration into account, the surface area of Ladoga was split into 183 cells (10 km per side) (Fig. 1). Each ice chart, from either an aircraft or a satellite image, was converted into a geographic projection. The average percentage ice cover was estimated for each cell with 10% accuracy. The ice concentration was averaged over all grid cells to calculate the lake-averaged ice concentration on each composite chart for a given winter season. Till date, about 1,000 surveys have been processed for the 63-year period. We took account only of the spatial ice concentration, not its thickness.

Results

In order to calculate the average long-term seasonal ice cover pattern, all available data for October through May were plotted without regard to year of observation. The data were approximated by a fifth order polynomial fit (Eq. 1). The main criteria of fit order choice were the maximum coefficient of determination and curve smoothness. In our case, the coefficient of determination was 64% (Fig. 3).

$$\begin{aligned} IC = & -0.11 + 0.05 \cdot D - 0.008 \cdot D^2 \\ & + 0.00026 \cdot D^3 - 1.64 \cdot 10^{-6} \cdot D^4 \\ & + 2.93 \cdot 10^{-9} \cdot D^5 \end{aligned} \quad (1)$$

where IC is the lake ice concentration (percentage) and D represents days since October 1st. The curve application is valid for the period November 5th through May 29th.

The first feature we should mention is that the mean curve does not extend to 100% ice concentration. The same result was cited by Tikhomirov (1966). The absence of the 100% value from the average ice-cover curve is due to a scatter of years

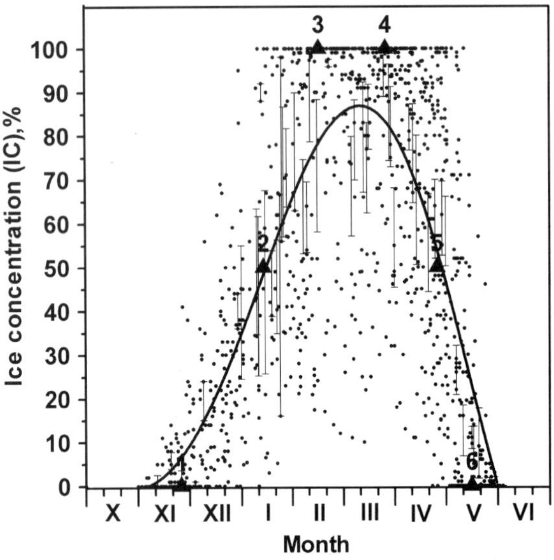

Fig. 3 The mean curve of ice concentration and average dates of basic winter stages (Δ) 1—freeze-up starting date, 2—50% ice cover date, 3—100% ice cover, 4—100% ice cover breakup, 5—50% ice cover breakup, 6—end of breakup (see Table 1)

with less than full ice cover. The second feature is that the average period of ice-cover formation is up to 1.5 times longer than the period of ice–cover destruction. This can be explained by wind action: wind prevents freeze-up and promotes ice breakup.

All changes in ice concentration were plotted for each year from the time of appearance of the first ice, through its maximum spatial distribution, and up to its complete disappearance from the water surface (Fig. 4). It is sometimes difficult to establish the time of appearance and disappearance of the lake ice using information from remote sensing surveys. In such cases, we used dates that were either taken from the coastal hydrometric network or obtained by extrapolating the chronological ice concentration curve to zero.

We propose the RICI to characterize each year in terms of a single dimensionless value. The RICI represents the area under the actual chronological ice concentration curve for a given year divided by the area under the mean seasonal curve (Fig. 3). The RICI allows years to be compared with each other. Figure 4 shows examples of different ice concentration curves with corresponding RICI values.

Data obtained throughout the continuous period of observation from 1943 to 2006 were used to estimate the main statistical characteristics of ice event dates and the durations of ice cover. Mean $X = (\sum x_i)/N$, median M, standard deviation $\sigma = \left[\sum (x_i - x)^2/N\right]^{1/2}$ and standard error of mean $E = \sigma \cdot (N)^{-1/2}$ were calculated for each winter season, where N is number of seasons (Table 1). The date of freeze-up was taken as the date of appearance of the first ice. The date of ice breakup was considered as the date of last ice set in the lake. The total ice cover period was calculated as the number of days between freeze-up and breakup (Assel et al. 2003). The complete ice cover was found to be unstable in some years, so the complete ice cover period was taken as 95% ice concentration. The date of 95% ice cover freeze-up represents the first date of complete ice cover, and the date of 95% ice cover breakup represents the last (Table 2).

Inter-annual changes in the first and last ice dates and in the duration of total and complete ice cover in the lake were analyzed together with the RICI. We calculated the linear trends of these parameters for the whole period of observation to investigate the inter-annual changes of ice conditions. All parameters under consideration showed significant (95%) trends. The slopes of regression lines and coefficients of determination were rather small for all trends.

We estimated the linear freeze-up date trend (Fig. 5c, curve 2) and found that freeze-up starts earlier every year (−0.14 days/year); however, the coefficient of determination for this trend is only 3%.

The dates of ice break-up (Fig. 5c, curve 1) have very small inter-annual variability (Table 1) and have no significant tendency at all (0.01 days/year), with a coefficient of determination of 0.3%.

The inter-annual changes in the duration of total ice cover for the period investigated (Fig. 5b, curve 1) have a minor tendency to growth (0.06 days per year), mostly due to later freeze-up dates.

The duration of complete ice cover on the lake (if it takes place) (Fig. 5b, curve 2) can reach 4 months (e.g. winter 1986–1987). The inter-annual change over the study period shows a minor tendency to growth (0.04 days per year) with a very small coefficient of determination (0.03%).

The RICI was calculated for all winters for which there were sufficient data to build up a

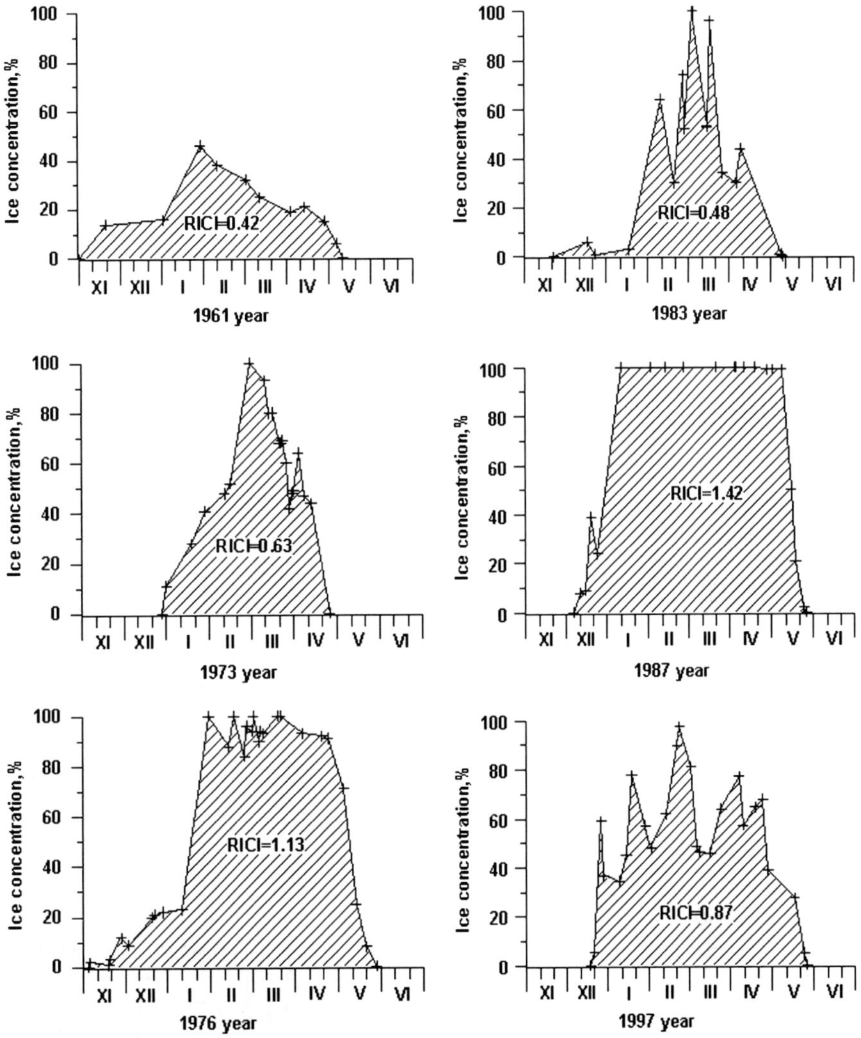

Fig. 4 Examples of various types of ice cover patterns during the winter period in selected years. Numbers within plot are RICI

Table 1 The statistical characteristics of dates of ice cover stages in Lake Ladoga

Date	Freeze-up	50% ice cover freeze-up	95% ice cover freeze-up	95% ice cover break-up	50% ice cover break-up	Break-up
Mean X	November 24	January 11	February 13	March 24	April 23	May 13
Std.err. E. days	1.8	2.6	3.2	3.7	2.5	1.3
Median M	November 24	January 9	February 11	March 22	April 26	May 12
Std.dev.σ, days	12.7	18.4	20	23.1	18.3	10.2
Minimum	03/11/75	09/12/55	08/01/87	18/01/50	25/02/90	20/04/75
Maximum	26/12/72	19/02/01	28/03/57	09/05/85	21/05/56	29/05/74,79
Range, days	53	72	79	111	85	39
Total number of winter seasons used, N	50	52	38	39	52	59

complete course of ice cover conditions. The linear trend of the RICI (Fig. 5a) shows a negative tendency (−0.002 unit/year). The coefficient of determination of this trend, however, is only 3%. The decrease in RICI was no more than 12% over this time.

Table 2 The statistical characteristics of the duration of ice cover in Lake Ladoga

Duration, (days)	95% ice cover	Total ice cover
Mean X	34	172
Std.err. E	4.6	2.5
Median M	27	172
Std.dev. σ	34.2	14.7
Minimum (years)	0 (1945, 1961, 1971, 1975, 1989, 1990, 2000, 2004)	118 (1973)
Maximum (years)	117 (1987)	203 (1976)
Range, days	117	85
Total number of winter seasons used, N	55	53

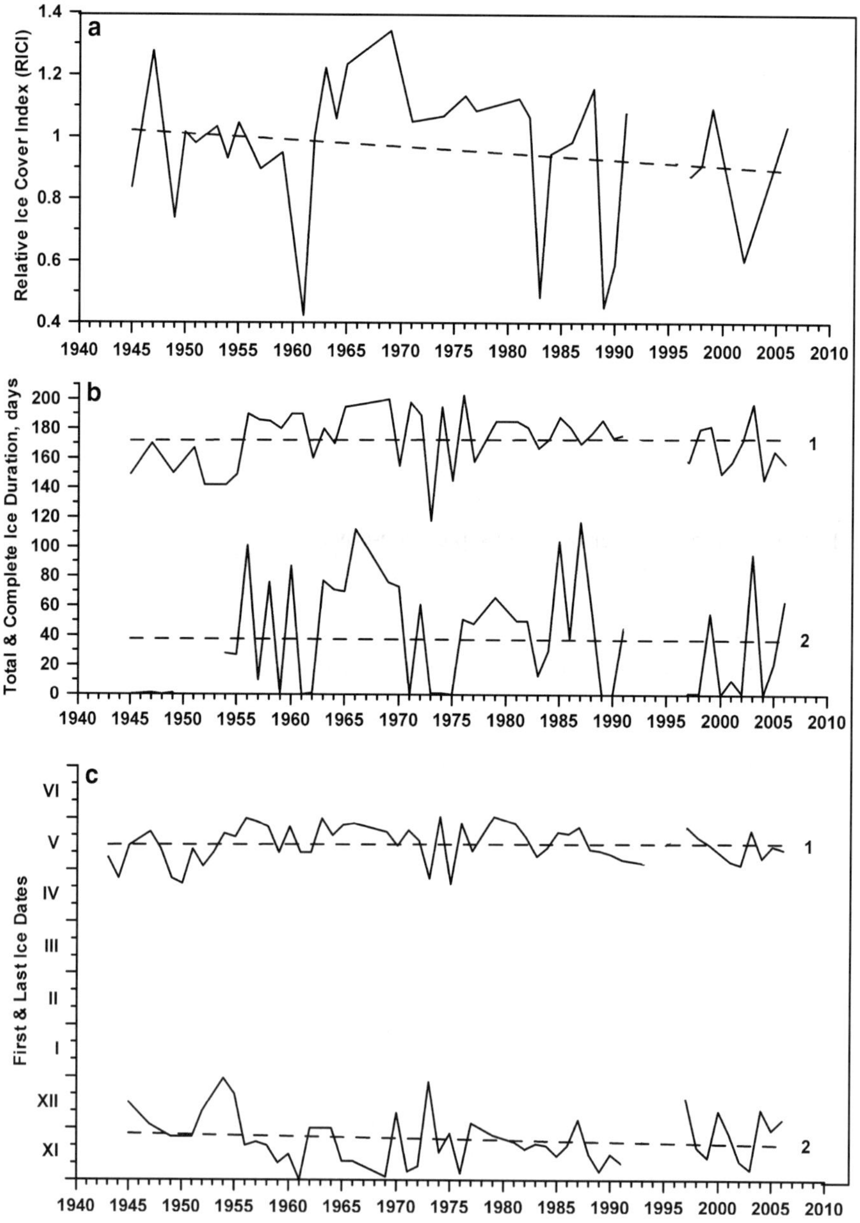

Fig. 5 (**a**) Relative Ice Cover Index (RICI), (**b**) total (curve 1) and compliete (curve 2) ice duration, and (**c**) dates of last (curve 1) and first (curve 2) ice with linear trends (dashed lines)

Discussion

As a rule, the first ice on Ladoga is formed in the shallow southern Petrokrepost Bay. It forms later in Volkhov Bay and Svir Bay, then along the eastern coast up to Mantsinsari Island. The water along the west coast and the northern deep-water zone are the last to become ice-covered. Water remains ice-free for longest over the deepest areas of the lake, and sometimes this area does not become ice-covered. An example of a MODIS Lake Ladoga ice-cover image (for 2 February 2006) is presented in Fig. 1. Stable ice cover along each shore (white color) is seen in the image; moveable ice is seen to the south from the Valaam Archipelago (grey color with crack), and there is a wide area of ice-free water (black) with clouds over it.

In spring, the ice in the central area of Ladoga can drift under the wind that breaks the floating ice. Depending on the wind direction, the ice is driven to inshore regions, where it melts. A small portion is driven to Petrokrepost Bay and reaches the Neva River.

In order to interpret the inter-annual changes of ice cover in Lake Ladoga, the RICI was compared with the NAO index (http://www.cru.uea.ac.uk/cru/data/nao.htm), which was calculated as an index sum for the period from October to May (Fig. 6). The coefficient of determination of this correlation was 18%. Importantly, the points on the graph (Fig. 6) were higher for winters with periods of complete ice-cover exceeding 77 days (1956, 1958, 1960, 1966, 1967, 1970, 1972, 1979, 1985, 1987, and 2003), while for winters with periods of total ice cover less than 148 days (1973, 1975, 2000, 2001, 2004 and 2005) the points were below the regression line. These years are marked with ellipses in Fig. 6. If we exclude them from the correlation analysis, the coefficient of determination increases to 58%, indicating that the NAO index cannot fully explain the extreme ice-covering years. This implies that years with extreme ice cover in the north-western region of Russia are influenced not only by the North Atlantic.

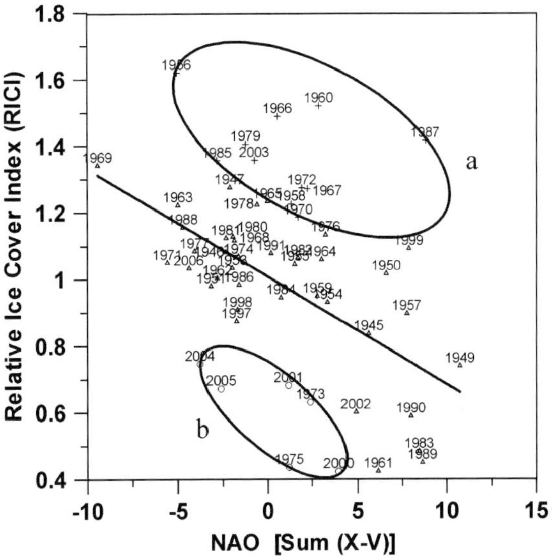

Fig. 6 The relationship between RICI and the North Atlantic Oscillation (NAO) Index. Numbers denote years, and ellipses indicate the extreme years

Conclusion

A number of important results follow from this analysis. Information about the percentage ice cover on Ladoga obtained from aircraft and satellite surveys between 1943 and 2006 has been collected and organized into a database. The database was used to derive the mean seasonal curve of ice cover and a formula to represent ice conditions in Lake Ladoga. The mean curve shows asymmetric branches: ice formation on the lake is slower than its destruction. Basic statistical characteristics for the dates of the main ice phenomena and the duration of coverage have been determined.

Analysis of the inter-annual variation in ice phenomena dates and the duration of coverage shows very small but significant trends (1–6 days per 100 years). The trend is most marked for the freeze-up dates (14 days per 100 years), but the coefficient of determination is no more than 3%. The RICI was derived to characterize each winter by a single dimensionless value. It shows a significant negative climatic trend. The decrease in RICI was no more than 12% over the 63-winter periods, with a rather small coefficient of determination. The RICI for Lake Ladoga correlates well with the winter NAO Index for average ice conditions. Nevertheless, the NAO Index cannot explain years with extreme ice cover.

References

Assel, R., K. Cronk & D. Norton, 2003. Recent trends in Laurentian Great Lakes ice cover. Climatic Change 57: 185–204.

Lebedev, V. V. & P. L. Medres, 1966. Ledovyi rezhim Ladozhskogo ozera po materialam aviarazvedok. Sbornik rabot Leningradskoi GMO 3: 135–182. (Lake Ladoga ice covers condition by the aircraft surveys data. In Russian).

Marszelewski, W. & R. Skowron, 2006. Ice cover as an indicator of winter air temperature changes: case study of Polish Lowland lakes. Hydrological Sciences Journal 51: 336–349.

Medres, P. L., 1957. Ledovyi rezhim Ladozhskogo ozera po materialam aviarazvedok, Vol. 66. Trudy GGI, Gidrometeoizdat, Leningrad, 92–140. (Lake Ladoga ice cover conditions by the aircraft surveys data. In Russian).

Nõges, T., 2004. Reflection of the changes of the North Atlantic Oscillation Index and Gulf Stream Position Index in the Hydrology and phytoplankton of Võrtsjarv, a large, shallow lake in Estonia. Boreal Environment Research 9: 401–407.

Palecki, D. E. & R. G. Barry, 1986. Freeze-up and break-up of lakes as an index of temperature changes during the transition seasons: a case study for Finland. Journal of Climate and Applied Meteorology 25: 893–902.

Tikhomirov, A. I., 1966. Raschet sredniykh mesyachnykh temperatur poverkhnosti vody Ladozhskogo ozera. In Malinina, T. I. (ed.), Gidrologicheskii rezhim i vodnyi balans Ladozhskogo ozera. Lengosuniversitet, 279–323. (Lake Ladoga mean monthly water surface temperatures calculation. In Russian.).

Usachev, V. F., V. G. Prokacheva & V. V. Borodulin, 1985. Otsenka dinamiki ozernykh l'dov, snezhnogo pokrova i rechnykh razlivov distantsionnymi sredstvami. Gidrometeoizdat, Leningrad, 103 pp. (The lake ice dynamics, snow covering and river floods estimation by remote sensing. In Russian).

Wang, K., M. Leppäranta & A. Reinart, 2006. Modeling ice dynamics in Lake Peipsi. Verhandlungen der internationalen Vereinigung fur Limnologie 29: 1443–1446.

Hydrobiologia (2008) 599:49–58
DOI 10.1007/s10750-007-9199-6

ELLS 2007

History of anthropogenically mediated eutrophication of Lake Peipsi as revealed by the stratigraphy of fossil pigments and molecular size fractions of pore-water dissolved organic matter

Aina Leeben · Ilmar Tõnno · Rene Freiberg · Viia Lepane · Nicolas Bonningues · Natalja Makarõtševa · Atko Heinsalu · Tiiu Alliksaar

Abstract We investigated stratigraphic changes in fossil pigments and the molecular structure of the UV-absorbing fraction of pore-water dissolved organic matter in a sedimentary record from Lake Peipsi (Estonia/Russia) temporally covering the 20th century. The aims of the study were to define the onset of eutrophication in the lake and to track its course. An attempt was also made to reconstruct lake conditions before the intensive nutrient loading began. Fossil pigment analysis indicated that the eutrophication of the lake started in the 1960s and accelerated in the 1970s. Sedimentary pigments also indicate a continuing tendency of the lake ecosystem towards eutrophy in the 1980s and 1990s. However, changes in the molecular size structure of pore-water dissolved organic matter indicated that the contribution of autochthonous matter to the organic pool of the lake ecosystem had already started to increase around the end of the 1930s. We conclude that this rise was generated by a coincidence of several anthropogenic and natural factors. The pore-water data also show that a slight relative reduction in the autochthonous organic matter took place in the 1990s. A discordance in the paleodata obtained for the beginning of the 20th century complicates clear conclusions about earlier conditions in the lake. On the one hand, the qualitative characteristics of pore-water dissolved organic matter and the low concentration of chlorophyll *a* indicate that the phytoplankton biomass was low in Lake Peipsi during that period. On the other hand, the concentrations of marker pigments of specific phytoplankton groups are high, comparable with the values in the recent sediments. Possible reasons for the high levels of these pigments in the early 1900s sediments, such as a shift in the preservation conditions of organic substances and their transport from the lake's catchment, are discussed.

Guest editors: T. Nõges, R. Eckmann, K. Kangur, P. Nõges, A. Reinart, G. Roll, H. Simola & M. Viljanen
European Large Lakes—Ecosystem changes and their ecological and socioeconomic impacts

A. Leeben (✉)
Marine Systems Institute, Tallinn University of Technology, Akadeemia tee 21, 12618 Tallinn, Estonia
e-mail: aina.leeben@phys.sea.ee

I. Tõnno · R. Freiberg
Centre for Limnology, Institute of Agricultural and Environmental Sciences, Estonian University of Life Sciences, Rannu, Tartu County 61101, Estonia

V. Lepane · N. Bonningues · N. Makarõtševa
Institute of Chemistry, Tallinn University of Technology, Akadeemia tee 15, 12618 Tallinn, Estonia

N. Bonningues
Joseph Fourier University, DSU, 480 Avenue Centrale, 38400 Saint-Martin d'Heres, France

A. Heinsalu · T. Alliksaar
Institute of Geology, Tallinn University of Technology, Ehitajate tee 5, 19086 Tallinn, Estonia

Keywords Paleolimnology · Lake sediments · Sedimentary pigments · Pore water · Dissolved organic matter · Lake Peipsi

Introduction

Increased nutrient loading due to intensified agriculture, industry and urbanization has been one of the forcing factors influencing European lakes during recent centuries. In particular, the period since the Second World War has been characterized by the enrichment of water bodies with biogenic substances. However, initial changes and the course of eutrophication in large water ecosystems are often either poorly documented or information on them is entirely missing. In such cases, a paleolimnological approach is often the only way to reconstruct the history of a lake.

Lake Peipsi is the fourth largest lake in Europe by surface area. The catchment consists of fertile soils; there are several municipalities with up to 200,000 inhabitants and numerous industrial establishments. The lake has been subjected to increasing anthropogenic inputs of nutrients from the 1950s onwards, as demonstrated by paleoecological analyses of diatom assemblages (Heinsalu et al., 2007). Monitoring data collected since the early 1960s reveal subsequent nutrient inputs and changes in the lake ecosystem (Nõges, 2001; Pihu & Haberman, 2001).

In this study, we present a reconstruction of the recent eutrophication history of Lake Peipsi at a high-resolution scale based on fossil pigment records and stratigraphic changes in the molecular size structure of pore-water dissolved organic matter (*p*DOM). Two methods were employed in the analyses of pigments and *p*DOM: first, crude samples were analysed spectrophotometrically; thereafter, compounds in the sample were separated and analysed by high-performance liquid chromatography (HPLC). Although measurements of *p*DOM are not yet widely employed in paleoinvestigations, several studies have demonstrated the capacity of these methods to provide insight into the origin and character of organic matter (OM) (Peuravuori & Pihlaja, 1997; Chin et al., 1998; Lepane et al., 2004). Moreover, results from a recent study have proved that the absorbance characteristics of *p*DOM have high potential for tracking the sources of organic carbon and for assessing the paleoproductivity of a lake (Leeben et al., 2005).

Study site

Lake Peipsi is a large (3,555 km^2), shallow (mean depth 7.1 m, max depth 15.3 m), unstratified and eutrophied water body, which consists of two extensive basins (Peipsi proper and Lake Pihkva) joined by a narrow strait (Lämmijärv). The lake receives its water from a catchment area of 47,800 km^2 and has an outflow into the Gulf of Finland. The mean residence time of the water is about 2 years. A characteristic feature of the lake is the fluctuation of the water level (mean annual range 1.2 m). Approximately 40% of the lake's catchment area is covered by coniferous and mixed forests, and approximately another 40% is used for agricultural purposes. The soils of the catchment are very diverse (loamy Cambisols, Luvisols, Podzoluvisols, sandy Podzols, Planosols, Gleysols, Histosols). During wintertime, the lake is covered by ice for an average of 114 days. A comprehensive overview of the catchment and limnology of Lake Peipsi can be found in Nõges (2001) and Pihu & Haberman (2001).

Materials and methods

Sediment coring and dating

Sediment samples were obtained from the middle of the broadest part of the lake (58°47′14″ N, 27°19′20″ E; water depth 9.2 m) with a freeze corer (Wright, 1980) in winter 2002 and 2006. The cores were sectioned into consecutive 1-cm sub-samples. The 2002 core was subjected to dating and pore-water analyses; pigment analyses were performed on the 2006 core. The two cores were correlated through their loss-on-ignition curves.

For dating, the sediment samples were analysed for ^{210}Pb, ^{226}Ra and ^{137}Cs by gamma spectrometry using a well-type coaxial low background intrinsic germanium detector (Appleby et al., 1986). For the ^{210}Pb chronology, the CRS model (Constant Rate of Supply; Appleby et al., 1986) was applied. The methodology, results and reliability of the chronology are presented and discussed in detail in Heinsalu et al. (2007).

Analyses of pore-water dissolved organic matter (*p*DOM)

The frozen sediment samples were thawed at 4°C. The pore water was collected and clarified by centrifugation and filtration (pore size 0.2 μm). Absorption spectra were recorded over the range 200–800 nm with a Cadas 100 scanning spectrophotometer (Dr. Lange) using Milli-Q water as a blank. The ratio of the absorbances at 250 and 365 nm was calculated to evaluate the character and molecular structure of *p*DOM: a high ratio indicates organic substances with a low degree of aromaticity and small molecular size (Peuravuori & Pihlaja, 1997). The absorbance values were converted into carbon concentrations using a linear relationship between the absorbance values of *p*DOM at 250 nm and the concentration of dissolved organic carbon (DOC) in the pore water ($r = 0.87$, $n = 10$). The DOC concentration was determined by the standard method ISO 8245 (1999) using a DC-80 Total Organic Carbon Analyzer (Rosemount-Dohrmann): the organic carbon is converted to CO_2 by ultraviolet-promoted persulphate oxidation, and the CO_2 formed is detected in the infrared part of the spectrum.

High-performance size-exclusion chromatography (HPSEC) of pore-water samples was carried out using an HPLC system with a UV detector (Knauer) coupled to a biocompatible BIOSEP-SEC-S2000 column (7.5 × 300 mm, Phenomenex). A 100-μl sample was injected into the column. Compounds were eluted with 0.02 M phosphate buffer (pH 6.8) at a flow rate of 1 ml min^{-1} and detected by absorbance at 254 nm. The HPLC system was calibrated using protein standards, which were run individually; the corresponding elution times were recorded and plotted against their molecular weights. Full details of the HPSEC equipment and procedures are described in Lepane et al. (2004). The total peak areas were calculated from the chromatograms, representing the total UV-absorbing fraction of the DOM in the pore-water sample. To obtain the percentages of molecular fractions, the chromatograms were divided into three molecular size fractions: 0.01–0.2 kDa (low molecular weight, LMW); 0.2–100 kDa (medium molecular weight, MMW) and 100–2,000 kDa (high molecular weight, HMW). The areas of the respective fractions were calculated and divided by the total peak area. Weight-average and number-average molecular weights of *p*DOM (M_w and M_n, respectively) were determined using the formulae

$$M_w = \Sigma(h_i \cdot M_i)/\Sigma h_i \text{ and } M_n = \Sigma(h_i)/\Sigma(h_i/M_i),$$

where h_i is the detector output and M_i is the molecular weight, both at the *i*th retention time (Mori & Barth, 1999). Thus, M_n is the weight of an 'average' molecule in the mixture, while M_w is the weight of the molecule to which the 'average' atom belongs (Zhou et al., 2000).

Analyses of fossil pigments

Sediment pigments were extracted and analysed following the recommendations of Leavitt & Hodgson (2001). The frozen sediment samples were freeze-dried and pigments were extracted with an acetone-methanol mixture (80:20 v:v) at −20°C for 24 h under a N_2 atmosphere. Thereafter, the extracts were clarified by filtration through a 0.45 μm pore-size filter (Millex LCR, Millipore). The absorption spectra of the pigment extracts were recorded with a Hitachi U-3010 spectrophotometer over the wavelength range 400–800 nm. The ratio between the absorbances at 410 and 665 nm was calculated from the spectra as an indicator of chlorophyll (Chl) *a* preservation. The degradation products of Chl *a* have an absorbance peak at 410 nm, while the absorbance peak around 665 nm is due to Chl *a* and its derivatives (Rowan, 1989).

The pigments were separated with an HPLC installation consisting of two pumps (CE 1100 Cecil Instruments Ltd, UK), a dynamic mixer (Cecil), an injection valve (Rheodyne 7125) with a 100-μl injection loop and a UV detector (CE1200, Cecil). A Spherisorb ODS2 column (4.6 × 150 mm, 3 μm particle size, Waters) was used for the separation. Prior to the HPLC run, an ion-pairing solution was added to the sample extract. The pigments were eluted using a non-linear binary gradient at a constant flow rate of 1.5 ml min^{-1} (Table 1) and identified by their retention times reported in the literature and

Table 1 Elution scheme and solvents used in the separation of pigments by HPLC

Solvent (%)	Time (min)				
	0	2	15	23	25
A	100	50	0	0	100
B	0	50	100	100	0

Solvent A = 80% methanol : 20% 1 M ammonium acetate (pH 7.2) (v:v)

Solvent B = 80% methanol : 20% acetone (v:v)

standards provided by DHI (Denmark). The chromatograms were quantified by absorbance at 450 nm.

Results

Pigment stratigraphy

According to Laugaste et al. (1996), the assemblage of dominant species in Lake Peipsi has not changed during the last 100 years: the main portion of phytoplankton biomass in the lake is formed by diatoms [*Aulacoseira islandica* (O. Müller) Sim., *Aulacoseira. granulata* (Ehr.) Sim., *Stephanodiscus binderanus* (Kütz.) Krieger], followed by cyanobacteria [*Gloeotrichia echinulata* (J. E. Smith) Richter, *Aphanizomenon*, *Microcystis* and *Anabaena* species]. Third come green algae, but their biomass is considerably smaller than that of the two predominant groups. Unfortunately, because the peaks overlap with those of other pigments, the HPLC method we used failed to separate fucoxanthin (a marker pigment for diatoms) and myxoxanthophyll (a marker pigment for cyanobacteria) properly. Therefore, Chl c_2 (for diatoms) and echinenone (for N_2-fixing filamentous cyanobacteria) were chosen to track changes in the dominant phytoplankton groups in the sediments (Leavitt & Hodgson, 2001; Bianchi et al., 2002). We also present zeaxanthin and canthaxanthin data, because the actual pigment composition of cyanobacteria species occurring in Lake Peipsi is not fully known. Zeaxanthin has been detected in *Microcystis* species (Bianchi et al., 2002; Kupperstein & Boyer, 2005), but it is also a common pigment in vascular plants. Canthaxanthin has been used as a marker pigment for colonial and filamentous cyanobacteria (Lami et al., 2000; Leavitt & Hodgson, 2001); however, sometimes herbivore tissues also contain this xanthophyll (Leavitt, 1993). Chlorophyll *b* was chosen as a proxy for green algae and Chl *a* as a proxy for total phytoplankton. However, one should keep in mind that both these are also major pigments in higher plants.

The quantities and profiles of fossil pigments in the sediment record of Lake Peipsi vary greatly. The concentrations of Chl *a* and Chl c_2 in the lower part of the record (50 and 1 nmol g^{-1} OM, respectively) are approximately one quarter of those in the upper part (Fig. 1a, c; Table 2). Their concentrations started to increase in the 1970s. The distribution of canthaxanthin exhibits a reverse pattern, having a slightly higher concentration (around 50 nmol g^{-1} OM) before the 1920s, after which it diminishes somewhat (to 40 nmol g^{-1} OM) and remains stable up to the top of the core (Fig. 1f). The concentrations of echinenone and zeaxanthin vary over the ranges 15–30 and 30–45 nmol g^{-1} OM, respectively (Fig. 1d, e). Their lowest concentrations are detected around the mid-20th century. The concentration of Chl *b* varies over the range 5–10 nmol g^{-1} OM (Fig. 1g) and its stratigraphy appears similar to those of echinenone and zeaxanthin. The lowest concentrations of Chl *b* occur in the 1960s and 1970s.

Temporal changes in the quantity and structure of *p*DOM

The concentration of *p*DOM (expressed as carbon) in the sediment varies from 2 to 3 mM (Fig. 1k, Table 2). Its highest values are around 1900 and at the end of the 20th century. The weight-average molecular weight was less than 7 kDa, varying from 3.5 to 5 kDa (Fig. 1l). The number-average molecular weight of *p*DOM was less than 4 kDa and mainly in the range 1–3 kDa (Fig. 1m). In two sediment layers, one dated to the late 1960s and the other to the very early 1990s, the pore-water organic substances have very high molecular weights (Fig. 1l, m). We would point out that the values for *p*DOM obtained here are not 'real' concentrations and molecular weights of organic substances in the pore water. Freezing and subsequent thawing of organic material in aqueous solution disintegrates particulate OM and disrupts dissolved organic molecules. Comparison of the molecular characteristics of *p*DOM in fresh and frozen samples showed that the molecular weights are approximately 2–3 times higher in unfrozen samples than in frozen samples (Lepane et al., 2006).

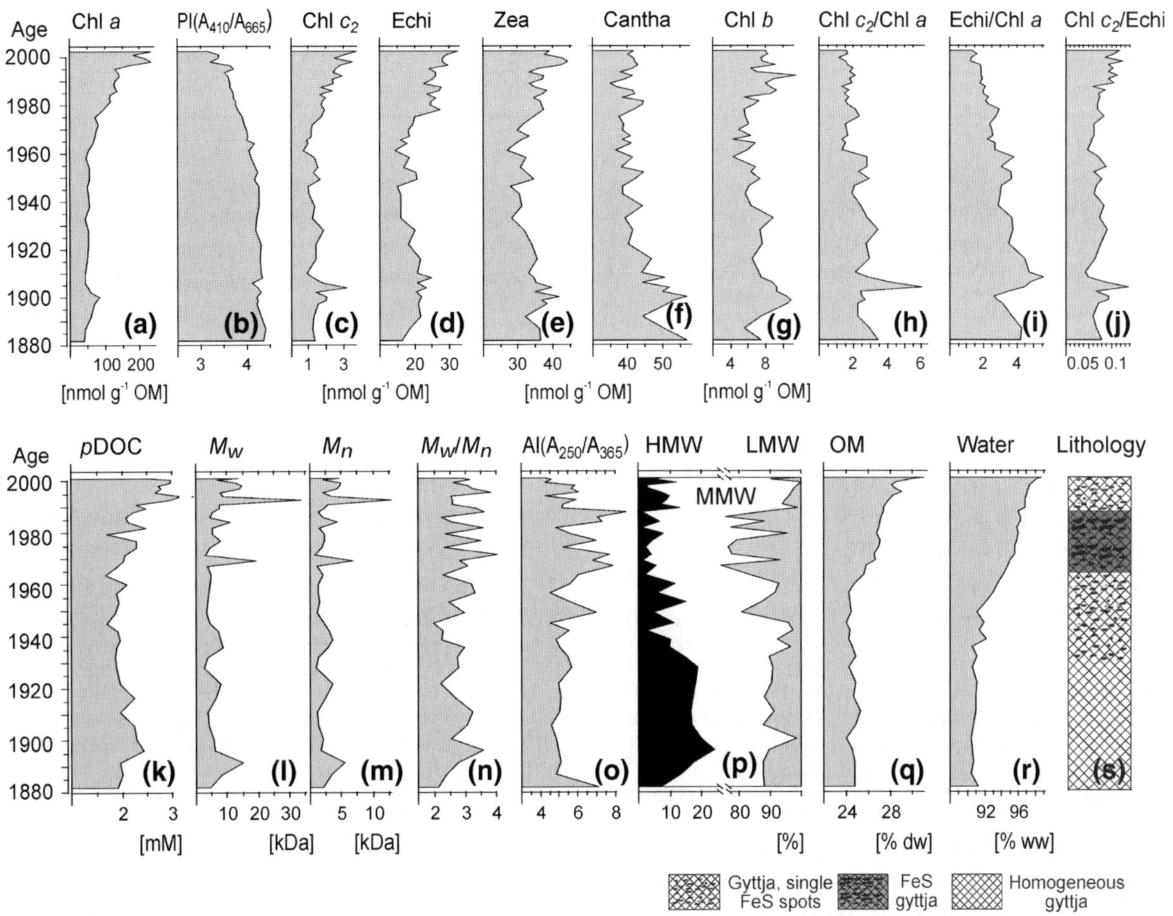

Fig. 1 (*Upper panel*) Age-resolved profiles of the concentrations of selected fossil pigments (**a**, **c**–**g**) and their ratios (**h**–**j**) in the sediments of Lake Peipsi. Preservation index of chlorophyll *a* (PI), calculated as the ratio between absorbances at 410 and 665 nm (A_{410}/A_{665}), is also shown (**b**). (*Lower panel*) Age-related changes in the characteristics of pore-water dissolved organic matter (**k**–**p**). For comparison, the profiles of major sediment constituents (**q**, **r**) and lithology (**s**) (reprinted with permission from Heinsalu et al., 2007) are also shown. Abbreviations: Chl—Chlorophyll; Echi—Echinenone; Zea—Zeaxanthin; Cantha—Canthaxanthin; *p*DOC—Pore-water dissolved organic carbon; M_w and M_n—Weight-average and number-average molecular weight, respectively; AI—Aromaticity index calculated as the ratio between absorbances at 250 and 365 nm (A_{250}/A_{365}); LMW—Low molecular weight; MMW—Medium molecular weight; HMW—High molecular weight; OM—Organic matter; dw and ww—Dry and wet weight of sediments, respectively. For details see Materials and methods

The ratio of M_w to M_n, called polydispersity, has been used as a quantitative measure of the molecular weight distribution of OM in a mixture (Chin et al., 1998). For pure organic substances the ratio is equal to 1; for a mixture of molecules it is greater than 1. In the sediments of Lake Peipsi, the polydispersity of *p*DOM varies over a very narrow range (Fig. 1n), indicating that temporal changes in the molecular weight distribution are small. However, the aromaticity index estimates of *p*DOM demonstrate slightly elevated values from the 1960s to the late 1980s (Fig. 1o, Table 2), implying that the proportion of substances of smaller size and lower aromaticity has increased in the OM accumulated during this period. Ordinarily, lower aromaticity and molecular weight (hence also size) are characteristic of autochthonous OM, while compounds with higher aromaticity and molecular weight predominate in OM derived from terrestrial sources (McKnight et al., 2001). Stratigraphic changes in the distribution of different molecular weight fractions confirm that the proportion of high molecular weight *p*DOM has decreased

Table 2 Spearman rank order correlations for selected characteristics of pore-water dissolved organic matter (*p*DOM) and fossil pigments obtained by different methods (given in parentheses) for the analysed sediments of Lake Peipsi

Pair of characteristics	Spearman	*P*-level
Fossil pigments		
A_{665} (Abs1) and Chlorophyll *a* (HPLC)	0.846	0.000
A_{665} (Abs1) and Chlorophyll *b* (HPLC)	0.247	0.095
A_{665} (Abs1) and Chlorophyll c_2 (HPLC)	0.691	0.000
A_{665} (Abs1) and Echinenone (HPLC)	0.804	0.000
A_{665} (Abs1) and Zeaxanthin (HPLC)	0.484	0.000
A_{665} (Abs1) and Canthaxanthin (HPLC)	−0.257	0.081
*p*DOM		
*p*DOC (PO + Abs2) and Total area (HPSEC)	0.882	0.000
A_{250}/A_{365} (Abs2) and HMW (HPSEC)	−0.545	0.000
A_{250}/A_{365} (Abs2) and MMW (HPSEC)	−0.257	0.109
A_{250}/A_{365} (Abs2) and LMW (HPSEC)	0.274	0.087
M_w/M_n (HPSEC) and HMW (HPSEC)	−0.011	0.947
M_w/M_n (HPSEC) and MMW (HPSEC)	0.440	0.004
M_w/M_n (HPSEC) and LMW (HPSEC)	0.169	0.327

A_{665}—Absorbance of crude pigment extract at 665 nm; *p*DOC—Concentration of pore-water dissolved organic carbon; Total area—Total area of chromatogram; A_{250}/A_{365}—Ratio of absorbances of pore water at 250 nm and 365 nm; HMW, MMW, LMW—Respective areas of the high, medium and low molecular weight fractions in the chromatograms; M_w/M_n—Ratio of weight-average molecular weight and number-average molecular weight; PO—Persulphate oxidation method; Abs1—Absorption spectroscopy of pigment extracts; Abs2—Absorption spectroscopy of pore water; HPLC—High-performance liquid chromatography; HPSEC—High-performance size exclusion chromatography. For details see Materials and methods

somewhat since the 1940s and the proportion of low molecular weight substances has increased (Fig. 1p, Table 2).

Discussion

Onset and course of eutrophication

Qualitative phytoplankton proxies—the concentrations of Chl *a*, Chl c_2, echinenone and zeaxanthin—indicate that a shift of the lake ecosystem towards eutrophy started in the 1970s. From that time onwards the concentrations of the pigments are higher than in earlier years, or are gradually increasing. Statistical analysis reveals that changes in the concentration of Chl *a* correlate well with those of echinenone and Chl c_2 (Table 2), suggesting that the main contributors to the phytoplankton biomass in the lake have been cyanobacteria and diatoms. However, it is difficult to estimate which of these has made the larger contribution. On the one hand, chlorophylls are more photosensitive than carotenoids. On the other, diatom cells sink more rapidly than cyanobacteria. Moreover, many cyanobacteria species form surface blooms and even after the bloom collapse, their colonies continue to float on the surface, prolonging their exposure to sunlight. Different sinking rates of cyanobacteria, and hence different times of exposure to sunlight, could also explain why the concentration of zeaxanthin was higher than that of echinenone in the sediments of Lake Peipsi, whereas according to the monitoring data, the biomass of filamentous cyanobacteria in the lake has been higher than that of unicellular *Microcystis* species (Nõges et al., 1996). The findings of Bianchi et al. (2002) suggest that unicellular cyanobacteria (including picocyanobacteria) may have higher sedimentation rates than the N_2-fixing filamentous cyanobacteria.

Canthaxanthin is the only pigment whose concentration shows no change during the 1970s. As mentioned previously, canthaxanthin has been widely detected also in herbivores. Moreover, when Quiblier-Llobéras et al. (1996) observed the summer planktonic community in an oligo-mesotrophic lake, canthaxanthin was found only in zooplankton, but it was totally absent from phytoplankton. In the case of Lake Peipsi, further investigations should be conducted to determine whether canthaxantin is a marker pigment of specific cyanobacteria taxa or whether it is confined to herbivores.

The changes in the pigment ratios in the sediment record imply that the ecosystem changed in the 1960s (Fig. 1h–j). The sediment composition data suggest that the sediment deposited since the 1960s clearly differs from the underlying layers: it is richer in OM and less compacted (Fig. 1q, r). The preservation index of Chl *a* declines steadily, simultaneously with changes in the pigment ratios, and this also implies that conditions in the water column or/and sediments have changed since the 1960s (Fig. 1b).

On the basis of long-term investigations of Lake Peipsi, it is generally considered that the lake trophy started to increase in the 1960s. In the 1970s, the process accelerated and a high level was reached in the 1980s. From the early 1990s until 1996 an improvement in the lake water quality was observed. However, since 1997 there has been a continuous and even accelerating deterioration of the lake status (Nõges et al., 1996; Nõges & Nõges, 2006). Changes in fossil pigments accord well with the monitored course of eutrophication, except for the transient recovery of the lake in the 1990s, which is not clear in the fossil pigment stratigraphy. However, limnological data of phytoplankton biomass from that period suggest that despite a lowered nutrient level in 1988–1994 (Loigu & Leisk, 1996), the average phytoplankton biomass remained high (Laugaste et al., 1996).

As shown by the *p*DOM data, changes in the lake OM occurred even before the 1960s. The distributions of molecular weight fractions of *p*DOM indicate a change in the late 1930s to early 1940s (Fig. 1p), suggesting a rise in the proportion of autochthonous matter in the OM pool. The increase in the autochthonous contribution to the bulk of OM becomes more and more evident over the following years. According to the *p*DOM characteristics, the autochthonous component in the sediments was highest during the 1970s and 1980s (Fig. 1o, p). A subsequent reduction in the proportion of the LMW fraction and an increase in the proportion of OM with higher aromaticity in the *p*DOM implies a somewhat decreased contribution of autochthonous matter in the bulk of OM during the 1990s, thus indicating a slight deceleration of eutrophication. The stratigraphic changes in the molecular size structure of *p*DOM (Fig. 1o, p) coincide well with the zonal changes in the sediment lithology (Fig. 1s), demonstrating that the dissolved organic fraction in the sedimentary OM has changed in concert with changes in the particulate fraction.

There is no clear explanation for the autochthonous shift in the character of *p*DOM in the late 1930s to early 1940s. A similar qualitative shift in *p*DOM was observed in the measurements of the fluorescence of *p*DOM from the same core in an earlier study (Heinsalu et al., 2007: Figs. 5, 6c, d). We suppose that the cumulative effects of several anthropogenic and natural factors (pre-war increases in agriculture and urban sewage discharges, intensive fishing, establishment of the Zebra mussel *Dreissena polymorpha* Pallas in the lake, numerous extensive forest fires and low water levels) may have triggered these changes in OM.

According to monitoring observations, very large fluctuations in the summer phytoplankton biomass are characteristic of the Peipsi ecosystem: the interannual differences can be up to 10-fold (Laugaste et al., 1996). Somewhat unexpectedly, we could detect no prominent changes in the productivity proxies in the sediments. However, the two abrupt peaks of *p*DOM molecular weights (Fig. 1l, m) appear to coincide temporally with the two densest phytoplankton bloom events registered by monitoring observations in Lake Pihkva, in 1972 and 1991 (Laugaste et al., 1996: Fig. 9b), when the biomass of cyanobacteria exceeded the mean annual biomass level by factors of 10 and 22, respectively. However, these bloom events are not detectable in other constituents of the sediments. Moreover, we are unable to surmise whether (1) the blooms extended to Lake Peipsi proper and/or (2) the OM produced in Lake Pihkva could be carried to and settle in Lake Peipsi proper. Therefore, the coincidence may be just accidental.

Conditions prior to the 1930s

The paleolimnological inferences about the nutrient loading response in Peipsi over the past 40 years, based on the stratigraphic changes in fossil pigments and molecular structure of *p*DOM, are consistent with

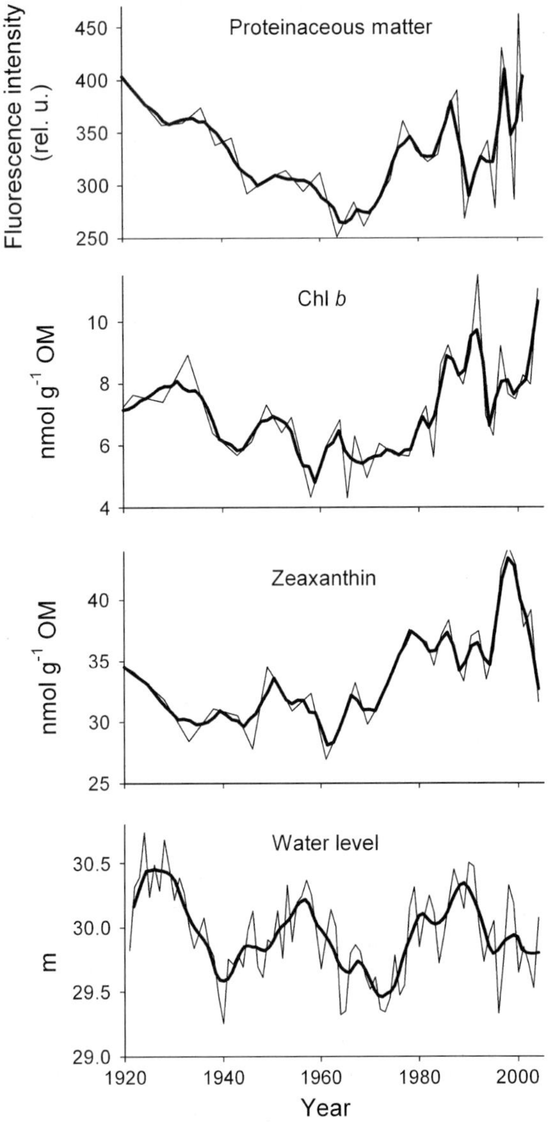

Fig. 2 Comparative curves (fine line—raw data, solid line—smoothed data, Loess curve fitting) of the annual water-level dynamics (elevation, m above sea level) and temporal distribution of proteinaceous matter, zeaxanthin and chlorophyll *b* (Chl *b*) in the settled material of Lake Peipsi. The initial unsmoothed curve of the proteinaceous matter is redrawn with permission from Heinsalu et al. (2007). OM—Organic matter

the monitoring results for the lake. Thus, we conclude that even the pre-eutrophication conditions in Peipsi can be assessed by the sedimentary organics.

Comparison of the fossil pigment chromatograms suggests that the pigment composition of phytoplankton has remained relatively unchanged over the entire time span investigated. This finding coincides well with the results of Laugaste et al. (1996). A rather stable, mesotrophic species composition characterizes the sedimentary diatom assemblage for the period 1880–1950 (Heinsalu et al., 2007). The sedimentary profile and preservation index of Chl *a* obtained in the present study indicate a low algal standing crop and high stability of the lake conditions at that time (Fig. 1a, b).

Besides the low Chl *a* concentrations, the high levels of echinenone, zeaxanthin and Chl *b* in the sediments dated to the 1900–1920s are noteworthy (Fig. 1d, e, g). A slight concurrent increase in the *p*DOC concentration is also discernible (Fig. 1k). Constituents of fluorescent *p*DOM were also present in greater quantities in the lower part of the sediment record than in the sediment layers dated to the mid-20th century (Heinsalu et al., 2007: Fig. 6a, b). This may be explained by the fact that the environmental conditions in the lake during that period were favourable for the preservation of OM and also other pigments, e.g. canthaxanthin (Fig. 1f), diatoxanthin and β-carotene (the last two are not shown). Another possible reason is revealed by comparing the temporal changes of these OM constituents with the changes of water level (Fig. 2). Keeping in mind that the dating of the lower part of the core involves a greater error, one could speculate that the runoff of OM from the lake catchment was higher in that period than in previous and subsequent decades. The presence of a greater proportion of large aromatic organic substances supports the hypothesis of intensive terrestrial transport of OM, as they indicate that allochthonous OM predominated in the lake ecosystem before the 1930s. Unfortunately, the water level observation series is too short to detect any further regularities, especially since the influence of water-level fluctuations is severely masked in the upper part of the sediment record by progressive eutrophication of the water body. Therefore, the extent of external loading of OM to the Lake Peipsi ecosystem at the beginning of the 20th century still remains ambiguous.

Conclusions

The first symptoms of environmental disturbance of Lake Peipsi appeared in the settled matter dated to the late 1930s, where changes in the molecular size composition of *p*DOM indicate an incipient relative increase of autochthonous matter and a corresponding reduction of allochthonous OM. In subsequent decades, this trend becomes more and more obvious. In the fossil pigments, an increase in water column production appears in the 1960s and since then the OM constituents indicate a gradual trend towards eutrophy. The stratigraphic changes in *p*DOM size fractions imply that the nutrient availability for phytoplankton might have declined somewhat in the 1990s. However, these changes are minor and subsequent persistently high concentrations of fossil pigments in the early 2000s do not suggest that the eutrophication level of the lake had been reduced.

Acknowledgements The study was funded by the Estonian Science Foundation (grants Nos. 5582, 5923, 6720, 6741 and G6020PKPK06). We are indebted to Mr. T. Laur for his technical help, Dr. J. Klõšeiko for software support, Dr. T. Nõges for the water-level data and Dr. H. Simola for his criticism, comments and suggestions on the manuscript.

References

Appleby, P. G., P. J. Nolan, D. W. Gifford, M. J. Godfrey, F. Oldfield, N. J. Anderson & R. W. Battarbee, 1986. ^{210}Pb dating by low background gamma counting. Hydrobiologia 141: 21–27.

Bianchi, T. S., C. Rolff, B. Widbom & R. Elmgren, 2002. Phytoplankton pigments in Baltic Sea seston and sediments: seasonal variability, fluxes, and transformation. Estuarine, Coastal and Shelf Science 55: 369–383.

Chin, Y.-P., S. J. Traina, C. R. Swank & D. Backhus, 1998. Abundance and properties of dissolved organic matter in pore waters of a freshwater wetland. Limnology and Oceanography 43: 1287–1296.

Heinsalu, A., T. Alliksaar, A. Leeben & T. Nõges, 2007. Sediment diatom assemblages and composition of pore-water dissolved organic matter reflect recent eutrophication history of Lake Peipsi (Estonia/Russia). Hydrobiologia 584: 133–143.

ISO 8245, 1999. Water quality. Guidelines for the determination of total organic carbon (TOC) and dissolved organic carbon (DOC).

Kupperstein, R. & G. Boyer, 2005. Carotenoid signatures as indicators of cyanobacteria. Retrieved from http://www.esf.edu/chemistry/reu/2005/Russ.Kupperstein.pdf (last accessed 05.05.2007).

Lami, A., A. Marchetto, R. Lo Bianco, P. G. Appleby & P. Guilizzoni, 2000. The last *ca* 2000 years paleolimnology of Lake Candia (N. Italy): inorganic geochemistry, fossil pigments and temperature time-series analyses. Journal of Limnology 59: 31–46.

Laugaste, R., V. V. Jastremskij & I. Ott, 1996. Phytoplankton of Lake Peipsi-Pihkva: species composition, biomass and seasonal dynamics. Hydrobiologia 338: 49–62.

Leavitt, P. R., 1993. A review of factors that regulate carotenoid and chlorophyll deposition and fossil pigment abundance. Journal of Paleolimnology 9: 109–127.

Leavitt, P. R. & D. A. Hodgson, 2001. Sedimentary pigments. In Smol, J. P., H. J. B. Birks & W. M. Last (eds), Tracking Environmental Change Using Lake Sediments, Volume 3: Terrestrial, Algal and Siliceous Indicators. Kluwer Academic Publishers, Dordrecht, 295–325.

Leeben, A., A. Heinsalu, T. Alliksaar & L. Saarse, 2005. A high-resolution spectroscopic study of pore-water dissolved organic matter in annually laminated lake sediments: a new tool for reconstructing eutrophication history. Proceedings of the International Association of Theoretical and Applied Limnology 29: 465–468.

Lepane, V., A. Leeben & O. Malachenko, 2004. Characterization of sediment pore-water dissolved organic matter of lakes by high-performance size exclusion chromatography. Aquatic Sciences 66: 185–194.

Lepane, V., N. Makarõtševa, N. Bonningues & A. Leeben, 2006. High-performance size exclusion chromatography analysis of lake sediment pore-water dissolved organic matter. In Frimmel, F. H. & G. Abbt-Braun (eds), Humic Substances: Linking Structure to Functions. Proceedings of the 13th Meeting of the International Humic Substances Society, July 30—August 4, 2006. Universität Karlsruhe, Karlsruhe, 209–212.

Loigu, E. & Ü. Leisk, 1996. Water quality of rivers in the drainage basin of Lake Peipsi. Hydrobiologia 338: 25–35.

McKnight, D. M., E. W. Boyer, P. K. Westerhoff, P. T. Doran, T. Kulbe & D. T. Andersen, 2001. Spectrofluorometric characterization of dissolved organic matter for indication of precursor organic material and aromaticity. Limnology and Oceanography 46: 38–48.

Mori, S. & H. G. Barth, 1999. Size Exclusion Chromatography. Springer Verlag, Berlin, Heidelberg, New York.

Nõges, T. (ed.), 2001. Lake Peipsi: Meteorology, Hydrology, Hydrochemistry. Sulemees Publishers, Tartu.

Nõges, T., J. Haberman, A. Jaani, R. Laugaste, S. Lokk, A. Mäemets, P. Nõges, E. Pihu, H. Starast, T. Timm & T. Virro, 1996. General description of Lake Peipsi-Pihkva. Hydrobiologia 338: 1–9.

Nõges, T. & P. Nõges, 2006. Indicators and criteria to assess ecological status of the large shallow temperate polymictic lakes Peipsi (Estonia/Russia) and Võrtsjärv (Estonia). Boreal Environmental Research 11: 67–80.

Peuravuori, J. & K. Pihlaja, 1997. Molecular size distribution and spectroscopic properties of aquatic humic substances. Analytica Chimica Acta 337: 133–149.

Pihu, E. & J. Haberman (eds), 2001. Lake Peipsi: Flora and Fauna. Sulemees Publishers, Tartu.

Quiblier-Llobéras, C., G. Bourdier, C. Amblard & D. Pepin, 1996. A qualitative study of zooplankton grazing in an oligo-mesotrophic lake using phytoplanktonic pigments as organic markers. Limnology and Oceanography 41: 1767–1779.

Rowan, K., 1989. Photosynthetic Pigments of Algae. Cambridge University Press, New York.

Wright H. E. Jr., 1980. Cores of soft lake sediment. Boreas 9: 107–114.

Zhou, Q., S. E. Cabaniss & P. A. Maurice, 2000. Considerations in the use of high-pressure size exclusion chromatography (HPSEC) for determining molecular weights of aquatic humic substances. Water Research 34: 3505–3514.

Hydrobiologia (2008) 599:59–65
DOI 10.1007/s10750-007-9198-7

ELLS 2007

Seasonality and trends in the Secchi disk transparency of Lake Ladoga

Mikhail A. Naumenko

Abstract Data on Secchi disk transparency in Lake Ladoga, the largest lake in Europe, collected between 1905 and 2003, were used to detect climatic (interannual) trends for lake regions with various depths. The seasonal variations in Secchi depth (D_s) during the ice-free period both for limnetic regions with large differences in bathymetric characteristics and for the whole lake were estimated by more than 7000 transparency measurements. The two-dimensional data sets have a spatial resolution of approximately 20 km and are geo-referenced by latitude and longitude in Lake Ladoga. Monthly mean spatial transparency distributions and their variances were calculated from May to October. The spatial distributions of the transparency for each month are discussed within the context of lake bathymetric patterns. The maximum values of Secchi depth (more than 4 m) occur during May and October in deep regions. Both the minimum mean value of water transparency and minimum horizontal gradients of D_s for the lake occur in August. The regions with significant interannual (climatic) decreasing trends of D_s have been identified. These areas increase in summertime and cover approximately half the lake area. In spring and autumn the areas decrease and occur in the southern near-shore regions. The mean downward climatic trend of water transparency in Lake Ladoga is 0.02 m/year.

Keywords Secchi disk transparency · Climatic trends · Lake Ladoga

Introduction

Secchi disk transparency observations have become an integral component of large lake surveillance strategies. These measurements form the only major "optical history" for water bodies on a global scale. The transparency of the surface water layer in large lakes defines the depth of light penetration, which is essential for the growth of green plants as sunlight provides the energy for photosynthesis. Water transparency decreases as color, suspended sediments and algae increase. The basic standard integrated parameter of surface water layer transparency is the depth (D_s) at which the white disk (Secchi disk) disappears from the surface observer's view (Preisendorfer, 1986).

Moreover, the depth of disappearance of the Secchi disk is a visual index of the trophic state of

Guest editors: T. Nõges, R. Eckmann, K. Kangur, P. Nõges, A. Reinart, G. Roll, H. Simola & M. Viljanen.
European Large Lakes—Ecosystem changes and their ecological and socioeconomic impacts.

M. A. Naumenko (✉)
Institute of Limnology of the Russian Academy of Sciences, ul. Sevast'yanova 9, St. Petersburg 196105, Russia
e-mail: naumenko@limno.org.ru

a lake. Secchi disk transparency is the most frequently measured characteristic of lake waters and can be regarded as a parameter for analyzing interannual (climatic) changes in the lake. For example, climatic trends in Secchi disk transparency have been detected in various North American lakes (Francis et al., 1994; Siver et al., 1996; Lathrop et al., 1999).

Morphometrical features of the world's large lakes along with their thermal and dynamic regimes determine the distribution of total suspended matter (organic and inorganic) within a water body, which ultimately regulates the optical properties of the lake water surface layer. The purpose of the present paper is to obtain a monthly mean spatial distribution of the transparency in Lake Ladoga for the period of open water within the context of the lake's bathymetric patterns. In addition, the climatic trends in D_s for the period 1905–2003 were analyzed for every month from May to October. The biological features determining interannual variations in D_s over the period examined were not considered in the present paper.

Materials and methods

The first measurements of water transparency in the largest European lake, Lake Ladoga (surface area: 17,800 km, mean depth: 47 m, maximum depth: 230 m), were carried out at the end of the 19th century. The most comprehensive water transparency and color data for Lake Ladoga, obtained using the white Secchi disk, were generalized in the monograph "Lake Ladoga" published in 1945 (Molchanov, 1945). Later, Petrov (1967) investigated the vertical transparency distribution by analyses of water samples. In the last quarter of the 20th century, eutrophication of Lake Ladoga resulted in a decline in Secchi disk depths (Petrova, 1982). Modern generalizations of water transparency distributions for spring, summer, and autumn are given in the Atlas of Lake Ladoga (Hvorov & Utin, 2002).

Data were assembled using the Lake Ladoga database, including 7,085 measurements of transparency by the white Secchi disk for the periods of open water since 1905. The database includes all the data obtained by the Institute of Limnology of the Russian Academy of Sciences and the State Hydrometeorological Service. The locations of the observation stations on Lake Ladoga at which Secchi disk transparencies were recorded between 1905 and 2003 are shown in Fig. 1.

Basic statistical analysis and linear regression analysis were used to detect climatic trends for every month of individual measurements.

It is evident that the seasonal changes in water transparency differ among different areas of the lake owing to the morphological features of the coastal zone, distributions of inflow waters, and degrees of biological activity. Therefore, to calculate the monthly mean distribution of water transparency in Lake Ladoga, the total surface area of the lake was divided into 20*20 km grid cells. This division is optimum for analyzing both spatial distribution and interannual trends. The specified values were referred to the average coordinates of each grid cell. All water transparency measurements were analyzed on a monthly basis (from May to October) for each of the 71 cells. The monthly mean values, dispersions, and trends were calculated. Climatic trends were analyzed by a linear model for every month for each grid cell. The cells with significant trends (at a significance level of 5%) were identified for each

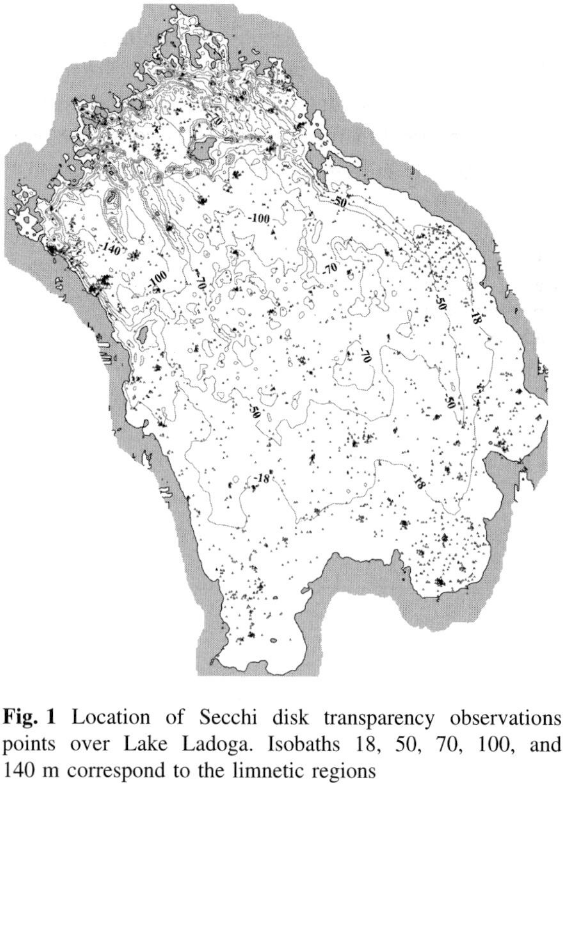

Fig. 1 Location of Secchi disk transparency observations points over Lake Ladoga. Isobaths 18, 50, 70, 100, and 140 m correspond to the limnetic regions

month. The coefficient of determination (R^2) was used to estimate how well the data are explained by the best-fit line. 'Zero' means that the fitted line explains no data variation. $R^2 = 1$ means that the fitted line explains all the data variation. For linear regression, R^2 is the square of the correlation coefficient.

Results and discussion

The penetration of sunlight into a water body determines the depth of distribution and quantity of phytoplankton, macroalgae, and other underwater plants. The variability of water transparency during the season depends on the periods of the greatest runoff from the catchment basin, temperature, and wind regime, as well as conditions of development of lake microorganisms (plankton). In Fig. 2 the quantitative characteristics of the data for all periods of observations on the lake are presented. It is obvious that only since 1955 measurements have been conducted regularly. The linear trend (represented by the thick line), which indicates the decrease in water transparency for all the data combined, does not exceed 0.73 m per hundred years. The slope of the regression line, 0.0073 m/year, is statistically significant. We accept this value as a limiting value from which the trend is apparent.

The empirical distribution of D_s is consistent with the normal distribution function: the mean, modal, and median values of D_s for the whole observation period practically coincide with each other and are equal to 2.9–3.0 m with a standard deviation of 0.9 m. As the data are normally distributed, the parametric method of trend analysis was used.

Figure 3 shows the seasonal course of Lake Ladoga water transparency for limnetic regions, which were described by Naumenko (1995). Each limnetic region is characterized by light, thermal, hydrological, and hydrochemical features. D_s in a shallow area (to a depth of 18 m) hardly changes during the ice-free period because of the constant intensive wind activity and wave motion, which stir up the bottom sediments. Volkhov Bay, included in this area, is the bay of Lake Ladoga with the lowest surface transparency in all seasons. The Volkhov River brings turbid water to the lake with a significant amount of suspended particulate matter.

Strongly marked minimum in water transparency occurs during August in the regions with bottom depth more than 50 m. The greatest differences between maximal and minimal monthly mean water transparencies are observed in the very deep-water area with water depth more than 100 m.

It is necessary to mention that the maximal water transparency never exceeded 6.6 m throughout the whole period. Molchanov's assumption (1945) that the transparency can be 7–8 m was not corroborated.

The monthly mean spatial distributions of transparency and their trends for each month of six-month open water period are presented in Fig. 4. In May–June,

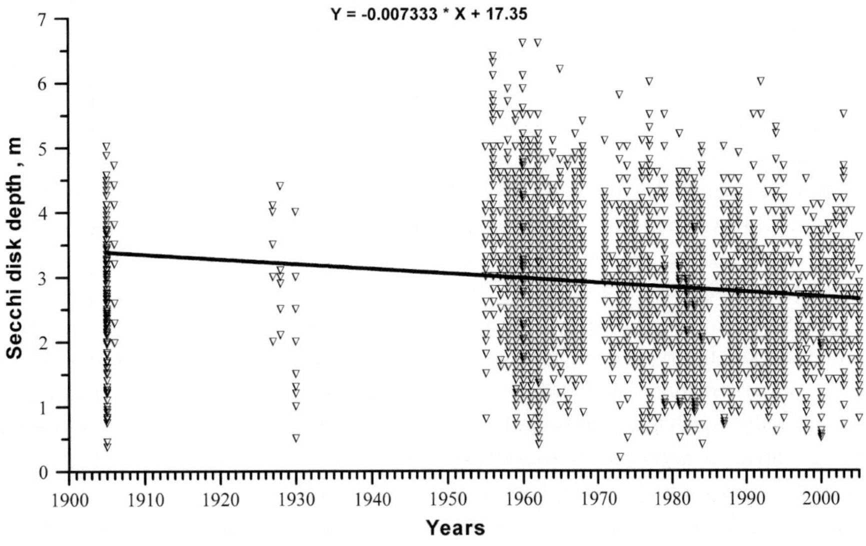

Fig. 2 Changes in the Secchi disk transparency (m) with linear trend over the whole observation period

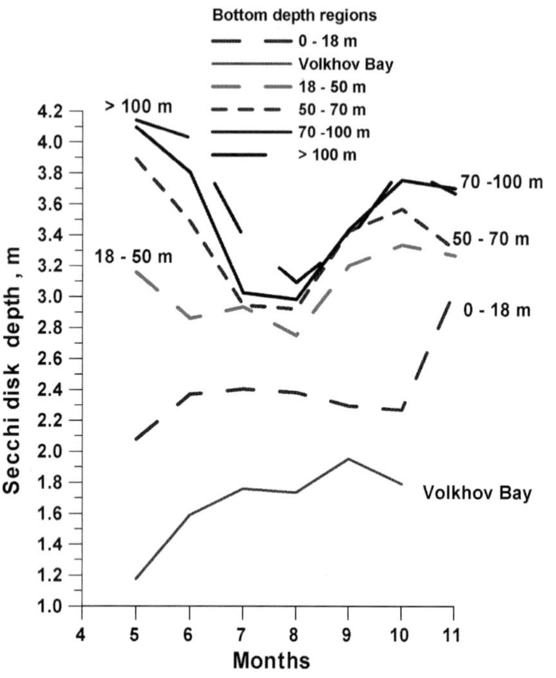

Fig. 3 Monthly mean water transparency for limnetic regions with various depths in Lake Ladoga

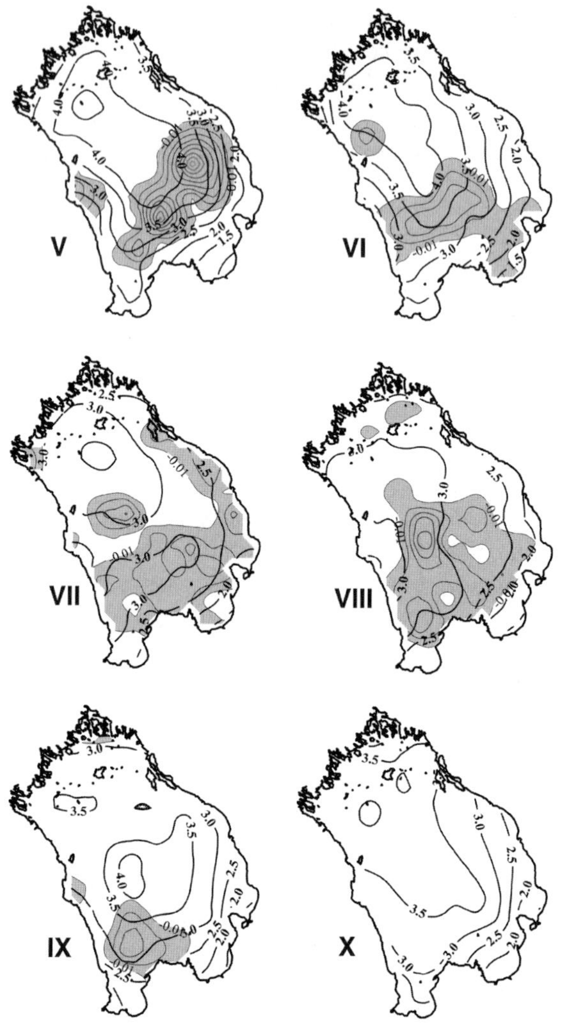

Fig. 4 The spatial distributions of mean monthly transparencies in Lake Ladoga and their trend

variations in spatial transparency correspond to the distribution of lake depths. Water transparency increases with the distance from the shore. The maximum horizontal gradients between the deep-water part of the lake and the shallow Volkhov Bay are observed in May. In this month the surface temperatures are less than 4°C and the total suspended matter concentration is low in regions with depths exceeding 100 m, where the D_s reaches 4.5 m. The Secchi depth in Volkhov Bay at this time is less than 1.5 m. When the water becomes warmer in June–July both the water transparency and the horizontal gradients of D_s decrease. During this period the spatial distribution of transparency does not follow the distribution of lake depths; it is mainly defined by the level of development of biological communities. Water transparency minima over Lake Ladoga as well as minimum horizontal gradients of D_s occur in August. This is related to phytoplankton development, which leads to an increased amount of suspended organic matter in the lake. The eastern part of the lake has the greatest number of river inflows rich in colored dissolved organic matter. In these areas the transparency is less than 2.5 m, several kilometres away from the coast. This value (D_s = 2.5 m) can apparently be regarded as the limit for separating river plumes from the open lake waters and shows no particular change in either space or time. The western part of Lake Ladoga has greater transparency than the eastern part (D_s = 3 m).

To estimate the relationship between bottom depth distribution and spatial distribution of D_s, a third-degree polynomial approximation was used. The Secchi disk transparency was plotted as a function of bottom depth for the grid points (Fig. 5) using a digital morphometric model (Naumenko, 1995). The value of the determination coefficient R^2 evidently varies during the period of open water. The largest share of the dispersion (i.e., maximum R^2 = 64%) explained by this model was in May and decreased slightly in June (Fig. 5). The same trend in the

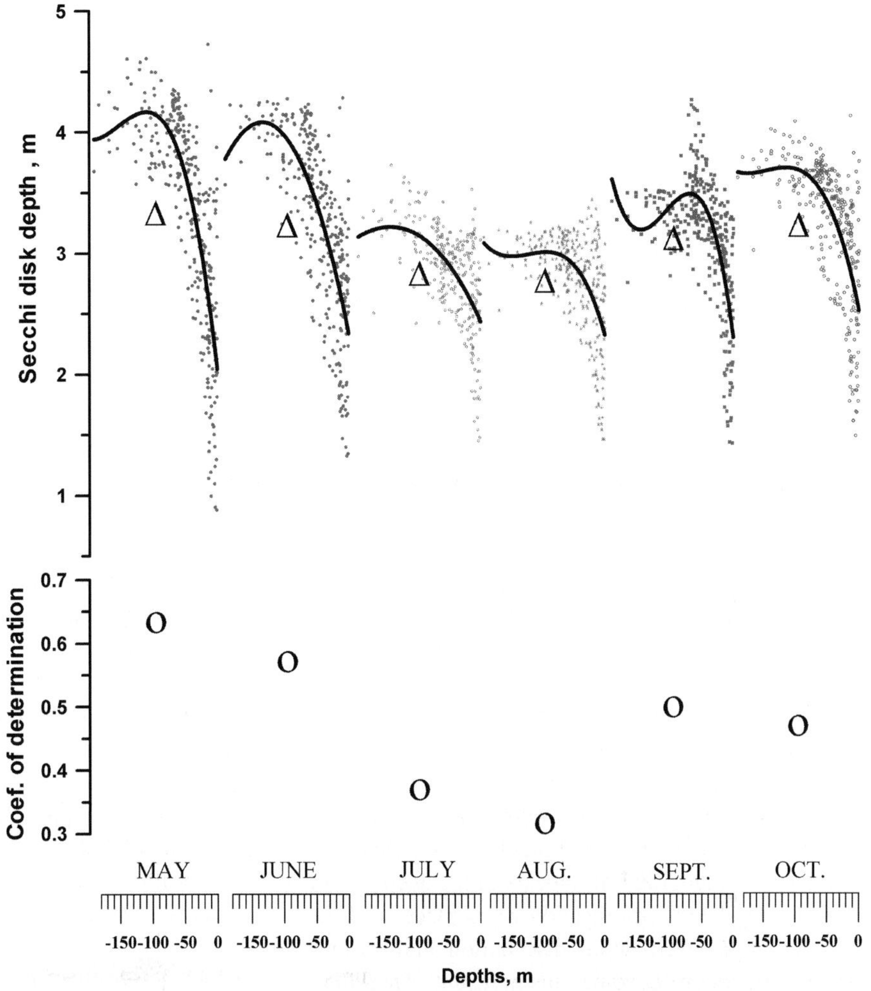

Fig. 5 Correlations between transparency and bottom depth (upper) and mean coefficient of determination (lower) for May–October. (Curves are third-degree polynomial. Triangles denote monthly mean water transparency over the lake)

Table 1 Portion of Lake Ladoga area with significant decreasing trends in water transparency and $R^2 \frac{\text{min–max}}{\text{mean}}$

	Month				
	May	June	July	August	September
Area (%)	32.7	29.8	48	44.9	12.7
Coefficient of determination R^2 (%)	$\frac{22\text{–}92}{48}$	$\frac{7\text{–}55}{24}$	$\frac{14\text{–}45}{25}$	$\frac{12\text{–}66}{31}$	$\frac{30\text{–}40}{34}$

determination coefficient typified the spatial distribution of temperature in Lake Ladoga (Naumenko & Karetnikov, 2002). In August, when the surface water temperature had reached its maximum, the determination coefficient fell to less than 40%. This indicates that the spatial D_s distribution during this month was controlled more by hydrobiological and hydrodynamic factors than by bottom depth distribution. In August both the mean value of D_s and its spatial gradients were minimal in the lake. After September the determination coefficient increased but did not reach the spring value. The polynomial approximation of the spatial transparency distribution allows the location of maximum water transparency to be determined for each month. In May and June, water transparency with a value greater than 4 m occurred above the bottom depths over 100 m. In August–September the maximum appeared over

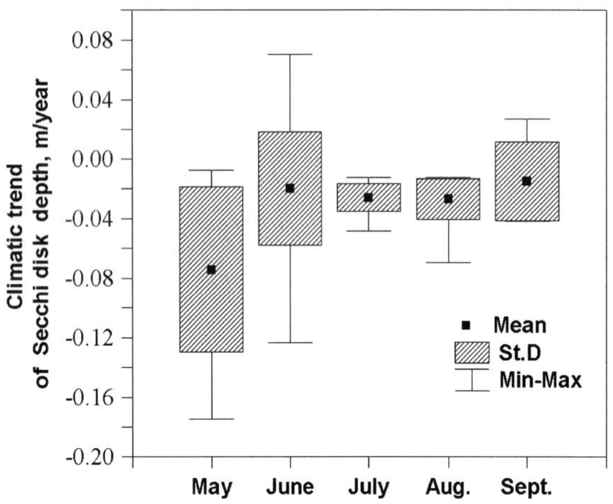

Fig. 6 Climatic (interannual) trend of Secchi disk transparency in Lake Ladoga from 1905 to 2003

depths of 60–80 m, i.e., in the central part of the lake. As Naumenko & Karetnikov (2002) reported, the surface water temperature in May–June correlates strongly with bottom depth. This correlation results from continuous water warming and the appearance of vertical temperature stratification near the shore. In spring the thermobar phenomenon occurs, which divides the stratified water from the cold central water (Naumenko, 1994, Naumenko et al., 1996). In mid-July the thermobar disappears in the deep part of Lake Ladoga. Full stratification takes place, and the correlation between depth distribution and surface water temperature becomes insignificant. Therefore, Secchi disk transparency is significantly correlated with the bottom depth in May–June when inshore surface water temperature and nutrients strongly affect the spatial phytoplankton distribution and, in turn, the water transparency.

The analysis of spatial distribution of water transparency dispersions as a measure of interannual variability shows that during the period of open water the greatest dispersions take place in the southern part of lake.

Practically all significant interannual trends of D_s were negative, indicating a distinct decline in surface water transparency over the study period. The determination coefficients R^2 of the linear trends describe about 20–50% of the total dispersions (Table 1). The spatial distribution of the areas with significant trends in D_s varies depending on the month.

In summer, the area with a significant trend in D_s is larger than in spring or autumn (Fig. 4). Because of the lack of data, the spatial D_s trend distribution for October is not presented. The marked areas have a significant D_s trend with a regression line gradient of more than 0.01 m/year. The greatest area, about half the Lake Ladoga surface, is subject to a negative D_s trend in July (Table 1). In spring and autumn this area is much smaller and occurs in the southern parts of the lake. Using the comparison of monthly spatial distributions of water transparency trends shown in Fig. 4, the area in which the trend of water transparency always exceeds 0.01 m/year was identified. The area is situated in the southern part of the lake with depths from 10 to 30–40 m, and its location practically coincides with the September distribution of D_s trends presented in Fig. 4.

The results obtained demonstrate the Secchi disk transparency trends for the period from 1905 to 2003 in Lake Ladoga (Fig. 6). The average values of the negative summer and autumn D_s trends are approximately identical and vary from 0.027 to 0.015 m/year (on average, 0.02 m/year). As an exception, the largest value is −0.07 m/year in May.

This research provides an opportunity to determine the areas of Lake Ladoga with significant long-term variability in water transparency during different seasons.

Acknowledgments I am grateful to Mrs Tatiana Podsetchine for linguistic revision of the text.

References

Francis, J. C., M. A. Poirrier, D .E. Barbe, V. Wijesundera & M. M. Mulino, 1994. Historic trends in the Secchi disk transparency of Lake Pontchartrain. Gulf Research Reports 9: 1–16.

Hvorov, G. V. & G. N. Utin (eds), 2002. Atlas of Lake Ladoga. Limnology Institute of the Russian Academy of Science, Saint-Petersburg. (In Russian).

Lathrop, R. C., S. R. Carpenter & D. M. Robertson, 1999. Summer water clarity responses to phosphorus, *Daphnia* grazing, and internal mixing in Lake Mendota. Limnology and Oceanography 44: 137–146.

Molchanov, N. V., 1945. Ladozhskoe ozero. Hydrometeoizdat Publishing House, Leningrad-Moscow. 557 pp. (Lake Ladoga. In Russian).

Naumenko, M. A., 1994. Some aspects of large lakes thermal regime (Ladoga, Onega). Water Pollution Research Journal of Canada 29: 423–439.

Naumenko, M.A., 1995. Novoe opredelenie morfometricheskikh khararteristik Ladozhskogo ozera. Reports of the Academy of Sciences 345: 514–517. (New estimation of Lake Ladoga morphometrical characteristics. In Russian).

Naumenko, M. A., S. G. Karetnikov & A. I. Tikhomirov, 1996. Main features of the thermal regime of Lake Ladoga during ice-free period. Hydrobiologia 322: 69–73.

Naumenko, M. A. & S. G. Karetnikov, 2002. Sezonnaya evolutsiya prostranstvennogo raspredeleniya temperaturi poverkhnosti Ladozhskogo ozera v svyazi s ego morfometriei. Reports of the Academy of Sciences 386: 250–253. (Seasonal evolution of spatial distribution of temperature of Lake Ladoga surface water in connection with its morphometry. In Russian).

Petrov, B. M., 1967. Prozrachnost i tsvet vody Ladozhskogo ozera. In Alekin, O. A. (ed.), Gidrokhimiya i gidrooptika Ladozhskogo ozera. Nauka Publishing House, Leningrad: 156–216. (Transparency and colour of water of Lake Ladoga. In Russian).

Petrova, N. (ed.), 1982. Antropogennoe evtrofirovanie Ladozhskogo ozera. Nauka Publishing House, Leningrad. (Anthropogenic eutrophication of Lake Ladoga. In Russian).

Preisendorfer, R. W., 1986. Secchi disk science: visual optics of natural waters. Limnology and Oceanography 31: 909–926.

Siver, P. A., R. W. Canavan IV, C. K. Field, L. J. Marsicano & A.-M. Lott, 1996. Historical changes in Connecticut Lakes over a 55 year period. Journal of Environmental Quality 25: 334–345.

Hydrobiologia (2008) 599:67–76
DOI 10.1007/s10750-007-9194-y

Silicon load and the development of diatoms in three river-lake systems in countries surrounding the Baltic Sea

Peter Nõges · Tiina Nõges · Rita Adrian · Gesa A. Weyhenmeyer

Abstract We studied the formation of dissolved silicon loads from rivers to lakes, the development of diatoms in lakes and the role of climatic forcing on the silicon cycle in three river-lake systems in Sweden, Estonia and Northern Germany. We found coherent seasonality in the silicon loads of the two northern rivers, which was probably caused by the common snow-type hydrology of the catchments as distinct from the rain-type hydrology of the catchment, further south. The similarity among lakes in the dynamics of the Si-related variables studied resulted from similarities in mean lake depth and mixing type rather than the climatic regime. Among the variables measured at the three sites, river water discharge responded most coherently to climatic forcing as synchronized by the North Atlantic Oscillation winter index (NAOw). Water discharge and Si load were strongly linked variables and showed coherent patterns among the river systems. We found significant season-specific correlations of the NAOw with either the biomass or the share of diatoms in each lake, but no coherent pattern among the lakes. Our results indicate that processes driven by water discharge are more coherent across regions than in-lake processes.

Keywords Silicon · Lakes · Rivers · Diatoms · Climate · NAO · Discharge

Guest editors: T. Nõges, R. Eckmann, K. Kangur, P. Nõges, A. Reinart, G. Roll, H. Simola & M. Viljanen
European Large Lakes—Ecosystem changes and their ecological and socioeconomic impacts

P. Nõges · T. Nõges
Institute of Agricultural and Environmental Sciences, Centre for Limnology, Estonian University of Life Sciences, Rannu, Tartumaa 61101, Estonia

P. Nõges (✉)
Institute for Environment and Sustainability, European Commission – Joint Research Centre, Ispra (VA) 21020, Italy
e-mail: peeter.noges@jrc.it

R. Adrian
Leibniz-Institute of Freshwater Ecology and Inland Fisheries, Müggelseedamm 301, 12587 Berlin, Germany

G. A. Weyhenmeyer
Department of Environmental Assessment, Swedish University of Agricultural Sciences (SLU), Box 7050, Uppsala 750 07, Sweden

Introduction

Human activities are mainly responsible for changes in the riverine delivery of nitrogen (N) and phosphorus (P). Unlike N and P, the source of dissolved silicon (Si) in aquatic systems is mainly the chemical weathering of silicate minerals and the delivery of Si by rivers is influenced by lithology, weathering intensity, climatic variation and diatom production (Conley, 1997; Conley et al., 2000). Weathering of silicate rocks is considered to be a sink for atmospheric carbon dioxide (Ittekkot et al., 2000). Another process in which silicon plays a key role is anthropogenic CO_2 sequestration: silicate inputs fertilize the seas by stimulating the production of diatoms, which play a crucial role in the biological uptake of CO_2 by the ocean through the so-called biological pump (Smetacek, 1998). As rapidly sinking particles, diatoms play a major role in exporting organic carbon to deep ocean layers and thus in sequestering CO_2 (Tréguer & Pondaven, 2000).

Anthropogenic activities have significantly increased the riverine loading of N and P (Smith et al., 1999), while their effects on Si loading have been minor (Berner & Berner, 1996). The essential difference in behaviour among N, P and Si lies in their rates of regeneration and recycling. Si is released back to the water column via the dissolution of biogenic silica, which is a considerably slower process than the remineralisation of N and P. If Si influx were constant, the Si:N and Si:P ratios would change due to changes in N and P loadings (Ptacnik et al., 2005). However, the Si load may also decrease as a result of increasing Si retention in watersheds, e.g. because of dam construction (Humborg et al., 2000; Ittekkot et al., 2000). Large-scale hydrological alterations of rivers change the ratios of N:P:Si and these changes affect phytoplankton species composition and diversity, and hence carbon sequestration (Ittekkot, 2003). The excess delivery of N and P relative to Si by rivers affects the development of diatoms. If the spring diatom bloom is terminated by exhaustion of Si while N and P are still present, the residual N and P in the water column are subsequently used by non-siliceous organisms, among which cyanobacteria may form harmful and nuisance blooms in lakes (Huisman & Hulot, 2005). In view of the role of Si in global carbon sequestration as well as its importance in the structuring of biological communities, long-term changes in Si concentrations should be followed through the aquatic continuum comprising rivers, lakes, wetlands and estuaries where intense physico-chemical and biological processes occur.

We have analyzed long-term data on Si concentration, Si loading and the development of diatoms in three systems, each composed of a lake and its main inflowing river, located in Sweden, Estonia and Northern Germany. We address the following questions:

1. What are the similarities and differences in Si load formation between catchments of different hydrological regimes (snow-type versus rain-type hydrology)?
2. Are Si concentrations and the development of diatoms in the three lakes related to the riverine Si load?
3. How similar are the dynamics of different Si cycle variables—Si concentrations in rivers and lakes, Si loadings to the lakes, Si loss rates during diatom development and diatom biomass—among the three river/lake systems, and how coherent is the response to climatic forcing?

Study sites

The study encompasses three well-studied lakes together with their main inflowing rivers (Table 1) in three countries surrounding the Baltic Sea. Considered in pairs, these river-lake systems show some common features. The two northern systems, L. Ekoln (basin of Lake Mälaren) in Sweden and L. Võrtsjärv in Estonia, are characterized by the northern snow regime of their catchments, while the runoff from the Müggelsee catchment is smoother with a less-developed spring peak. Müggelsee and Võrtsjärv are shallow polymictic lakes, differing in this respect from the dimictic Ekoln basin. Müggelsee and the Ekoln basin both have rather short water residence times, 0.1 and 0.5 y, respectively, compared to 1 year in Võrtsjärv. Võrtsjärv exhibits large changes in water level. The average discharge of the Fyris River is about 1.5 times higher than those of the other two rivers. Despite differences in trophic status, size and mixing regime among the lakes, there are similarities in their dominating diatom species.

Materials and methods

We used three-letter codes for the sites comprising the letters R or L indicating river or lake, and the two-letter ISO code for the name of the country: SE for Sweden, EE for Estonia, DE for Germany. For example, REE stands for the Estonian River Väike Emajõgi and LEE for the Estonian Lake Võrtsjärv (Table 1).

Table 1 General descriptors for the study sites

Descriptor	Sweden (SE)	Estonia (EE)	Germany (DE)
Lake: Name	Ekoln (=LSE)	Võrtsjärv (=LEE)	Müggelsee (=LDE)
Coordinates	59°45′ N	57°50′–58°30′ N	52°26′ N
	17° 30′ E	25°35′–26°40′ E	13°39′ E
Area (km^2)	30	270	7.3
Mean depth (m)	15.7	2.8	4.9
Maximum depth (m)	37	6	8
Water retention time (y)	0.5	1	0.1
Ice cover (months)	4–5	4–5	0–4
Mixing type	Dimictic	Polymictic	Polymictic
Trophic type	Eutrophic	Eutrophic	Hypertrophic
Dominating diatom taxa	*Aulacoseira* spp.	*Aulacoseira ambigua* (Grun.) Simonsen,	*Aulacoseira granulata* (Ehr.) Simonsen,
	Stephanodiscus hantzschii Grun.	*A. granulata* (Ehr.) Simonsen,	*Stephanodiscus hantzschii* Grun., *Synedra acus* Kütz.
	Diatoma tenuis Agardh	*Synedra acus* Kütz.	
River: Name	Fyris (=RSE)	Väike Emajõgi (=REE)	Spree (=RDE)
Catchment area (km^2)	1982	1271	7000
Length (km)	80	82	400
Proportion in total water discharge to the lake (%)	62	41	97
Precipitation to the catchment (mm y^{-1})	563	680	568
Hydrological regime	Northern snow regime with low winter- and relatively high spring snowmelt runoff	Northern snow regime with low winter- and high spring snowmelt runoff	Mixed snow- and rain regime with higher winter runoff

We based our study on long-term records of monthly mean water discharges from our three rivers (Q), the dissolved silicon concentrations in the rivers (SiRiver) and lakes (SiLake), and the biomasses of diatoms (Bbac) and total phytoplankton (Bphy). The time series covered periods from 12 to 40 years (Table 2). The data covered the full 12-month period from all sites except for LSE, for which data were available from March to October. The derived variables were the silicon loading (Si*Q) and the percentage of diatoms in the total phytoplankton biomass (Bac%). Monthly sampling frequencies were used in LSE and LEE, while LDE was sampled weekly during the growing season and biweekly during winter.

Dissolved Si was determined by a colorimetric method based on the ammonium molybdate reaction. Phytoplankton biomass was determined according to Utermöhl (1958). As a proxy indicator for climate forcing we used the North Atlantic Oscillation winter index (NAOw; December–March http://www.cru.uea.ac.uk/cru/data/nao.htm), which shows strong relationships with a number of natural phenomena in the northern hemisphere in winter and spring (Weyhenmeyer et al., 1999; Gerten & Adrian, 2000; Ottersen et al., 2001; Hurrell et al., 2003). Prior to correlation analyses, the NAOw series were detrended and all the other data series were log-transformed and detrended. STATISTICA for Windows (v 6.0) was used for statistical analyses.

We used a two-step approach to study coherence in the responses to climatic forcing by Si-cycle variables such as Si concentrations in rivers and lakes, Si loadings to the lakes and diatom biomass. First, we correlated the NAOw data to the monthly data of our variables. Second, we analysed the

Table 2 General statistics of the data used in the analyses

Variable	Water body	Period	Months	*n*	Mean	Median	Min	Max
Discharge ($m^3 s^{-1}$)	RSE	1965–2004	Jan–Dec	480	12.9	8.5	0.4	68.0
	REE	1965–2004	Jan–Dec	480	8.2	6.3	1.3	38.3
	RDE	1979–2000	Jan–Dec	263	8.6	8.2	2.2	20.1
Silicon concentration ($mg l^{-1}$)	RSE	1965–2004	Jan–Dec	480	3.98	4.01	0.12	8.32
	REE	1993–2004	Jan–Dec	142	3.13	3.10	0.95	6.00
	RDE	1991–2004	Jan–Dec	159	4.26	4.55	0.06	7.55
	LSE	1965–2004	March–Oct	226	2.49	2.46	0.11	5.73
	LEE	1993–2004	Jan–Dec	145	2.13	2.00	0.10	5.40
	LDE	1979–2004	Jan–Dec	312	3.79	3.90	0.00	8.90
Phytoplankton biovolume ($mm^3 l^-$)	LSE	1965–2004	March–Oct	236	1.70	0.71	0.00	20.5
	LEE	1965–2004	Jan–Dec	440	14.5	10.4	0.0	100
	LDE	1979–2004	Jan–Dec	312	8.25	5.74	0.12	50.6
Biovolume of diatoms ($mm^3 l^-$)	LSE	1965–2004	March–Oct	236	0.80	0.07	0.00	16.5
	LEE	1965–2004	Jan–Dec	440	3.23	2.21	0.00	33.5
	LDE	1979–2004	Jan–Dec	312	4.21	2.48	0.00	35.4
Share of diatoms (%)	LSE	1965–2004	March–Oct	232	33	27	0	100
	LEE	1965–2004	Jan–Dec	440	25	19	0	100
	LDE	1979–2004	Jan–Dec	312	47	47	0	97

similarities among the monthly correlograms by Pearson correlations. This kind of "metacorrelation" enabled us to identify sites at which the variables analysed showed similar types of seasonal response to the climatic signal despite the often weak and noisy character of the correlations among single monthly values.

Results

Silicon concentration was highest in the German system and lowest in the Estonian one, but the differences among the regions were not very great (Table 2). Si load, determined as a product of SiRiver and water discharge, showed a strong hysteresis caused by higher SiRiver values during periods in which the limbs of the hydrographs were rising (Fig. 1). The graph shows that in RSE, Q_{max} and Q_{min} differed by a factor of nine, while the concentrations differed by a factor of five only. In RDE, characterized by the widest range of SiRiver values (factor of five across seasons), the seasonal variability of Q was the lowest (factor of two). The correlation between Q and SiRiver was positive for all months in RSE (Fig. 2) with a significant relationship in January and

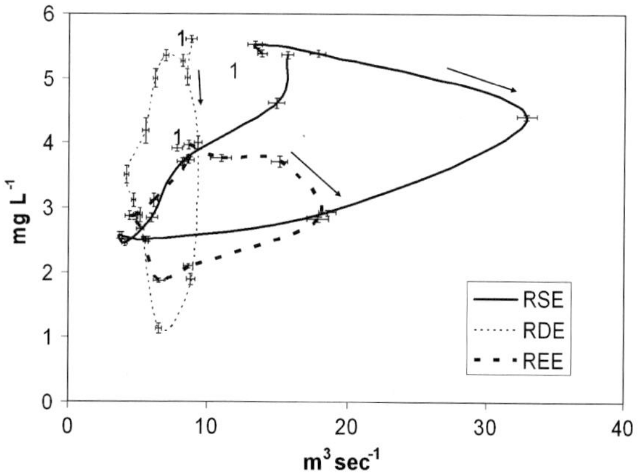

Fig. 1 Seasonal cycle of the relationship between monthly average discharge (Q) and silicon concentration (Si) in the Swedish River Fyris (RSE), Estonian River Väike Emajõgi (REE) and German River Spree (RDE). Error bars indicate the standard error, number 1 indicates January and the arrows show the direction of the seasonal cycle

from August to November. Also in REE this relationship was positive and significant in August and September, but revealed a dilution effect (negative relationship) at the time of the flood peak in April. In RDE no significant correlation between Q and SiRiver was found.

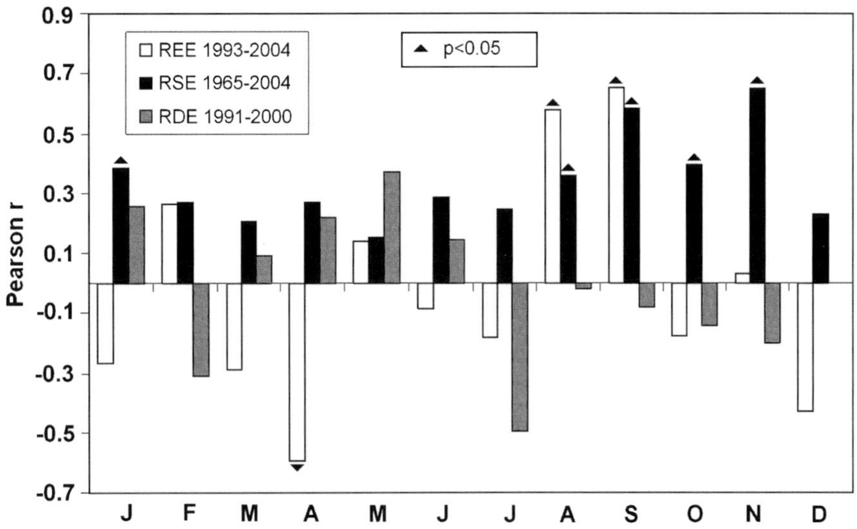

Fig. 2 Relationship between monthly average water discharge (Q) and silicon concentration in the Estonian River Väike Emajõgi (REE), Swedish River Fyris (RSE), and German River Spree (RDE). All data series were log-transformed and detrended prior to analysis

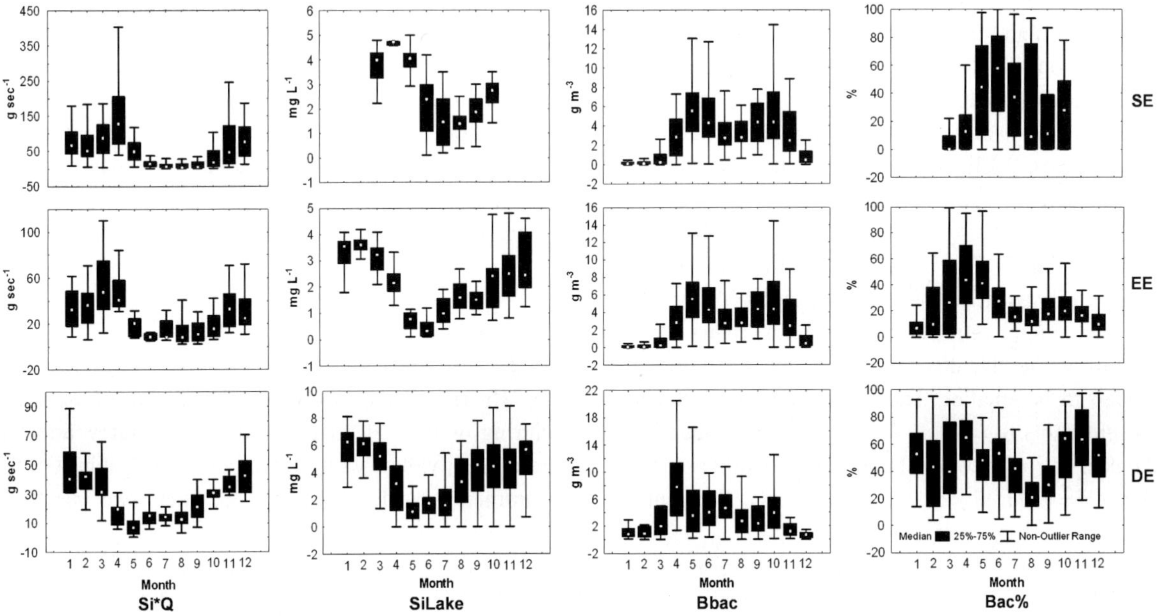

Fig. 3 Seasonal dynamics of dissolved silicon loads (Si*Q), in-lake silicon concentrations (SiLake), diatom biomass (Bbac), and diatom percentage in the total phytoplankton biomass (Bac%) at the study sites in Germany (DE), Sweden (SE) and Estonia (EE)

The resulting silicon loads (Si*Q; Fig. 3) had a V-shaped dynamic in RDE with the highest loadings in winter and the lowest in May. The loading patterns in the northern rivers followed more or less the annual hydrograph with peak values during the spring flood and an extended summer low-load period. From June to September the silicon loadings remained very low in all three systems.

The dissolved silicon concentrations had a V-shaped seasonal dynamic in all three lakes (SiLake; Fig. 3). There were strong correlations between the monthly average silicon loads and in-lake concentrations (r from 0.3 to 0.6, $P < 0.001$) but not between the annual values (not shown). The median Si concentration was highest in LDE, being nearly double of that in LEE (Table 2). In the latter, the annual minimum Si concentations also reached significantly lower values (0.5 mg l^{-1} on average) than those in LDE and LSE (0.9 and 1.1 mg l^{-1}, respectively).

The average phytoplankton biomass (Bphyto) and the biomass of diatoms (Bbac) were several times higher in LEE and LDE compared to LSE, while the average Bac% was rather similar in LEE and LSE, and 1.5 times higher in LDE (Table 2).

All river-related variables (discharges, Si concentrations and Si loads) were significantly correlated among all three rivers (Table 3). All lake-related variables were significantly correlated between the two shallow lakes, LEE and LDE, while phytoplankton biomass was the only variable that correlated between LSE and LDE. No correlations between the Estonian and Swedish lakes were found for any of the lake variables (Table 3).

On average, the rate of Si decrease in spring ranged from −1 to −2 mg l^{-1} $month^{-1}$ in all three lakes (Fig. 4). However, the increase in diatom biomass was nearly four times higher in the nonstratified lakes (LEE, LDE) than in the stratified lake LSE.

Seasonal correlograms relating the NAOw index and Q (Fig. 5) had similar dynamics in the two northern rivers, REE and RSE. In high NAOw years, Q was significantly higher in both rivers during winter, while the discharge peak in April or May tended to be smaller. The correlation between these correlograms ($r = 0.91$; $P < 0.01$) was stronger than the correlation between the monthly Q values at these sites ($r = 0.71$; $P < 0.01$; Tables 3, 4). We found a significantly similar response of Q to the NAOw for RSE and RDE ($r = 0.71$; $P < 0.01$; Table 4), while the pair REE-RDE ($r = 0.56$; $P > 0.05$) was the least coherent. Despite strong correlations between the silicon concentrations in the rivers (Table 3), no significant relationship between SiRiver and the NAOw index could be found in any river for any given month (not shown). Given the linkage between the NAOw and Q, silicon loads still showed a certain coherence between two pairs of rivers (Table 4).

In years with high NAOw, the silicon concentration was significantly lower in both shallow lakes (LEE, LDE) in February and March (not shown), but later the coherence between these lakes disappeared and as a whole the similarity between the correlograms was non-significant. Total phytoplankton biomass was positively related to NAOw from February to May in LEE and in June and July in LDE. The same relationship in LSE was negative in July (not shown). The biomass of diatoms showed a significant positive relationship with NAOw only in LEE from February to April. The percentage of diatoms in the total phytoplankton biomass showed some significant relationships at all three sites but there was no coherence in their occurrence.

Discussion

In all rivers the highest SiRiver values occurred during winter at medium discharge rates and the minimum values were found during or just before the summer low-flow period. In the two northern

Table 3 Pearson correlation coefficients between the log-transformed and detrended monthly average water discharges (Q), silicon concentrations in the rivers (SiRiver) and lakes (SiLake), silicon loads (Q*Si), biomasses of phytoplankton (Bphy) and diatoms (Bbac), and the share of diatoms in total phytoplankton biomass (Bac%) at the three study sites in Sweden (SE), Estonia (EE) and Germany (DE)

Variable	EE-DE	*N*	EE-SE	*N*	SE-DE	*N*
Q	0.46**	264	0.71**	480	0.52**	264
SiRiver	0.41**	144	0.41**	144	0.18*	161
Q*Si	0.24*	96	0.74**	144	0.31**	113
SiLake	0.43**	145	0.15	58	−0.16	150
Bphyto	0.34**	300	0.44**	217	−0.03	158
Bbac	0.33**	300	0.13	216	−0.14	158
Bac%	0.14*	300	0.03	212	0.05	157

N, number of monthly mean values used for each analysis

*$P < 0.05$, **$P < 0.01$

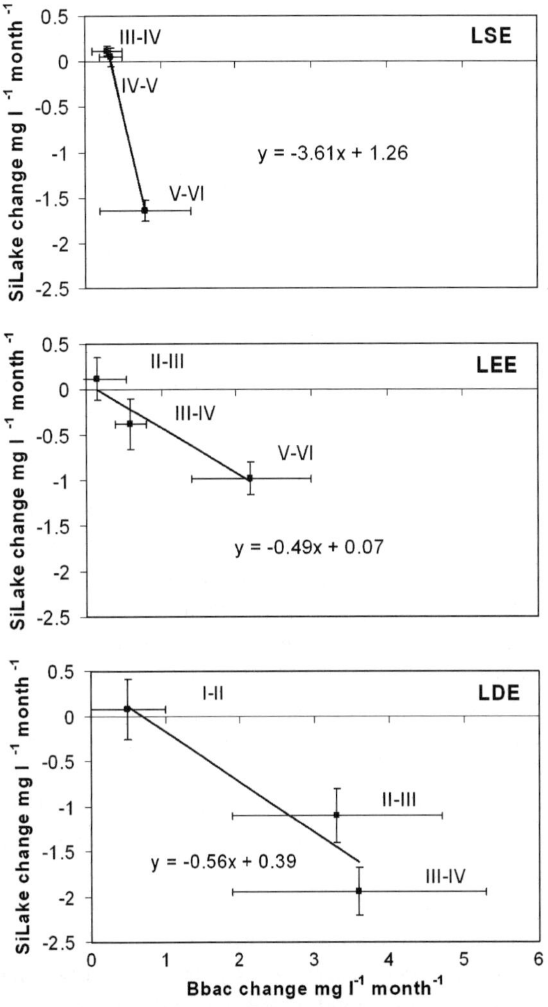

Fig. 4 Relationship between the changes in diatom biomass (Bbac) and dissolved silicon concentration (SiLake) in the lakes studied during the spring diatom growth phase. LSE—Ekoln, LEE—Võrtsjärv, LDE—Müggelsee. I–II etc. stand for months

catchments, the hysteresis cycle was stretched along the discharge axis, implying that changes in discharge were mostly responsible for the variability in loading. However, the relationship between SiRiver and Q differed between basins being always positive only in Sweden where a more continuos supply of silicon from the weathering of silicate minerals can be speculated. Unusually high loadings of reactive silica to Lake Mälaren during high water discharges in the year 2000 were described also by Weyhenmeyer et al. (2004). In Estonia high water discharge in spring caused a dilution effect of silicon in the river.

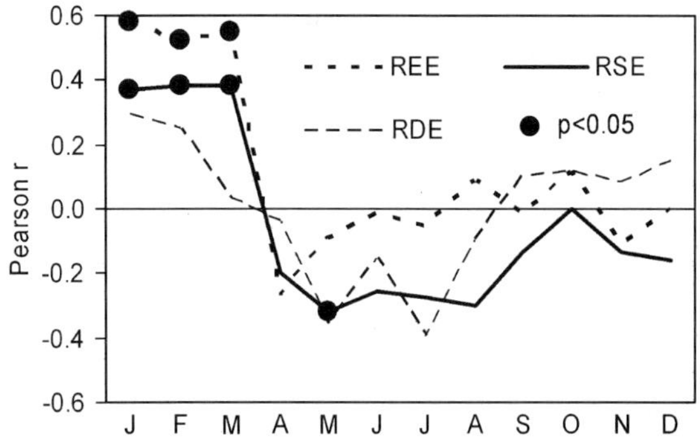

Fig. 5 Correlograms between river discharge and the NAO winter index at the study sites. The discrete values of monthly correlation coefficients are connected by lines for easier visualisation

The V-shaped seasonal dynamic of dissolved silicon in all three lakes concentrations was probably caused by large loadings at the beginning and at the end of the year and by losses due to the uptake by diatoms in spring/summer. The strong correlations between the monthly silicon loads and in-lake concentrations arose mainly from a similar seasonality as there was no correlation between the annual values.

The higher yield of diatom biomass per unit loss of reactive silicon from the water column in nonstratified lakes can be explained by smaller sedimentation losses in these systems. As planktonic diatoms are balanced between sinking losses and mixing-induced resuspension (Raubitschek et al., 1999), most of the biomass produced remains suspended in the water column in polymictic lakes while a considerable part settles to the lake bottom in deeper lakes once they sink below the thermocline. Loss of silica from the trophogenic layer in LSE was probably one reason why diatoms developed only one seasonal peak. A distinct autumn peak of diatoms with biovolumes above 10 $mm^3 l^{-1}$ was observed in LDE in several years during the 1980s (Köhler & Hoeg, 2000), while an autumn peak occurred regularly in LEE with diatom biomasses comparable to the spring maximum. González Sagrario et al. (2005) pointed out that in shallow lakes, in contrast to deep lakes, silicon limitation is often a short-lived phenomenon in spring, because silicate is returned to the water column via remineralisation shortly after sedimentation. The discrepancy between the decrease in silicon concentrations and the increase in planktonic diatom

Table 4 Pearson correlations between the seasonal correlograms relating monthly average water discharges (Q), silicon concentrations in the rivers (SiRiver) and lakes (SiLake), silicon loads (Q*Si), biomasses of phytoplankton (Bphy) and diatoms (Bbac), and the shares of diatoms in phytoplankton biomass (Bac%) to the North Atlantic Oscillation winter index (NAOw) in the three study sites in Sweden (SE), Estonia (EE) and Germany (DE)

Correlogram	EE-DE	*N*	EE-SE	*N*	SE-DE	*N*
Q–NAOw	0.56	12	0.91**	12	0.71**	12
SiRiver–NAOw	0.05	12	−0.40	12	0.22	12
Q*Si–NAOw	0.06	12	0.64*	12	0.72**	12
SiLake–NAOw	0.49	12	−0.38	8	0.23	8
Bphyto–NAOw	0.16	12	0.57	8	−0.25	8
Bbac–NAOw	0.35	12	0.21	8	0.16	8
Bac%–NAOw	0.57	12	−0.33	8	−0.06	8

N, the number of cases

$*P < 0.05$, $**P < 0.01$

biomass may be related to lake-specific differences in Si-loading and biomass flushing, zooplankton grazing (Poister & Armstrong, 2003), silicon uptake by benthic or periphytic diatoms, sorption of dissolved silicon on to clay particles and surficial iron hydroxides (Mortlock & Frohlich, 1987), and to differences in Si cell quota. Different diatom species have different cell quotas of biogenic silica (Sommer & Stabel, 1983; Conley et al., 1989) but the cell quota of a single species might also decrease under Si-limiting conditions (Harrison et al., 1977). The overall range of silicification in diatoms measured under different conditions was 4.7-fold, which is much greater than variations in the C or N cell quota (Hildebrand, 2002). Moreover, in lakes with rather short residence times such as LDE and LSE, in-lake processes may be masked by water exchange rates, which are of minor importance in lakes with long residence times.

The river and lake variables showed different levels of coherence in their seasonal changes and responses to the NAOw signal. In the three rivers, water discharge, silicon concentration and silicon loading showed a rather similar seasonality. Si concentrations were highest in winter and Si load into the lakes was most pronounced during spring.

The lower silicon concentration observed in February and March of high NAOw years in both shallow lakes (LEE, LDE), revealed either a diluting effect by the increased winter discharge or an earlier onset of diatom development, or both. As no clear dilution effect caused by the enhanced winter discharge was apparent in the river data (Fig. 2), the earlier uptake of silicon by diatoms seems the more likely cause of the lower SiLake values. Early spring diatom development in high NAOw years has indeed been described for LDE (Gerten & Adrian, 2000), LSE (Weyhenmeyer, 2004) and other lakes (Adrian et al., 1995, Weyhenmeyer et al., 1999).

In general, on the basis of the correlograms, there was no significant similarity among the three lakes in the responses of phytoplankton to the NAOw, indicating that lake-internal biological processes are rather lake-specific.

Our findings, which showed high coherence in the discharge and loading variables in contrast to the low or absent coherence in the in-lake and biological variables, are in good agreement with results from lakes and terrestrial ecosystems from the UK, the US and Sweden, where high levels of coherence were recorded for physical variables but lower levels for biological variables (Magnuson et al., 1990; Kratz et al., 1997; George et al., 2000; Blenckner & Hillebrand, 2002; Weyhenmeyer, 2004).

The results of our study showed that the three rivers studied, being mostly physically driven systems, were supra-regionally more coherent in their responses to climatic forcing than the patterns observed in the three lakes. The discharge regime of rivers is largely determined by precipitation and temperature regimes, i.e. by climatic variables. In lakes the climate signal is transformed by the lake morphology and subsequently by the mixing regime and the water residence time. The nature of catchment and lake features that may modify the climate signal has led to the development of an "effect filter" concept, in terms of which the different pathways of climate impact on lakes are described (Blenckner, 2005). Even in lakes with similar mixing regimes, such as LEE and LDE in our study, which showed highly correlated seasonality of several variables (SiLake, Bphyto, Bbac), we found only very limited coherence in the responses to climatic forcing expressed by NAOw. However, these two lakes differ considerably in size, and LEE is mostly driven by changes in water level, unlike LDE (Nõges & Nõges, 1998; Nõges et al., 2003).

Conclusions

1. In all three rivers, the silicon concentration–water discharge relationship showed a strong seasonal hysteresis with higher dissolved concentrations during periods in which the limbs of the hydrographs were rising. In RDE, characterised by rain-type hydrology, the resulting silicon loads had a V-shaped dynamic with the highest loadings in winter and the lowest in May. In the northern rivers, chacterized by snow-type hydrology, the loading patterns followed the annual hydrograph with peak values during the spring flood and with an extended summer low-load period. Of the two components of silicon load formation (discharge and Si concentration), river discharge was the more variable in the two northern rivers while silicon concentration varied most markedly in RDE.
2. Strong correlations between monthly silicon loads and in-lake concentrations were most probably caused by similar seasonality; there were no correlations on an annual basis. The rates of decrease of the in-lake silicon concentration in spring were fairly comparable in all three lakes, but obviously because of a large sedimentation loss, the visible increase in diatom biomass was nearly four times smaller in the stratified lake than in the nonstratified lakes.
3. The high seasonal correlation in the two northern rivers went along with the common snow-type hydrology of their catcments. The similarity between lakes arose more from lake morphometry than from the climatic regime: the shallow lakes in Estonia and Germany had more similarities than the Estonian and the deeper Swedish lake because they share a similar type of catchment hydrology.
4. Among the variables measured at the three sites, river discharge responded most coherently to climatic forcing as represented by the NAOw. Because of the hydraulic linkage, silicon loads also demonstrated a certain coherence. Phytoplankton variables had several significant correlations with NAOw in different seasons but they were mostly non-coherent among the sites. Our study confirmed the view that rivers, as mostly physically driven systems, are likely to show more supra-regionally coherent reactions to climatic forcing than lakes.

Acknowledgements Funding was provided by the Estonian Ministry of Education (grant 0362480s03), the Estonian Science Foundation (grant 5738), the EU projects CLIME (EVK1-CT-2002-00121) and Euro-limpacs (GOCE-CT-2003-505540), the Royal Swedish Academy of Sciences (Knut and Alice Wallenberg grant) and the Swedish Research Council (621-2005-4335). We thank the Swedish Environmental Protection Agency and all laboratories for financing, sampling and analysing numerous water samples.

References

Adrian, R., R. Deneke, U. Mischke, R. Stellmacher & P. Lederer, 1995. A long term study of the Heiligensee (1975–1992). Evidence for effects of climatic change on the dynamics of eutrophied lake ecosystems. Archiv für Hydrobiologie 133: 315–337.

Berner, E. K. & R. A. Berner, 1996. Global Envrironment. Water, Air, and Geochemical Cycles. Prentice Hall, Upper Saddle River, New Jersey.

Blenckner, T., 2005. A conceptual model of climate-related effects on lake ecosystems. Hydrobiologia 533: 1–14.

Blenckner, T. & H. Hillebrand, 2002. North Atlantic Oscillation signatures in aquatic and terrestrial ecosystems—a meta-analysis. Global Change Biology 8: 203–212.

Conley, D. J., 1997. Riverine contribution of biogenic silica to the oceanic silica budget. Limnology and Oceanography 42: 774–777.

Conley, D. J., S. S. Kilham & E. Theriot, 1989. Differences in silica content between marine and freshwater diatoms. Limnology and Oceanography 34: 205–213.

Conley, D. J., P. Stålnacke, H. Pitkänen & A. Wilander, 2000. The transport and retention of dissolved silicate by rivers in Sweden and Finland. Limnology and Oceanography 45: 1850–1853.

George, D. G., J. F. Talling & E. Rigg, 2000. Factors influencing the temporal coherence of five lakes in the English Lake District. Freshwater Biology 43: 449–461.

Gerten, D. & R. Adrian, 2000. Climate driven changes in spring plankton dynamics and the sensitivity of shallow polymictic lakes to the North Atlantic Oscillation. Limnology and Oceanography 45: 1058–1066.

González Sagrario, M. A., E. Jeppesen, J. Gomà, M. Søndergaard, J. P. Jensen, T. Lauridsen & F. Landkildehus, 2005. Does high nitrogen loading prevent clear-water conditions in shallow lakes at moderately high phosphorus concentrations? Freshwater Biology 50: 27–41.

Harrison, P. J., H. L. Conway, R. W. Holmes & C. O. Davis, 1977. Marine diatoms grown in chemostats under silicate or ammonium limitation. III. Cellular chemical composition and morphology of *Chaetoceros debilis*, *Skeletonema costatum*, and *Thalassiosira gravida*. Marine Biology 43: 19–31.

Hildebrand, M., 2002. Lack of coupling between silicon and other elemental metabolisms in diatoms. Journal of Phycology 38: 841–843.

Huisman, J. & F. D. Hulot, 2005. Population dynamics of harmful cyanobacteria. Factors affecting species composition. In Huisman, J., H. C. P. Mattehijs & P. M Visser (eds), Harmful Cyanobacteria. Springer, 143–176.

Humborg, C., D. J. Conley, L. Rahm, F. Wulff, A. Cociasu & V. Ittekkot, 2000. Silica retention in river basins: far-reaching effects on biogeochemistry and aquatic food webs in coastal marine environments. Ambio 29: 45–50.

Hurrell, J. W., Y. Kushnir, G. Ottersen & M. Visbeck (eds), 2003. The North Atlantic Oscillation: Climatic Significance and Environmental Impact. American Geophysical Union, Washington.

Ittekkot, V., 2003. A new story from the Ol' Man River. Science 301: 56–58.

Ittekkot, V., C. Humborg & P. Schafer, 2000. Hydrological alterations and marine biogeochemistry: a silicate issue? BioScience 50: 776–782.

Köhler, J. & S. Hoeg, 2000. Phytoplankton selection in a river-lake system during two decades of changing nutrient supply. Hydrobiologia 424: 13–24.

Kratz, T. K., K. E. Webster, C. J. Bowser, J. J. Magnuson & B. J. Benson, 1997. The influence of landscape position on lakes in Northern Wisconsin. Freshwater Biology 37: 290–317.

Magnuson, J. J., B. J. Benson & T. K. Kratz, 1990. Temporal coherence in the limnology of a suite of lakes in Wisconsin, U.S.A. Freshwater Biology 23: 145–149.

Mortlock, R. A. & P. N. Frohlich, 1987. Continental weathering of germanium: Ge/Si in the global river discharge. Geochimica et Cosmochimica Acta 51: 2075–2082.

Nõges, P. & T. Nõges, 1998. The effect of fluctuating water level on the ecosystem of Lake Võrtsjärv, Central Estonia. Proceedings of Estonian Academy of Sciences. Biology. Ecology 47: 98–113.

Nõges, T., P. Nõges & R. Laugaste, 2003. Water level as the mediator between climate change and phytoplankton composition in a large shallow temperate lake. Hydrobiologia 506–509: 257–263.

Ottersen, G., B. Planque, A. Belgrano, E. Post, P. C. Reid & N. C. Stenseth, 2001. Ecological effects of the North Atlantic Oscillation. Oecologia 128: 1–14.

Poister, D. & D. E. Armstrong, 2003. Seasonal sedimentation trends in a mesotrophic lake: influence of diatoms and implications for phosphorus dynamics. Biogeochemistry 65: 1–13.

Ptacnik, R., G. D. Jenerette, A. M. Verschoor, A. F. Huberty, A. G. Solimini & J. D. Brookes, 2005. Applications of ecological stoichiometry for sustainable acquisition of ecosystem services. Oikos 109: 52–62.

Raubitschek, S., A. Lücke & G. H. Schleser, 1999. Sedimentation patterns of diatoms in Lake Holzmaar, Germany—(on the transfer of climate signals to biogenic silica oxygen isotope proxies). Journal of Paleolimnology 21: 437–448.

Smetacek, V., 1998. Diatoms and the silicate factor. Nature 391: 224–225.

Smith, V. H., G. D. Tilman & J. C. Nekola, 1999. Eutrophication: impacts of excess nutrient inputs on freshwater, marine, and terrestrial ecosystems. Environmental Pollution 100: 179–196.

Sommer, U. & H.-H. Stabel, 1983. Silicon consumption and population density changes of dominant planktonic diatoms in Lake Constance. The Journal of Ecology 71: 119–130.

Tréguer, P. & P. Pondaven, 2000. Global change. Silica control of carbon dioxide. Nature 406: 358–359.

Utermöhl, H., 1958. Zur Vervollkommnung der quantitativen Phytoplankton-Methodik. Mitteilungen der internationalen Vereinigung für Limnologie 9: 1–38.

Weyhenmeyer, G. A., 2004. Synchrony in relationships between the North Atlantic Oscillation and water chemistry among Sweden's largest lakes. Limnology and Oceanography 49: 1191–1201.

Weyhenmeyer, G. A., T. Blenckner & K. Pettersson, 1999. Changes of the plankton spring outburst related to the North Atlantic Oscillation. Limnology and Oceanography 44: 1788–1792.

Weyhenmeyer, G. A., E. Willén & L. Sonesten, 2004. Effects of an extreme precipitation event on water chemistry and phytoplankton in the Swedish Lake Mälaren. Boreal Environment Research 9: 409–420.

Hydrobiologia (2008) 599:77–86
DOI 10.1007/s10750-007-9195-x

ELLS 2007

Critical N:P ratio for cyanobacteria and N_2-fixing species in the large shallow temperate lakes Peipsi and Võrtsjärv, North-East Europe

Tiina Nõges · Reet Laugaste · Peeter Nõges · Ilmar Tõnno

Abstract In the 1990s a sharp decrease in nitrogen loading occurred in Estonian rivers, bringing about a reduction of the nitrogen-to-phosphorus ratio (N:P ratio) in the large shallow lakes, Peipsi (3,555 km^2, mean depth 7.1 m) and Võrtsjärv (270 km^2, 2.8 m). The average mass ratio of total nitrogen (TN) and total phosphorus (TP) in Võrtsjärv (45) was about twice as high as that in Peipsi (22). In Peipsi, the N_2-fixing *Gloeotrichia echinulata, Aphanizomenon flos-aquae* and *Anabaena* species prevailed in the summer phytoplankton, while in Võrtsjärv the dominant cyanobacteria were *Limnothrix planktonica, L. redekei* and *Planktolyngbya limnetica*, which cannot fix N_2; the main N_2-fixing taxa *Aphanizomenon skujae* and *Anabaena* sp. seldom gained dominance. In May–October the critical TN:TP mass ratio, below which N_2-fixing cyanobacteria (Nfix) achieved high biomasses, was $\sim$40 in Võrtsjärv and $\sim$30 in Peipsi. The percentages of both total cyanobacteria (CY) and Nfix (CY% and Nfix%) in Peipsi achieved their maximum values at an N:P mass ratio at or below 20 for both TN:TP and Nmin:SRP. In Võrtsjärv, the TN:TP supporting a high Nfix% was between 30 and 40 and the N_{min}:SRP supporting this high percentage was in the same range as that in Peipsi (<20), though the maximum Nfix% values in Võrtsjärv (69%) were much lower than in Peipsi (96%). The Nmin:SRP ratio explained 77% of the variability in Nfix% in May–October. The temperature dependence of Nfix% approximated to the maximum function type, with an upper limiting value at a certain water temperature, and this was most distinct in May–October. The critical TN:TP ratios obtained from our study (roughly 30 for Peipsi and 40 for Võrtsjärv) are much higher than the Redfield N:P mass ratio routinely considered (7). Our results represent valuable guidelines for creating effective management strategies for large shallow lakes. They provide a basis for stressing the urgent need to decrease phosphorus loading and to keep the in-lake P concentration low, and *not* to implement nitrogen reduction measures without a simultaneous decrease of phosphorus concentration.

Keywords Large and shallow lake · Nitrogen–phosphorus ratio · Cyanobacteria · N_2-fixing cyanobacteria

Guest editors: T. Nõges, R. Eckmann, K. Kangur, P. Nõges, A. Reinart, G. Roll, H. Simola & M. Viljanen
European Large Lakes—Ecosystem changes and their ecological and socioeconomic impacts

T. Nõges (✉) · R. Laugaste · P. Nõges · I. Tõnno
Institute of Environmental and Agricultural Sciences, Centre for Limnology, Estonian University of Life Sciences, 61101 Tartu County, Estonia
e-mail: tiina.noges@emu.ee

P. Nõges
Institute of Environment and Sustainability, European Commission – Joint Research Centre, 21020 Ispra, VA, Italy

Introduction

Understanding the causes of cyanobacterial dominance has been a focal point of classical and contemporary limnological research (Havens et al., 2003). Schindler (1977) noticed that high concentrations of P and a low N:P supply ratio favour the development of cyanobacteria blooms. Focus on N and P is fundamental because these macronutrients most frequently limit primary production in aquatic ecosystems (Hall et al., 2005). The critical N:P ratio at which plankton switches over from P limitation to N limitation, favouring cyanobacteria, has often been considered close to the 'Redfield ratio' (N:P 16:1 by atoms or 7:1 by mass), which has been found to be characteristic of the plankton and seawater in oceans throughout the world (Redfield, 1958; Falkowski & Davis, 2004). Cyanobacterial dominance at lower N:P ratios has been linked to the fact that cyanobacteria compete better for nitrogen than other phytoplankton. Because many cyanobacterial species are capable of fixing N_2, they usually dominate in lakes with low N:P ratios (Huisman & Hulot, 2005). According to Howarth et al. (1988), planktonic N_2-fixation can occur if the N:P ratio is equal to or lower than the Redfield mass ratio of seven. However, the critical N:P mass ratios at which cyanobacteria, both N_2-fixing and non-fixing, tend to dominate in lakes are found to range from 5 to 10 (Schindler, 1977; Seip, 1994; Michard et al., 1996; Bulgakov & Levich, 1999) up to 29 (Smith, 1983). Smith et al. (1995) showed that an N:P mass ratio of 22:1 provides a rather distinct boundary between lakes dominated by N_2-fixing cyanobacteria and lakes in which few of these algae occur. Cyanobacterial blooms in moderately deep, stratified eutrophic lakes are typically composed of N_2-fixing taxa, including *Anabaena* and *Aphanizomenon* (Paerl et al., 2001). In contrast, shallow eutrophic lakes are typically dominated by cyanobacterial taxa that do not fix N_2, in particular the Oscillatoriaceae family, including *Oscillatoria, Planktothrix* and *Limnothrix* (Havens et al., 2003). Oscillatoriaceae can attain high biomasses in shallow eutrophic lakes, but do not form surface blooms. They continue to grow even when the biomass and light attenuation become extremely high, possibly by setting up a stable feedback mechanism that maintains their dominance by creating shady conditions (Scheffer et al., 1997); it has been shown that non-N_2-fixing cyanobacteria can maintain net growth at low underwater irradiance (van Duin et al., 1995).

In their treatise on ecological stoichiometry, Sterner and Elser (2002) hypothesize that the N:P content of producers should closely match the ratio of N and P supplied to ecosystems. In contrast, Hall et al. (2005) provided evidence that the cellular stoichiometry of primary producers in nature is much less responsive to variations in the N:P supply ratios than previously proposed. They stressed that other mechanisms potentially influence the N:P content of producers. For instance, N_2 fixation by cyanobacteria could elevate producer N content and thus increase the N:P ratio of the whole ecosystem; in shallow lakes, phytoplankton may contain resuspended benthic algae that receive nutrients from sediments, and the N:P supply from these sediments may differ from that measured in the water column.

According to many authors (Smith et al., 1987; Willén, 1992; Lathrop et al., 1998; Downing et al., 2001), cyanobacterial dominance and blooms may be connected more strongly to the variations in P and N concentrations than to changes in the N:P ratio. Other factors such as water temperature, pH, light intensity and total carbon dioxide concentration are also important.

To obtain a better understanding of the causes of cyanobacteria dominance in large shallow temperate lakes, and to develop an effective nutrient management strategy, the aim of the present study was to estimate the critical N:P ratio favouring cyanobacterial development and nitrogen fixation in two large shallow lakes in Estonia (North West Europe).

Study site

Lake Peipsi (3,555 km^2) is the fourth largest lake in Europe by surface area. Being shared between Estonia and Russia, Peipsi is the largest transboundary lake in Europe. The lake consists of three parts: Lake Peipsi s.s. (the largest; 2,611 km^2, mean depth 8.4 m, maximum depth 12.9 m), Lake Pihkva and the river-shaped Lake Lämmijärv connecting two larger basins. The outflowing River Narva runs into the Gulf of Finland, the Baltic Sea.

Lake Võrtsjärv (270 km^2, mean depth: 2.8 m, maximum depth: 6 m) is a large shallow lake located in the watershed of Peipsi (for the location map see Nõges et al., 2007). The water residence time in Peipsi is about 2 years and in Võrtsjärv about 1 year. Both lakes are typically ice-covered from mid-November until mid-April. According to Nõges et al. (2007) the loading of nitrogen to Lakes Peipsi (from Estonian catchment) and Võrtsjärv decreased substantially from 1980–1991 to 1992–2004 (by 39% and 51%, respectively), whereas phosphorus loading decreased much less (by 13% and 35%, respectively), so the N:P ratio was considerably lowered in both the loading (30% and 33%, respectively) and the water (18% and 31%, respectively) of these lakes.

Materials and methods

We used nutrient concentration, phytoplankton biomass and species composition data for Lakes Võrtsjärv and Peipsi gathered within different monitoring programmes and research projects in Estonia in 1985–2004. Monthly data were collected throughout the year in Võrtsjärv and from March to November in Peipsi s.s. Total N (TN), total P (TP), soluble reactive phosphorus (SRP) and the forms of mineral nitrogen (Nmin = nitrates + nitrites + ammonium) were analysed by the methods described by Grasshoff et al. (1983) against the corresponding ISO standards at the accredited laboratory, Tartu Environmental Research Centre Ltd. Phytoplankton samples were preserved with formaldehyde (1962–1995) and Lugol's iodine (1996 onwards). Microscopic counting was performed at 400× magnification in a Fux-Rosenthal chamber in 1962–1999 and from 2000 onwards by the Utermöhl (1958) method. Intercalibration of the different preservation and counting methods revealed good correspondence among the results obtained using all the methods applied (R. Laugaste & P. Nõges, unpublished data).

We used the program Statistica for Windows version 7.0 for statistical analyses. We used Spearman rank correlation throughout the analysis to avoid the influence of non-normality in data distributions.

Results

In Võrtsjärv, the average total nitrogen concentration (TN) has been about twice as high as that in Peipsi, while the total phosphorus (TP) concentration has remained in roughly the same range in both lakes. The TN:TP mass ratio in Võrtsjärv has been about twice as high as in Peipsi, and the mass ratio of the mineral forms of nitrogen (Nmin) to the soluble reactive phosphorus (SRP) 3–4 times higher (Table 1). In both lakes, changes in the TN:TP ratio correlate significantly with changes in Nmin:SRP (Spearman 0.48 and 0.59 for Peipsi and Võrtsjärv, respectively; $P < 0.0001$). The N:P ratio in both lakes has typically been quite high in spring, achieving its lowest values in the period July–November (Fig. 1a and b).

The N_2-fixing *Gloeotrichia echinulata* (J. E. Smith) Richter, *Aphanizomenon flos-aquae* (L.) Ralfs and *Anabaena* spp., and non N_2-fixing *Microcystis* species, prevail in the summer phytoplankton in Peipsi. In Võrtsjärv, the dominant cyanobacterial species *Limnothrix planktonica* (Wolosz.) Meffert, *L. redekei* (Van Goor) and *Planktolyngbya limnetica* (Lemm.) Kom.-Legn. cannot fix N_2; the main N_2-fixing taxa, *Aphanizomenon skujae* Kom.-Legn. & Cronb. and *Anabaena* sp., seldom achieve dominant status. In both lakes, the average share of cyanobacteria and N_2-fixing taxa in the phytoplankton biomass in May–October increased substantially from 1980–1991 to 1992–2004 (in Peipsi by 31% and 66%, and in Võrtsjärv by 13% and 783%).

Even though the share of cyanobacteria in the phytoplankton in Võrtsjärv has remained twice as high as that in Peipsi (Table 1), the share of N_2-fixing species in Võrtsjärv has never achieved such high values as in Peipsi (maximum values 69% and 96%, respectively).

The biomasses of all cyanobacteria (CY) and N_2-fixing cyanobacteria (Nfix), and the percentages of the total phytoplankton biomass represented by these two groups (CY% and Nfix%, respectively), correlated strongly with water temperature in both lakes (Fig. 2). As regards nutrients, the biomass and the share of CY and Nfix correlated positively with phosphorus concentration in Peipsi, whereas in Võrtsjärv there was only a weak positive correlation with TP and the correlation with SRP was negative. The biomass and share of CY and Nfix correlated negatively with nitrogen concentrations in Võrtsjärv, the strongest negative correlations occurring with Nmin; the respective correlations with N compounds were much weaker in Peipsi. In Peipsi, the TN:TP

Table 1 Average values of Secchi depth, biomass of phytoplankton (Bphyto), share of cyanobacteria and share of N_2-fixing species in total phytoplankton biomass (CY% and Nfix%, respectively), concentration of total nitrogen (TN), total phosphorus (TP), and the mass ratios of TN:TP, and mineral nitrogen (Nmin) to soluble reactive phosphorus (SRP), in Peipsi s.s. and Võrtsjärv in 1985–2004

	Peipsi s.s. 1985–2004					Võrtsjärv 1985–2004				
	n	Mean	Median	Std	CV (%)	*n*	Mean	Median	Std	CV (%)
Secchi, m	627	1.9	1.8	0.7	37	590	1	0.9	0.4	40
Bphyto, gWW m^{-3}	619	6.9	4.6	11.5	167	438	13.1	9.7	12.5	95
CY% in biomass	593	29	23	26	90	430	58	67	29	50
Nfix% in biomass	563	10.6	0.4	19	179	435	3.2	0.6	6.8	213
TN, g m^{-3}	654	0.73	0.64	0.34	47	786	1.62	1.4	0.9	56
TP, mg m^{-3}	658	43	40	21	49	859	53	45	46	87
TN:TP (mass)	647	20	17	13	65	785	45	31	53	118
Nmin:SRP (mass)	447	22	11	29	132	529	76	45	93	122
Chl a, mg m^{-3}	696	19	16	14	74	897	27	26	19	70

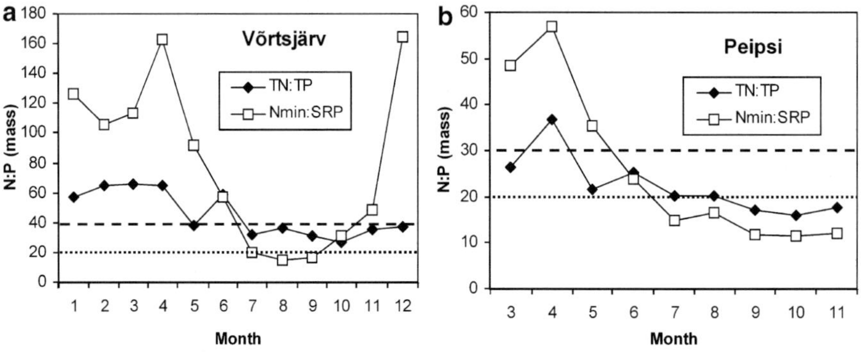

Fig. 1 Seasonal cycle of monthly average mass ratios of total nitrogen to total phosphorus (TN:TP) and of mineral nitrogen to soluble reactive phosphorus (Nmin:SRP) in Võrtsjärv (**a**) and Peipsi s.s. (**b**) in 1985–2004. Critical ratios—TN:TP (30 for Peipsi and 40 for Võrtsjärv) and Nmin:SRP (20 for both lakes)—are denoted with dashed lines

ratio was more closely related to cyanobacterial indices, while in Võrtsjärv these indices corresponded slightly more strongly with the Nmin:SRP ratio (Fig. 2).

The temperature (Temp) dependence of CY% and Nfix% showed a seasonal hysteresis: the water temperatures corresponded better with lower CY% and Nfix% during the increasing phase in spring and summer than during the decreasing phase in autumn. The maximum values of CY% and Nfix% coincided with the highest water temperatures in August (around 20°C) in both lakes (Fig. 3).

To elucidate the temperature-dependence of algal blooming, we ran a further set of analyses focusing on the period May–October, because our lakes are completely ice-free in this period and the water temperature is high enough to support the development of cyanobacteria. In May–October the critical TN:TP mass ratio below which N_2-fixing cyanobacteria potentially achieve high biomasses was ~40 in Võrtsjärv (Fig. 4a and b) and ~30 in Peipsi (Fig. 4c–e). TN:TP ratios below 40 favoured an increased biomass of *Microcystis* sp. in Peipsi (Fig. 4f) and total cyanobacterial biomass in both lakes (Fig. 4g and h).

CY% was persistently higher and Nfix% lower in Võrtsjärv than in Peipsi, and during the increasing phase of CY% and Nfix% (June–August in both lakes) the unit decrease of Nmin:SRP brought about a much higher increase of both indices in Peipsi than in Võrtsjärv (Fig. 5). In Peipsi both CY% and Nfix% achieved their highest values at N:P mass ratios at or below 20 for both TN:TP and Nmin:SRP. In Võrtsjärv the CY% remained high over a wide range (20–60) of

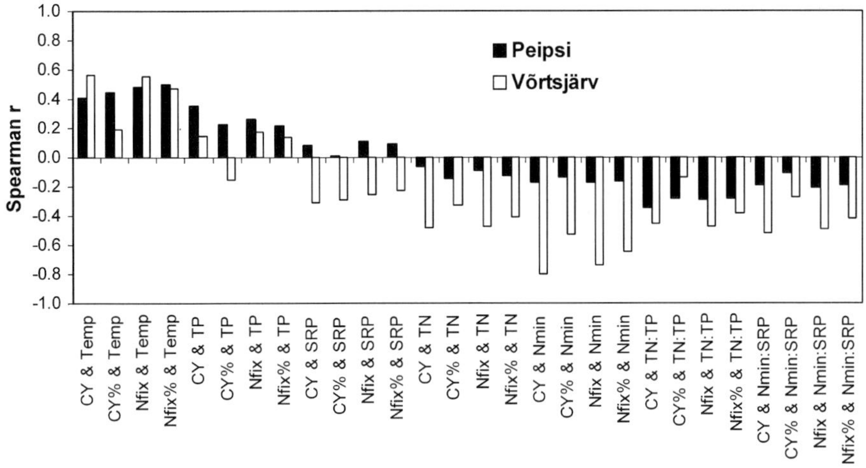

Fig. 2 Spearman correlations between the biomass of all cyanobacteria and share of cyanobacteria in total phytoplankton biomass (CY and CY%, respectively), and the biomass of N_2-fixing cyanobacteria and share of N_2-fixing cyanobacteria in total phytoplankton biomass (Nfix and Nfix%, respectively), with water temperature (Temp), concentration of total phosphorus (TP), soluble reactive phosphorus (SRP), total nitrogen (TN), mineral nitrogen (Nmin) and the TN:TP and Nmin:SRP ratios in Peipsi s.s. and Võrtsjärv based on all data in 1985–2004. All $r > 0.12$ and $r < -0.12$ are significant at the $P < 0.01$ level

Fig. 3 Seasonal cycle of the relationship between monthly average values of the share of cyanobacteria and share of N_2-fixing species in total phytoplankton biomass (CY% and Nfix%, respectively) and water temperature in Võrtsjärv (**a** and **c**) and Peipsi s.s. (**b** and **d**) in 1985–2004

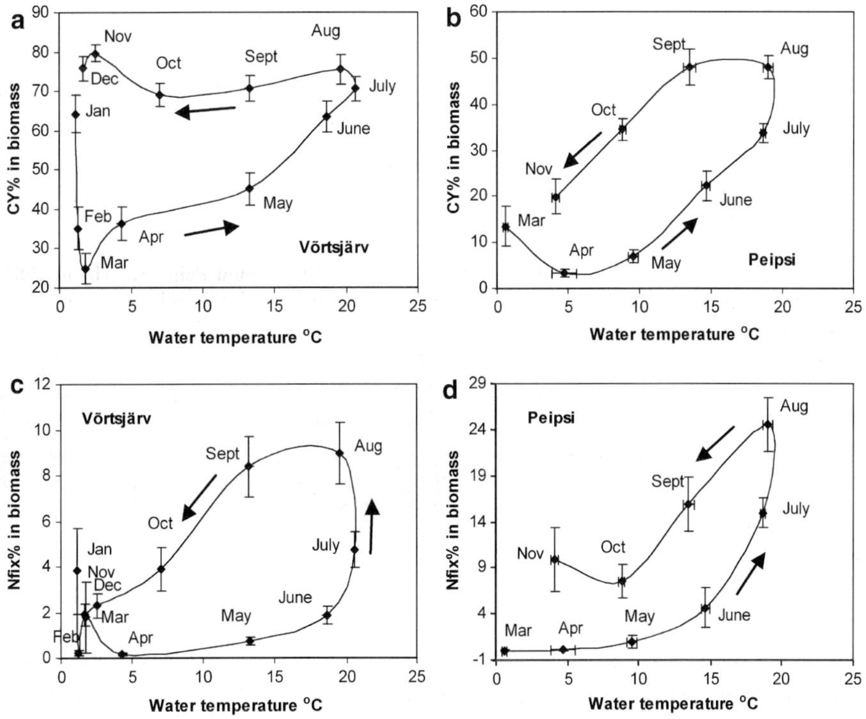

TN:TP (Fig. 5a and b) while high Nfix% occurred at TN:TP ratios between 30 and 40 (Fig. 5c). The Nmin:SRP values connected with high Nfix% were in the same range as those in Peipsi (<20), even though the maximum Nfix% values in Võrtsjärv were much lower (Fig. 5d). On the basis of the monthly average Nmin:SRP ratios, a common power function could be developed for both lakes, which explained 42% of the variability of Nfix% when all monthly averages were considered and as much as 77% when the period May–October was considered (Fig. 6a). The temperature dependence of Nfix% approximated to the maximum

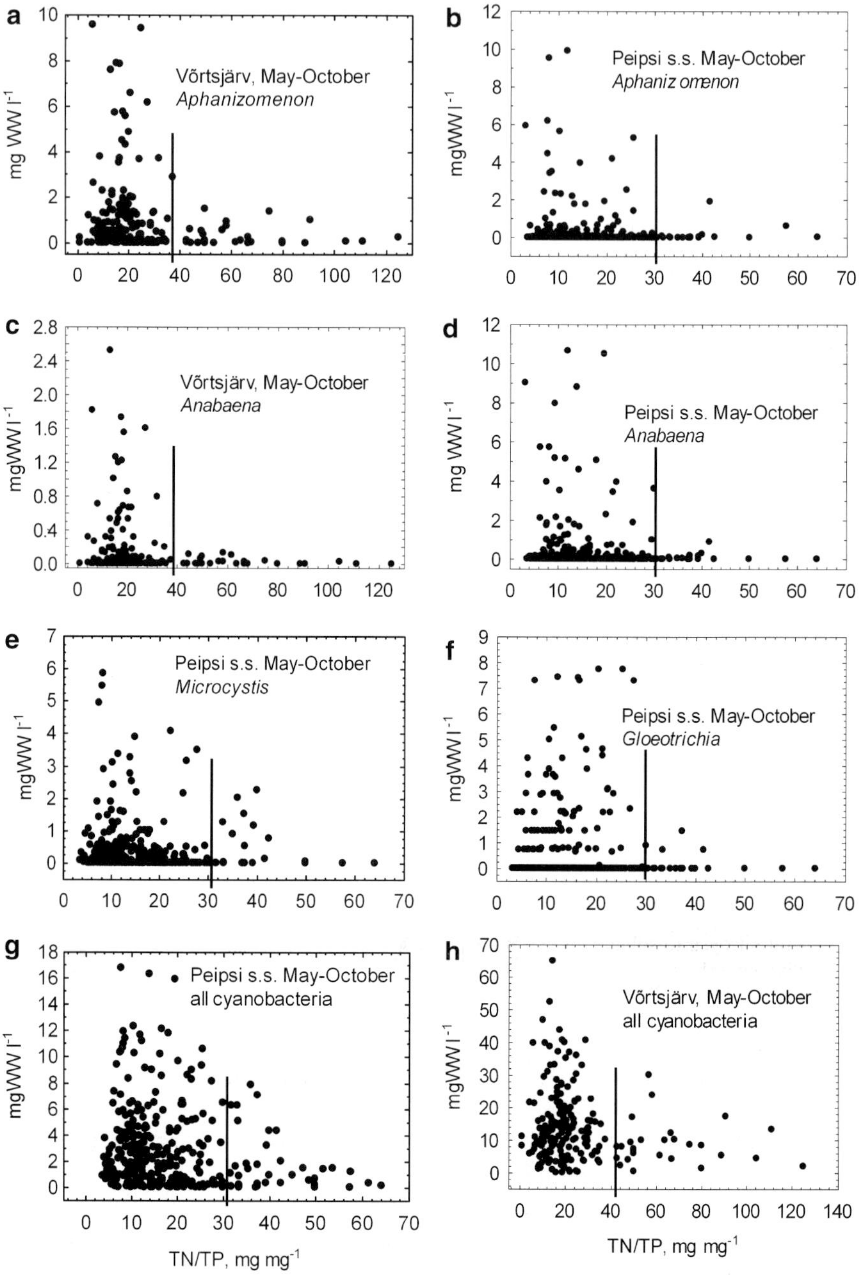

Fig. 4 The relationships between TN:TP ratio and the biomass of N_2-fixing cyanobacteria *Aphanizomenon* (**a**, **b**), *Anabaena* (**c**, **d**), *Microcystis* (**e**), *Gloeotrichia* (**f**) and all cyanobacteria (**g**, **h**) in Peipsi s.s (**b**, **d**–**g**) and Võrtsjärv (**a**, **c**, **h**) in May–October, 1985–2004

function type, with an upper limit for Nfix% at certain water temperatures, and this was again most evident in the period May–October (Fig 6b).

Discussion

In both lakes, cyanobacterial blooms were already a common phenomenon at the beginning of the 20th century (Mühlen & Schneider, 1920; Laugaste et al., 2001). No blooming was observed under the heavy nitrogen loading conditions in the 1980s, but started again in Peipsi in the 1990s when the TN:TP ratio in the inflows and in the lake decreased (Nõges et al., 2007). Between the 1980s and 2001, the average TN:TP mass ratio in Peipsi decreased from 38 to 14, mainly because the nitrogen loading was reduced. In Peipsi, the declining TN:TP ratio favoured the

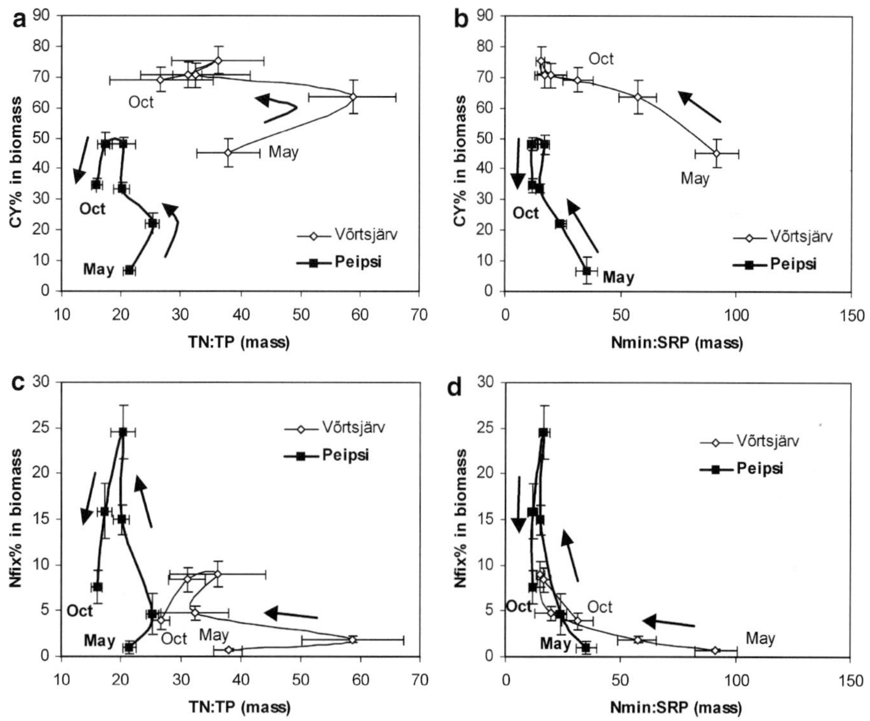

Fig. 5 Seasonal cycle of the relationship between monthly average values of the share of cyanobacteria and share of N_2-fixing species in total phytoplankton biomass (CY% and Nfix%, respectively) with (**a** and **c**) the mass ratio of total nitrogen to total phosphorus (TN:TP), and (**b** and **d**) the ratio of mineral nitrogen to soluble reactive phosphorus (Nmin:SRP) in Peipsi s.s. and Võrtsjärv in 1985–2004

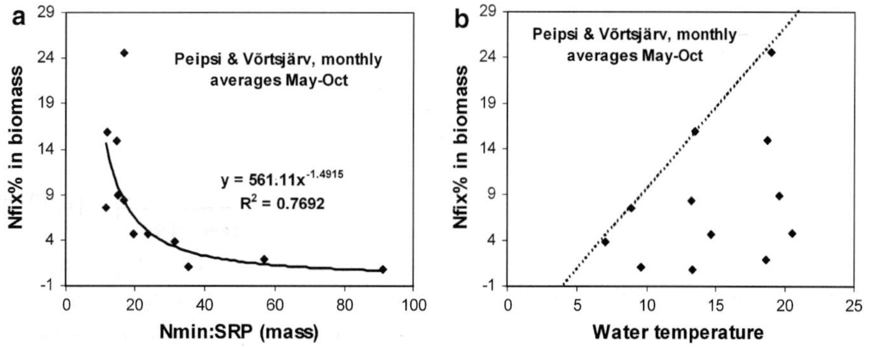

Fig. 6 The relationship between the monthly average values of the share of N_2-fixing species in total phytoplankton biomass (Nfix%) and the mass ratio of the concentration of mineral nitrogen to soluble reactive phosphorus (Nmin:SRP, **a**), and to water temperature (**b**), in Peipsi s.s. and Võrtsjärv in May–October 1985–2004

development of cyanobacteria including N_2-fixing species (Nõges et al., 2004). In Võrtsjärv, the TN:TP mass ratio has remained above 30 until the present and massive blooms have not reappeared. Smith et al. (1995) showed that an N:P mass ratio of 22:1 defines a rather distinct boundary between lakes dominated by N_2-fixing cyanobacteria and lakes in which these algae are relatively rare. Our results are consistent with this finding, especially regarding the Nmin:SRP ratio. In both lakes, an Nmin:SRP ratio of 20 appeared to be a boundary at and below which the share of N_2-fixing algae increased sharply (Fig. 5d). The Nmin:SRP ratio appears to be a more straightforward indicator than the TN:TP ratio for the resource supply stoichiometry for phytoplankton, but the latter should not be neglected. During the period of phytoplankton growth, the concentrations of inorganic forms of N and P may be close to the detection limits, and consequently the precision with which their ratio can be measured may be quite low. Furthermore, concerning N and P loading in lakes, the total amounts of N and P and the TN:TP ratio reproduce the supply ratio more reliably than the amounts of inorganic forms or the Nmin:SRP loading ratio. Havens et al. (2003) described a relationship between CY% and Nfix% and water column TN:TP, comparing the data of Smith (1985) with records from the shallow subtropical Lake Okeechobee (Fig. 9 in Havens et al., 2003). These graphs showed that at TN:TP mass ratios below 30, CY%, and especially Nfix%, increases.

Comparing our data with these graphs, we can see that Nfix% in both lakes and also CY% in Peipsi correspond closely to the data presented by Havens et al. (2003), but CY% in Võrtsjärv (60%) is exceptionally high given the TN:TP mass ratio of 40 in this lake. The high CY% in Võrtsjärv could be explained by the species composition of cyanobacteria in this lake. Võrtsjärv is a typical light-limited lake (Nõges et al., 1998) with a mean depth of <3.0 m and a low Secchi depth, dominated by Oscillatoriaceae, which are adapted to harvest light effectively under low irradiance conditions (Berger, 1989). These ''shade algae'' (Reynolds, 1984) may achieve near-complete dominance of the plankton in strongly light-limited lakes (Hall et al., 2005), while the share of N_2-fixing cyanobacteria remains low because of the poor light climate, high N:P ratio, high shear stress and low water column stability (Havens et al., 2003), which is also the case in Võrtsjärv.

Earlier studies on Võrtsjärv showed that N_2 fixation starts when the TN:TP mass ratio decreases to about 20. However, throughout the period in which N_2-fixation has been observed to occur, this ratio has never declined below 17 (Tõnno & Nõges, 2003). The present study revealed that the critical TN:TP mass ratios for the cyanobacterial communities in Lakes Peipsi and Võrtsjärv are around 30 and 40, respectively, and the critical Nmin:SRP mass ratios in both lakes are around 20. The probable reason why the symptoms of N-limitation appear at higher N:P ratios than predicted by Redfield's ratio is that P is recycled more rapidly than N and the ratio of recycled N to recycled P is lower than the N:P found in the lake water. In principle, the faster turnover of P compared to N in lakes is a widely acknowledged phenomenon (Harris, 1986). In shallow lakes such as Võrtsjärv and Peipsi the turnover rate of P could be much higher than in deep stratified lakes. Nõges et al. (2007) found a significant positive relationship between the P content of sediments and the relative depth of the lake, supporting this possibility. In deep lakes, most of the bound phosphorus, which settles out from the euphotic zone during the growing season, first becomes available for photosynthetic organisms after the turnover in the following spring. In shallow lakes where sediments are in permanent contact with the water, the same portion of phosphorus could be recycled and brought back to water column much more rapidly and may support planktonic photosynthesis several times during one vegetation period. The finding that N:P ratios are an important determinant of cyanobacterial dominance is of major practical importance from a water management perspective, because N:P ratios can be manipulated by sewage diversion and nutrient removal. However, several studies have also shown that the absolute levels of nutrients, primarily P, may be even more important for cyanobacterial dominance than the nutrient ratios. Cyanobacteria are rather poor competitors for phosphorus and do not prevail at low P concentrations. In Peipsi, where the cyanobacterial biomass and the dominance of potentially toxic and bloom-forming N_2-fixing cyanobacteria increase with increasing TP (Fig. 2) and decreasing TN:TP ratio, reduction of P loading would appear to be the most effective measure for reducing cyanobacterial blooms in the lake. Nutrient supply should be kept strictly under control at the drainage basin level and also reduced in Võrtsjärv, because without high nutrient levels no eutrophication problems will exist. However, in Võrtsjärv, where the development of cyanobacteria is much less correlated with the P concentration and the N:P ratio, other factors such as the water level obviously also have a strong influence on the composition of the phytoplankton community. In earlier studies of Võrtsjärv, light limitation was shown to be an important factor controlling the phytoplankton community (Nõges et al., 1998, 2003).

The temperature dependence of cyanobacterial development and N_2 fixation should be taken into account in the context of global warming. Since a higher water temperature supports both cyanobacterial growth (Huisman et al., 2005) and P recycling from sediments (Genkai-Kato & Carpenter, 2005), and since the temperature dependence of Nfix% proves to be a maximum function (Fig. 6) determining the highest possible Nfix% at certain water temperatures, there are management implications: to avoid blue-green blooms in a warmer world, the phosphorus supply has to be reduced even further than predicted on the basis of existing data sets.

Conclusions

Our study revealed that the critical TN:TP mass ratios for the cyanobacterial communities in Lakes Peipsi and Võrtsjärv are around 30 and 40, respectively, and the critical Nmin:SRP mass ratios in both lakes are around 20. The critical N:P ratios deduced from our study are much higher than the routinely considered Redfield N:P mass ratio (7) and represent valuable guidelines for creating effective management strategies for large shallow lakes. Our results provide a basis for stressing the urgent need to decrease phosphorus loading and to keep the in-lake P concentration low, and *not* to implement nitrogen reduction measures without simultaneously decreasing the phosphorus concentration.

Acknowledgements Funding for this research was provided by target funding project 03962480s03 by the Estonian Ministry of Education and by Estonian Science Foundation grant 5738. We used data obtained from frames of the State Monitoring Program of the Estonian Ministry of Environment. The authors are grateful to the guest editor Heikki Simola and to two anonymous reviewers for valuable advice and comments for improving the manuscript.

References

Berger, C., 1989. In situ primary production, biomass and light regime in the Wolderwijd, the most stable Oscillatoria agardhii lake in The Netherlands. Hydrobiologia 185: 233–244.

Bulgakov, N. G. & A. P. Levich, 1999. The nitrogen:phosphorus ratio as a factor regulating phytoplankton community structure. Archiv für Hydrobiologie 146: 3–22.

Downing, J. A., S. B. Watson & E. McCauley, 2001. Predicting Cyanobacteria dominance in lakes. Canadian Journal of Fisheries and Aquatic Sciences 58: 1905–1908.

Falkowski, P. G. & C. S. Davis, 2004. Natural proportions. Redfield ratios, the uniformity of elemental ratios in the oceans and the life they contain underpins our understanding of marine biogeochemistry. Nature 431: 131.

Genkai-Kato, M. & S. R. Carpenter, 2005. Eutrophication due to phosphorus recycling in relation to lake morphometry, temperature, and macrophytes. Ecology 86: 210–219.

Grasshoff, K., M. Ehrhardt & K. Kremling (eds), 1983. Methods of Seawater Analysis. Verlag Chemie, Weinheim, 419 pp.

Hall, S. R., V. H. Smith, D. A. Lytle, & M. A. Leibold, 2005. Constraints on primary producer N:P stoichiometry along N:P supply ratio gradients. Ecology 86: 1894–1904.

Harris, G. P., 1986. Phytoplankton Ecology. Chapman and Hall, USA, 137–165.

Havens, K. E., R. T. James, T. L. East & V. H. Smith, 2003. N:P ratios, light limitation, and cyanobacterial dominance in a subtropical lake impacted by non-point source nutrient pollution. Environmental Pollution 122: 379–390.

Howarth, R. W., J. J. Cole & R. Marino, 1988. Nitrogen fixation in freshwater, estuarine and marine ecosystems. 2. Biogeochemical controls. Limnology and Oceanography 33: 688–701.

Huisman, J. & F. Hulot, 2005. Population dynamics of harmful cyanobacteria. Factors affecting species composition. In Huisman, J., H. C. P. Matthijs & P. M. Visser (eds), Harmful Cyanobacteria. Springer-Verlag, Dordrecht, The Netherlands, 143–176.

Huisman, J., H. C. P. Matthijs & P. M. Visser (eds), 2005. Harmful Cyanobacteria. Springer-Verlag, Dordrecht, The Netherlands, 241 pp.

Lathrop, R. C., S. R. Carpenter, C. A. Stow, P. A. Soranno & J. C. Panuska, 1998. Phosphorus loading reductions needed to control blue-green algal blooms in Lake Mendota. Canadian Journal of Fisheries and Aquatic Sciences 55: 1169–1178.

Laugaste, R., T. Nõges, P. Nõges, V. V. Jastremskij, A. Milius & I. Ott, 2001. Algae. In Pihu, E. & J. Haberman (eds), Lake Peipsi. Flora and Fauna. Sulemees Publishers, Tallinn, 31–49.

Michard, M., L. Aleya & J. Verneaux, 1996. Mass occurrence of the Cyanobacteria *Microcystis aeruginosa* in the hypereutrophic Villerest Reservoir (Roanne, France): Usefulness of the biyearly examination of N/P (nitrogen phosphorus) and P/C (protein/carbohydrate) couplings. Archiv für Hydrobiologie 135: 337–359.

von zur Mühlen, M. & G. Schneider, 1920. Der See Wirzjerw in Livland. Archiv für die Naturkunde des Ostbaltikums 14: 1–156.

Nõges, T., V. Kisand, P. Nõges, A. Põllumäe, L. Tuvikene & P. Zingel, 1998. Plankton seasonal dynamics and its controlling factors in shallow polymictic eutrophic Lake Võrtsjärv, Estonia. International Review of Hydrobiology 83: 279–296.

Nõges, T., P. Nõges & R. Laugaste, 2003. Water level as the mediator between climate change and phytoplankton composition in a large shallow temperate lake. Hydrobiologia 506: 257–263.

Nõges, T., I. Tõnno, R. Laugaste, E. Loigu & B. Skakalski, 2004. The impact of changes in nutrient loading on cyanobacterial dominance in Lake Peipsi (Estonia/Russia). Archiv für Hydrobiologie 160: 261–279.

Nõges, T., A. Järvet, A. Kisand, E. Loigu, B. Skakalski & P. Nõges, 2007. Reaction of large and shallow lakes Peipsi and Võrtsjärv to the changes of nutrient loading. Hydrobiologia 584: 253–264.

Paerl, H. W., R. S. Fulton, P.H. Moisander & J. Dyble, 2001. Harmful freshwater algal blooms, with an emphasis on cyanobacteria. The Scientific World Journal 1: 76–113.

Redfield, A. S., 1958. The biological control of chemical factors in the environment. American Scientist 46: 205–211.

Reynolds, C. S., 1984. The Ecology of Freshwater Phytoplankton. Cambridge University Press, UK.

Scheffer, M., 1998. Ecology of Shallow Lakes. Chapman and Hall, London, UK.

Scheffer, M., S. Rinaldi, A. Gragnani, L. R. Mur & E. H. van Nes, 1997. On the dominance of filamentous cyanobacteria in shallow, turbid lakes. Ecology 78: 272–282.

Schindler, D. W., 1977. Evolution of phosphorus limitation in lakes. Science 195: 260–262.

Seip, K. L., 1994. Phosphorus and nitrogen limitation of algal biomass across trophic gradients. Aquatic Sciences 56: 16–28.

Smith, V. H., 1983. Low nitrogen to phosphorus ratios favour dominance by blue-green algae in lake phytoplankton. Science 221: 669–771.

Smith, V. H., 1985. Predictive models for the biomass of blue-green algae in lakes. Water Resources Bulletin 21: 433–439.

Smith, V. H., E. Willén & B. Karlsson, 1987. Predicting the summer peak biomass of four species of blue-green algae (Cyanophyta/cyanobacteria) in Swedish lakes. Water Resources Bulletin 23: 397–402.

Smith, V. H., V. J. Bierman, B. L. Jones & K. E. Havens, 1995. Historical trends in the Lake Okeechobee ecosystem IV. Nitrogen:phosphorus ratios, cyanobacterial dominance, and nitrogen fixation potential. Archiv für Hydrobiologie, Monographische Beiträge 107: 71–88.

Sterner, R. W. & J. J. Elser, 2002. Ecological Stoichiometry: The Biology of Elements from Molecules to the Biosphere. Princeton University Press, Princeton, New Jersey, USA.

Tõnno, I. & T. Nõges, 2003. Nitrogen fixation in a large shallow lake: rates and initiation conditions. Hydrobiologia 490: 23–30.

Utermöhl, H., 1958. Zur Vervollkomnung der quantitativen Phytoplanktonmethodik. Mitteilungen der Internationalen Vereinigung für Limnologie 9: 1–38.

van Duin, E. H. S., R. H. Aalderink & L. Lijklema, 1995. Light adaptation of *Oscillatoria agardhii* at different time scales. Water Science and Technology 32: 35–48.

Willén, E., 1992. Long-term changes in the phytoplankton of large lakes in response to changes in nutrient loading. Nordic Journal of Botany 12: 575–587.

Hydrobiologia (2008) 599:87–95
DOI 10.1007/s10750-007-9191-1

Phytoplankton nitrogen demand and the significance of internal and external nitrogen sources in a large shallow lake (Lake Balaton, Hungary)

Mátyás Présing · Tom Preston · Anikó Takátsy · Péter Sprőber · Attila W. Kovács · Lajos Vörös · Gyöngyi Kenesi · István Kóbor

Abstract Since the middle of 1990s the trend of Lake Balaton towards an increasingly trophic status has been reversed, but N_2-fixing cyanobacteria are occasionally dominant, endangering water quality in summer. The sources of nitrogen and its uptake by growing phytoplankton were therefore studied. Experiments were carried out on samples collected from the middle of the Eastern (Siófok) and Western (Keszthely) basins between February and October 2001. Ammonium, urea and nitrate uptake and ammonium regeneration were measured in the upper 5-cm layer of sediment using the ^{15}N-technique. Ammonium was determined by an improved micro-diffusion assay. N_2 fixation rates were measured by the acetylene-reduction method. Ammonium regeneration rates in the sediment were similar in the two basins. They were relatively low in winter (0.13 and 0.16 μg N cm^{-3} day^{-1} in the Eastern and Western basin, respectively), increased slowly in the spring (0.38 and 0.45 μg N cm^{-3} day^{-1}) and peaked in late summer (0.82 and 1.29 μg N cm^{-3} day^{-1}, respectively). Ammonium uptake was predominant in spring in the Eastern basin and in summer in the Western basin, coincident with the cyanobacterial bloom. The amount of N_2 fixed was less than one third of the internal load during summer when external N loading was insignificant. Potentially, the phytoplankton N demand could be supported entirely by the internal N load via ammonium regeneration in the water column and sediment. However, the quantity of N from ammonium regeneration in the upper layer of sediment combined with that from the water column would limit the standing phytoplankton crop in spring in both basins and in late summer in the Western basin, especially when the algal biomass increases suddenly.

Keywords Ammonium regeneration · Microdiffusion · Nitrogen uptake · ^{15}N-technique

Guest editors: T. Nõges, R. Eckmann, K. Kangur, P. Nõges, A. Reinart, G. Roll, H. Simola & M. Viljanen
European Large Lakes—Ecosystem changes and their ecological and socioeconomic impacts

M. Présing (✉) · P. Sprőber · A. W. Kovács · L. Vörös · Gy. Kenesi
Balaton Limnological Research Institute of the Hungarian Academy of Sciences, Klebelsberg K. u. 3, 8237 Tihany, Hungary
e-mail: matyas@tres.blki.hu

T. Preston
Isotope Biochemistry Laboratory, Scottish Universities Environmental Research Centre, East Kilbride, Glasgow G75 0QF, UK

A. Takátsy
Janus Pannonius University, Szigeti u. 13, 7643 Pécs, Hungary

I. Kóbor
Central-Transdanubian Environmental and Water Authority, Balatoni u. 1, 8000 Székesfehérvár, Hungary

Introduction

Lake Balaton is the largest shallow lake in Central Europe with a surface area of 594 km^2 and a mean depth of 3.5 m. The lake is of international significance as an environmental resource and of great importance to the Hungarian economy as a recreational resort, visited by four million tourists annually; apart from Budapest it is the most important source of tourist income in Hungary. Because of increased nutrient loading during the 1980s the Western part became hypertrophic, the Central and Eastern basins eutrophic (Herodek, 1986). Several measures have been taken to reduce the nutrient load, particularly phosphorus. Reservoirs have been built on the main tributaries feeding Lake Balaton. Phosphorus precipitation has been applied at the sewage treatment plants, and a sewage transfer pipe system diverts wastewater from the catchment area of the lake. Pollution from liquid manure has been stopped and fertiliser use has been reduced. The biologically available phosphorus loads in the Eastern and Western basins have decreased by 60 and 30%, respectively (Somlyódy et al., 1997). Since the most recent bloom in 1994 the phytoplankton biomass has been low, the Eastern basin of the lake has become mesotrophic, while the Western basin remains eutrophic. However, since the algal biomass has decreased, the presence and often the dominance of nitrogen-fixing and potentially toxic cyanobacteria in summer give rise to concern. The most probable reason for their success is that they benefit from extremely warm summers, they are able to take up P very rapidly and they may be able to store it. They also benefit from the low N to P ratio and ability to fix molecular nitrogen as an additional N source if necessary. Due to nitrogen deficiency may contribute to the development of these species (Blomqvist et al., 1994), it seemed worthwhile to study the importance of the different nitrogen sources used by the phytoplankton. Following earlier studies of ammonium regeneration in the water column and sediment (Présing et al., 2001b), the present study was designed to estimate ammonium regeneration in sediment over several seasons, using a newly established measurement method and comparing it with the external nitrogen load in the lake.

Materials and methods

Experiments were carried out between February and October 2001 (Fig. 1) on samples collected from the middle of the Western and Eastern basins of the lake. Details of the methods used for sampling, chlorophyll-*a* and nutrient analysis of integrated water samples were published previously (Présing et al., 2001b). Ambient nutrient and chlorophyll-*a* concentrations were determined in three replicates with coefficients of variation less than 10 and 5%, respectively. The mean values of ammonium, urea and nitrate concentrations were compared with paired sample *t*-test. Ammonium and nitrate concentrations in interstitial water and extractable ammonium in the sediment layers were determined as follows: for each experiment, 10 sediment cores were sliced at 5 cm intervals and the corresponding layers were mixed. From each homogenised sediment layer, 40 g was immediately centrifuged for 10 min at 4,000 rpm in a Hermle Z 320 centrifuge. Extractable ammonium was analysed in another 40 g sample of the same layers. The sediment was mixed with 10 ml of 2 M KCl, incubated for 1 h and centrifuged. The centrifugation supernatants were used for chemical analysis as described above. Ammonium standards were prepared in an equivalent concentration of KCl.

Phytoplankton samples taken from the integrated water samples from the basins were preserved in Lugol's solution. Algal species were enumerated with an inverted plankton microscope (Utermöhl, 1958). The wet weight of each species was calculated from cell volumes (Németh & Vörös, 1986). At least 25 cells (or filaments) of each species were measured to determine biomass and at least 400 were counted. N_2 fixation in both basins was measured by a modified acetylene-reduction technique (Présing et al., 2001a) in the middle of June, July and August. Carbon uptake was estimated using a model developed for the lake (Vörös & V.-Balogh, 1998) based on water temperature and chlorophyll-*a* concentration. The methods for measuring phytoplankton ammonium, urea and nitrate uptake and the calculations of nitrogen uptake rates were detailed previously (Présing et al., 2001a). The nitrogen contents and $^{15}N/^{14}N$ ratios of dried seston samples (60°C) were determined by an automated elemental analyser interfaced with an Isotope Ratio Mass Spectrometer (ANCA-MS, Europa Scientific Ltd., UK). Ammonium, urea

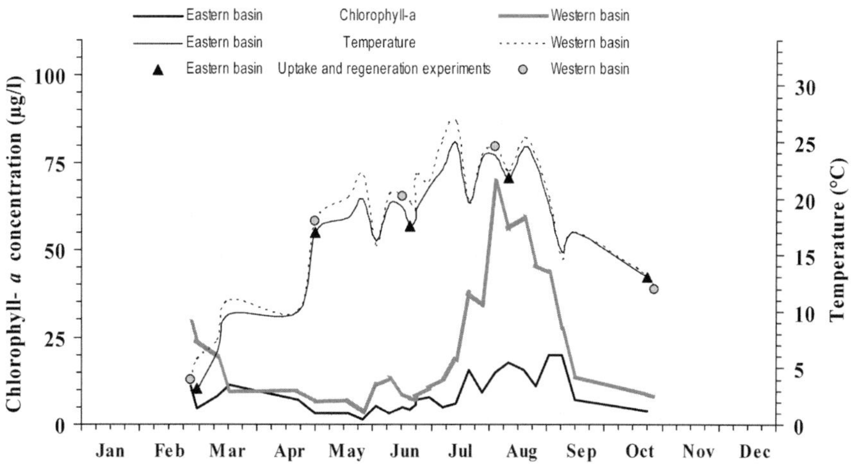

Fig. 1 Water temperature and chlorophyll-*a* concentration in Eastern and Western basins of Lake Balaton in 2001

and nitrate uptake per unit surface area was calculated from the uptake velocities measured at optimal light intensity (110 µmol photon m^{-2} s^{-1}) at the average depth of the basins. In order to obtain daily uptake, the conversion factors described previously (Présing et al., 2001a) were used, taking account of global irradiation, vertical light attenuation and day–night rhythm.

Ammonium regeneration in the sediment was measured by ^{15}N dilution. The upper 5 cm layers from 10 sediment cores were mixed and supplemented with $^{15}NH_4Cl$. Filtered lake water (20 ml) was pipetted on top of the mixed, labelled 40 g samples in centrifuge tubes. The samples were incubated in the dark for 1–12 days at the lake temperature. After incubation, the samples were mixed with an equal volume of 2 M KCl for 1 h and then centrifuged. The supernatant was used for microdiffusion and chemical analysis. Ammonium was measured in the supernatant by an improved microdiffusion method after Brooks et al. (1989) and Slawyk & Raimbault (1995) using our modified ammonium trap. Supernatant (10–30 ml) was pipetted into incubation bottles (Duran Schott, 100 ml volume) and 10 mg of MgO and 5 glass beads (4 mm) were added. An aluminium capsule was fixed under the acidified (25 µl of 0.5 N H_2SO_4) filter trap, separated from the top of the incubation bottle with a specially bent stainless steel wire, to gather the occasional drops and avoid loss of ammonium. The bottles were gently shaken with a vertical shaker for 10 days at 24°C, then the filters were dried for 1–2 days in a helium atmosphere in desiccators containing H_3PO_4. The dried filters were placed in the same capsule. The regeneration rate was calculated on the basis of the ^{15}N dilution of ammonium at T_0, i.e. the start of the experiment, when the KCl was added. ^{15}N enrichment was analysed using the same isotope ratio mass spectrometer used to measure nitrogen uptake.

Results

Temperature and phytoplankton succession in Lake Balaton

After a relatively mild winter the water temperature of the lake rose to 10°C in March, and a cold April 2001 was followed by a second period of increasing temperature in May (Fig. 1). The water temperature reached 20°C by the middle of June and remained above this level until the beginning of September. In July and August the water temperature reached or exceeded 24–25°C often and for long periods. From September to December the lake cooled rapidly. In the Eastern basin, chlorophyll-*a* concentrations were low, ranging between 2 and 20 µg l^{-1}. There were two small maxima, one in early spring and the other in late summer–autumn (August–September). In the Western basin, chlorophyll-*a* concentrations were higher and peaked at around 30 µg l^{-1} at the end of February and at 75 µg l^{-1} in August–September. During the period under investigation the phytoplankton biomass varied between 0.5 and 6.1 mg l^{-1} in the Eastern basin, and Heterocontophyta dominated in spring (Fig. 2a). Of the two classes included in this division, Bacillariophyceae formed the overwhelming majority with a predominance of centric diatoms (*Cyclotella* spp.), while Chrysophyceae were negligible. In early summer, the phytoplankton community became more diverse with dinoflagellates increasingly noteworthy: (*Ceratium hirundinella* (O.F.M.) Schrank) and the N_2-fixing cyanobacterium (*Aphanizomenon flos-aquae* (L.) Ralfs). In August and September, N_2-fixing cyanobacteria (*A. flos-aquae* and *Cylindrospermopsis raciborskii* (Wołosz.) Seenayya and Subba Raju) dominated. In October, filamentous cyanobacteria disappeared; centric diatoms, cryptophytes and picocyanobacteria became the dominant species. Total phytoplankton biomass was much higher in the Western basin of the lake, varying between 1.4 and 21 mg l^{-1} (Fig. 2b). In April, centric diatoms prevailed, but towards summer they were slowly replaced by cryptophytes (*Rhodomonas minuta* Skuja), dinoflagellates (*C. hirundinella*) and N_2-fixing blue-greens (*Anabaena aphanizomenoides* Forti, *A. flos-aquae*). Phytoplankton biomass was highest (14–21 mg l^{-1}) in August and was dominated by N_2-fixing cyanobacteria, *A. flos-aquae*, *Aphanizomenon issatschenkoi* (Ussach.) Prosk.-Lavr. and *C. raciborskii.* From September, the total biomass decreased markedly and cryptophytes along with chlorophytes were the most abundant organisms.

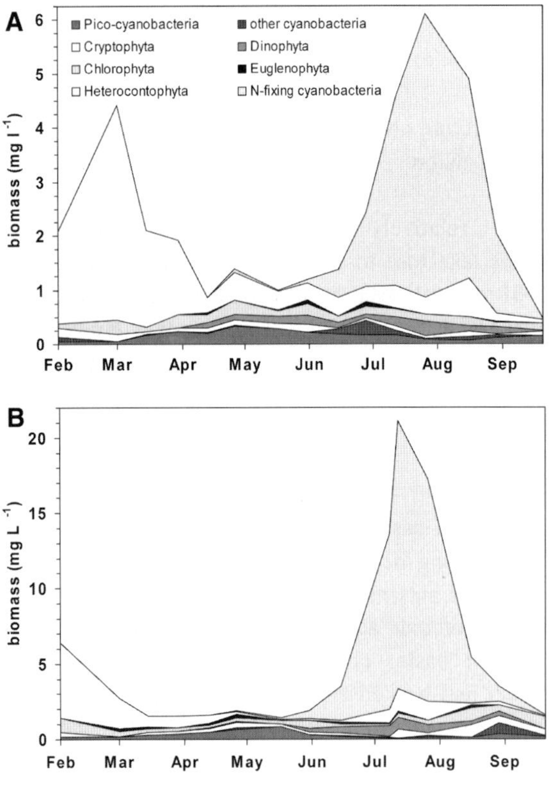

Fig. 2 Seasonal changes in phytoplankton composition and biomass in Eastern (**A**) and Western (**B**) basins of Lake Balaton in 2001

Pelagic inorganic nitrogen and urea concentration, uptake rates and nitrogen fixation

Annual mean concentrations of ammonium (8.01 ± 3.98 μg N l^{-1}) and urea (21.98 ± 10.95 μg N l^{-1}) in pelagic water of both basins were low the entire year and did not follow a seasonal pattern. Nitrate concentrations changed during the year in both basins with higher values from late autumn to spring (55.6 ± 21.7 μg N l^{-1}) and significantly ($P < 0.001$) lower ones during summer (5.7 ± 2.53 μg N l^{-1}). Nitrate and ammonium concentrations were similarly low and did not differ significantly ($P > 0.05$) in summer. However annual mean concentrations of urea were also relatively low, significantly ($P < 0.001$) exceeded those of ammonium and summer values of nitrate. Nitrate concentrations from autumn to spring were far above ($P < 0.001$) the highest ambient nitrogen concentrations of other N-forms during the whole year.

The results of the nitrogen uptake experiments are shown in Table 1. In the Eastern basin the uptake rates were mostly below 1 μg N l^{-1} h^{-1} with peaks of 3.24 μg N l^{-1} h^{-1} for ammonium in August. Because of the higher algal biomass, uptake rates were higher in the Western than the Eastern basin, especially for ammonium at the time of the cyanobacterial bloom in August (6.18 μg N l^{-1} h^{-1}). For the most part, ammonium assimilation was the most pronounced, followed by nitrate and/or urea uptake. However, the nitrate uptake rate exceeded that of ammonium in the Western basin in February and in the Eastern basin in June. In the Eastern basin with only one exception, and in the Western basin always, the V_{max} was highest for ammonium. The greatest values were determined in May and August in both basins. On the basis of the calculated daily uptake per unit surface area, ammonium was taken up in highest amount in both basins between March and October. Its contribution to the daily nitrogen supply of the

Table 1 Characteristics of ammonium, urea and nitrate uptake in Eastern and Western basin of Lake Balaton in 2001

Date	N-forms	K_s	v (μg N l^{-1} h^{-1})	V_{max} (μg N l^{-1} h^{-1})	V_{max}/Chl-*a*	α V_{max}/K_s	mg N m^{-2} day^{-1}
Eastern basin							
2 Mar	NH_4	10.00	0.39	0.65	0.14	0.06	22.96
	NO_3		0.13				5.92
	Urea		0.19				9.98
2 May	NH_4	7.80	0.71	1.75	0.57	0.22	42.13
	NO_3	2.98	0.21	0.26	0.09	0.09	9.38
	Urea	2.87	0.22	0.25	0.08	0.09	11.74
22 Jun	NH_4	15.6	0.18	0.56	0.13	0.04	10.87
	NO_3	4.86	0.53	0.80	0.18	0.16	24.03
	Urea	7.81	0.77	1.16	0.27	0.15	40.30
	N_2-fixation						1.33
16 Aug	NH_4	1.09	3.24	3.99	0.47	3.66	192.87
	NO_3		0.30	0.35	0.04		13.65
	Urea						144.65
	N_2-fixation						18.64
25 Oct	NH_4	15.97	0.21	0.49	0.12	0.03	12.78
	NO_3		0.03		0.00		1.14
	Urea	24.8	0.01	0.02	0.01	0.00	0.53
Western basin							
27 Feb	NH_4	2.87	1.17	1.50	0.05	0.52	49.53
	NO_3		1.45				47.13
	Urea		0.75				28.13
3 May	NH_4	3.06	2.78	4.22	0.62	1.38	118.17
	NO_3		0.50				16.25
	Urea	1.74	2.36	2.75	0.40	1.58	88.58
18 Jun	NH_4	12.50	0.86	2.38	0.54	0.19	36.58
	NO_3	8.48	0.13	0.13	0.03	0.02	4.23
	Urea						27.43
	N_2-fixation						10.23
8 Aug	NH_4	16.20	6.18	15.20	0.41	0.94	262.66
	NO_3	2.10	0.11	0.15	0.004	0.07	3.47
	Urea						196.99
	N_2-fixation						50.68
29 Oct	NH_4	3.17	0.97	1.58	0.19	0.50	41.10
	NO_3	9.71	0.16	0.32	0.04	0.03	5.28
	Urea	3.87	0.25	0.45	0.05	0.12	9.51

v = uptake velocity at ambient concentration, K_s = half saturation constant, V_{max} = uptake velocity at the saturation concentration

phytoplankton usually exceeded 50% (up to 66% in May). The contribution of urea to the daily nitrogen uptake was between 18 and 53% and was the largest portion of daily nitrogen demand in June. While ammonium was the preferred nitrogen source of phytoplankton, urea sometimes contributed more than ammonium to daily total nitrogen uptake during 2001 because the ambient urea concentrations were higher. Nitrate uptake was not dominant in the nitrogen supply to the phytoplankton. In the Eastern basin, its highest value (24 mg N m^{-2} day^{-1} in June) represented 31% of total daily uptake. At other times

Table 2 External load of Lake Balaton from the River Zala in 2001

2001	NH_4 (ton)	NO_2 (ton)	NO_3 (ton)	Total N (ton)	Water inflow ($m^3 \times 10^6$)
Jan	0.33	0.07	3.34	16.04	17.53
Feb	0.38	0.08	4.62	20.75	17.78
Mar	0.61	0.10	1.84	19.58	16.85
I.	1.32	0.25	9.80	56.37	52.16
Apr	0.41	0.08	1.46	20.26	18.58
May	1.12	0.06	0.75	16.18	10.41
June	0.35	0.03	0.44	5.77	5.22
II.	1.88	0.17	2.65	42.21	34.21
July	0.24	0.01	0.15	2.58	3.32
Aug	0.07	0.01	0.14	2.79	1.59
Sept	0.19	0.01	0.16	2.26	1.70
III.	0.50	0.03	0.45	7.63	6.61
Oct	0.38	0.03	0.33	5.99	3.63
Nov	1.33	0.10	0.88	11.55	6.97
Dec	2.26	0.12	1.40	15.07	7.58
IV.	3.97	0.25	2.61	32.61	18.18
Total	7.67	0.70	15.51	138.82	111.16

it contributed only 4–15%. N_2 fixation was negligible in June and July (only 2–5%) and reached only 5% even when 70% of the phytoplankton biomass comprised N_2-fixing cyanobacteria. The amount of fixed nitrogen was only a few percent of the total nitrogen demand of the phytoplankton. In the Western basin the ratios of daily ammonium, urea and nitrate uptake per unit surface area to the total nitrogen utilisation by phytoplankton were similar to those in the Eastern basin. Nitrogen fixation played a more important role in the Western than the Eastern basin. Considering the measured N fixation rate (50.68 mg N m^{-2} day^{-1}) and the surface area of the Western basin (38 km^2), the amount of nitrogen fixed during August may be about 60 ton.

External nitrogen load and ammonium generation in the sediment

It was assumed that the external load of the Western basin was determined by the single significant inflow, the River Zala, which transports up to 90% of the total nutrient load of the basin. The load was calculated from daily water discharge and nutrient concentrations and was summarised monthly (Table 2). It showed a clear annual trend, with higher quantities in autumn and winter and lowest in the summer months. The total DIN charged by Zala to the lake was only 0.22 ton in August.

At the end of February, when the lake temperature was 5°C, the daily ammonium productions per cm^3 wet sediment were 0.13 and 0.16 μg in the Eastern and Western basins, respectively. This means that ammonium regeneration in the upper 5-cm layer of 1 m^2 sediment can produce 6.6 and 8 mg nitrogen daily in the Eastern and Western basins, respectively (Fig. 3). By the time of the experiments in spring, the water temperature had increased to 10°C in the Eastern basin and to 15°C in the shallower Western basin. The daily ammonium production increased to 0.4 μg N cm^{-3} wet sediment, which meant around 20 mg m^{-2} of ammonium nitrogen production in both basins. At the end of June the water temperature in both basins was 20°C. Ammonium regeneration per cm^3 wet weight was 1.15 and 0.82 μg, and 57.3 and 41.2 mg m^{-2}, in the Eastern and Western basins, respectively. At the time of the experiments in August the water in the Eastern basin was at 23°C while that in the shallower Western basin was at

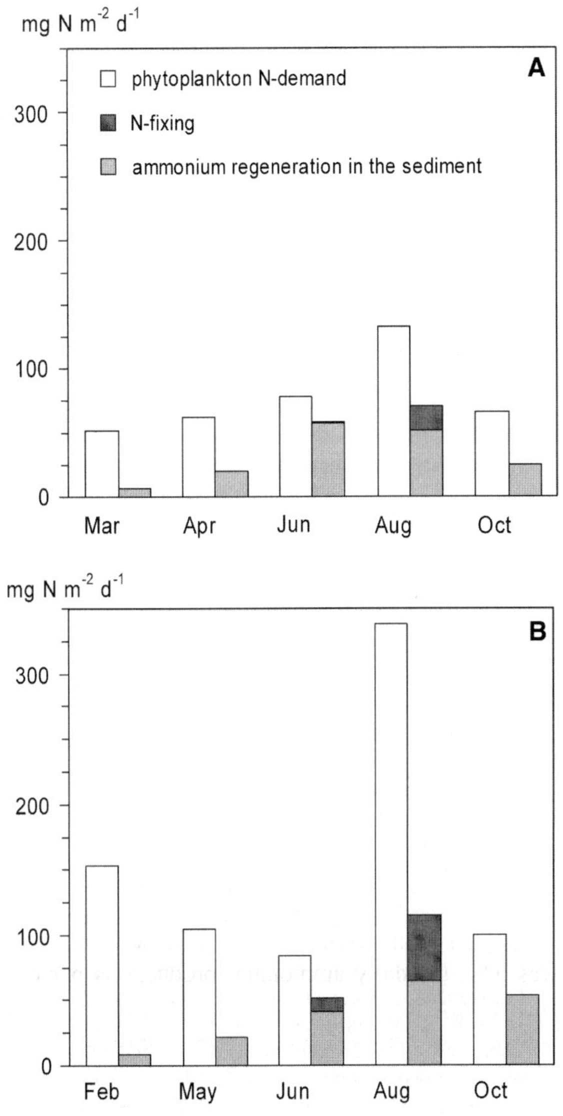

Fig. 3 Phytoplankton nitrogen demand (primary production divided by 5.6), the amount of regenerated ammonium in the sediment and fixed N_2 in the Eastern (**A**) and Western (**B**) basins of Lake Balaton in 2001

26°C. Ammonium regeneration values in the Eastern basin were close to those measured in June. In the Western basin the ammonium production had almost doubled and reached its yearly maximum of 1.3 μg cm^{-3} and 64.2 mg m^{-2}. Scaling this over the 38 km^2 surface area of the Western basin, the amount of nitrogen originating from the sediment via ammonium regeneration may reach 75 ton during August. By the end of October, the water temperature had dropped to around 10°C. In the Eastern part of the lake the ammonium regeneration fell to half of that measured in the summer. In the Western basin the decrease was slower and the values remained around 1 μg cm^{-3} and 53.6 mg m^{-2}.

Discussion

Kinetic parameters (K_s and V_{max}) determined in our experiments are conventionally used to evaluate the possible nutrient limitation or substrate affinity of phytoplankton. Maximal uptake velocity (V_{max}) rather than the half saturation constant (K_s) was used to characterise nitrogen uptake; determination of K_s is less accurate and the value generally increases with increasing V_{max} (Takamura et al., 1987), which is a more independent indicator and usually increases with increasing nitrogen limitation (Lomas et al., 1996). The initial slope of the uptake rate versus nutrient concentration curve (α) is also thought to be a better indicator of nutrient competition than K_s alone (Cochlan & Harrison, 1991). Based on these kinetic parameters the order of preference of phytoplankton for nitrogen forms in both basins of Lake Balaton was ammonium > urea ≫ nitrate uptake. The highest affinity of algae for nitrogen especially for ammonium occurred in spring and coinciding with the cyanobacterial blooms in summer in both basins. Generally, the contribution of urea to the total N utilised by the phytoplankton is intermediate between those of ammonium and nitrate, somewhat closer to the former (Takamura et al., 1987; Mitamura et al., 1995). The contribution of urea can be as much as 70–80% of the daily nitrogen demand (Glibert et al., 1991). Measurements during 1998 showed that the contribution of urea to phytoplankton nitrogen supply in Lake Balaton reached 50–70% of ammonium uptake (Présing et al., 2001a). The contributions of reduced nitrogen forms (ammonium and urea) were far more important than that of nitrate, which was some 10% of the daily nitrogen supply to the algae. However, the contribution of nitrogen fixation was also around 10% but this can be a critically important additional nitrogen source for cyanobacteria, conferring a competitive advantage.

The measured ammonium regeneration rates are close to the very few values in the literature (Bowden, 1984). One reason why there are so few ammonium regeneration results based on direct measurements is the lack of appropriate method. Our modified microdiffusion method has proved suitable for both extraction and concentration of ammonium. It does not need specialised laboratory equipment and gives 80–100% recoveries.

We assumed that the nitrogen assimilated by phytoplankton can be calculated from its primary production and Redfield's classical C to N (weight:-weight) ratio of 5.6. By this calculation we estimated the total phytoplankton N demand (open columns in Fig. 3) independently of our different N uptake measurements. In the water column, approximately 50% of primary phytoplankton production can be utilised by bacteria (Vörös et al., 1996). On average, a similar proportion of the phytoplankton nitrogen demand was provided by ammonium regeneration in the water column (Présing et al., 2001b). The balance of the nitrogen demand may be supplied by ammonium regeneration in the sediment together with external sources, such as inflowing water and fixation of atmospheric nitrogen. From autumn to spring, nitrogen demand can be met by the relatively high inflowing nitrogen load (Table 2). In summer, especially in the Western part of the lake, phytoplankton nitrogen demand is met only by adding the input from nitrogen fixation to the ammonium supplied by elevated regeneration. It should be noted that these calculations do not consider N loss via denitrification. A portion of the regenerated ammonium will be oxidised to nitrate. This, together with exogenous nitrate, will be subject to loss via denitrification, which will result in even less available combined nitrogen to support the standing crop.

Potentially, more of the phytoplankton nitrogen demand could be met by inflowing external load and by internal sources via ammonium regeneration, which occurs in the water column and the upper layer of sediment. These sources could barely supply enough combined nitrogen to maintain the standing phytoplankton crop in late summer, especially in the Western basin, when the algal biomass increases rapidly, and at the same time compensate for nitrogen loss by denitrifcation. External nitrogen load is insignificant in summer, when the atmospheric N_2 can supplement available nitrogen, but only for cyanobacteria capable of nitrogen fixation. The impact of nuisance algae may possibly be minimised by managing the nutrient loading: decreasing the external P while not reducing the external N load.

Acknowledgements This work was supported by the Office of the Prime Minister of Hungary (MeH). Support from NATO in the form of travelling expenses to Tom Preston is gratefully acknowledged. We are very grateful to Gerd Slawyk for consultation about the microdiffusion method. We wish to thank the Western-Transdanubian Environmental and Water Authority for nutrient loading data. We are extremely grateful for Terézia Horváth, Erika Kozma and Géza Dobos for their assistance with the sampling and laboratory experiments.

References

Blomqvist, P., A. Peterson & P. Hyenstrand, 1994. Ammonium nitrogen: A key regulatory factor causing dominance of non-nitrogen-fixing cyanobacteria in aquatic systems. Archiv für Hydrobiologie 132: 141–164.

Bowden, W. B., 1984. A nitrogen-15 isotope dilution study of ammonium production and consumption in a marsh sediment. Limnology and Oceanography 29: 1004–1015.

Brooks, P. D., J. M. Stark, B. B. McInteer & T. Preston, 1989. Diffusion method to prepare soil extracts for automated nitrogen-15 analysis. Soil Science Society of America Journal 53: 1707–1711.

Cochlan, W. P. & P. J. Harrison, 1991. Kinetics of nitrogen (nitrate, ammonium and urea) uptake by the picoflagellate *Micromonas pusilla* (Prasinophyceae). Journal of Experimental Marine Biology and Ecology 153: 129–141.

Glibert, P. M., C. Garside, J. A. Fuhrman & M. R. Roman, 1991. Time-dependent coupling of inorganic and organic nitrogen uptake and regeneration in the plume of the Chesapeake Bay estuary and its regulation by large heterotrophs. Limnology and Oceanography 36: 895–909.

Herodek, S., 1986. Phytoplankton changes during eutrophication and P and N metabolism. In Somlyódy, L. & G. van Straten (eds), Modeling and Managing Shallow Lake Eutrophication. Springer Verlag, Berlin, 183–204.

Lomas, M. W., P. M. Glibert, G. M. Berg & M. Burford, 1996. Characterization of nitrogen uptake by natural populations of *Aureococcus anophagefferens* (Chrysophyceae) as a function of incubation duration, substrate concentration, light and temperature. Journal of Phycology: 32: 907–916.

Mitamura, O., Y. Saijo, K. Hino & F. A. R. Barbosa, 1995. The significance of regenerated nitrogen for phytoplankton productivity in the Rio Doce Valley Lakes, Brazil. Archiv für Hydrobiologie 134: 179–194.

Németh, J. & L. Vörös, 1986. Koncepció és módszertan felszíni vizek algológiai monitoringjához. OKTH, Budapest, 135 pp. (in Hungarian).

Présing, M., S. Herodek, T. Preston & L. Vörös, 2001a. Nitrogen uptake and the importance of internal nitrogen loading in Lake Balaton. Freshwater Biology 46: 125–139.

Présing, M., T. Preston, A. Kovács & P. Sprőber, 2001b. Internal and external nitrogen supply of phytoplankton in Lake Balaton (Hungary). Biwako 2001, 9th International Conference on the Conservation and Management of Lakes, Kyoto, Japan. Partnerships for Sustainable Life in Lake Environments, Conference Proceedings, Session 5: 272–275.

Slawyk, G. & P. Raimbault, 1995. Simple procedure for simultaneous recovery of dissolved inorganic and organic nitrogen in ^{15}N-tracer experiments and improving the isotopic mass balance. Marine Ecology Progress Series 124: 289–299.

Somlyódy, L., S. Herodek, Cs. Aradi, A. Clement, Gy. Dévai, V. Istvánovics & Gy. Varga, 1997. Revision of the Lower Kis-Balaton Reservoir. Synthesis Report, Technical University, Budapest.

Takamura, N., T. Iwakuma & M. Yasuno, 1987. Uptake of ^{13}C and ^{15}N (ammonium, nitrate and urea) by *Microcystis* in Lake Kasumigaura. Journal of Plankton Research 9: 151–165.

Utermöhl, H., 1958. Zur Vervollkommnung der quantitativen Phytoplankton-Methodik. Mitteilungen Internationale Vereinigung für Theoretische und Angewandte Limnologie 9: 1–38.

Vörös, L. & K. V.-Balogh, 1998. A Balaton Keszthelyimedencéjének szénforgalma. Hidrológiai Közlöny 78: 385–386 (in Hungarian).

Vörös, L., K. V.-Balogh & S. Herodek, 1996. Microbial food web in a large shallow lake (Lake Balaton, Hungary). Hydrobiologia 339: 57–65.

Hydrobiologia (2008) 599:97–104
DOI 10.1007/s10750-007-9192-0

ELLS 2007

Changes in the water level of Lake Peipsi and their reflection in a sediment core

Jaan-Mati Punning · Galina Kapanen · Tiit Hang · Natalia Davydova · Mihkel Kangur

Abstract A comprehensive study (chronological, lithological and geochemical) of an 8.5 m postglacial sediment sequence from Lake Peipsi was conducted to elucidate the effects of lake-level changes on the sedimentary environment and biogeochemical dynamics in a large lake. Four lithological units were distinguished in the sediment sequence studied: clayey silt, slightly laminated greyish carbonaceous gyttja, brownish-grey gyttja and dark gyttja. These units indicate that large shifts in sedimentation processes occurred in the past. The sediment data show that fluctuations in water depth had a profound impact on the lake environment, recorded as changes in the lithological composition, phosphorus content of sediments and composition of diatom assemblages. The corresponding changes are best reflected in sediments accumulated during phases of regression when the lake area was also smaller. As the water depth and area of the lake increased, wave-induced erosion and resedimentation smoothed or even disturbed the initial information. The increase of phosphorus content in the unconsolidated, high-porosity surface sediments (0.5 m) is most probably connected with active diffusion and matter exchange between the water-sediment pools.

Keywords Lake sediments · Lake Peipsi · Sediment lithology · Water-level fluctuations · Diatom assemblages · Phosphorus content

Guest editors: T. Nõges, R. Eckmann, K. Kangur, P. Nõges, A. Reinart, G. Roll, H. Simola & M. Viljanen.
European Large Lakes—Ecosystem changes and their ecological and socioeconomic impacts.

J.-M. Punning (✉) · G. Kapanen · M. Kangur
Institute of Ecology, Tallinn University, Uus-Sadama 5, 10120 Tallinn, Estonia
e-mail: punning@tlu.ee

T. Hang
Institute of Geology, University of Tartu, Vanemuise 46, 51014 Tartu, Estonia

N. Davydova
Institute of Limnology, Russian Academy of Sciences, Sevastjanova 9, 196105 St. Petersburg, Russia

Introduction

Knowledge of the dynamics of the environmental parameters affecting the lakes is a prerequisite for appropriate interpretation of palaeolimnological information. Thus, studies of lakes with different temporal scales and spatial sizes are important. In Estonia, many reconstructions have been made on the basis of sediment data from small lakes (Punning et al., 2004, 2005), whereas data from large lakes are still limited. Palaeoecological studies of large lakes are more complex than those of small lakes because: (1) the high energy of waves in large lakes may significantly affect the sedimentation process; (2) recent geochemical studies of bottom sediments have shown that changes in the concentrations of mobile elements such as phosphorus are strongly determined by physico-chemical changes in the water–sediment interface, and evidence of these changes does not necessarily provide direct information about specific events in the past.

Despite long-term and interdisciplinary investigations, very little is known about the sediments of Lake Peipsi. Initial studies on these sediments were undertaken in the 1970s (Raukas, 1981), while research on the curative mud of Värska Bay was conducted by Pirrus and Tassa (1981). These data resist serious palaeoecological interpretations because no chronostratigraphic background is available. The Holocene water level changes in Lake Peipsi have been reconstructed mainly from the sediment sequences in the central part of the lake and studies on the lake's catchment area (Orviku, 1960; Sarv & Ilves, 1975; Miidel et al., 1995; Hang et al., 2001). These data showed significant lake-level fluctuations during the Holocene, but little is known about their impact on the ecological history of the lake. The Holocene development of the lake ecosystem described previously through studies of the diatom flora of Lake Lämmijärv (Davydova, 1981; Davydova & Kimmel, 1991), the central part of Lake Peipsi, showed that throughout the Holocene the lake depression was occupied by a body of freshwater in which eutrophication started during the Atlantic Chronozone and has persisted to the present.

Numerous recent studies (e.g. Kozerski et al., 1994, 1999; Søndergaard et al., 2003; Battarbee et al., 2005) have demonstrated that, in addition to external phosphorus loading, changes in the hydrological regime and physico-chemical conditions within the lake play an important role in the development of ecological conditions. The main aim of the present study was to estimate the impact of water level fluctuations on the lithology of sediments and on their diatom and phosphorus contents.

Study area

The submeridionally elongated Lake Peipsi (surface area 3555 km^2, maximum length approximately 150 km, width 42 km) is a shallow body of water with a mean depth of about 7 m (max 15 m). Lake Peipsi (30 m a.s.l.) is surrounded by a flat lowland area 30–45 m a.s.l. with peculiar landforms rarely higher than 80 m. Till covers the bedrock over the entire lake bottom and is overlain by a ca 10 m thick bed of glaciolacustrine clay or silt. The distribution of lake marl and gyttja is limited to the deeper central part of the lake, while the thickness of the calcareous layer varies from a few centimetres to 1.9 m. Gyttja deposits, up to 5 m thick, are greenish in their lower parts and change to dark black towards the top (Hang et al., 2001).

Deglaciation began in the Lake Peipsi basin ca 13,500 ^{14}C year BP (Hang & Miidel, 1999a) when glacial Lake Peipsi was located in front of the retreating ice margin. The glacioisostatic factor has been the main trigger of changes in lake level and the corresponding shore displacement throughout the postglacial period (Hang & Miidel, 1999b). Owing to the higher uplift rate in the northern part, Lake Peipsi is steadily spreading southwards and inundates low coastal areas.

Currently, eutrophication is a major problem in Lake Peipsi. Several studies have been undertaken to track the water quality trends of the lake, but important questions remain unsolved, particularly concerning the role of internal and external phosphorus loading. Although nutrient loading decreased during the 1990s (Blinova, 2001; Kangur et al., 2002), phosphorus concentrations in Lake Peipsi remain high, partly because phosphorus is released from a large pool in the bottom sediments.

Methods

Sampling

The sediment sequence was obtained in March 2004 by coring from the ice at a water depth of 10 m in the central part of Lake Peipsi (N 58°41′53″, E 27°23′53″). A Livingstone–Vallentyne piston corer (diameter 7 cm) operated by rods was used to obtain a 0.5 m long core of unconsolidated surface sediments. In the field, the uppermost 10 cm of this core was divided into 1 cm and the rest of the core into 2 cm sections. The lower part of the sediments, from 0.5 to 8.5 m depth, was sampled with a Russian type peat sampler. This part of the core was divided into 1 m sections that were wrapped in plastic and placed in PVC tubes for transport. In the laboratory, the samples were sliced into 20 cm subsamples, packed into plastic boxes and kept frozen prior to analysis.

Analysis

To determine the water content, the samples were dried to constant weight at 105 °C. The organic matter content was measured by loss-on-ignition (LOI) at 550°C for 2 h and expressed as the percentage dry matter. The carbonate content was calculated as the loss of weight after burning the LOI residue at 950°C for 30 min (Heiri et al., 2001).

Total (TP) and inorganic phosphorus (inorg-P) were measured at the Institute of Limnology of the Russian Academy of Sciences according to Finnish Standard Methods (SFS 3025, 1986; SFS 3026, 1986). The inorg-P compounds were fractionated with hydrochloric acid, sulphuric acid and potassium hydroxide, and analysed according to SFS 3025 (1986). To determine TP, inorganic phosphorus compounds and organically bound P were oxidized with peroxodisulphate according to the ammonium molybdate method (SFS 3026, 1986).

The samples for diatom analysis were treated using standard procedures (Battarbee et al., 2001). Valve concentrations per 1 g of dry sediment were calculated by the method of Davydova (1985). The diatom taxonomy used here follows that of Krammer & Lange-Bertalot (1986–1991) and Simonsen (1987). Quantitative reconstruction of the diatom-inferred TP for Lake Peipsi was performed by a weighted averaging transfer function, using a combined TP dataset in The European Diatom Database, a web-based information system (http://craticula.ncl.ac.uk/Eddi/jsp/index.jsp).

Three bulk gyttja samples were dated by the ^{14}C method using accelerator-mass-spectrometry (AMS) at the Institute of Physics at the University of Erlangen-Nürnberg (Germany).

Results

Lithology and chronology

The sediment sequence studied can be divided into four lithostratigraphic units (Table 1). Unit 1 (8.5–7.0 m) consists of fine-grained clayey silt, which corresponds to the unit described by Hang et al. (2001) as glaciolacustrine silt covering the proglacial varved clay. This demonstrates that the coring had penetrated the whole Holocene sequence. The organic matter content of the clayey silt is ca 5% and the water content is up to 50% (Fig. 1). The upper contact of glaciolacustrine sediments is sharp at a depth of 7.0 m. Unit 2 (7.0–5.2 m) is dark-grey slightly laminated gyttja characterized mainly by a continuous increase in carbonate from 4% up to 25%. The organic matter and water contents of this unit increase continuously up to 15% and 60%, respectively. Unit 3 (5.2–0.4 m) consists of homogeneous greenish-grey gyttja characterized by a slight increase in water content (from 60% to 85%). The organic matter content reaches its maximum value (28%) at a depth of 3.9 m, varying upwards to around 25%. The topmost 0.4 m of the sediment (unit 4) consists of unconsolidated homogeneous dark gyttja with water content >90% (Fig. 1).

Phosphorus content

Phosphorus compounds (total and inorganic) were measured in 37 samples from the upper 6.0 m of the sediments (Fig. 1). The TP concentrations are relatively stable from 6 to 0.6 m (0.7 mg g^{-1} DM) with an increase in the depth interval from 4.5 to 2.5 m (up to 0.8 mg g^{-1} DM). The concentrations of inorganic phosphorus are somewhat more variable (from 0.5 to 0.6 mg g^{-1} DM) and are smaller in the lower part of the sequence (Fig. 1). In the near-surface layers (from a depth of 0.5 m), both types of phosphorus compounds (total and inorganic) increase towards the surface, reaching 1.6 mg g^{-1} and 1.17 mg g^{-1} DM, respectively.

Diatom-inferred TP

There was little similarity between the diatom species of Lake Peipsi and the combined TP dataset. There are two reasons for this: first, the number of taxa observed in the sediment sequence of Lake Peipsi is relatively low; second, some of the diatom flora of the lake are unique. It includes diatom species characteristic of small lakes along with those typical of

Table 1 Lithostratigraphic subdivision of the Lake Peipsi sediment sequence studied

Depth from the sediment surface (m)	Unit	Lithological description
0.4–0	4	Unconsolidated homogeneous dark gyttja
5.2–0.4	3	Homogeneous greenish-grey gyttja, lower interval (5.2–4.7 m) laminated
7.0–5.2	2	Carbonaceous dark-grey gyttja, slightly laminated
8.5–7.0	1	Clayey silt

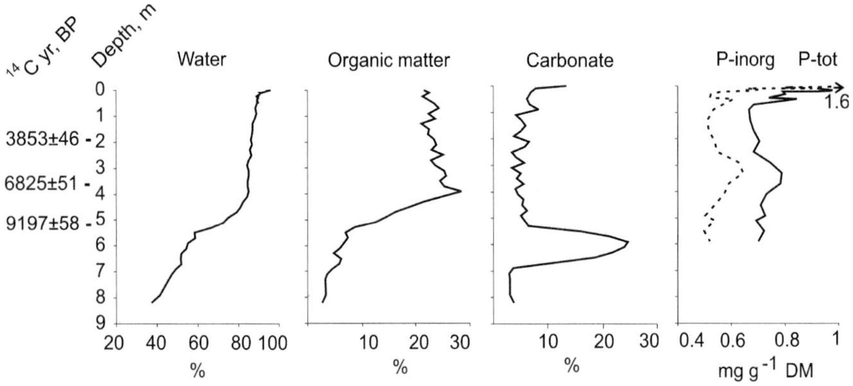

Fig. 1 Water, organic matter and carbonate contents, and concentrations of phosphorus compounds in the sediment core from Lake Peipsi

large lakes such as Lake Ladoga and Lake Onega (Davydova, 1981). Therefore diatom-inferred TP values do not realistically reflect the phosphorus content of the lake and they have not been used in further discussion.

Diatom assemblages

In total, 57 diatom species were found, out of which 22 are planktonic, 23 benthic and 12 bottom-living. Valve concentration per 1 g of dry sediment varied significantly (Fig. 2). Five diatom assemblage zones were identified on the basis of the composition and abundance of diatoms. In the clayey silt and in most of the carbonaceous gyttja, only some fragments of diatom shells were found.

In zone P4-04-I (7.3–5.5 m) the diatom abundance is very low (ca 50×10^3 g^{-1} DM). Single planktonic alkaliphilous diatoms *Aulacoseira granulata* (Ehr.) Sim., *A., ambigua* (Grun.) Sim., *Fragilaria capucina* Desm. and less often benthic *F. brevistriata* Grun. and *F. leptostauron* var. *martyi* (Hreib.), were observed.

In the overlying zone P4-04-II (5.5–3.9 m), the diatom concentration increases to 120×10^6 g^{-1} DM. Benthic diatoms are the most abundant making up 65–82% of the frustules; planktonic diatoms constitute 16–33%. The dominant species in this zone are the benthic species *Fragilaria* (*F. brevistriata*, *F. construens* (Ehr.) Grun. et var., *F. heidenii* Oestr.), and subdominants are the planktonic *Aulacoseira islandica* (O. Mill.) Sim. and *A. ambigua.*

In the zone P4-04-III (3.9–2.2 m), the diatom concentration varies from 30 to 80×10^6 g^{-1} and the species diversity increases. The proportion of planktonic species increases and that of benthic species decreases. The dominant taxa in these layers are the planktonic *Aulacoseira granulata* and the benthic *Fragilaria heidenii* (up to 25%). Some planktonic *Stephanodiscus* (*S. rotula* (Kutz.) Hendey, *S. minutulus* (Kutz.) Cl. and Moll., *S. niagarae* Ehr.) appeared in this zone.

Zone P4-04-IV (2.2–0.5 m) is characterized by a slight decrease in diatom concentration (25–85×10^6 g^{-1} DM) compared to the previous zone. The main dominants here are two planktonic *Aulacoseira* species (*A. granulata* and *A. islandica*) and subdominants are represented by benthic *Fragilaria* (*F. brevistriata, F. heidenii*).

In the uppermost zone, P4-04-V (0.5–0 m), the diatom concentration varies from 18 to 85×10^6 g^{-1}

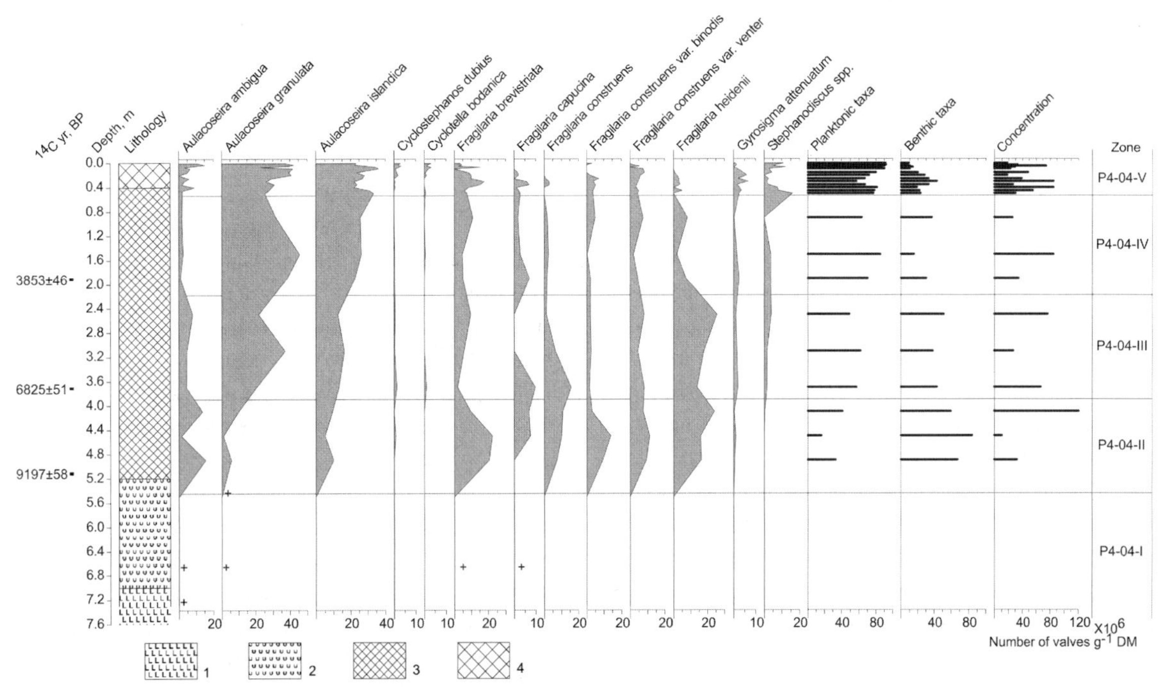

Fig. 2 Diatom diagram of the core studied indicating the main diatom assemblage zones. Relative abundances of major taxa are given, along with percentages of planktonic and benthic taxa in the totality of diatoms, and concentration of valves. Lithology: 1—clayey silt; 2—carbonaceous gyttja; 3—homogeneous greenish-grey gyttja; 4—unconsolidated dark gyttja

DM with planktonic diatoms prevailing (up to 92% in the uppermost subsamples). The same two *Aulacoseira* species as in zone P4-04-IV are the main dominants here. A third dominant (*A. ambigua*) appears in the samples from the uppermost 10 cm. The single frustules of planktonic *Cyclostephanos dubius* (Fricke) Round, which is a typical inhabitant of mesotrophic waters, were found regularly within this zone.

Discussion

Clear lithostratigraphic units could be distinguished in the Holocene sediment sequence of Lake Peipsi. These form a good basis for correlation with cores from the southern areas of the lake system and enable some conclusions to be made about changes in water depth during the Holocene. The low organic and carbonate contents of clayey silt in unit 1 (Fig. 1) reflect the sedimentary environment characteristic of a deep proglacial lake. A sharp transition to the overlying calcareous gyttja deposits (unit 2) is most probably connected with a marked lowering of water level at the transition from late-glacial to Holocene prior to 9200 14C year BP. Previously, Hang et al. (2001) described an erosional discontinuity at this boundary in the near-shore areas of the recent lake, which they explain in terms of high wave erosion or even emergence of areas caused by a fall in lake level. In the sequence studied, an increase in the carbonate content from 5% to ca 30% followed by a decrease to 8% in unit 2 indicates cyclic changes in the sedimentary environment. It is known that the carbonate content of sediments depends on water depth. Dean (1999) showed that the rates of organic matter and $CaCO_3$ accumulation in lake sediments are balanced by the rates of biomass production in the epilimnion and of organic matter decomposition and $CaCO_3$ dissolution in the hypolimnion and sediments. It was also demonstrated by Punning et al. (2005) that the high $CaCO_3$ concentrations (35–40%) in sediments of small lakes suggest sedimentation at a water depth of about 2–4 m. It is worth mentioning that a water depth of no more than 4 m in Preboreal Lake Peipsi was also previously suggested by ostracod studies (Niinemets, 1999). Therefore, we can conclude that during the accumulation of the carbonaceous gyttja, the water depth in the lake was around 2–4 m. According to pollen and radiocarbon data from the Saviku section (Sarv & Ilves, 1975) in the mouth of the Emajõgi River, the water level in Lake Peipsi began to rise during the Atlantic Chronozone and has continued.

Figure 3 presents composite data about the most significant changes in sediment composition and suggested water depth in Lake Peipsi. In some sediment intervals, most of the records show changes in trends. The late-glacial-Early-Holocene lowering of the water level and accumulation of carbonaceous gyttja (7.0–5.2 m) induced the most significant changes in the lake ecosystem. The absence of diatoms from the corresponding sediments precludes conclusions about the trophic state of the lake at that time. The next well-expressed interval in the development of Lake Peipsi is recorded in the sediment interval ca 5.2–5.0 m (accumulated ca 9200 ^{14}C year BP). This interval is characterized by a decrease in the carbonate content and a rapid increase in the organic content. As demonstrated previously (Dean, 1999), a gradual increase in organic productivity would result in an increased oxygen demand in the hypolimnion, making it anoxic. Decomposition of organic matter in the hypolimnion decreases the pH and enhances the dissolution of $CaCO_3$. Obviously, the increase in water depth in Lake Peipsi discussed here was rather slow, because the diatom data show a continuous dominance of benthic forms in the sediments from a depth of 5.5–3.9 m. The dominance of the periphythic diatoms *Fragilaria brevistriata*, *F. construens* and *F. heidenii* indicates mesotrophic conditions in the lake. It has been shown (Barker et al., 1994; Yang & Duthie, 1995; Punning et al., 2004) that differences in the types of diatom taxa (ratio of planktonic to non-planktonic diatoms) may be used as an indicator of past water depths. At high water levels, the sediments are dominated by planktonic forms, while benthic forms are expected to increase at low water levels.

The sediments above 3.9 m (accumulated after ca 6000 ^{14}C year BP) reflect important changes in the development of Lake Peipsi. From this level upwards, the number of planktonic diatoms continuously increases, which may indicate a further increase of water depth at the site studied. However, the dominance of planktonic diatoms can also often be ascribed to an increased nutrient supply and the consequent increase in productivity of the aquatic ecosystem. At a depth of 3.9 m, the content of organic matter reaches a maximum (28%). The phosphorus content of the sediments also starts to increase, which might be explained by a shift in the phosphorus pool towards the sediments under the influence of a pH decrease in the hypolimnion caused by organic matter decay (Golterman, 2004).

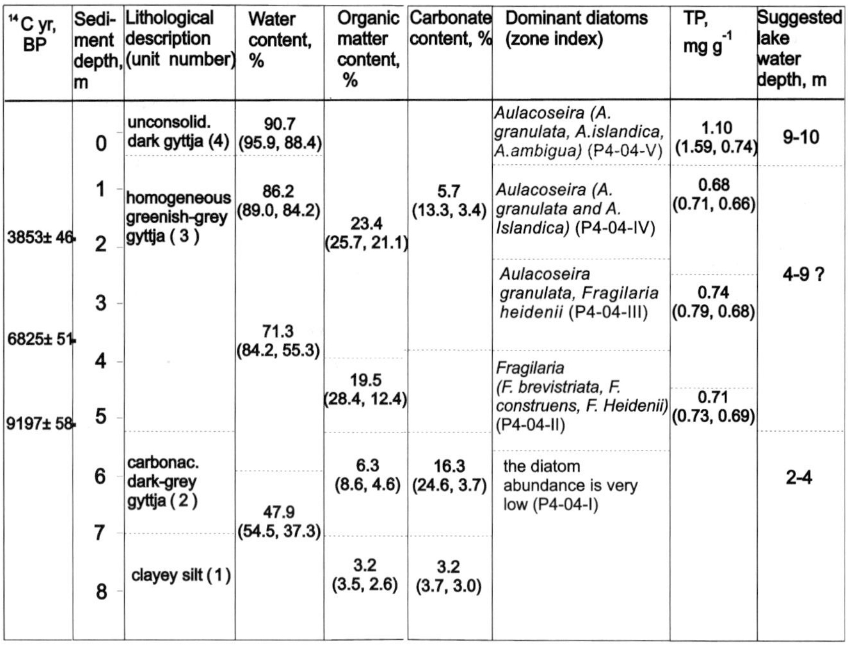

Fig. 3 Comparison of the changes in sediment lithology, water, organic matter and carbonate content (mean value (min, max)), dominant diatoms, diatom zones, total phosphorus concentration and suggested water depths. The most significant changes in values are marked by lines

Striking changes in the colour and composition of sediments are reported in the unconsolidated surface layer (0.5–0 m) accumulated during recent centuries (Nõges et al., 2006; Kangur et al., 2007). Most noteworthy is the rapid increase of phosphorus concentration towards the surface. Recent geochemical studies of the surface sediments have proved that changes in the concentrations of mobile elements such as phosphorus, and pore water, are conditioned by physico-chemical changes in the water–sediment interface, while their distribution in the sediments provides no direct information about specific past events (Ciglenečki et al., 2005). In the sediments studied, the water content record shows that an active process of diffusion, and exchange of chemical elements in sediments reaches to a depth of 0.5 m, where the porosity of the sediments exceeds 89%.

The reasons for the relatively conservative behaviour of the phosphorus concentration throughout the sediment sequence studied remain to be elucidated. In small lakes, it has been possible to estimate the relationship between the known lake-level fluctuations and phosphorus concentrations (Punning & Leeben, 2003; Punning et al., 2003). Palaeolimnological reconstructions for large and shallow lakes are generally much less credible, mostly because of possible resuspension caused by the greater influence of waves on sediment accumulation (Håkanson & Jansson, 1983). The waves and near-bottom currents may have caused resedimentation of fine-grained deposits.

Conclusions

The temporal changes in the lithological composition of sediments in the core studied support the conclusion that profound changes in water level have occurred in Lake Peipsi during the postglacial period. The clearly represented lithological units and their correlation with the sequences from the southernmost areas of the lake correspond well with the earlier views about the significant lowering of water level in Lake Peipsi at the late-glacial-Holocene transition. At that time Lake Peipsi was relatively shallow (2–4 m) in the area represented by the sequence studied. As suggested by the increase in planktonic diatoms, the water level started to rise after 9200 ^{14}C year BP when the accumulation of calcareous gyttja was replaced by the accumulation of gyttja. The fluctuations in water depth had a profound impact on the lake environment and are recorded as changes in lithological composition, phosphorus contents of sediments and composition of diatom assemblages. The corresponding changes are most clearly reflected in sediments accumulated during the phases of regression when the lake area was also smaller. As the water depth and area of the lake increased, wave-induced erosion and resedimentation smoothed or even disturbed the information. The increase in phosphorus content in the unconsolidated, high-porosity surface sediments (0.5 m) is connected with active diffusion and matter exchange between the water–sediment pools.

Acknowledgements The study was supported by the Estonian target financed project SF0280016s07 and the Estonian Science Foundation (grant no. 5370). This research was partly supported by the Estonian Environmental Investment Centre (contract no. 06-06-3/381). We thank our colleagues Jaanus Terasmaa for assistance in the fieldwork and Liisa Puusepp and Anna Ludikova for the help in compiling the diatom diagram. We thank the Estonian Border Guard for supporting field work in Lake Peipsi and Mrs Tiia Kaare for improving the language. The journal reviewers and the guest editor of the special issue are acknowledged for constructive comments.

References

Barker, P. A., N. Roberts, H. F. Lamb, S. van der Kaars & A. Benkaddour, 1994. Interpretation of Holocene lake-level changes from diatom assemblages in Lake Sidi Ali, Middle Atlas, Morocco. Journal of Paleolimnology 12: 223–234.

Battarbee, R. W., V. J. Jones, R. J. Flower, N. G. Cameron & H. Bennion, 2001. Diatoms. In Smol, J. P., H. J. B. Birks & W. M. Last (eds), Tracking Environmental Change Using Lake Sediments, Vol. 3: Terrestrial, Algal and Siliceous Indicators. Kluwer Academic Publishers, Dordrecht, 155–202.

Battarbee, R. W., N. Anderson, E. Jeppesen & P. Leavitt, 2005. Combining paleolimnological and limnological approaches in assessing lake ecosystem response to nutrient reduction. Freshwater Biology 50: 1772–1780.

Blinova, I., 2001. Riverine load into L. Peipsi. In Nõges, T. (ed.), Peipsi. Meteorology, Hydrology, Hydrochemistry. Sulemees Publishers, Tartu, 94–96.

Ciglenečki, I., M. Carić, F. Kršinić, D. Viličić & B. Ćosović, 2005. The extinction by sulphide – turnover and recovery of a naturally eutrophic, meromictic seawater lake. Journal of Marine Systems 56: 29–44.

Davydova, N., 1981. Diatoms in sediment cores from Lake Peipsi–Pskov. In Raukas, A. (ed.), Bottom Deposits of

Lake Peipsi. Estonian Academy of Sciences, Tallinn, 74–81, (In Russian).
Davydova, N., 1985. Diatoms – Indicators of Natural Conditions of Water Reservoir in Holocene. Nauka, Leningrad, (In Russian).
Davydova, N. & K. Kimmel, 1991. Palaeogeography of Lake Peipsi on the basis of biostratigraphical studies of bottom sediments. Proceedings of the Estonian Academy of Sciences, Geology 40: 16–23, (In Russian).
Dean, W. E., 1999. The carbon cycle and biogeochemical dynamics in lake sediments. Journal of Paleolimnology 21: 375–393.
Golterman, H. L., 2004. The Chemistry of Phosphate and Nitrogen Compounds in Sediments. Kluwer Academic Publishers, Dordrecht/Boston/London.
Håkanson, L. & M. Jansson, 1983. Principles of Lake Sedimentology. Springer-Verlag, Berlin.
Hang, T. & A. Miidel, 1999a. Bedrock topography. In Miidel, A. & A. Raukas (eds), Lake Peipsi: Geology. Sulemees Publishers, Tallinn, 27–28.
Hang, T. & A. Miidel, 1999b. Holocene history of the lake. In Miidel, A. & A. Raukas (eds), Lake Peipsi: Geology. Sulemees Publishers, Tallinn, 131–135.
Hang, T., A. Miidel, V. Kalm & K. Kimmel, 2001. New data on the distribution and stratigraphy of the bottom deposits of Lake Peipsi. Proceedings of the Estonian Academy of Sciences, Geology 50: 233–253.
Heiri O., A. F. Lotter & M.-J. Lemcke, 2001. Loss on ignition as a method for estimating organic and carbonate content in sediments: reproducibility and comparability of results. Journal of Paleolimnology 25: 101–110.
Kangur, K., T. Möls, J. Haberman, K. Kangro, R. Laugaste, A. Milius, T. Nõges, H. Timm, T. Timm & P. Zingel, 2002. Lake Peipsi change of ecological state during 1992–2001. In Roose, A. (ed.), Estonian Environmental Monitoring 2001. Tartu, 57–64, (In Estonian).
Kangur, M., K. Kangur, R. Laugaste, J.-M. Punning & T. Möls, 2007. Combining limnological and palaeolimnological approaches in assessing degradation of Lake Pskov. Hydrobiologia 584: 121–132.
Kozerski H. P., H. Behrendt & J. Köhler, 1994. Possibilities and limitations of sediment traps to measure sedimentation and resuspension. Hydrobiologia 284: 93–100.
Kozerski H. P., H. Behrendt & J. Köhler, 1999. The N and P budget of the shallow, flushed lake Müggelsee: retention, external and internal load. Hydrobiologia 408/409: 159–166.
Krammer, K. & H. Lange-Bertalot, 1986–1991. Bacillariophyceae. In Susswasserflora Mitteleuropas 2/1–4. Stuttgart.
Miidel, A., T. Hang, R. Pirrus & A. Liiva, 1995. On the development of the southern part of Lake Peipsi in the Holocene. Proceedings of the Estonian Academy of Sciences, Geology 44: 33–44.
Niinemets, E., 1999. Ostracods. In Miidel, A. & A. Raukas (eds), Lake Peipsi: Geology. Sulemees Publishers, Tallinn, 90–97.
Nõges, T., A. Heinsalu, T. Alliksaar & P. Nõges, 2006. Paleolimnological assessment of eutrophication history of large transboundary Lake Peipsi, Estonia/Russia. Verhandlungen der Internationalen Vereinigung für Limnologie 29: 1135–1138.
Orviku, K., 1960. Geomorphology. In Orviku, K. (ed.), Geology of U.S.S.R., 28. Gosgeoltekhizdat, Moscow, 209–230, (In Russian).
Pirrus, R. & V. Tassa, 1981. On the sapropel layer at Värska. In Raukas, A. (ed.), Bottom Deposits of Pihkva-Peipsi Lake. Estonian Academy of Sciences, Tallinn: 82–93, (In Russian).
Punning, J.-M. & A. Leeben, 2003. A comparison of sediment and monitoring data: implications for paleomonitoring a small lake. Environmental Monitoring and Assessment 89: 1–13.
Punning, J.-M., M. Kangur, T. Koff & G. Possnert, 2003. Holocene lake-level changes and their reflection in the paleolimnological records of two lakes in northern Estonia. Journal of Paleolimnology 29: 167–178.
Punning, J.-M., L. Puusepp & T. Koff, 2004. Spatial variability of diatoms, subfossil macrophytes and OC/N values in surface sediments of Lake Väike Juusa (southern Estonia). Proceedings of the Estonian Academy of Sciences. Biology, Ecology 53: 147–160.
Punning, J.-M., T. Koff, E. Kadastik & A. Mikomägi, 2005. Holocene lake level fluctuations recorded in the sediment composition of Lake Juusa, southeastern Estonia. Journal of Paleolimnology 34: 377–390.
Raukas, A., 1981. On the lithological composition of bottom deposits in Pihkva-Peipsi Lake. In Raukas, A. (ed.), Bottom Deposits of Pihkva-Peipsi Lake. Estonian Academy of Sciences, Tallinn, 23–41, (In Russian).
Sarv, A. & E. Ilves, 1975. On Holocene deposits in the delta of River Emajõgi (Saviku). Proceedings of Estonian Academy of Sciences. Chemistry, Geology 24: 64–69, (In Russian).
SFS 3025: 1986. Determination of Phosphorus in Water. Finnish Standards Association SFS, Helsinki.
SFS 3026: 1986. Determination of Phosphate in Water. Finnish Standards Association SFS, Helsinki.
Simonsen, R., 1987. Atlas and Catalogue of the Diatom Types of Friederich Hustedt, Vol. I–III. Stuttgart.
Søndergaard, M., J. P. Jensen & E. Jeppesen, 2003. Role of sediment and internal loading of phosphorus in shallow lakes. Hydrobiologia 506/509: 135–145.
Yang, J.-R. & H. C. Duthie, 1995. Regression and weighted averaging models relating surficial sedimentary diatom assemblages to water depth in Lake Ontario. Journal of Great Lakes Research 21: 84–94.

Hydrobiologia (2008) 599:105–110
DOI 10.1007/s10750-007-9193-z

Rates of change in physical and chemical lake variables – are they comparable between large and small lakes?

Gesa A. Weyhenmeyer

Abstract Changes over time in 16 physical and chemical variables were analysed and compared between Sweden's largest lakes, Vättern and Vänern, and 48 smaller Swedish reference lakes during spring over the period 1984–2003. The rates of changes varied substantially among lakes and among variables, and they were clearly influenced by changes in both climate and atmospheric deposition. Rates of change of variables associated with atmospheric deposition such as sulphate concentrations were dependent on lake morphometry. This also applied to the rates of change of variables associated with climate change effects in the catchment such as calcium and magnesium concentrations. However, climate change effects could also be comparable between large and small lakes. Rates of change in physical and chemical variables directly influenced by the climate via the lake water surface, e.g., surface water temperature, and variables associated with the spring phytoplankton development such as phosphate–phosphorus and nitrate–nitrogen concentrations, were similar and therefore independent of lake morphometry. This study shows that climate change effects that act via the lake surface can be of the same order of magnitude among large and small lakes, but climate change effects that act via the catchment differ substantially in large lakes. It is essential to differentiate between these two types of climate effects in order to assess the impacts of climate change and the adaptation and vulnerability of lake ecosystems.

Keywords Climate · Atmospheric deposition · Water chemistry · Spring phytoplankton development · Lake volume

Guest editors: T. Nõges, R. Eckmann, K. Kangur, P. Nõges, A. Reinart, G. Roll, H. Simola & M. Viljanen
European Large Lakes - Ecosystem changes and their ecological and socioeconomic impacts

G. A. Weyhenmeyer (✉)
Department of Environmental Assessment, Swedish University of Agricultural Sciences (SLU), Box 7050, 750 07 Uppsala, Sweden
e-mail: Gesa.Weyhenmeyer@ma.slu.se

Introduction

The quality of water changes frequently in lakes, depending on natural and anthropogenic changes in the catchment and the atmosphere. Climate and atmospheric deposition are two examples of drivers that influence the quality of water. How quickly the water quality changes in response to changes in the climate and atmospheric deposition is related to various factors including lake morphometry. However, some lake variables show rates of change that are quite independent of lake morphometry, one being the timing of ice breakup. Rates of change in the timing of lake ice breakup were found to depend on the geographical region rather than lake morphometry (Weyhenmeyer et al., 2004, 2005), implying that small and large lakes can respond quite similarly to climatic changes.

A variety of processes are related to the timing of lake ice breakup, e.g., light penetration into the water (Leppäranta et al., 2003), nutrient recycling (Järvinen et al., 2002) and the occurrence of the spring phytoplankton bloom (Weyhenmeyer et al., 1999), and the question arises whether rates of change in these ice breakup-related processes are dependent or independent on lake morphometry. An answer to this question might reveal the lake variables for which we have to expect time lags in response to natural and anthropogenic changes. Here, I hypothesize that rates of change of climate-induced direct alterations of lake surface waters, i.e., ice-breakup-related lake internal processes, are independent on lake morphometry; whereas rates of change in catchment-related processes are dependent on lake morphometry. To test this hypothesis, long-term data on 50 Swedish lakes with various morphometries were analyzed in this study.

Data and methods

From the Swedish national freshwater database, available from the Department of Environmental Assessment, Swedish University of Agricultural Sciences at http://www.ma.slu.se, complete data series were downloaded for surface water temperature and 15 surface (0.5 m) water chemical variables (sulphate, pH, conductivity, calcium, magnesium, sodium, potassium, chloride, ammonium–nitrogen, nitrate–nitrogen, total nitrogen, phosphate–phosphorus, total phosphorus, water color [estimated as light absorption at 420 nm by 0.45 μm-filtered water in a 5-cm cuvette], and reactive silica) for 50 lakes (Fig. 1) during May over the period 1984–2003. The month of May in these lakes corresponds to the spring turnover time when the lake ice has disappeared. Since data series for other months were less complete, and since the focus of the study was on processes influenced by the timing of lake ice breakup, the spring turnover period was used to compare physical and chemical changes between large and small lakes. All physical and chemical lake variables were sampled and analyzed by the same laboratory using standard limnological methods, described at http://info1.ma.slu.se/ma/www_ma.acgi $Analysis?ID=AnalysisList.

Winter (December through February) air temperature data from Mariestad at Lake Vänern from 1984

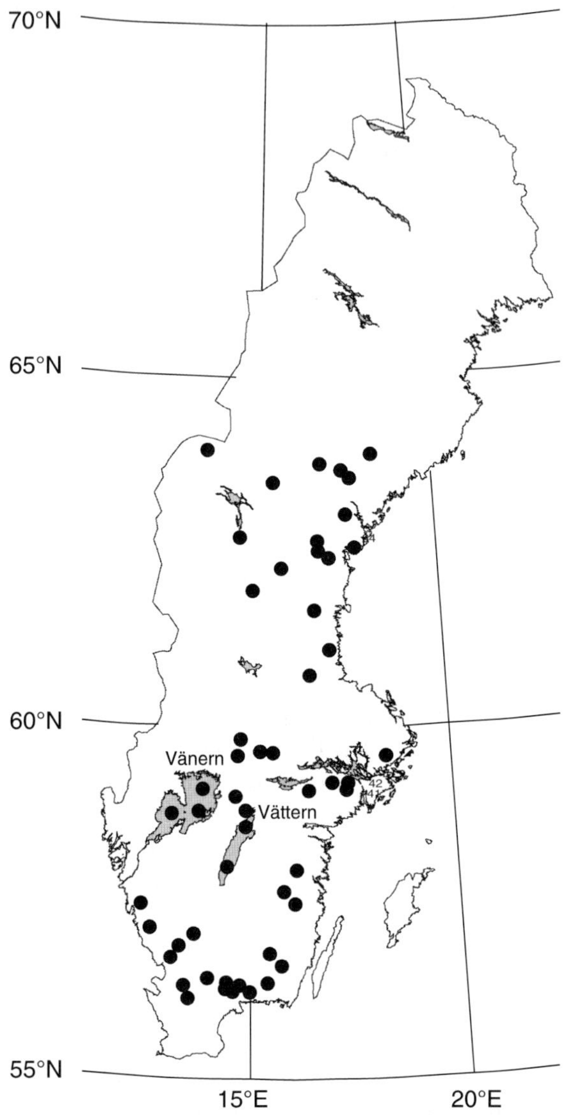

Fig. 1 Map of Sweden showing the geographical distribution of the 48 reference lakes and Sweden's largest lakes Vänern and Vättern

to 2003 were obtained from the Swedish Meteorolgical and Hydrological Institute, and yearly mean sulphate concentrations in the Lake Vänern area were downloaded from the Swedish Environmental Research Institute at http://www.ivl.se/miljo/db/statkart.asp.

About 48 of the 50 lakes are classified as nutrient-poor reference lakes, implying that anthropogenic changes in the catchments other than climatic and atmospheric deposition changes are negligible. The remaining two lakes are Sweden's largest lakes,

Vänern and Vättern. Two sampling sites were available for Lake Vättern and three for Lake Vänern. Both lakes have been subjected to drastic reductions in phosphate loading from the catchment, but since the late 1970s the phosphorus concentrations have stabilized after earlier reductions (Wilander & Persson, 2001). We therefore assume that changes in these large lakes during 1984–2003 were primarily driven by changes in climate and atmospheric deposition.

The surface areas of the lakes ranged from 0.03 to 5,648 km^2 (Lake Vänern) and the mean lake depths from 0.9 to 40 m (Lake Vättern). Both Lake Vänern and Lake Vättern are at the extremes of area and mean depth (Fig. 2); moreover, the water retention times of these two lakes (10 and 58 years, respectively) are exceptionally long compared to the 48 reference lakes. The study therefore focuses primarily on comparing differences between the patterns of change in Vänern and Vättern with those in the remaining 48 lakes.

Changes over time of a variable were determined from the slopes of its linear regression to time, i.e., rates of change correspond to trends over time. For this study, relative and not absolute rates of change were considered so that changes could be compared among the lake variables. All statistical tests were carried out using the JMP programme, version 5.0.1.2 (SAS Institute, 2002).

Results

Analysis of the trends over time in 16 variables from 50 Swedish lakes revealed substantial differences in the rates of change among lakes and among variables during 1984–2003. Two groups of variables could be differentiated: group 1, in which the rates of change in Lakes Vänern and Vättern fell within the 10–90% range of the rates of change in the 48 smaller reference lakes; and group 2, in which at least one of the large lakes fell outside this percentile range (Fig. 3). The variables surface water temperature, ammonium–nitrogen, nitrate-nitrogen, total nitrogen, chloride, phosphate–phosphorus and pH belonged to group 1; sulphate, calcium, magnesium, conductivity, sodium, potassium, water color, total phosphorus, and reactive silica belonged to group 2. Taking Lake Vänern as an example, surface water temperatures (group 1) increased on average by 0.7°C per year during 1984–2003, while sulphate concentrations (group 2) decreased on average by 1.3 % per year. The surface water temperature increase in May in Lake Vänern was related to a 1.9°C winter air temperature increase in the lake during 1984–2003 ($R^2 = 0.45$, $P < 0.01$, $n = 20$) and the decrease in sulphate concentrations in May was related to a 82% decrease in the yearly mean sulphate deposition in the Vänern area over this same period ($R^2 = 0.52$, $P < 0.001$, $n = 20$). Temporal variations in surface water temperatures and sulphate concentrations in Lake Vänern were not related to each other ($P > 0.05$). Moreover, rates of change in surface water temperatures were not related to rates of change in sulphate concentrations over the whole sample of 50 lakes ($P > 0.05$), making it possible to identify and distinguish climate and atmospheric deposition effects.

The rates of change in Lake Vättern, the lake with the longest water retention time of 58 years, deviated most from the rates of change of the 48 small reference lakes. A clear outlier was the rate of change in sulphate concentrations. Sulphate concentrations in Lake Vättern continued to increase during 1984–2003 while they decreased in all the other lakes. Seven other variables showed rates of change in Lake Vättern that deviated from those in the other lakes (Fig. 3). In Lake Vänern, the lake with the second longest water retention time of 10 years, rates of change deviating from those in the 48 reference lakes were only observed for water color and reactive silica concentrations.

Differences in the rates of change between the two large lakes and the 48 small reference lakes could

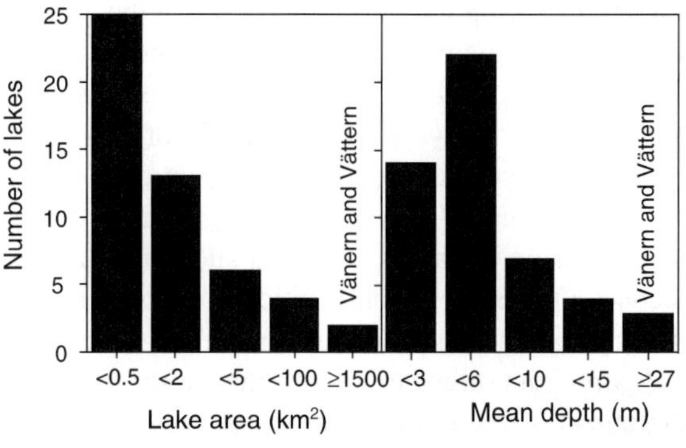

Fig. 2 Frequency distribution of area and mean depth of the 50 lakes studied, including different sites in Sweden's largest lakes, Vänern and Vättern

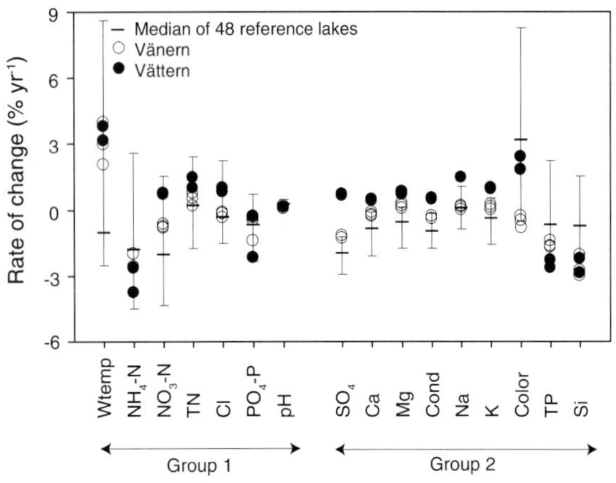

Fig. 3 Relative rates of change of 16 lake variables in the 50 lakes during 1984–2003. The rates of change of sites in Sweden's largest lakes are indicated by circles; lines indicate the 10, 50 and 90% of the rates of change in the 48 reference lakes. Group 1 describes variables for which the rates of change from Sweden's largest lake sites fall within the 10 and 90% of the rates of change from the reference lakes; group 2 describes variables for which the rates of change from at least one large lake fall outside this range of percentiles

imply that these rates of change depend on lake morphometric variables. The available morphometric variables were mean depth, area and volume of the lake and the size of the drainage area. Lake volume was the variable that explained most of the differences in rates of change (Table 1). Significant relationships between lake volume and rate of change could be established for sulphate, magnesium, calcium, total phosphorus, water color, and reactive silica (Table 1). The ratio between drainage area and lake area was used as a measure of the lake water retention time, but this estimate was probably too inaccurate to reveal significant relationships.

Discussion

In both of Sweden's largest lakes and the 48 smaller reference lakes, an impact of climate on, e.g., water temperature and an effect of atmospheric deposition on, e.g., sulphate concentration became apparent. The impact of climate on water temperature was independent of lake morphometry, and the rates of change in water temperature did not differ between Sweden's largest lakes, Vänern and Vättern, and the smaller reference lakes. Since changes in the timing of lake ice breakup are also independent of lake morphometry (Weyhenmeyer et al., 2004, 2005), this pattern might have been expected. It is most likely that climate effects that directly alter the physical properties of lake surface waters are as rapid in large lakes as in small ones. It has been shown that changes in physical variables are most coherent among lakes (e.g., Magnuson et al., 1990; Baines et al., 2000), indicating that lake morphometry is of only minor importance. However, direct climate-induced alterations of lake surface waters can also include changes in biological variables, e.g., spring phytoplankton development. A variety of studies have shown that an earlier timing of lake ice breakup leads to an earlier spring phytoplankton development in both large and small lakes (Weyhenmeyer et al., 1999; Gerten & Adrian, 2000; Weyhenmeyer, 2001). In this study, no biological data were available, but data existed on chemical variables that are associated with the spring phytoplankton bloom such as phosphate–phosphorus and nitrate–nitrogen concentrations (Eriksson & Forsberg, 1992). Changes in these nutrients in Sweden's largest lakes were quite comparable to changes in the smaller reference lakes, indicating that these nutrients might be influenced by lake internal processes such as the spring phytoplankton bloom. Reactive silica concentrations are also associated with the spring phytoplankton bloom, especially with a spring diatom bloom (e.g., Eriksson & Forsberg, 1992). Here, Sweden's largest lakes showed a substantially greater rate of change than the smaller reference lakes. This could indicate that reactive silica is influenced by catchment processes rather than by lake internal processes, or it could be explained by the fact that only Sweden's largest lakes show a clear dominance of diatoms during the spring bloom while the smaller nutrient-poor reference lakes are dominated by flagellated species such as chrysophytes (Andersson & Willén, 1999). Chrysophytes are more or less unrelated to reactive silica concentrations but diatoms are strongly associated with them. Changes in the spring phytoplankton bloom as described above may consequently change the reactive silica concentrations in the large lakes but not so much in the small ones.

Apart from climate effects that lead to direct alterations of lake surface waters and are associated with the timing of lake ice breakup, there are climate effects that lead to alterations in physical and chemical conditions in the catchment. Small

Table 1 Relationships (R^2-values) between the rates of change in lake variables in 50 lakes during 1984–2003 and log-transformed lake morphometric variables

Variable	Lake area	Mean depth	Volume	Ratio drainage area/lake area
Water temperature (W_{temp})	–	–	–	–
Sulphate (SO_4)	0.22***	0.31***	0.30***	0.10*
pH	–	–	–	–
Conductivity (Cond)	–	–	–	–
Calcium (Ca)	0.16**	0.24***	0.24***	–
Magnesium (Mg)	0.18**	0.25***	0.24***	–
Sodium (Na)	–	–	–	–
Potassium (K)	–	–	–	–
Chloride (Cl)	–	–	–	–
Ammonium-nitrogen (NH_4–N)	–	–	–	–
Nitrate-nitrogen (NO_3–N)[a]	–	–	–	–
Total nitrogen (TN)	–	–	–	–
Phosphate–phosphorus (PO_4–P)	–	–	–	–
Total phosphorus (TP)	0.12*	0.11*	0.13*	–
Water color (color)[b]	0.14**	0.14**	0.16**	–
Reactive silica (Si)	0.13*	–	0.11*	–

Significance levels are given as three stars ($P < 0.001$), two stars ($P < 0.01$) and one star ($P < 0.05$). Short lines indicate non-significant relationships ($P > 0.05$)

[a] Nitrate–nitrogen includes nitrite–nitrogen, which is less than 5% of the sum of nitrate + nitrite–nitrogen (Wilander pers. comm.)

[b] Water color was estimated as light absorption at 420 nm of 0.45 μm filtered water in a 5-cm cuvette

alterations in the catchment might not affect concentrations in lakes with large water volumes because of the strong dilution effect. Greater alterations in the catchment will affect concentrations even in lakes with a large water volume, but there will be a substantial time lag, depending on the water retention time of the lake, before these effects can be detected in the water column. Deviating rates of change in large lakes are expected for all catchment-driven variables. Calcium and magnesium might represent such variables. Calcium and magnesium concentrations are known to be influenced by climate-driven weathering processes and water flow characteristics in the catchment (Kalff, 2002). Total phosphorus concentrations and water color are also known to be influenced by hydrological, chemical and biological alterations in the catchment, which Håkanson (2005) was able to demonstrate by relating catchment variables to such catchment-driven variables. The rates of change of the catchment-driven variables were particularly dependent on the lake water volume, implying that changes in the catchment and consequently in the loading are damped by a large water volume.

Like changes in the climate, changes in atmospheric deposition can also directly influence water quality via the lake water surface or indirectly via the catchment. It is suggested that the indirect effects of atmospheric deposition on water quality are more pronounced than direct effects, since catchment areas are usually larger than lake surface areas, so the amounts of substances deposited are higher in the catchment than on the lake surface. It is likely that indirect effects of atmospheric deposition are responsible for the deviant rates of change in atmospheric deposition-driven variables such as sulphate concentration observed in Sweden's large Lake Vättern. In Lake Vättern, with the longest water retention time of 58 years, the deviation in this pattern was so pronounced that sulphate concentrations increased, while they decreased in all the other lakes. The decrease in sulphate concentrations is most probably a result of decreased atmospheric deposition, as also observed in other lake systems (Moldan et al., 2001; Forsius et al., 2003), while the increase in Vättern indicates a substantial time lag before decreasing atmospheric deposition can be detected in the water columns of lakes with very long water retention times. Of all the variables tested, rates of change in sulphate concentrations were most closely related to lake morphometric variables, implying that changes in this variable are very little influenced by climate effects that act via the lake water surface. Most of the other variables were influenced by changes in both the climate and atmospheric deposition. In these cases, it becomes important to differentiate between effects acting via the lake water surface and effects acting via the catchment.

Conclusion

According to this study, direct climate-induced alterations of lake surface waters, in this case ice breakup-related processes, are independent of lake morphometry, while rates of change of catchment-related processes are dependent. Variables primarily associated with atmospheric deposition are mainly influenced by catchment processes. For these variables, and also for lake variables indirectly affected by climate via the catchment, substantial time lags in lakes with a long water retention time have to be expected, and models simulating the effects of global change on these variables need to consider lake morphometry. In contrast, lake morphometry can be neglected for variables that are affected by climate directly via the lake water surface. These findings have important implications for lake management and simulations of the effects of global change.

Acknowledgments This work was partly funded by the European Union within the framework of the European Commission projects CLIME ("Climate and Lake Impacts in Europe", EVK1-CT-2002-00121 and Euro-limpacs ("Integrated Project to Evaluate the Impacts of Global Change on European Freshwater Ecosystems", GOCE-CT-2003-505540). The author is a research fellow of the Royal Swedish Academy of Sciences supported by a grant from the Knut and Alice Wallenberg Foundation. Funding was also received by the Swedish Research Council (621-2005-4335). The author is grateful to the Swedish Environmental Protection Agency and the IMA laboratory for financing, sampling and analyzing numerous water samples.

References

Andersson, B. & E. Willén, 1999. Lakes. In Rydin, H., P. Snoeijs & M. Diekmann (eds), Swedish plant geography. Acta Phytogeographica Suecica 84: 149–168.

Baines, S. B., K. E. Webster, T. K. Kratz, S. R. Carpenter & J. J. Magnuson, 2000. Synchronous behavior of temperature, calcium, and chlorophyll in lakes of northern Wisconsin. Ecology 81: 815–825.

Eriksson, C. & C. Forsberg, 1992. Nutrient interactions and phytoplankton growth during the spring bloom in Lake Erken, Sweden. Internationale Revue der gesamten Hydrobiologie 77: 517–551.

Forsius, M., J. Vuorenmaa, J. Mannio & S. Syri, 2003. Recovery from acidification of Finnish lakes: regional patterns and relations to emission reduction policy. Science of the Total Environment 310: 121–132.

Gerten, D. & R. Adrian, 2000. Climate-driven changes in spring plankton dynamics and the sensitivity of shallow polymictic lakes to the North Atlantic Oscillation. Limnology & Oceanography 45: 1058–1066.

Håkanson, L., 2005. The importance of lake morphometry and catchment characteristics in limnology – ranking based on statistical analyses. Hydrobiologia 541: 117–137.

Järvinen, M., M. Rask, J. Ruuhijärvi & L. Arvola, 2002. Temporal coherence in water temperature and chemistry under the ice of boreal lakes (Finland). Water Research 36: 3949–3956.

Kalff, J., 2002. Limnology. Prentice Hall. .

Leppäranta, M., A. Reinart, A. Erm, H. Arst, M. Hussainov & L. Sipelgas, 2003. Investigation of ice and water properties and under-ice light fields in fresh and brackish water bodies. Nordic Hydrology 34: 245–266.

Magnuson, J. J., B. J. Benson & T. K. Kratz, 1990. Temporal coherence in the limnology of a suite of lakes in Wisconsin, U.S.A. Freshwater Biology 23: 145–159.

Moldan, F., R. F. Wright, S. Löfgren, M. Forsius, T. Ruoho-Airola & B. L. Skjelkvale, 2001. Long-term changes in acidification and recovery at nine calibrated catchments in Norway, Sweden and Finland. Hydrology and Earth System Sciences 5: 339–349.

SAS Institute, 2002. JMP statistics and graphics guide. Version 5, SAS Institute.

Weyhenmeyer, G. A., 2001. Warmer winters – are planktonic algal populations in Sweden's largest lakes affected? Ambio 30: 565–571.

Weyhenmeyer, G. A., T. Blenckner & K. Pettersson, 1999. Changes of the plankton spring outburst related to the North Atlantic Oscillation. Limnology & Oceanography 44: 1788–1792.

Weyhenmeyer, G. A., M. Meili & D. M. Livingstone, 2004. Nonlinear temperature response of lake ice breakup. Geophysical Research Letters 31: L07203. doi: 10.1029/2004GL019530.

Weyhenmeyer, G. A., M. Meili & D. M. Livingstone, 2005. Systematic differences in the trend towards earlier ice-out on Swedish lakes along a latitudinal temperature gradient. Verhandlungen der Internationalen Vereinigung der Limnologie 29: 257–260.

Wilander, A. & G. Persson, 2001. Recovery from eutrophication: experiences of reduced phosphorus input to the four largest lakes of Sweden. Ambio 30: 475–485.

Hydrobiologia (2008) 599:111–118
DOI 10.1007/s10750-007-9188-9

ELLS 2007

Increasingly ice-free winters and their effects on water quality in Sweden's largest lakes

Gesa A. Weyhenmeyer · Anna-Karin Westöö ·
Eva Willén

Abstract Mean global air temperatures have steadily increased during recent decades, resulting in an earlier timing of lake ice breakup. In Sweden's largest lakes, Vänern and Vättern, the breakup of ice has occurred considerably earlier since 1979 and ice-free winters have become more frequent. Comparison between the years when the lakes were ice covered with those when they remained ice-free in terms of 37 lake variables revealed significant differences in water temperatures, sulphate concentrations and the biomass of diatoms in May after ice breakup ($P < 0.01$). In particular, the biomass of the genus *Aulacoseira* increased significantly, which may explain increasing complaints about algae that clog fishing-nets, filter-beds and micro-strainers in waterworks in Vänern and Vättern. We assume that *Aulacoseira* is mainly affected by changes in climate-driven water circulation patterns. In contrast, other observed water quality changes such as changes in sulphate concentration might rather be attributed to changes in atmospheric deposition. To explain water quality changes in Sweden's largest lakes it is important to consider changes in both climate and atmospheric deposition as well as catchment measures.

Keywords Climate · Atmospheric deposition · Ice · Water Chemistry · Phytoplankton

Guest editors: T. Nõges, R. Eckmann, K. Kangur, P. Nõges, A. Reinart, G. Roll, H. Simola and M. Viljanen
European Large Lakes—Ecosystem changes and their ecological and socioeconomic impacts

G. A. Weyhenmeyer (✉) · E. Willén
Department of Environmental Assessment, Swedish University of Agricultural Sciences (SLU), Box 7050, 750 07 Uppsala, Sweden
e-mail: Gesa.Weyhenmeyer@ma.slu.se

Present Address:
A.-K. Westöö
Department of Regions and Environment, Statistics Sweden, Box 24 300, 104 51 Stockholm, Sweden

Introduction

Regional climate models for Sweden simulate increasing air temperatures over the next decades, especially during winters (Rummukainen et al., 2001). Changes in winter air temperatures can be linked directly to changes in ice cover dynamics (e.g. Palecki & Barry, 1986; Vavrus et al., 1996; Livingstone, 1997). As a response to a 1.2°C air temperature increase, Magnuson et al. (2000) reported a 6.3-day earlier ice breakup, on average, over the northern hemisphere during the past 100 years. However, Weyhenmeyer et al. (2005) found systematic differences in the trend towards earlier ice breakup along a latitudinal temperature gradient. Taking Sweden as an example, an annual mean air temperature increase of 1°C caused up to a 35-day earlier ice breakup in the relatively warm southern part of Sweden while the same air temperature increase caused only a 4-day earlier ice breakup in the relatively cold northern part of the country (Weyhenmeyer et al., 2004). The reason for this difference is that the timing of lake ice breakup responds non-linearly to changes in air temperature (Weyhenmeyer et al., 2004). Sweden's largest lakes, Vänern and Vättern, are located in the southern part of Sweden, where considerable changes in the timing of lake ice breakup in response to increasing air temperatures can be expected. Since Sweden's largest lakes already remain occasionally ice-free during winter periods, the incidence of ice-free winters is likely to increase in future in response to air temperature increases as simulated for that region (Rummukainen et al., 2001). To quantify the effects of ice-free winters on water quality and phytoplankton development in the large lakes we used long-term physical, chemical and biological data collected since 1979. Studies on other lake systems have shown that ice cover has impacts on, for example, underwater light conditions (Leppäranta et al., 2003), nutrient recycling (Järvinen et al., 2002), oxygen conditions (Stewart, 1976; Livingstone, 1993), the production and biodiversity of phytoplankton (Rodhe, 1955; Weyhenmeyer et al., 1999; Phillips & Fawley, 2002) and the occurrence of winter fish kill (Greenbank, 1945; Barica & Mathias, 1979). We therefore hypothesize that water quality, phytoplankton composition and phytoplankton biomass in Sweden's largest lakes, Vänern and Vättern, are significantly different after ice-free winters than after winters that have been ice-covered.

Study sites and methods

For this study we used long-term data from Lakes Vänern and Vättern (Fig. 1). In terms of area, Vänern is one of the 30 largest lakes in the world (Table 1). The area of Vättern is only one-third of that of Vänern but because it is deep it has an unusually long water retention time of 58 years (Table 1). For more hydrological information about the large lakes, see Kvarnäs (2001).

Water samples have been taken frequently at three stations in Lake Vänern (since 1973) and at two stations in Lake Vättern (since 1966) (Fig. 1). Sampling has been conducted in the middle of each month during the main growing season from May to October (i.e. ice-free period). Water sampling and determination of all chemical and biological variables have been carried out by the same laboratory using standard limnological methods. Both lakes were subjected to drastic reductions in phosphorus loading from the catchment during the early 1970s. By the late 1970s the phosphorus concentrations had stabilized (Wilander & Persson, 2001). We therefore used data since 1979 on water temperature, 22 chemical variables and phytoplankton during May when the water column of the lakes is mixed (Table 2). For phytoplankton we compared the biomass changes of six groups. Since diatoms are the dominant phytoplankton group during May in both large lakes (Willén, 2001) we further investigated changes in the biomass of the diatom genera *Asterionella, Aulacoseira, Cyclotella, Diatoma, Stephanodiscus, Synedra* and *Tabellaria*. For water chemistry, surface (0.5 m) water samples were taken. Phytoplankton was collected with a tube sampler and represented the upper 0–8 m mixed water layer.

In addition to lake variables, we received monthly mean air temperature data from two sites from the Swedish Meteorological and Hydrological Institute (SMHI): Karlstad located at Lake Vänern and Jönköping at Lake Vättern (Fig. 1). For this study we used the mean monthly air temperatures from January to May. SMHI also provided us with ice cover data from Lakes Vänern and Vättern from 1979 to 2002. SMHI defines the date of ice-on when the main lake area is first covered by ice and the ice cover lasts for at least 3 days. The date of ice-off is defined when the main lake area is ice-free, implying that some small bays may still have an ice cover. When the lake freezes again, the last date when it opens is taken (Eklund, 1999). According to SMHI's definition, several years have been reported in which the main lake area did not freeze during a whole winter period. We consider these to be years when the lakes lack an ice cover. For winters when the lakes are ice covered we use the abbreviation W_{ice} and for winters when the lakes remain ice-free we use $W_{\text{ice-free}}$.

In our study, we compared physical, chemical and biological lake variables during the spring turnover in May after W_{ice} and $W_{\text{ice-free}}$. For this comparison we used the non-parametric Wilcoxon test. At least 5% of the Wilcoxon tests carried out for each lake must be regarded as significant ($P < 0.05$) by pure chance for stochastic reasons. Rice (1989) described

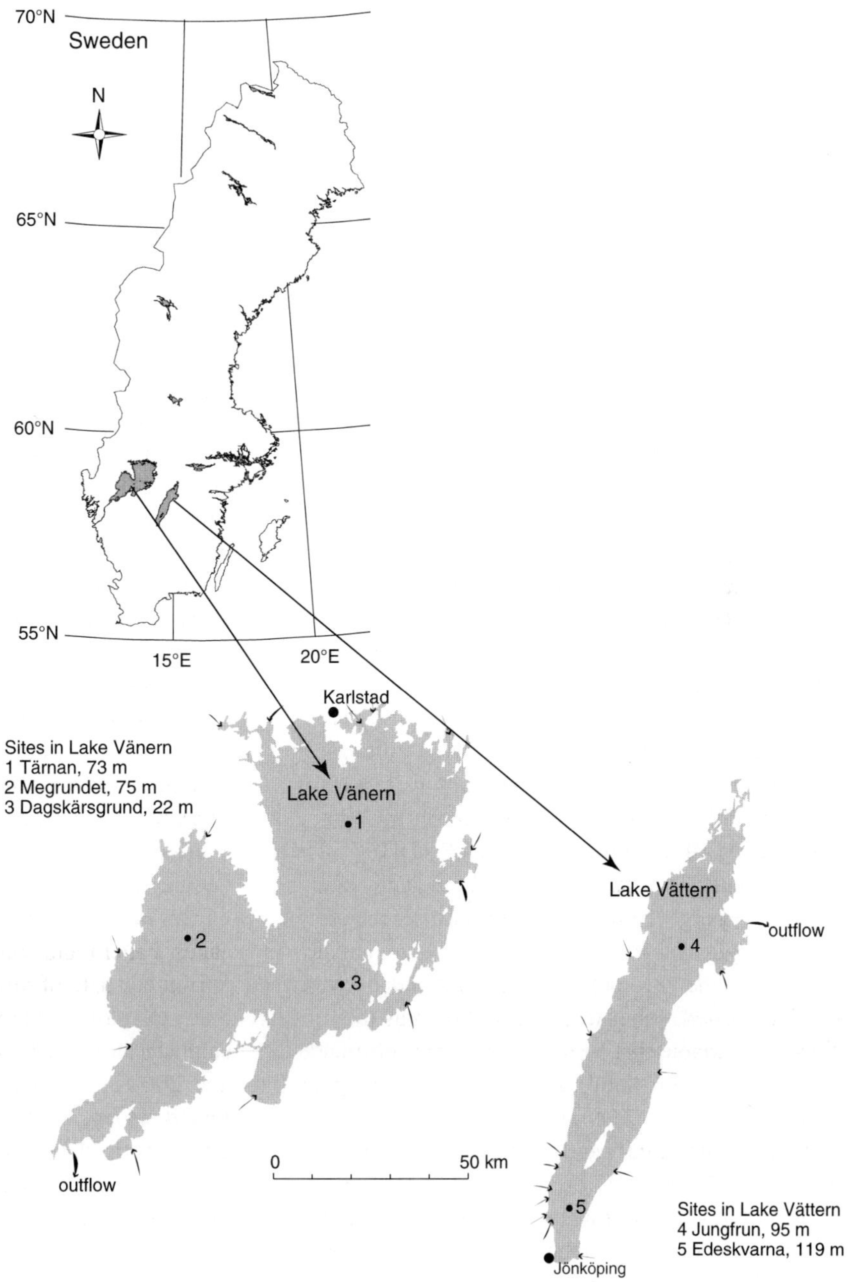

Fig. 1 Sampling sites in Lakes Vänern and Vättern, locations of the main inflowing and outflowing rivers (indicated by arrows) and the locations of the two meteorological stations at the lakes

a sequential Bonferroni correction procedure to control this problem. A Bonferroni correction in this study, however, would decrease the *P*-value to 0.001. To reach such low *P*-values with our available data is not reasonable, so we decided to consider *P*-values less than 0.01 as significant.

For the non-parametric Wilcoxon test we considered data from the different sites in the same lake to obtain results for the whole water column and to increase the number of data. All statistical tests were carried out using the JMP program, version 5.0.1.2 (SAS Institute, 2002).

Results

Within the 24-year period, 1979–2002, Vänern was ice-covered during nine and Vättern during eight

Table 1 Morphometric and hydrological data on Lake Vänern and Lake Vättern (after Kvarnäs, 2001)

	Vänern	Vättern
Area (km^2)	5,648	1,856
Volume (km^3)	153	74
Mean depth (m)	27	40
Maximum depth (m)	106	128
Theoretical turnover time (years)	10	58
Catchment area (km^2)	46,830	6,359

winters (Fig. 2). There was consistency between the two lakes in the occasions of ice cover except for 1990–1991 when Vänern was ice covered and Vättern was not (Fig. 2). From winter 1996–1997 onwards both lakes remained ice-free, indicating a positive trend in the incidence of ice-free winters between 1979 and 2002 (Fig. 3)

Comparing all years in which the lakes remained ice-free (15 for Vänern and 16 for Vättern; $W_{\text{ice-free}}$) with years in which the lakes were covered by ice (9 for Vänern and 8 for Vättern; W_{ice}), the mean monthly air temperatures at the lakes were significantly higher in January and February during $W_{\text{ice-free}}$ than during W_{ice} (Wilcoxon test, $P < 0.01$). The surface water temperatures in Vänern and Vättern were on average 2 and 1.3°C warmer, respectively, after $W_{\text{ice-free}}$ (Table 2).

Among the 23 physical and chemical water variables tested, sulphate concentration was the only chemical variable that showed significant differences after $W_{\text{ice-free}}$ in both lakes (Table 2). Sulphate concentrations in May were significantly lower in Vänern and significantly higher in Vättern after $W_{\text{ice-free}}$. In Vänern, $W_{\text{ice-free}}$ gave additional signals in the physical and chemical variables: Secchi depth, pH, nitrate-nitrogen, silica and chlorophyll *a* (Table 2).

Phytoplankton biomass also showed significant differences after $W_{\text{ice-free}}$. In Vänern, the biomasses of total phytoplankton, cyanobacteria, diatoms, cryptophytes and chlorophytes were significantly greater after $W_{\text{ice-free}}$ than after W_{ice} (Table 2). In Vättern, only the biomass of diatoms increased significantly after $W_{\text{ice-free}}$. Changes in the biomass of diatoms after $W_{\text{ice-free}}$ resulted in a different phytoplankton composition in both lakes, with diatoms out-competing other phytoplankton groups (Fig. 2).

Of the seven main diatom genera examined only the genus *Aulacoseira* showed a significantly different biomass after $W_{\text{ice-free}}$ (Table 2). *Aulacoseira* increased its biomass more than fourfold after $W_{\text{ice-free}}$. In both Vänern and Vättern the diatom species *Aulacoseira islandica* is dominant during the spring turnover.

Discussion

Our results indicate that surface water temperatures, Secchi depth, pH, concentrations of sulphate, nitrate-nitrogen (NO_3-N), silica and chlorophyll *a* and the phytoplankton biomass showed significant differences after ice-free winters ($W_{\text{ice-free}}$) in at least one of the lakes studied (Table 2). Increases in surface water temperatures after $W_{\text{ice-free}}$ can easily be attributed to global warming (reviewed in, e.g. Gerten & Adrian, 2002). Global warming, however, may be reflected in the positive trend in occurrence of $W_{\text{ice-free}}$ between 1979 and 2002. This positive trend causes difficulties in interpreting our results since all variables showing a strong upward or downward trend will appear significantly different after $W_{\text{ice-free}}$. Well-known climate-independent trends include the decline in sulphate concentrations in various Scandinavian lake ecosystems since the 1990s (e.g. Moldan et al., 2001; Forsius et al., 2003). These trends can be attributed to decreasing atmospheric deposition rather than to global warming. We observed declining sulphate concentrations in Vänern between 1979 and 2002. In Vättern, sulphate concentrations continued to increase over this period even though atmospheric deposition had decreased. Probably the very long water turnover time of 58 years causes a time lag of several years or even decades before the effects of decreasing atmospheric deposition can be detected in the water body of Lake Vättern.

Many results of this study might be attributed to decreasing atmospheric deposition; examples are increasing pH and decreasing NO_3-N concentrations. However, most lake variables are probably influenced by both atmospheric deposition and global warming. According to Weyhenmeyer (2001, 2004) a warmer winter climate, in this study defined as $W_{\text{ice-free}}$, can lead to an increase in total phytoplankton biomass in May, resulting in an increased uptake of the bioavailable nutrients, NO_3-N and silica. We suggest

Table 2 List of lake variables in surface waters/upper mixed layer used in this study and comparison of the median values of the variables between years with (W_{ice}) and without ($W_{ice\text{-}free}$) ice cover (Fig. 2)

Variable	Lake Vänern		Lake Vättern	
	W_{ice}	$W_{ice\text{-}free}$	W_{ice}	$W_{ice\text{-}free}$
Water temperature (°C)	**3.2**	**5.2**	**3.4**	**4.7**
Oxygen (mg l^{-1})	13.1	12.9	13.2	12.6
Secchi depth (m)	**5.0**	**4.5**	10.8	11.0
Sulphate (mEq l^{-1})	**0.32**	**0.29**	**0.36**	**0.38**
pH	**7.2**	**7.3**	7.7	7.7
Alkalinity (mEq l^{-1})	0.2	0.2	0.5	0.5
Conductivity (mS l^{-1})	9.4	9.1	13.4	13.5
Calcium (mEq l^{-1})	0.4	0.4	0.7	0.7
Magnesium (mEq l^{-1})	0.1	0.1	0.2	0.2
Sodium (mEq l^{-1})	0.3	0.3	0.3	0.3
Potassium (mEq l^{-1})	0.03	0.03	0.04	0.04
Chloride (mEq l^{-1})	0.2	0.2	0.3	0.3
Ammonium-nitrogen (μg l^{-1})	4	7	6	7
Nitrate-nitrogen[a] (μg l^{-1})	**580**	**550**	453	496
Total nitrogen (μg l^{-1})	861	838	688	715
Phosphate-phosphorus (μg l^{-1})	2	1	1	1
Other phosphorus (μg l^{-1})	7	6	4	4
Total phosphorus (μg l^{-1})	8	8	6	5
Turbidity[b] (m^{-1})	0.01	0.01	0.005	0.004
Water colour[c] (m^{-1})	0.04	0.04	0.008	0.009
Potassium permanganate (mg l^{-1})	18.8	17.6	7.3	6.7
Reactive silica (mg l^{-1})	**0.35**	**0.25**	0.21	0.25
Chlorophyll *a* (μg l^{-1})	**1.0**	**2.5**	1.1	1.1
Total phytoplankton biomass (mg l^{-1})	**0.11**	**0.37**	0.06	0.10
Cyanobacteria (mg l^{-1})	**0.000**	**0.008**	0.000	0.000
Diatoms (mg l^{-1})	**0.06**	**0.26**	**0.03**	**0.05**
Asterionella (mg l^{-1})	0.002	0.003	0.007	0.004
Aulacoseira (mg l^{-1})	0.019	0.080	0.002	0.009
Cyclotella (mg l^{-1})	0.011	0.008	0.006	0.004
Diatoma (mg l^{-1})	0.003	0.002	0.003	0.001
Stephanodiscus (mg l^{-1})	0.003	0.002	0.003	0.001
Synedra (mg l^{-1})	0.0003	0.0001	0.002	0.001
Tabellaria (mg l^{-1})	0.004	0.006	0.002	0.002
Cryptophytes (mg l^{-1})	**0.016**	**0.038**	0.008	0.012
Chrysophytes (mg l^{-1})	0.011	0.014	0.013	0.014
Chlorophytes (mg l^{-1})	**0.000**	**0.004**	0.004	0.007
Dinoflagellates (mg l^{-1})	0.000	0.001	0.001	0.002

Variables that showed significant differences between W_{ice} and $W_{ice\text{-}free}$ (Wilcoxon-test, $P < 0.01$) are shown in bold

[a] Nitrate-nitrogen includes nitrite-nitrogen, which is less than 5% of nitrate- + nitrite-nitrogen (Wilander, pers. comm.)

[b] Turbidity was estimated as the difference between light absorption at 420 nm of unfiltered and filtered (0.45 μm) water in a 5-cm cuvette

[c] Water colour was estimated as light absorption at 420 nm of 0.45 μm filtered water in a 5-cm cuvette

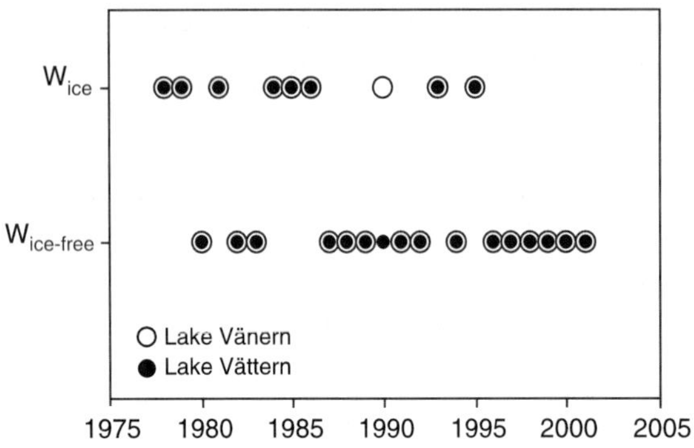

Fig. 2 Winters with (W_{ice}) and without ($W_{ice\text{-}free}$) an ice cover on the Lakes Vänern and Vättern

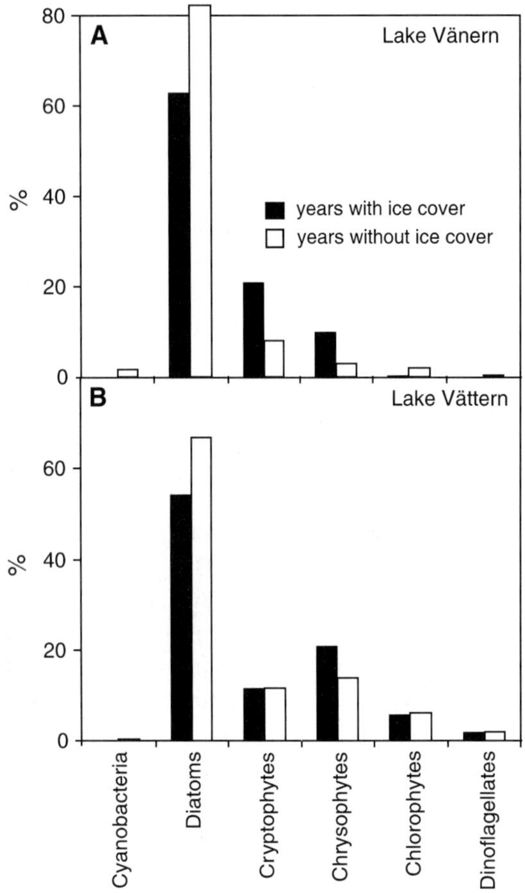

Fig. 3 Percentage contribution of different taxonomic groups to the total phytoplankton biomass in Lakes Vänern and Vättern in May during ice-covered and ice-free years (median values of the different years were taken)

that such a global warming effect can be seen in Vänern where the total phytoplankton biomass and chlorophyll *a* concentrations were significantly higher, and NO_3-N and silica concentrations significantly lower, after $W_{ice\text{-}free}$ (Table 2). In Vättern, changes in the biomass and composition of the spring phytoplankton bloom after $W_{ice\text{-}free}$ were less apparent than in Vänern. One reason might be that chemical changes in the water were also less pronounced; another might be a stronger nutrient limitation in Vättern, implying that warmer waters could not be efficiently employed for the growth of phytoplankton other than diatoms. Diatoms are sensitive to physical conditions such as the presence of ice. Gerten & Adrian (2000) found, for example, that diatom biomass increased after winters with short ice-cover periods or ice-free winters, and Lotter & Bigler (2000) detected a relationship between the timing of lake ice breakup and diatom blooms as determined by sediment deposits in Swiss lakes. In both Vänern and Vättern we observed an increase in the biomass of the genus *Aulacoseira* after $W_{ice\text{-}free}$. *Aulacoseira* belongs to the genera that are frequently found in large lakes with deep-mixing such as Ladoga, Onega, Baikal, Michigan and several of the large Rift Valley lakes (Willén, 1991). *Aulacoseira* needs water mixing to develop in the water column, which occurs when lakes are not covered by ice. The dominant diatom species *Aulacoseira islandica* in Vänern and Vättern is a good competitor at water temperatures of about 5°C (Willén, 1991), which corresponds to the May water temperature after $W_{ice\text{-}free}$ (Table 2). Water mixing together with suitable water temperatures and sufficient nutrient availability in May after ice-free winters strongly favour the development of *Aulacoseira* in Vänern and Vättern.

A significantly increased *Aulacoseira* biomass in Vänern and Vättern after $W_{ice\text{-}free}$ may have far-reaching consequences. Complaints about algae that clog fishing-nets, filter-beds and micro-strainers in waterworks are already common. It is most likely that these problems are caused by the genus *Aulacoseira*, which is known to have substantial clogging abilities (Willén, 2001). The clogging problems may become cost-intensive if fish catches are reduced and boats, bridges, pipes, etc. require additional maintenance. However, an increase in the diatom biomass might also have positive effects on lake ecosystems because diatoms are a good food resource for the benthic fauna, as shown by a high positive relationship between diatom biomass and the growth of the benthic fauna with a time lag of 1 year (Johnson &

Wiederholm, 1992). Further possible effects of an increase in the *Aulacoseira* biomass on lake ecosystems remain to be determined.

According to this study, increases in water temperatures and in the *Aulacoseira* biomass in spring can be expected as a direct effect of warmer winters. Indirect effects that function via the catchment may need some time to become apparent in the water column, depending on the water residence time of the lake. Further studies are particularly needed to quantify the indirect effects of ice-free winters on lake ecosystems with different water residence times.

Acknowledgements This work was partly funded by the European Union within the framework of the European Commission projects CLIME ("Climate and Lake Impacts in Europe", EVK1-CT-2002-00121) and Euro-limpacs ("Integrated Project to Evaluate the Impacts of Global Change on European Freshwater Ecosystems", GOCE-CT-2003-505540). The first author is a research fellow of the Royal Swedish Academy of Sciences supported by a grant from the Knut and Alice Wallenberg Foundation. Additional funding was received from the Swedish Research Council (621-2005-4335). We thank the Swedish Environmental Protection Agency and the IMA laboratory for financing, sampling and analysing numerous water samples.

References

Barica, J. & J. A. Mathias, 1979. Oxygen depletion and winterkill risk in small Prairie lakes under extended ice cover. Journal of Fisheries Research Board Canada 36: 980–986.

Eklund, A., 1999. Ice-on and ice-off in Swedish lakes. SMHI-Report Hydrologi 81 (in Swedish).

Forsius, M., J. Vuorenmaa, J. Mannio & S. Syri, 2003. Recovery from acidification of Finnish lakes: regional patterns and relations to emission reduction policy. Science of the Total Environment 310: 121–132.

Gerten, D. & R. Adrian, 2000. Climate-driven changes in spring plankton dynamics and the sensitivity of shallow polymictic lakes to the North Atlantic Oscillation. Limnology and Oceanography 45: 1058–1066.

Gerten, D. & R. Adrian, 2002. Effects of climate warming, North Atlantic Oscillation, and El Nino-Southern Oscillation on thermal conditions and plankton dynamics in northern hemispheric lakes. The Scientific World 2: 586–606.

Greenbank, J., 1945. Limnological conditions in ice-covered lakes, especially related to winterkill of fish. Ecological Monographs 15: 343–392.

Järvinen, M., M. Rask, J. Ruuhijärvi & L. Arvola, 2002. Temporal coherence in water temperature and chemistry under the ice of boreal lakes (Finland). Water Research 36: 3949–3956.

Johnson, R. K. & T. Wiederholm, 1992. Pelagic-benthic coupling – the importance of diatom interannual variability for population oscillations of *Monoporeia affinis*. Limnology and Oceanography 37: 1596–1607.

Kvarnäs, H., 2001. Morphometry and hydrology of the four large lakes of Sweden. Ambio 30: 467–474.

Leppäranta, M., A. Reinart, A. Erm, H. Arst, M. Hussainov & L. Sipelgas, 2003. Investigation of ice and water properties and under-ice light fields in fresh and brackish water bodies. Nordic Hydrology 34: 245–266.

Livingstone, D. M., 1993. Lake oxygenation: Application of a one-box model with ice cover. International Revue der gesamten Hydrobiologie 78: 465–480.

Livingstone, D. M., 1997. Break-up dates of alpine lakes as proxy data for local and regional mean surface air temperatures. Climatic Change 37: 407–439.

Lotter, A. F. & C. Bigler, 2000. Do diatoms in the Swiss Alps reflect the length of ice-cover? Aquatic Sciences 62: 125–141.

Magnuson, J. J., D. M. Robertson, B. J. Benson, R. H. Wynne, D. M. Livingstone, T. Arai, R. A. Assel, R. G. Barry, V. Card, E. Kuusisto, N. G. Granin, T. D. Prowse, K. M. Stewart & V. S. Vuglinski, 2000. Historical trends in lake and river ice cover in the Northern Hemisphere. Science 289: 1743–1746.

Moldan, F., R. F. Wright, S. Löfgren, M. Forsius, T. Ruoho-Airola & B. L. Skjelkvale, 2001. Long-term changes in acidification and recovery at nine calibrated catchments in Norway, Sweden and Finland. Hydrology and Earth System Sciences 5: 339–349.

Palecki, M. A. & R. G. Barry, 1986. Freeze-up and break-up of lakes as an index of temperature changes during the transition seasons: A case study for Finland. Journal of Climate and Applied Meteorology 25: 893–902.

Phillips, K. A. & M. W. Fawley, 2002. Winter phytoplankton community structure in three shallow temperate lakes during ice cover. Hydrobiologia 470: 97–113.

Rice, W. R., 1989. Analyzing tables of statistical tests. Evolution 43: 223–225.

Rodhe, W., 1955. Can phytoplankton production proceed during winter darkness in subarctic lakes? Verhandlungen der Internationalen Vereinigung für Limnologie 12: 117–122.

Rummukainen, M., J. Räisänen, B. Bringfelt, A. Ullerstig, A. Omstedt, U. Willén, U. Hansson & C. Jones, 2001. A regional climate model for northeastern Europe: model description and results from the downscaling of two GCM control simulations. Climate Dynamics 17: 339–359.

SAS Institute, 2002. JMP Statistics and graphics guide. Version 5, SAS Institute.

Stewart, K. M., 1976. Oxygen deficits, clarity and eutrophication in some Madison lakes. Internationale Revue der gesamten Hydrobiologie 61: 563–579.

Vavrus, S. J., R. H. Wynne & J. A. Foley, 1996. Measuring the sensitivity of southern Wisconsin lake ice to climate variations and lake depth using a numerical model. Limnology & Oceanography 41: 822–831.

Weyhenmeyer, G. A., 2001. Warmer winters – are planktonic algal populations in Sweden's largest lakes affected? Ambio 30: 565–571.

Weyhenmeyer, G. A., 2004. Synchrony in relationships between the North Atlantic Oscillation and water chemistry among Sweden's largest lakes. Limnology & Oceanography 49: 1191–1201.

Weyhenmeyer, G. A., T. Blenckner & K. Pettersson, 1999. Changes of the plankton spring outburst related to the North Atlantic Oscillation. Limnology & Oceanography 44: 1788–1792.

Weyhenmeyer, G. A., M. Meili & D. M. Livingstone, 2004. Nonlinear temperature response of lake ice breakup. Geophysical Research Letters 31: L07203, doi: 10.1029/2004GL019530.

Weyhenmeyer, G. A., M. Meili & D. M. Livingstone, 2005. Systematic differences in the trend towards earlier ice-out on Swedish lakes along a latitudinal temperature gradient. Verhandlungen der Internationalen Vereinigung der Limnologie 29: 257–260.

Wilander, A. & G. Persson, 2001. Recovery from eutrophication: Experiences of reduced phosphorus input to the four largest lakes of Sweden. Ambio 30: 475–485.

Willén, E., 1991. Planktonic diatoms – an ecological review. Algological Studies 62: 69–106.

Willén, E., 2001. Four decades of research on the Swedish large lakes Mälaren, Hjälmaren, Vättern and Vänern: The significance of monitoring and remedial measures for a sustainable society. Ambio 30: 458–466.

Hydrobiologia (2008) 599:119–125
DOI 10.1007/s10750-007-9185-z

Phosphorus fractions and alkaline phosphatase activity in sediments of a large eutrophic Chinese lake (Lake Taihu)

Yiyong Zhou · Chunlei Song · Xiuyun Cao · Jianqiu Li · Guoyuan Chen · Zhuoying Xia · Pinghong Jiang

Abstract Spatial, vertical, and seasonal variations in phosphorus fractions and in alkaline phosphatase activity (APA) were investigated in sediments in a large-shallow eutrophic Chinese lake (Lake Taihu) in 2003–2004. The phosphorus content was highest in the most seriously polluted lake area. Iron-bound phosphorus (Fe(OOH) ~ P) dominated (47% on average) among the phosphorus fractions determined according to Golterman (Hydrobiologia 335:87–95, 1996). Notably, organically-bound P comprised a further significant additional portion (acid-soluble + hot NaOH-extractable organic P = 25%), which was highest at the most polluted sites. The Fe(OOH) ~ P content was the lowest in spring (April, 2004), suggesting that degradation of organic matter led to the release of iron-bound phosphates. Sediment APA showed a significant positive relationship with both organically-bound P and Fe(OOH) ~ P. Consequently, organically-bound P is an important portion of the sediment phosphorus in Lake Taihu. It is mainly derived from freshly-settled autochthonous particles and from external discharges. Organically-bound P induces APA and may lead to the release of bioavailable phosphates from the organic sediments, thereby accelerating lake eutrophication.

Keywords P fractionation · Alkaline phosphatase · Kinetics · Distributions · Sediments · Interstitial water · Dredging · Lake eutrophication

Introduction

Eutrophication appears to be a serious environmental problem in a large-shallow Chinese lake (Lake Taihu) in which the concentrations of nitrogen, phosphorus (P), organic pollutants, and water blooms are all increasing at an accelerating rate (Fan et al., 1997). Sediment serves as an important P reservoir, release from which depends on the composition of phosphorus forms (Zhang et al., 2001). Much attention has been paid to the release of inorganically-bound phosphates, but there have been fewer studies on the fate of organically-bound phosphorus. Hua et al. (2000) reported that organic phosphorus is more readily released from the sediment of Lake Taihu and is quickly bioavailable to algae.

Guest editors: T. Nõges, R. Eckmann, K. Kangur, P. Nõges, A. Reinart, G. Roll, H. Simola & M. Viljanen
European Large Lakes—Ecosystem changes and their ecological and socioeconomic impacts

Y. Zhou (✉) · C. Song · X. Cao · J. Li · G. Chen · Z. Xia · P. Jiang
Institute of Hydrobiology, The Chinese Academy of Sciences, Wuhan 430072, P.R. China
e-mail: zhouyy@ihb.ac.cn

G. Chen · Z. Xia · P. Jiang
Graduate School of the Chinese Academy of Sciences, Beijing 100039, P.R. China

Phosphatase (APA) plays an important role in the hydrolysis of organic phosphorus and consequent phosphate release. In six Finnish lakes with different trophic states, phosphatase activity and levels of organic substances were highest in the top layers of all sediment profiles (Matinvesi & Heinonen-Tanski, 1992), suggesting that microbiological phosphate mineralization from organic substances could determine the internal phosphorus load in those lakes. Phosphatase is an important factor in accelerating eutrophication in Lake Taihu (Song et al., 2006). In this study, the spatial and vertical distributions of P fractions and APA in sediments were investigated during different seasons in Lake Taihu, to address the following issues that have previously remained unclear: (1) Is organically-bound P an important phosphorus component in sediments of Lake Taihu? (2) Does extracellular phosphatase catalyzing the liberation of orthophosphate from organic phosphorus compounds play an important role in phosphorus cycling in Lake Taihu?

Materials and methods

Lake Taihu is situated in the Yangtze River delta (30°05′–32°08′N, 119°08′-121°55′E; mean depth 2.0 m; surface area 2338 km^2), with Wuli Bay being the most polluted area, surrounded by Wuxi City. It functions as a drinking water source, but also as a receiver of untreated domestic sewage (Chen et al., 2003). In 2002–2004, dredging was carried out to remove polluted sediment, using an IHC Holland suction hopper dredger (more details are given in the legend to Fig. 1). Six sites (Sites 1–6) were sampled thrice (in October 2002, January and July 2003) to determine phosphorus fractions and APA, using a 3.5-cm-diameter hand-driven corer. Three or four cores 30 cm in depth was taken from each site. Columns were sliced at 5 cm intervals and the corresponding layers from each core were mixed thoroughly. The samples were stored at 4°C in the dark for 2–4 days before analysis. Each mixed sample was analyzed in triplicate. In 2004, seasonal

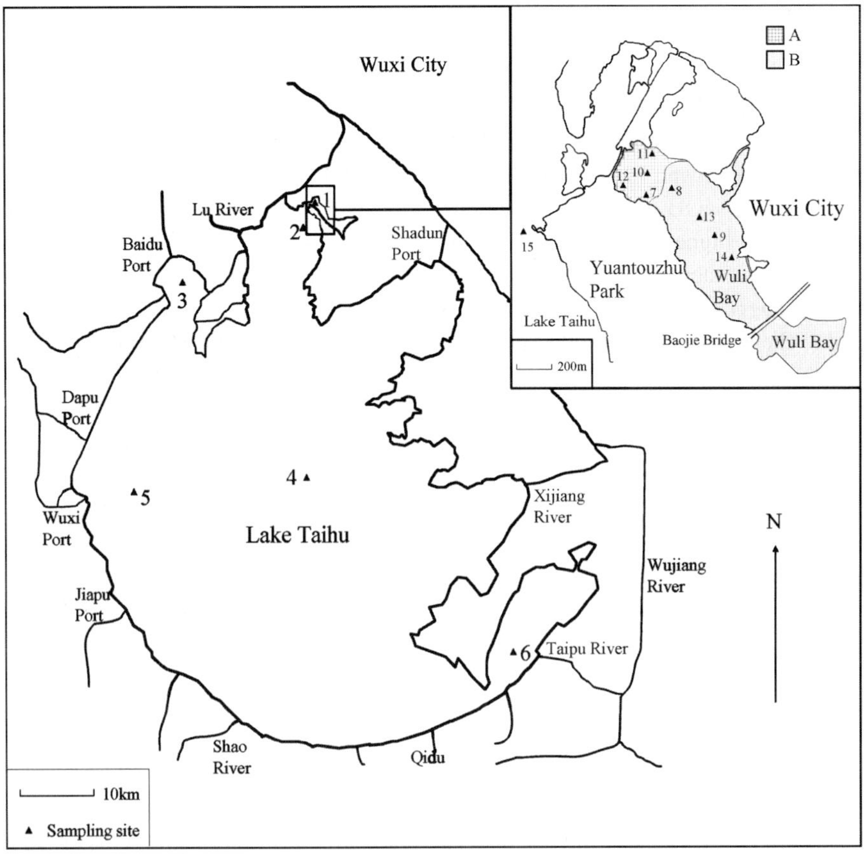

Fig. 1 A map of Lake Taihu showing the sampling sites. Almost the whole of Wuli Bay (Area B) with a total area of 8 km^2 was dredged in 2002–2003. From November 2003 to February 2004 the remaining part in the bay with a total area of 0.2 km^2 (Area A) was dredged. The dredging depth was 0.5 m on average

samples were collected for P fractionation by a Peterson dredge (Buresh & Patrick, 1981) from the area of sediment removal (Sites 7–14) and from Site 15 where sediment was not removed (Fig. 1).

Sediment P fractionation was carried out according to Golterman (1996). This method groups sediment P into iron-bound P (Fe(OOH)$\sim$P), calcium-bound P ($CaCO_3\sim$P), acid-soluble organic P (ASOP) and hot NaOH-extractable organic P (P_{alk}). Later in this article, the sum of ASOP and P_{alk} is considered to represent the organically-bound P.

APA was assayed by p-nitrophenylphosphate (pNPP), which is hydrolyzed by the enzyme to yield p-nitrophenol; with this system, enzyme activity is indicated by an increase in light absorbance (Sayler et al., 1979). Sediment samples were suspended in Tris buffer (pH 8.9). The pNPP was added to slurries at six final concentrations ranging from 0.0625 to 1.0 mmol l^{-1}. Samples were incubated at 37°C for 1 h, then 1.6 ml of slurry were centrifuged for 10 min at 3000 rpm. Supernatant (1 ml) was mixed with 4 ml 0.1 M NaOH to stop the reaction. The absorbance of the final solution was measured at 400 nm using a spectrophotometer (Model 721) made by the Shanghai Third Factory of Analytical Instrument, China. The pNPP was added to the reagent blanks after the NaOH. APA was converted to absolute units using a standard curve based on enzymatically hydrolyzed p-nitrophenol. V_{max} and K_m values were estimated by fitting the linearized Michaelis-Menten equation (Dick & Tabatabai, 1984), and the Lineweave-Burk plot was used (Zhou et al., 2001).

Results

For the samples collected in 2004, Fe(OOH)$\sim$P made the largest contribution to the sediment P pool, accounting for 22–67% of the sum of extracted fractions (47% on average), while $CaCO_3\sim$P

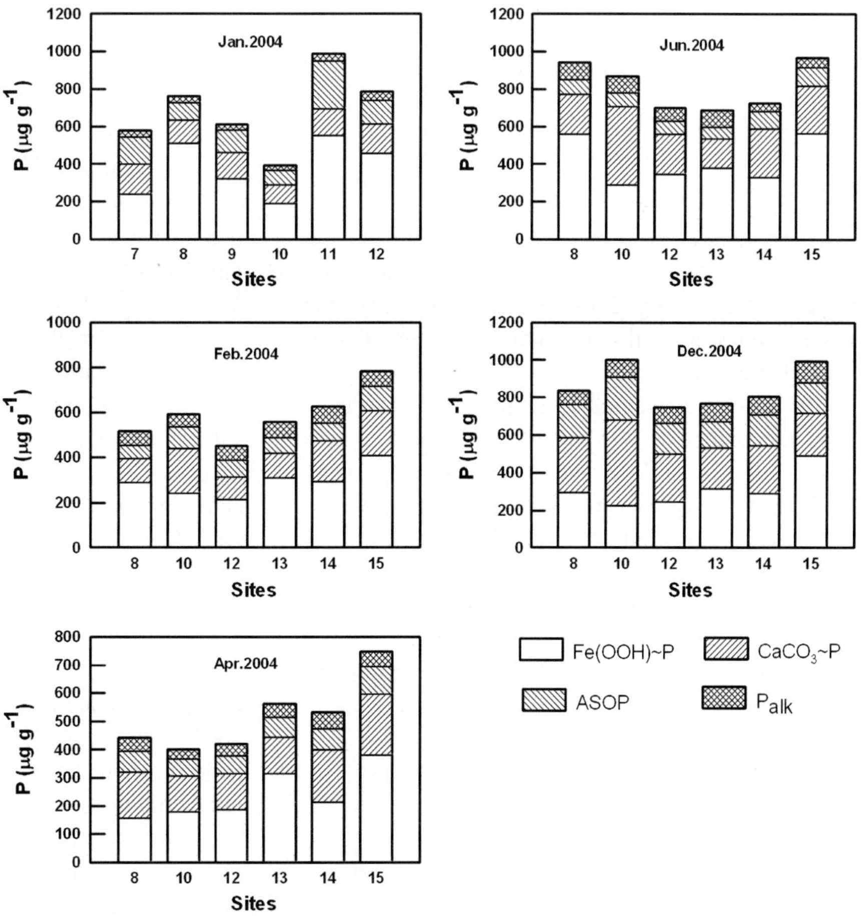

Fig. 2 Distribution of P fractions in surface sediments of Lake Taihu at different sites and in different seasons (the standard errors for each fraction were generally <5%)

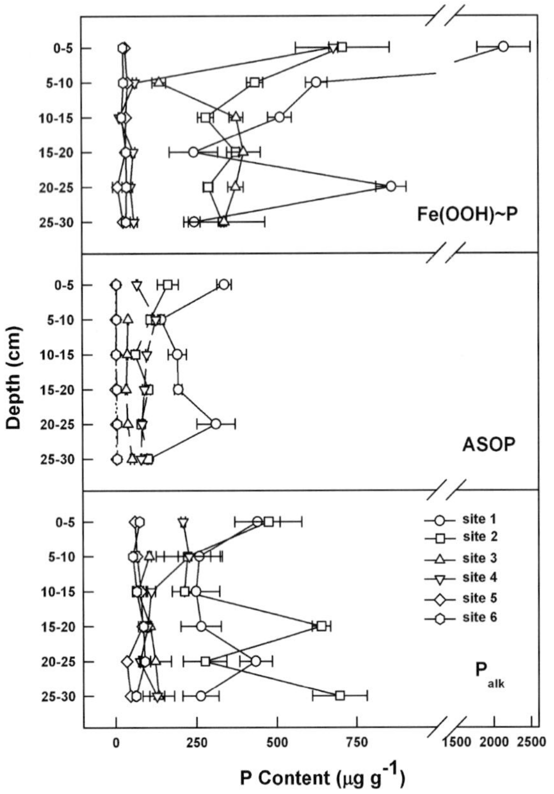

Fig. 3 Depth distribution of P fractions in sediment of Lake Taihu

averaged 28%. ASOP and P_{alk} were estimated on average at 16% and 9% of the total extracted P, respectively. During 2004, Fe(OOH)~P had the lowest values in April in the sediments from all sites studied (Fig. 2).

The highest concentration of phosphate extracted from the sediments was recorded at Site 1. It was greatest at the surface and decreased with depth (Fig. 3). This was a general distribution pattern in all seasons. Site 1 and the surficial sediment layer (0–5 cm) of site 2 showed markedly higher V_{max} values of APA (Table 1). There were significant correlations between the V_{max} values and P species (Fe(OOH)~P, ASOP and P_{alk}) in the sediments ($P < 0.01$, Fig. 4).

Discussion

Iron-bound phosphate accounted for most of the phosphorus extracted from the sediment of Lake Taihu (Fig. 2). Notably, the organically-bound P comprised a significant additional portion (ASOP + P_{alk} = 25%). It has two major sources, autochthonous and allochthonous. First, the concentrations of organically-bound P were always highest in the 0–5 cm sediment layer (Fig. 3), presumably reflecting the contribution of freshly-settled autochthonous organic particles. Furthermore, in 2004, ASOP concentrations in the intact sediment at the undredged site 15 were higher than those in the dredged sites 7–14, where the newly-formed surface had been a deeper layer prior to dredging (Fig. 2). Second, Site 1 generally showed the highest concentrations of organically-bound P in sediments at different depths (Fig. 3), attributable to the input of untreated domestic sewage from the large city of Wuxi (Qu et al., 2001).

Most of the organically-bound phosphorus in sediments is remineralized by springtime. This has been shown in the upper 10–20 cm of sediments from 32 shallow meso- to hyper-trophic Danish lakes (Sondergaard et al., 1996). In addition, in the Mejean Lagoon in France, the springtime rise in temperature reactivated the decomposition of the macroalgal biomass that had accumulated during autumn. The redox potential fell as a result of this biological activity, leading to the leakage of inorganically-bound P (Gomez et al., 1998). In our results, the content of Fe(OOH)~P was also lowest in spring (April, 2004; Fig. 2), suggesting that the degradation of organic matter led to phosphate release.

Extracellular phosphatase plays an important role in the degradation of organic matter in sediments. Our most eutrophic Site 1 generally had the highest values especially at the surface (Table 1). In the shallow eutrophic Chinese Lake Donghu, the most eutrophic basin also showed the highest sediment APA (Zhou et al., 2002). As shown in Fig. 4, there was a significant positive relationship between APA and organically-bound P (ASOP and P_{alk}). This may be explained by a mechanism of phosphatase induction by organic matter, as the total organic carbon content correlated significantly with APA (Su et al., 2005). The coincidence of APA with phosphatase-hydrolyzable phosphorus was also observed in the interstitial water of Lake Donghu (Zhou et al., 2002). Increased alkaline phosphatase V_{max} values were measured in the sediments, which were mainly formed from organic wastes of fish (*Oreochromis*

Table 1 Distribution of V_{max} values (μmol g^{-1} h^{-1}) of APA in sediments of Lake Taihu

Depth (cm)	Sites (October 2002)					
	1	2	3	4	5	6
0–5	963.43	1170.99	483.31		328.85	80.99
	247.86	*93.76*	*10.96*		*71.70*	*26.90*
5–10	1065.38	1037.29	250.87		37.04	62.75
	73.09	*201.85*	*72.80*		*16.92*	*39.52*
10–15	687.11	927.87	262.30	205.90	78.03	
	87.65	*28.88*	*30.91*	*22.72*	*9.71*	
15–20	688.88	526.54	169.16	164.17	93.41	115.65
	14.63	*134.34*	*36.74*	*35.65*	*12.11*	*36.96*
20–25	554.16	510.84	223.10		10.36	126.15
	138.90	*57.43*	*14.61*		*32.64*	*24.03*
25–30	581.76	795.38	401.55		71.56	9.21
	2.91	*75.12*	*35.31*		*3.33*	*8.63*
Depth (cm)	Sites (January 2003)					
	1	2	3	4	5	6
0–5	775.68	1308.63	481.12	260.85	494.87	237.08
	5.56	*59.19*	*40.04*	*45.58*	*25.49*	*18.89*
5–10	845.67	852.95	395.36		97.01	285.50
	75.81	*53.90*	*66.74*		*33.57*	*60.06*
10–15	659.97	284.94	218.85		205.93	119.76
	6.99	*76.70*	*40.63*		*20.56*	*32.67*
15–20	707.27	416.83	226.54	109.76	143.94	78.08
	109.60	*95.45*	*22.01*	*34.80*	*9.32*	*15.62*
20–25	530.76	315.22	31.47	99.72	332.47	23.59
	98.80	*44.52*	*16.50*	*24.86*	*34.58*	*6.63*
25–30	780.45	382.51	124.31	40.54	335.56	
	114.89	*41.54*	*21.65*	*21.61*	*2.89*	
Depth (cm)	Sites (July 2003)					
	1	2	3	4	5	6
0–5	426.28	557.58	612.57	475.28	472.16	127.57
	118.20	*59.85*	*100.34*	*60.88*	*3.64*	*27.79*
5–10	829.15	594.48	1604.53	157.04	283.54	167.89
	79.95	*128.47*	*210.35*	*6.34*	*24.15*	*45.89*
10–15	676.55	745.88	579.68	316.65	37.56	82.76
	162.66	*37.45*	*15.55*	*41.82*	*37.66*	*39.35*
15–20	942.78	249.31	609.40		103.06	13.84
	40.05	*76.79*	*15.02*		*26.24*	*25.76*
20–25	1101.12	223.61	254.76		70.52	104.14
	55.79	*29.75*	*21.12*		*25.62*	*14.28*
25–30	928.11	368.65	72.99		139.14	17.12
	113.44	*38.66*	*36.12*		*13.18*	*8.48*

Standard errors are given in italics

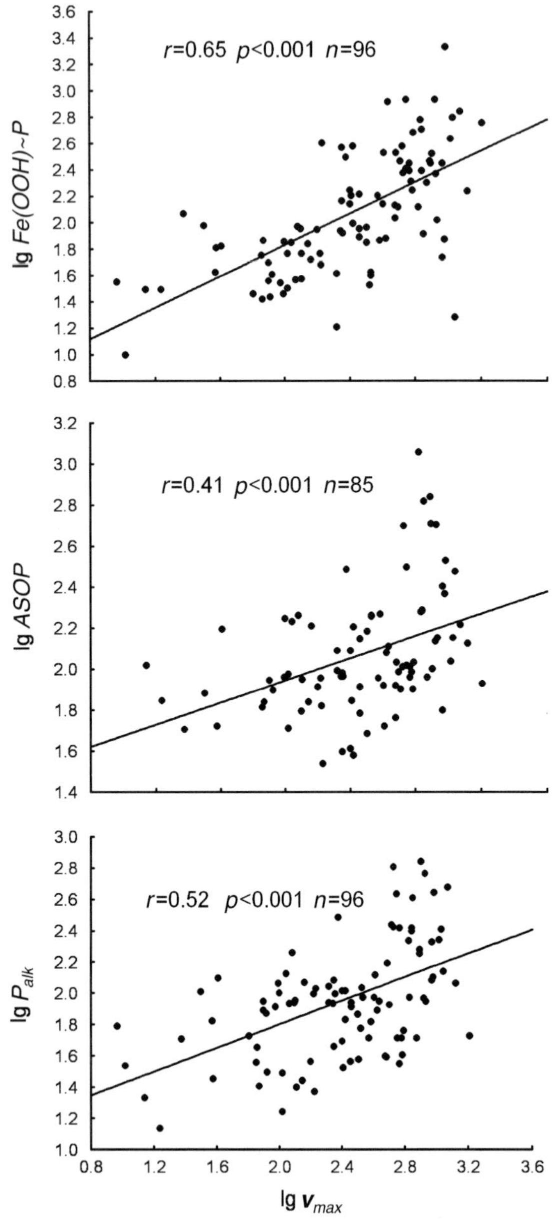

Fig. 4 Positive relationship between V_{max} of APA and P content of organically-bound fractions in sediments of Lake Taihu

niloticus) culture (Zhou et al., 2001). In a reservoir in Morocco, most of the total APA was produced by bacteria that were attached to organic matter (Mhamdi et al., 2003). All this information supports the idea that the activity of extracellular phosphatases is induced by an increased supply of organic matter rather than by a deficiency of phosphates. Nevertheless, the increased activity of alkaline phosphatase(s) enhances the availability of phosphates from both organically-and inorganically-bound sources, since there was also a significant positive relationship between APA and Fe(OOH)~P (Fig. 4), which is considered to be the most mobile fraction in the sediments (Taoufik et al., 2005).

In conclusion, organically-bound P constitutes an important portion of the sediment phosphorus in the large-shallow eutrophic Lake Taihu. It is mainly derived from freshly-settled autochthonous particles and from external discharges. Organically-bound P induces phosphatase activity and may lead to the release of bioavailable phosphate from the organic sediments, thereby accelerating lake eutrophication. The increased APA is induced to degrade the large amounts of organic matter rather than to compensate P deficiency.

Acknowledgments This work was supported by the National Basic Research Program of China (2002CB412304), through the grants KZCX2-YW-426 and 2002AA601013. The authors also gratefully acknowledge the support of the Mr. Li Lin, Zeng Hui, Chen Wei, Liu Bingqin and Peng Liang, and of Taihu Laboratory for Lake Ecosystem Research of the Chinese Academy of Sciences in sample collection, and of Mr. Liu Bingqin in sample analysis.

References

Buresh, R. J. & W. H. Patrick Jr., 1981. Nitrate reduction to ammonium and organic nitrogen in an estuarine sediment. Soil Biology and Biochemistry 13: 279–283.

Chen, Y., C. Fan, K. Teubner & M. Dokulil, 2003. Changes of nutrients and phytoplankton chlorophyll-a in a large shallow lake, Taihu, China: an 8-year investigation. Hydrobiologia 506–509: 273–279.

Dick, W. A. & M. A. Tabatabai, 1984. Kinetic parameters of phosphatases in soils and organic waste materials. Soils Science 137: 7–15.

Fan, C., J. Jiang, W. Zhang, Q. Wu, K. Chen & Y. Chen, 1997. Comprehensive evaluation and preliminary prediction for water quality and eutrophication of Gonghu Bay. Transactions of Oceanology and Limnology/Haiyang Huzhao Tongbao 3: 18–24.

Golterman, H. L., 1996. Fractionation of sediment phosphate with cheating compounds: a simplification, and comparison with other methods. Hydrobiologia 335: 87–95.

Gomez, E., M. Fillit, M. C. Ximenes & B. Picot, 1998. Phosphate mobility at the sediment-water interface of a Mediterranean lagoon (Etang du Mejean), seasonal phosphate variation. Hydrobiologia 373/374: 203–216.

Hua, Z., X. Zhu & X. Wang, 2000. Study on bioavailability of *Selenastrum capricornutum* influenced by released phosphorus. Acta scientiae circumstantiae/Huanjing Kexue Xuebao 20: 100–105.

Matinvesi, J. & H. Heinonen-Tanski, 1992. Biodegradable substances in lake sediments and their relation to sediment microbiological activity and phosphorus recycling. Aqua Fennica 22: 193–200.

Mhamdi, B. A., S. M. Raoui, M. A. Mhamdi & K. Derraz, 2003. Bacterial alkaline phosphatase activity at the water sediment interface in the Sahela reservoir. Journal of Water Science 16: 305–316.

Qu, W., M. Dickman & S. Wang, 2001. Multivariate analysis of heavy metal and nutrient concentrations in sediments of Taihu Lake, China. Hydrobiologia 450: 83–89.

Sayler, G., S. M. Puziss & M. Silver, 1979. Alkaline phosphatase assay for freshwater sediments: application to perturbed sediment system. Applied and Environmental Microbiology 38: 922–927.

Sondergaard, M., J. Windolf & E. Jeppesen, 1996. Phosphorus fractions and profiles in the sediment of shallow Danish lakes as related to phosphorus load, sediment composition and lake chemistry. Water Research 30: 992–1002.

Song, C., X. Cao, J. Li, Q. Li, G. Chen & Y. Zhou, 2006. Contributions of phosphatase and microbial activity to internal phosphorus loading and their relation to lake eutrophication. Science in China: Series D Earth Sciences 49(Supplement): 102–113.

Su, Y., S. Ma & S. Dong, 2005. Variation of alkaline phosphatase activity in sediments of shrimp culture ponds and its relationship with the contents of C, N and P. Journal of Ocean University of China 4: 75–79.

Taoufik, M., J. E. Dafir & S. Kemmou, 2005. Hydrogeochemical behaviour of phosphorus in the Imfout Reservoir (Oum Rabiaa, Morocco). Water Quality Research Journal of Canada 40: 202–210.

Zhang, L., C. Fan, B. Qin & L. Yang, 2001. Phosphorus release and absorption of surficial sediments in Taihu Lake under simulative disturbing conditions. Journal of Lake Sciences 13: 42–49.

Zhou, Y., J. Li, Y. Fu & M. Zhang, 2001. Kinetics of alkaline phosphatase in lake sediment associated with cage-culture of *Oceochrolmis miloticus*. Aquaculture 203: 23–32.

Zhou, Y., J. Li, Y. Fu & M. Zhang, 2002. Temporal and spatial variations in kinetics of alkaline phosphatase in sediments of a shallow Chinese eutrophic lake (Lake Donghu). Water Research 36: 2084–2090.

Hydrobiologia (2008) 599:127–134
DOI 10.1007/s10750-007-9209-8

ELLS 2007

The impact of the invasive Ponto-Caspian amphipod *Pontogammarus robustoides* on littoral communities in Lithuanian lakes

Simona Gumuliauskaitė · Kęstutis Arbačiauskas

Abstract The intentionally introduced *Pontogammarus robustoides* is the most successful amphipod invader of Lithuanian inland waters and has become established in large lakes. Its impact on littoral invertebrate communities was studied by comparing similar habitats across lakes that harbour or are devoid of the invader. In habitats where *P. robustoides* is well established and numerous, it significantly reduces species richness and community diversity. Moderate pontogammarid density in habitats that can sustain the native gammarid *Gammarus lacustris*, however, revealed no negative impact on diversity metrics. Among the lakes studied, the benthic biomass did not differ in invaded and uninvaded habitats. The biomass of indigenous invertebrates (excluding chironomids, which exhibited high lake-specific biomass variation) was lower in the places with well-established *P. robustoides*. A detrimental impact was observed upon the native isopod *Asellus aquaticus* and a negative correlation with most of the higher taxa of native invertebrates. In the invaded lake habitats that favour *P. robustoides*, a change in community structure and a decrease in diversity up to twofold or more are to be expected.

Keywords Pontogammarid · Diversity · Biomass · Native · Non-native · Invertebrates

Guest editors: T. Nõges, R. Eckmann, K. Kangur, P. Nõges, A. Reinart, G. Roll, H. Simola & M. Viljanen
European Large Lakes—Ecosystem changes and their ecological and socioeconomic impacts

S. Gumuliauskaitė (✉) · K. Arbačiauskas
Institute of Ecology, Vilnius University, Akademijos St. 2, 08412 Vilnius, Lithuania
e-mail: simonagum@ekoi.lt

K. Arbačiauskas
e-mail: arbas@ekoi.lt

Introduction

Among the most ecologically successful and aggressive immigrants into European inland waters are Ponto-Caspian amphipods of the family Pontogammaridae, i.e. pontogammarids (Bij de Vaate et al., 2002; Dick et al., 2002; Jazdzewski et al., 2004). In Lithuania, *Pontogammarus robustoides* (G. O. Sars, 1894) is the most widely distributed and abundant non-native amphipod in the inland waters and has become established in six large mesotrophic lakes (Arbačiauskas, 2002, 2005). Recently, its distribution area has been shown to occupy the southern and eastern parts of the Baltic Sea basin and it has already invaded the North Sea basin (Arbačiauskas & Gumuliauskaitė, 2007). *P. robustoides* together with two other Ponto-Caspian species *Obesogammarus crassus* (G. O. Sars, 1894) and *Chaetogammarus warpachowskyi* (G. O. Sars, 1894) were deliberately introduced into the Kaunas Water Reservoir located on the River Nemunas during 1960–1961 (Gasiūnas, 1972). This was the first point outside the Ponto-Caspian region from which the pontogammarid started to expand (primarily across Lithuanian waters) through deliberate or accidental introductions and natural dispersal.

The impacts of invading alien amphipods on native communities are usually significant and drastic. They have been shown to cause the native amphipod extinctions (Musko, 1994; Dick, 1996; Jazdzewski et al., 2004; Arbačiauskas 2005) and a decline in the diversity, abundance and biomass of indigenous fauna (Kelly et al., 2003). The effects that *P. robustoides* may have on invaded communities, however, have not so far been studied, except for the observation that in the stony littoral of Neva Bay, Gulf of Finland, this and another alien amphipod have together reduced the densities of benthic detritivores (Berezina & Panov, 2003). Outside the native range, successful establishments of *P. robustoides* are predicted in habitats with lentic or stagnant fresh waters including large temperate lakes (Arbačiauskas & Gumuliauskaitė, 2007). Therefore, understanding the potential impacts of the pontogammarid on invaded lake communities is relevant. The purpose of the current study was to determine the effects of *P. robustoides* on littoral communities by comparing invaded and uninvaded lake habitats.

Materials and methods

Littoral communities of macroinvertebrates in 10 natural lakes and the artificial Lake Elektrėnai, which are invaded by or devoid of *P. robustoides*, were compared (Table 1). In two lakes, we intended to identify littoral communities on which the impact of the alien amphipod varied according to its presence or abundance. Thus, two sites were selected for study in each of these lakes (and one site elsewhere): the open littoral and the closed bay in Lake Plateliai (hereafter denoted Plateliai1 and Plateliai2), and two sites with or without *P. robustoides* in Lake Asveja (Asveja1 and Asveja2). All littoral habitats were studied over a two-year period except in Lakes Nedingis and Akmena, which were investigated in the first and the second year, respectively. The three types of littoral communities were identified a priori with respect to the occurrence of alien and native amphipods: (1) communities containing only *P. robustoides*, (five sites); (2) communities containing *P. robustoides* and *Gammarus lacustris* G. O. Sars, 1863 (two sites, Plateliai2 and Asveja1) and (3) communities devoid of *P. robustoides* (six sites). Thus, 5, 2 and 5 communities of types 1, 2 and 3, respectively, were studied each year.

Possibly similar littoral biotopes in the selected study sites (sandy bottom with some macrophytes: *Chara* sp., *Potamogeton* sp. or *Phragmites australis* (Cav.) Trin. ex Steudel) were sampled in parallel thrice per year, between late June and October of 2004 and 2005. On each occasion, three quantitative samples of benthic macroinvertebrates were collected by a stovepipe sampler with a cross-sectional area of 0.1 m^2. Samples were preserved in 4% formaldehyde. All specimens (except Oligochaeta, which were not identified further, and Chironomidae and other

Table 1 Characteristics of the lakes studied: area, maximum depth, mean depth and annual water circulation

Lake	Area, km^2	Maximum depth, m	Mean depth, m	Annual water circulation, %
Akmena	2.76	30.2	11.2	8
Asveja	10.15	50.2	14.7	46
Baluošai	2.52	37.5	12.5	52
Beržoras	0.52	6.3	4.6	36
Daugai	9.11	44.0	13.2	12
Dusia	23.34	32.6	15.4	7
Elektrėnai	13.89	30.0	7.2	110
Lavysas	1.62	10.2	5.3	180
Nedingis	3.16	6.9	3.4	140
Plateliai	12.05	46.0	10.4	12
Seirijis	5.01	19.2	7.9	29

Lakes inhabited by *Pontogammarus robustoides* are in bold

Diptera, which were identified at eight lower taxa of subfamily/tribe rank and to the family level, respectively) were identified to species level and weighed (wet weight). When identification to species level was not possible (young developmental stage or injury during sampling), a higher taxon was identified. The three replicate samples were averaged to obtain an estimate of community characteristics on each sampling date.

Community metrics (species richness, Shannon-Wiener diversity index and wet weight biomasses of benthic macroinvertebrates) were calculated; molluscs were excluded because it is hardly likely that *P. robustoides* has a significant impact on them, especially large-sized species, and random variations in their biomass across the study sites and sampling dates may mask variation in biomass of other benthic animals (see results). These data were used to determine variations in the measured community characteristics. To test the impact of well-established *P. robustoides* on community metrics (see results), values measured at each sampling point were averaged for each study year. These data were further analysed by two methods. Initially, untransformed (species richness, Shannon-Wiener diversity index) or log-transformed (biomass) data were tested by a mixed model nested ANOVA with pontogammarid as a fixed factor and lake, i.e. study site, as a random nested factor. Type III sums of squares decomposition and Satterthwaite's method for synthesized error were applied. Secondly, untransformed data were subjected to a Kruskal-Wallis ANOVA presuming that each annual average represents a random estimate of the community metric within a group. The impact of the pontogammarid was considered significant if both tests showed significant effects. Spearman rank correlation was applied to evaluate relationships between the biomasses of *P. robustoides* and higher taxa of native invertebrates. The significance of the different patterns of population density change from summer to October between native and alien amphipods at the Plateliai2 site was tested by a three-way ANOVA with species, season and year as fixed factors. That would indicate the significance of the effect of interaction between species and season. Prior to analysis, individual estimates of density were coded by adding 0.5 to exclude zero values, and then log-transformed. All tests were performed using Statistica software.

Results

The occurrence of native and alien peracaridan species and species numbers of other higher invertebrate taxa in quantitative samples from the study sites are presented in Table 2. Estimates of mollusc biomass varied markedly between habitats and dates. For example, in Lakes Baluošai and Beržoras, the biomasses of molluscs were 191.6 and 0.1 g m^{-2}, respectively, whereas in Lake Elektrėnai, estimates from July and August of the same year were 4.3 and 91.1 g m^{-2}, respectively. Such patterns would definitely mask any variations in biomass of other littoral invertebrates. Thus, molluscs were excluded from further data analysis.

Well-established populations of *P. robustoides* were found in five out of the six lakes (excluding Lake Asveja) where aliens are present (Table 1). In the littoral habitats of these lakes (type 1 communities), the contribution of the pontogammarid to the total benthic animal biomass was 10–90%. In contrast, the share of the native amphipod *G. lacustris* in the total benthic biomass was always below 10%. The most numerous populations of *P. robustoides* inhabited Lakes Dusia and Elektrėnai. During bouts of reproduction, its density and biomass had maxima of 3347 and 2113 ind. m^{-2} and 25.4 and 19.6 g m^{-2}, but were never <407 and 313 ind. m^{-2} and 3.9 and 2.4 g m^{-2}, respectively.

Variations in the diversity and biomass characteristics of interest within the a priori distinguished community types are shown in Fig. 1. The metrics of the type 2 and 3 communities studied (containing alien and native amphipods or devoid of aliens, respectively) were quite similar and did not differ statistically (Kruskal-Wallis test, $P > 0.05$). Therefore, data for these community types were merged to test the impact of well-established *P. robustoides* on community metrics.

In the littoral habitats of the lakes investigated in which *P. robustoides* is well established and numerous, species richness and community diversity were significantly reduced (Table 3). On average, 1.5- and 1.6-fold differences in species richness (SR: 12.4 vs. 19.2) and Shannon-Wiener diversity index (H: 1.58 vs. 2.53) were recorded, respectively. When the littoral habitats sheltering the most numerous populations of the pontogammarid from Lakes Dusia and Elektrėnai (SR and H averaged 8.6 and 1.08) were

Table 2 Occurrence (+) of alien and native peracaridan species (*Pontogammarus robustoides*, *Obesogammarus crassus*, *Chaetogammarus warpachowskyi*, *Paramysis lacustris* (Czerniavsky, 1882) and *Limnomysis benedeni* Czerniavsky, 1882, and *Gammarus lacustris*, *Pallaseopsis quadrispinosa* (G. O. Sars, 1867) and *Asellus aquaticus* (Linne, 1758), respectively); species numbers of other higher macroinvertebrate taxa and total number of identified taxa in quantitative samples from lake habitats studied

Taxa	Akmena	Asveja1	Asveja2	Baluošai	Beržoras	Daugai	Dusia	Elektrėnai	Lavysas	Nedingis	Plateliai1	Plateliai2	Seirijis
P. robustoides		+				+	+	+			+	+	+
O. crassus							+	+					
C. warpachowskyi						+	+	+					+
P. lacustris							+						
L. benedeni						+		+					
G. lacustris	+	+	+	+	+					+		+	
P. quadrispinosa				+									
A. aquaticus		+	+	+	+	+			+	+		+	+
Oligochaeta	1	1	1	1	1	1	1	1	1	1	1	1	1
Hirudinea	4	5	5	7	4	4	0	4	6	4	4	5	3
Ephemeroptera	4	5	4	6	4	5	2	1	3	2	2	5	4
Odonata	4	12	10	2	1	6	0	3	5	2	1	7	1
Megaloptera	1	1	1	1	1	1	0	1	1	1	1	1	1
Coleoptera	4	5	6	5	3	6	3	2	5	2	5	1	5
Trichoptera	11	19	23	16	15	13	5	7	7	13	9	18	10
Lepidoptera	1	1	1	0	1	1	0	1	0	0	0	1	1
Hemiptera	0	0	2	0	0	0	0	0	0	0	0	0	0
Diptera	6	11	9	5	5	5	2	6	7	5	6	5	7
Mollusca	7	16	10	8	7	14	10	12	10	5	11	9	6
Total	44	79	74	54	43	60	27	42	45	36	41	56	41

Chironomids and other dipterans were identified at eight lower taxa of subfamily/tribe rank, and at the family level, respectively

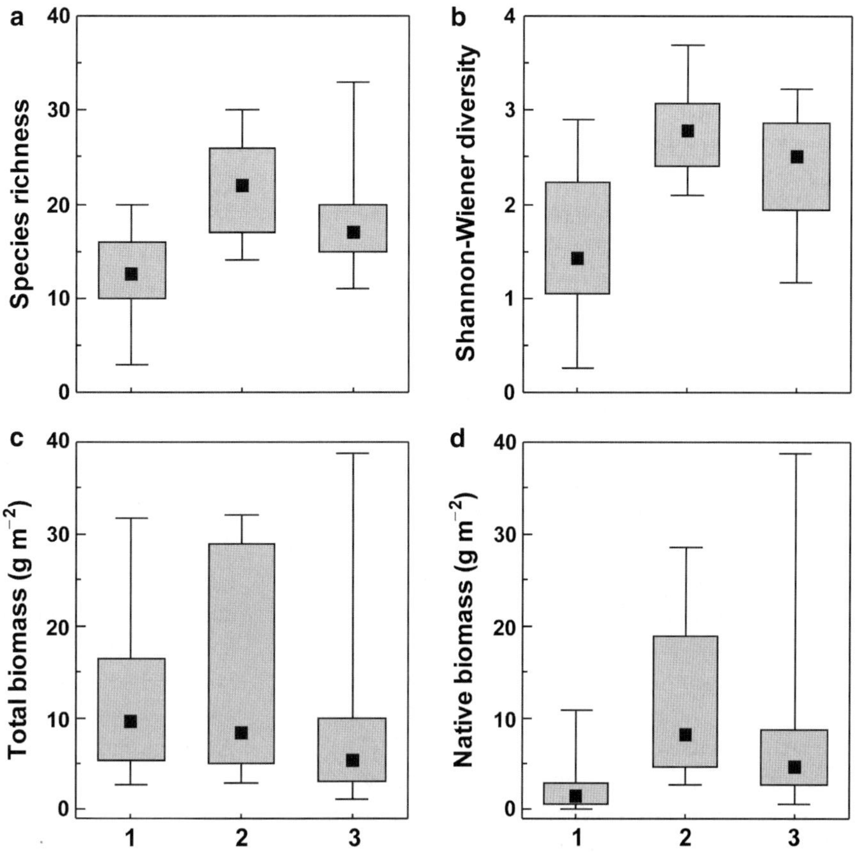

Fig. 1 Variation (median, quartiles and min-max range) of species richness (**a**), Shannon-Wiener diversity index (**b**), total benthic biomass (**c**) and biomass of native macroinvertebrates excluding chironomids (**c**) for the three types of lake littoral communities during the 2 year study period. 1—communities with well-established population of *Pontogammarus robustoides* (five study sites); 2—communities in which pontogammarid and indigenous gammarid *Gammarus lacustris* co-exist (two sites); 3—communities devoid of alien species (six sites)

Table 3 Summary of mixed-model nested ANOVAs testing for well-established *Pontogammarus robustoides* (PR) effect and lake (L) effect, and probabilities of Kruskal-Wallis tests (KWT) for PR effect on community metrics (see Methods for details): species richness (SR), Shannon-Wiener diversity index (H), total benthic biomass (B_{Total}), biomass of chironomids (B_{Chir}) and biomass of other native macroinvertebrates (B_{Other})

Metric	Source	Effect		Error		F	P	KWT, *P*
		df	MS	df	MS			
SR	PR	1	239.174	11.03	25.26	9.47	**0.011**	**0.003**
	L	11	25.321	11	9.25	2.74	0.055	
H	PR	1	4.637	11.03	0.535	8.67	**0.014**	**0.005**
	L	11	0.536	11	0.223	2.40	0.081	
B_{Total}	PR	1	0.787	11.01	0.673	1.17	0.303	0.219
	L	11	0.675	11	0.053	12.76	**<0.001**	
B_{Chir}	PR	1	10.619	11.01	3.800	2.79	0.123	**0.041**
	L	11	3.811	11	0.668	5.70	**0.004**	
B_{Other}	PR	1	10.059	11.01	1.167	8.62	**0.014**	**0.002**
	L	11	1.171	11	0.188	6.24	**0.003**	

Significant probabilities are in bold

compared with the type 2 and 3 communities, a more than two-fold difference was observed (2.2 and 2.3 times, respectively). Diversity metrics in type 2 communities, which include moderate numbers of *P. robustoides*, were on average higher than those in native communities (Fig. 1). Therefore, the difference between type 1 and type 3 communities also was tested. Analysis revealed a significant pontogammarid effect on both diversity indicators (mixed model ANOVAs, SR: $F_{1,9.3} = 5.83$, $P = 0.038$; H: $F_{1,9.3} = 5.64$; $P = 0.041$; Kruskal-Wallis ANOVAs, SR: $P = 0.013$, H: $P = 0.016$).

The benthic biomass varied significantly between the lakes (study sites) and did not depend upon the pontogammarid factor (Fig. 1, Table 3). No negative impact of well-established *P. robustoides* on the native biomass (excluding the biomass of *P. robustoides* and other alien peracaridans, which was always <12% that of the pontogammarid) was detected (mixed model ANOVA, $F_{1,11} = 0.83$; $P = 0.38$). However, there was a significant negative impact when the chironomid biomass was excluded from the native biomass. Variation in chironomid biomass (72% of random variation) depended significantly upon the study site and tended (as suggested by the Kruskal-Wallis test) to be larger in the lakes with well-established *P. robustoides* (Fig. 1, Table 3). Actually, the highest chironomid biomasses among the study points were in Lakes Dusia and Elektrėnai (5.2 and 7.0 g m^{-2} on average, respectively). The effect of the pontogammarid on the native biomass without chironomids was also significant when type 2 communities were excluded from the analysis (mixed model ANOVA, $F_{1,9.1} = 5.42$; $P = 0.045$; Kruskal-Wallis test, $P = 0.011$).

Across all lakes, the correlation in biomass between *P. robustoides* and most higher taxa of native invertebrates was significantly negative (Hirudinea −0.31, Ephemeroptera −0.48, Odonata −0.35, Trichoptera −0.35, Megaloptera −0.34, Diptera excluding chironomids −0.37). There was no correlation with Oligochaeta and Coleoptera, and a significantly positive correlation with chironomids (0.44). *P. robustoides* and *A. aquaticus* were also negatively related (−0.41). This native isopod occurred in quantitative samples together with the alien at only two sampling points, which suggests a negative impact of *P. robustoides* on *A. aquaticus* (Fisher exact test on occurrence frequencies in samples where pontogammarid is present or absent, $P < 0.001$).

Native *G. lacustris* and alien *P. robustoides* were found to co-exist at one site in Lake Asveja (Asveja1) and in the bay of Lake Plateliai (Plateliai2). At the former site, the densities of both amphipods were low and they appeared only occasionally in quantitative samples. The most numerous population of the native amphipod was discovered at the second site. Its density during a bout of reproduction was estimated at 1053 ind. m^{-2} (late June, 2005). At Plateliai2, *G. lacustris* dominated numerically over *P. robustoides* in summer, but towards October in both study years, its density declined substantially. In contrast, the abundance of *P. robustoides* was low during summer but increased towards October. The difference between species in the pattern of density dynamics was significant (Fig. 2).

Discussion

The results of the current study clearly show that in lake habitats where *P. robustoides* is well established and numerous, the newcomer significantly reduces species richness and community diversity. In large lakes, which among stagnant fresh waters are probably most vulnerable to *P. robustoides* invasion (see Arbačiauskas, 2005), a decline of more than twofold in littoral community diversity may occur. A negative impact on species diversity has also been recorded for other invading amphipod species (Dick, 1996; Dick et al., 2002; Kelly et al., 2003). On the other hand, a

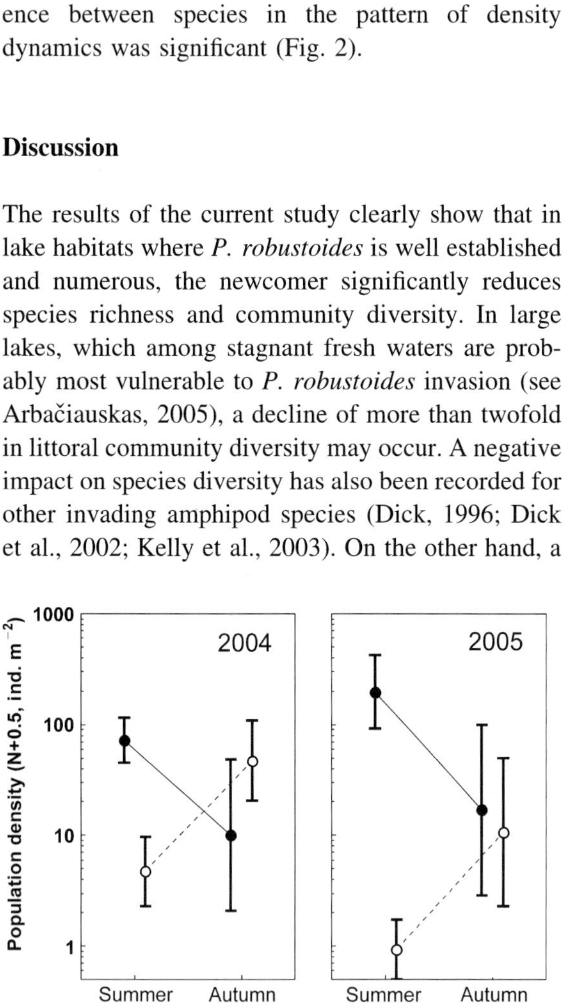

Fig. 2 Population density (mean ± SE) of indigenous *Gammarus lacustris* (closed circles) and alien *Pontogammarus robustoides* (open circles) in the closed bay of Lake Plateliai during the summer and autumn of 2004 and 2005. Note logarithmic scale. The density of indigenous species between June and October was higher, whereas the pattern of dynamics towards October differed significantly in species (three-way ANOVA on log-transformed (N + 0.5): species effect, $F_{1,28} = 6.6$; $P = 0.016$; species and season interaction effect, $F_{1,28} = 11.4$; $P = 0.002$)

moderate density of *P. robustoides* in lake habitats that limit the development of the invader population and can sustain the native gammarid *G. lacustris* had no negative impact upon invertebrate diversity.

When invaded and uninvaded habitats were compared among different lakes, the biomasses of littoral animals were similar. It is noteworthy that the intention behind the introduction of Ponto-Caspian amphipods into Lithuanian lakes was to supplement the food base for commercially important fish, and consequently to increase the benthic biomass. However, the current study does not support such an effect on littoral communities. When *P. robustoides* is well established, its contribution to the total invertebrate biomass is usually substantial, so a negative affect on the native benthic biomass can be expected. However, that was not detected in our data. The expected impact on native biomass was probably hidden by the high lake-specific variation in chironomid biomass, which tended to be greater in lakes sheltering numerous populations of aliens. A negative effect on native biomass emerged only when chironomids were excluded. The observed pattern of chironomid biomass variation across the lakes studied (and the positive correlation in biomass between pontogammarids and chironomids) must not be related to the pontogammarid factor. In contrast, a negative influence of *P. robustoides* on chironomids is to be expected, as this species in common with other amphipods preys on these dipterans (Berezina et al., 2005; Krisp & Maier, 2005).

Although correlation analysis is not sufficient to reveal causal relationships, the negative correlations between *P. robustoides* and most of the higher taxa of native invertebrates suggest that the invader may, directly or indirectly, affect various benthic animals. Invading amphipods have been shown to be capable of impacting negatively on the abundance and biomass of different indigenous invertebrates (Kelly et al., 2003). An adverse impact of the pontogammarid upon the native isopod *A. aquaticus*, however, is obvious in our data. Such an impact also has been indicated by other data and probably indicates direct predation (Arbačiauskas, 2005).

The alien and native amphipods in the closed bay of Lake Plateliai co-exist for a quite long time. Thus, the observed species differences in the patterns of density change from mid-summer towards October might indicate a temporal change in the competitive superiority of these species, which permits long-term co-survival. *G. lacustris* is probably superior to *P. robustoides* during the winter as it is more resistant to low oxygen concentrations, which are likely to occur under the ice cover (Arbačiauskas, 2005). Conversely, during the growing season, *P. robustoides* has an advantage, as it is more aggressive than the native species (Arbačiauskas & Gumuliauskaitė, 2007). Asymmetry in intraguild predation is believed to be responsible for the replacement of resident amphipod species by invaders (Dick et al., 1999, 2002). The conditions that allow amphipods of differing origins to co-exist, however, warrant a special study.

In conclusion, the invasive Ponto-Caspian amphipod *P. robustoides* is capable of affecting indigenous communities detrimentally. The magnitude of the impact on lake littoral communities probably depends upon environmental conditions. When a habitat is favourable for the pontogammarid and it can attain high population density, a drastic change in community structure and a decline in diversity of two-fold or more seem very probable. Such an impact is to be expected primarily in large lakes as their littorals provide benign conditions for the pontogammarids, which demand waters of high oxygen content.

Acknowledgements We thank anonymous peer reviewers for their comments on an early draft of the article.

References

Arbačiauskas, K., 2002. Ponto-Caspian amphipods and mysids in the inland waters of Lithuania: history of introduction, current distribution, relations with native malacostracans. In Leppakoski, E., S. Gollasch & S. Olenin (eds), Invasive Aquatic Species of Europe. Kluwer Academic Publishers, Dordrecht, Boston, London, 104–115.

Arbačiauskas, K., 2005. The distribution and local dispersal of Ponto-Caspian Peracarida in Lithuanian fresh waters with notes on *Pontogammarus robustoides* population establishment, abundance and impact. Oceanological and Hydrobiological Studies 34(Supplement 1), 93–111.

Arbačiauskas, K. & S. Gumuliauskaitė, 2007. Invasion of the Baltic Sea basin by Ponto-Caspian *Pontogammarus robustoides*, its ecological impact. In Gherardi, F. (ed.), Biological Invaders in Inland Waters: Profiles, Distribution and Threats. Springer, 463–477.

Berezina, N. A. & V. E. Panov, 2003. Establishment of new gammarid species in the eastern Gulf of Finland (Baltic Sea) and their effect on littoral communities. Proceedings of the Estonian Academy of Sciences. Biology Ecology 52: 284–304.

Berezina, N., S. Golubkov & J. Gubelit, 2005. Grazing effects of alien amphipods on macroalgae in the littoral zone of the Neva Estuary (Eastern Gulf of Finland, Baltic Sea). Oceanological and Hydrobiological Studies 34(Supplement 1): 63–82.

Bij de Vaate A., K. Jazdzewski, H. A. M. Ketelaars, S. Gollasch & G. Van der Velde, 2002. Geographical patterns in range extension of Ponto-Caspian macroinvertebrate species in Europe. Canadian Journal of Fisheries and Aquatic Sciences 59: 1159–1174.

Dick, J. T. A., 1996. Post-invasion amphipod communities of Lough Neagh, N. Ireland: influences of habitat selection and mutual predation. Journal of Animal Ecology 65: 756–767.

Dick, J. T. A., W. I. Montgomery & R. W. Elwood, 1999. Intraguild predation may explain an amphipod replacement; evidence from laboratory populations. Journal of Zoology (London) 249: 463–468.

Dick, J. T. A., D. Platvoet & D. W. Kelly, 2002. Predatory impact of the freshwater invader, *Dikerogammarus villosus* (Crustacea: Amphipoda). Canadian Journal of Fisheries and Aquatic Sciences 59: 1078–1084.

Gasiūnas, I., 1972. Enrichment of fodder basis of water bodies of Lithuania by acclimatized crustaceans-like organisms from the Caspian Sea complex. In Maniukas J. & J. Virbickas (eds), On the Breeding of Fish and Crustacean-like Organisms in the Water Bodies of Lithuania. Vilnius, 57–68 (in Russian).

Jazdzewski, K., A. Konopacka & M. Grabowski, 2004. Recent drastic changes in the gammarid fauna (Crustacea, Amphipoda) of the Vistula River deltaic system in Poland caused by alien invaders. Diversity and Distributions 10: 81–87.

Kelly, D. W., J. T. A. Dick, W. I. Montgomery & C. MacNeil, 2003. Differencies in the composition of macroinvertebrate communities with invasive and native *Gammarus* spp. (Crustacea: Amphipoda). Freshwater Biology 48: 306–315.

Krisp, H. & G. Maier, 2005. Consumption of macroinvertebrates by invasive and native gammarids: a comparison. Journal of Limnology 64: 55–59.

Muskó, I. B., 1994. Occurrence of Amphipoda in Hungary since 1853. Crustaceana 66: 144–152.

Hydrobiologia (2008) 599:135–141
DOI 10.1007/s10750-007-9203-1

Spatiotemporal and long-term variation in phytoplankton communities in the oligotrophic Lake Pyhäjärvi on the Finnish-Russian border

Anna-Liisa Holopainen · Liisa Lepistö · Riitta Niinioja · Anita Rämö

Abstract As part of the joint Finnish-Russian research and development project "Assessment of the ecological state of the transboundary waters", seasonal and spatial variations in the phytoplankton communities of Lake Pyhäjärvi were studied in order to identify possible long-term (1963–2002) changes and to present recent data (since 2002) on phytoplankton biomass and species composition. Some changes in both phytoplankton biomass and species composition, particularly in the littoral zone and northern basins of the lake, were obvious at the end of the 1980s; in particular, the density of blue-greens increased during that period. Biomasses increased five-fold in 1980, decreased after 1990 and have remained low (0.14 mg l^{-1}) to the present time. The effective water protection measures applied, especially reducing the phosphorus load from municipal wastewaters and airborne pollution, can now be seen to have improved the water quality. Although no significant temporal differences were found in phytoplankton biomass during the intensive biweekly sampling in 2002, considerable spatial variation was seen within the lake. An obvious change in the species composition of phytoplankton and an increase in biomass were seen even in the loaded parts of the lake. Despite the spatial differences in phytoplankton biomass and community structure, the changed species composition in the northern part of the lake indicated a clear recovery from the blue-green and Chrysophyceae dominated high biomass of the 1980s.

Keywords Phytoplankton · Water quality · Large lake · Long-term changes · Lake Pyhäjärvi

Guest editors: T. Nõges, R. Eckmann, K. Kangur, P. Nõges, A. Reinart, G. Roll, H. Simola & M. Viljanen
European Large Lakes—Ecosystem changes and their ecological and socioeconomic impacts

A.-L. Holopainen · R. Niinioja · A. Rämö
North Karelia Regional Environment Centre, P.O. Box 69, 80101 Joensuu, Finland

A.-L. Holopainen (✉)
Faculty of Biosciences, Ecological Research Institute, University of Joensuu, P.O. Box 111, 80101 Joensuu, Finland
e-mail: Anna-Liisa.Holopainen@Joensuu.fi

L. Lepistö
Finnish Environment Institute, P.O. Box 140, 00251 Helsinki, Finland

Introduction

Since phytoplankton is regarded as a sensitive and rapidly reacting indicator of changes in water quality (Heinonen, 1980; Reynolds, 1984; Willén, 2002), long-term monitoring of phytoplankton biomass and species composition has been used in many large lakes to evaluate environmental changes (e.g. Petrova, 1987; Willén, 1992; Talling, 1993; Makarewicz, 1993; Holopainen et al., 1996; Lepistö et al., 1999). The phytoplankton communities of Lake Pyhäjärvi have been monitored since 1960 as parts of national programmes. At the end of the 1980s, some changes were seen in the phytoplankton biomass and species composition in the loaded basins and littoral areas of Lake Pyhäjärvi, a large lake on the Finnish-Russian border. In particular, the density of blue-greens increased during that period. Local people were the first to notice the deterioration of their lake as an increased frequency of algal disturbances, turbid water and increasing vegetation in the littoral zone. Together with the authorities, the local citizens started continuous Secchi depth monitoring in different parts of the lake in 1997 (Niinioja et al., 2004).

Although algal blooms have been reported in some areas of the lake, the spatial and seasonal variation of phytoplankton in this oligotrophic lake has not previously been studied in detail. In this article, we present long-term results on seasonal and spatial variations in phytoplankton biomass and species composition. The aim of the work is to evaluate the recovery and the present state of Lake Pyhäjärvi.

Study area, materials and methods

Lake Pyhäjärvi is located in the Vuoksi river basin on the border between Finland and the Republic of Karelia, Russia (Fig. 1). The lake area is 248 km^2, of which 207 km^2 is situated in Finland. The mean depth of the lake is 7.6 m and the maximum depth is 27 m on the Finnish side. The catchment area of the lake is 1,026 km^2, the volume is 1.96 km^3 and the theoretical water retention time is 7 years.

Lake Pyhäjärvi is a valuable oligotrophic and clear-water lake that was included in the European Union's Natura 2000 programme in Finland, and in national and regional monitoring programmes since the 1960s. Many studies on the hydrology, water quality, hydrobiology and paleolimnology of the lake have been published previously (see Kukkonen et al., 2005, and references therein).

The main loading of Lake Pyhäjärvi derives from diffuse sources such as forestry and agriculture, but there are also point source loads from small villages (Kukkonen et al., 2006). Lake Pyhäjärvi is composed of several separate basins. Papinniemenselkä, Taipaleenselkä and Kalattomanselkä are situated in the southern part of the lake and Hiekanpäänselkä and Hummonselkä in the north, from whence the outlet discharges to L. Orivesi. The drainage basin is mainly forest (83%), with only small areas of fields and peat bogs.

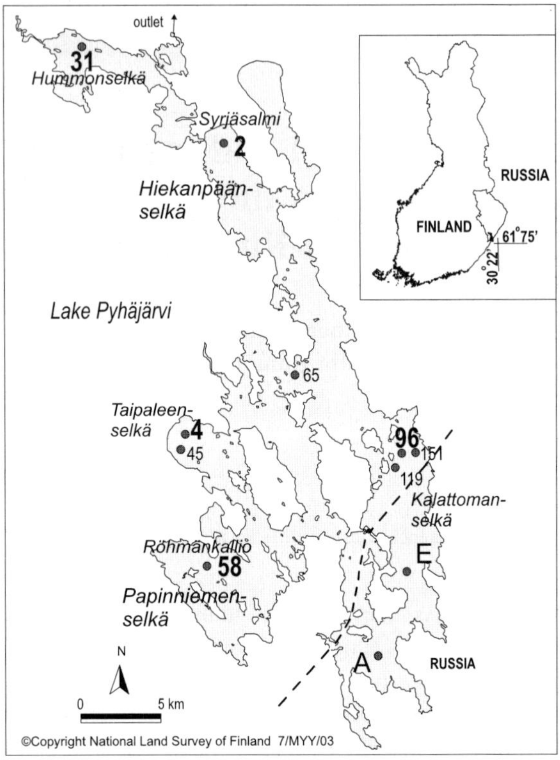

Fig. 1 Lake Pyhäjärvi and the sampling stations Röhmänkallio 58, Taipaleenselkä 4, Syrjäsalmi 2, Hummonselkä 31, Kalattomanselkä 96 and the stations A and E in Russia. The sampling stations 119, 151, 45 and 65 used for CCA-analysis are marked

The long-term (1963–2002) phytoplankton samples were collected from a depth of 0–2 metres in July to represent the stable summer situation. All samples were from the northern part of the lake (station 2). Samples for the seasonal succession of phytoplankton were taken biweekly from the deepest part of the lake (sampling station 2, Fig. 1) during the open water period of the intensive study year 2002. The spatial variation of phytoplankton was studied in five different basins of the lake at sampling stations 31, 2, 4, 58, 96, A and E (Fig. 1). As sampling was less frequent at the beginning of the study period, phytoplankton results from other sampling stations (65, 45, 119, 151) situated nearby were also used. Secchi depth values and samples for water chemical parameters were taken at the same time during 1970–2002. Chemical water quality was analysed according to Finnish standard methods and quality assurance (Niemi et al., 2001; Mitikka & Ekholm, 2003).

Samples were preserved using acidic Lugol solution. Phytoplankton biomass was estimated by microscopy using the Utermöhl technique (Utermöhl, 1958) and phase-contrast illumination. Cell counts were converted to biovolumes using the cell volumes in the phytoplankton database of the Finnish Environment Institute. The total phytoplankton biomasses are given as fresh weights (mg l^{-1}).

Canonical correspondence analysis (CCA) of the CANOCO computer programme was used for direct gradient analysis of the environmental and phytoplankton data (biomasses of phytoplankton taxa) (ter Braak, 1987, 1990). Secchi depth, total nitrogen, total phosphorus, NO_3–N, alkalinity, pH and colour of the water were used as environmental parameters.

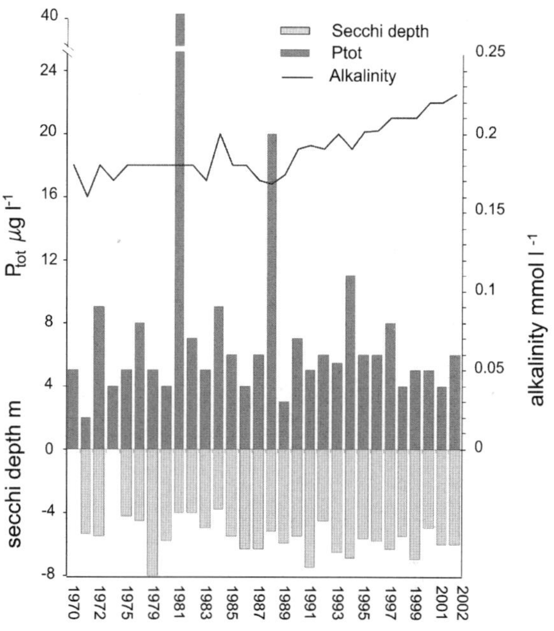

Fig. 2 Secchi depth, total phosphorus and alkalinity of water in July–August 1970–2002 at the Syrjäsalmi sampling station in Lake Pyhäjärvi

Results

The water quality of Lake Pyhäjärvi is characterised by clear water, low phosphorus and low chlorophyll *a* contents. However, increased phosphorus concentrations (up to 40 µg l^{-1}) and decreased Secchi depth values were occasionally noticed at the beginning of the 1980s. The oxygen concentration of surface water has been good, with only occasional decreased values in the deeps. Although variations in the water quality were negligible over the period 1970–2002 (Fig. 2), slight increases were seen in the Secchi depth and alkalinity values of the water. In 2002, the alkalinity values varied from 0.21 to 0.23 mmol l^{-1}; they were between 0.19 and 0.20 mmol l^{-1} in 1970–2002 (Fig. 2, Table 1).

In 2002, only minor differences were found in the water quality in different parts of this large lake (Table 1). Clear water (water colour ca. 10 mg l^{-1} Pt) with low phosphorus (mean 4–6 µg l^{-1}) and chlorophyll *a* (means 1.8–2.7 µg l^{-1}) contents was characteristic of the lake. Mean transparency or Secchi depth values varied from 4.9 m in Hummonselkä to 5.8 metres in Kalattomanselkä close to the Russian border. The most oligotrophic areas of the lake were found in Kalattomanselkä and in Papinniemenselkä (Röhmänkallio) in the south in 2002. The water was more coloured with slightly increased nutrient contents in the northern part of the lake in 2002.

In 1963–2002, the phytoplankton biomass in July varied from 0.12 to 0.66 mg l^{-1} (mean 0.31 mg l^{-1}) in the northern part of Lake Pyhäjärvi (Fig. 3). The lowest biomass (0.12 mg l^{-1}) was measured at the beginning of the period in 1960. Biomasses increased five-fold (0.66 mg l^{-1}) in 1980 and decreased again to 0.14 mg l^{-1} at the end of the period. The phytoplankton was dominated by small taxa such as the chrysophycean *Uroglena*, the Cryptophyceae flagellates (*Rhodomonas lacustris* Pascher and Ruttner and *Cryptomonas* sp.) and the diatom *Rhizosolenia longiseta* Zacharias. In the growing season of 2002, the phytoplankton biomass was mainly composed of small Cryptophyceae (*Rhodomonas lacustris, Katablepharis ovalis* Skuja and *Cryptomonas* sp.) and Chrysophyceae (*Dinobryon divergens* Imhof, *Uroglena* sp., *Dinobryon acuminatum* Ruttner, *Pseudopedinella* sp.) flagellates (Fig. 4).

Although no significant temporal differences were found in phytoplankton biomass in 2002, considerable spatial variation was seen within the lake (Fig. 5). Phytoplankton biomass was more than twice as high at station 2, in the northern part of Lake Pyhäjärvi, than in the other basins (Fig. 1). In addition to Cryptophyceae, the biomass of Cyanophyceae and Chysophyceae was higher in this area (Fig. 5).

Table 1 Secchi depth, colour, pH, chlorophyll *a*, total phosphorus, nitrogen and NO_3–N concentrations of the water in Lake Pyhäjärvi during the open water period 2002

		Kalattomanselkä 96	Röhmänkallio 58	Taipaleenselkä 4	Syrjäsalmi 2	Hummonselkä 31
Secchi depth, m	Mean	5.8	5.6	5.0	5.5	4.9
	Range	4.8–7.3	4.2–7.2	4.3–5.5	4.9–6.4	4.0–5.4
Colour, mg Pt l^{-1}	Mean	10	10	10	10	13
pH	Range	7.2–7.7	7.3–7.8	7.2–7.6	7.3–7.6	7.2–7.6
Chlorophyll *a*, μg l^{-1}	Mean	1.8	2.6	2.1	2.0	2.7
	Range	1.0–3.5	1.0–4.5	1.0–2.7	1.0–3.5	2.1–3.1
P_{tot}, μg l^{-1}	Mean	4.1	5.1	6.0	5.2	5.2
	Range	3–5	4–6	5–7	4–6	4–7
N_{tot}, μg l^{-1}	Mean	212	229	230	240	262
	Range	190–250	210–270	210–240	210–290	240–280
NO_3–N, μg l^{-1}	Mean	4.8	11.3	3.8	8.6	10.5
	Range	2.5–19	2.5–53	2.5–9	2.5–39	2.5–38

At Röhmänkallio and Syrjäsalmi $N = 9$; at Taipaleenselkä and Hummonselkä $N = 5$

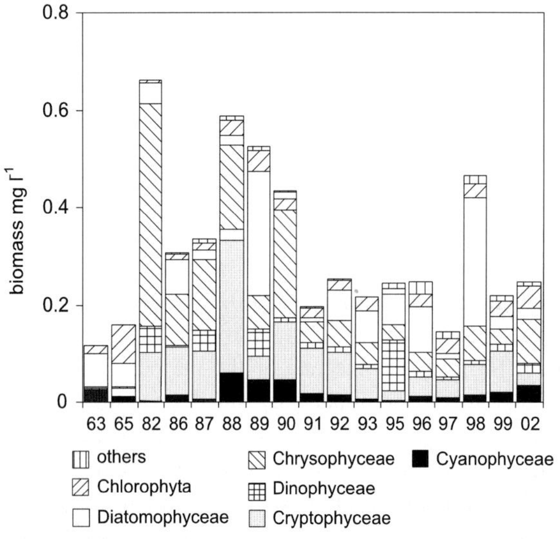

Fig. 3 Phytoplankton biomass and species composition in 1963–2002 at the Syrjäsalmi sampling station in Lake Pyhäjärvi

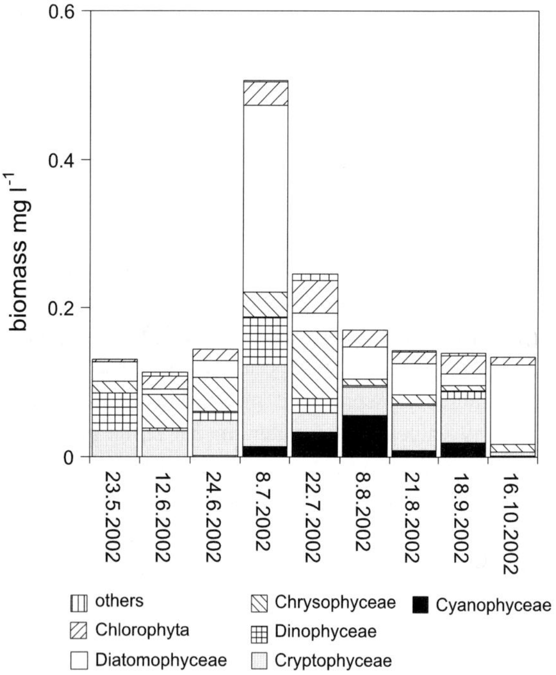

Fig. 4 Phytoplankton biomass and species composition in 2002 at the Syrjäsalmi sampling station in Lake Pyhäjärvi

The phytoplankton assemblages from years 1963 to 2002 were divided into three distinct groups by CCA-analysis (Fig. 6). The assemblages of the 1980s formed a group to the left side of axis 1. Typical phytoplankton taxa in that period were the chrysophyceans *Dinobryon sociale* Ehrenberg, *D. bavaricum* Imhof, *D. divergens* Imhof and *Uroglena* spp. and the dinophyte *Peridinium* spp. The samples of the 1990s and 2002 communities were grouped at the opposite end of axis 1, which correlated positively with alkalinity ($r = 0.931$). On axis 2 the Secchi depth and pH values of the water explained the changes in the phytoplankton community. Samples from the 1960s settled far from the other groups. The difference in this group might possibly be explained in terms of changes in the preservation and identification of

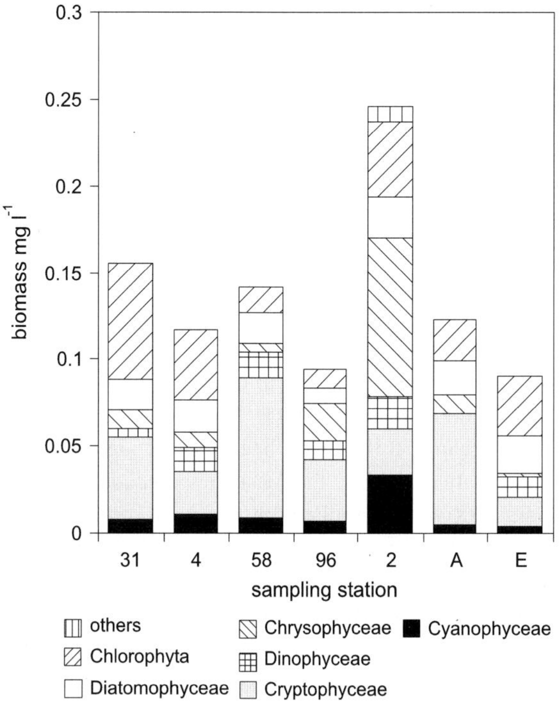

Fig. 5 Phytoplankton biomass and species composition in 2002 in Lake Pyhäjärvi at the sampling stations Röhmänkallio 58, Taipaleenselkä 4, Syrjäsalmi 2, Hummonselkä 31, Kalattomanselkä 96 and the stations A and E

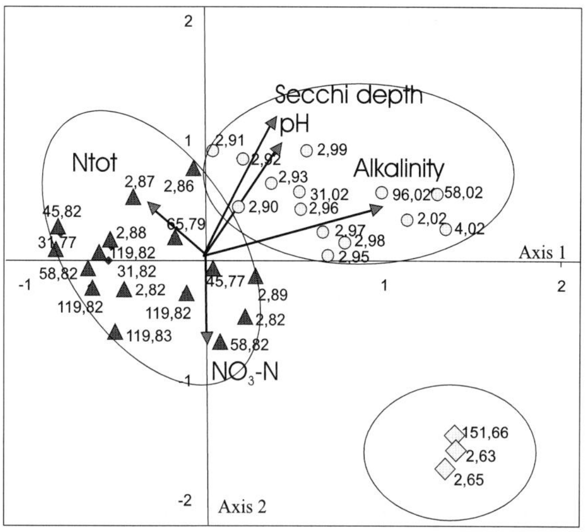

Fig. 6 CCA ordination of phytoplankton sampling stations with environmental variables in Lake Pyhäjärvi in 1963–2002. Sampling stations and years are expressed by numbers separated by a comma (2,65 = sampling station 2, year 1965). The eigenvalues of the axis are $Eig_1 = 0.374$, $Eig_2 = 0.274$. The axes explain 14% of the variation

phytoplankton or other methodological reasons. Typical species in the phytoplankton communities of the 1960s were the diatom *Stephanodiscus* cf. *hantzschii* Grunow, the greens *Eudorina elegans* Ehrenberg, *Xanthidium antilopeum* (Brébisson) Kützing and the blue-green *Gomphosphaeria aponina* Kützing. In 2002, at the end of this study period, the species composition had already changed in the northern part of Lake Pyhäjärvi. The CCA analysis in 2002 grouped the phytoplankton sample of the northern part of the lake together with samples from the other areas, thus indicating recovery of plankton communities from the situation in 1980.

Discussion

According to the long-term monitoring data, Lake Pyhäjärvi is an oligotrophic lake and its water quality is mostly good or excellent as indicated by the low phosphorus and chlorophyll *a* contents (Niinioja et al., 2004; Lepistö et al., 2006). Since only occasional decreases in oxygen concentration have been noticed in the deeps, the internal loading of the lake has been considered insignificant. Considerable long-term changes have been seen in Lake Pyhäjärvi in the increasing Secchi depth values and buffer capacity of the water. Similar changes have also been found in other lakes in Finland (Forsius et al., 2003; Mitikka & Ekholm, 2003).

During the open water season of 2002, the mean phytoplankton biomass in the surface water was low (0.20 mg l^{-1}, range 0.11–0.51 mg l^{-1}, Fig. 4). At the same time, the chlorophyll *a* concentration varied from 1.0 to 3.5 µg l^{-1} and was highest during the biomass maximum at the beginning of July (Holopainen et al., 2004). The phytoplankton biomass in the surface water was comparable to those in other large clear water lakes of North Karelia (Niinioja et al., 2005), but less than in large dark-coloured lakes, e.g. Lake Pielinen (Niinioja et al., 2000). However, when the phytoplankton biomass in the whole productive layer is compared, it may be higher in the clear-water lakes (Arvola et al., 1999). A deep-water biomass maximum can often be found, which increases the total biomass. This is also the case in Lake Pyhäjärvi, where a biomass maximum at 7–8 metres has been reported (Holopainen et al., 2004).

Some differences have been reported between the loaded and pristine basins of Lake Pyhäjärvi in the zooplankton communities (Rahkola-Sorsa et al., 2004) and the sedimented diatoms (Kukkonen et al., 2006). The changed zooplankton community in the northern part of the lake and the diatom species composition of the loaded bay areas agree particularly well with the spatial differences in phytoplankton. Although some spatial differences were found in the lake in 2002, the low biomass and the phytoplankton species composition indicated a clear oligotrophic state for Lake Pyhäjärvi (Heinonen, 1980; Lepistö et al., 2006).

In the growing season of 2002, the phytoplankton biomass was mainly composed of small Cryptophyceae and Chrysophyceae flagellates. This type of community seems to be typical of pristine areas and is recorded in many oligotrophic lakes in the Nordic countries (Lepistö & Rosenström 1998; Brettum & Halvorsen, 2004). The biomass of blue-greens was highest (30% of total biomass) at the beginning of August, and the numerically most abundant species were *Merismopedia warmingiana* Lagerheim and *Snowella lacustris* (Chodat) Komárek et Hindák, both typical of oligotrophic lakes with low phosphorus and nitrogen contents (Brettum, 1989; Lepistö, et al. 2006). In particular, the small chrysophyceans that were typical of this lake are known to be able to live in waters with low phosphorus content (Brettum, 1989).

The increasing biomass of chrysophyceans in the phytoplankton communities of the 1980s seems to indicate slight eutrophication of the lake. Abundance of the genus *Uroglena* is especially considered a sign of advancing eutrophication (Talling, 1993). According to Rosén (1981) and Brettum (1989), *Dinobryon divergens* in particular shows a tendency to thrive in more nutrient-rich lakes. Small Chroococcales were typical blue-greens in the communities of the 1980s, whereas the amount of *Anabaena lemmermannii* P. Richter increased later, after 1990. According to Lepistö et al. (2004), this massive bloom-forming species is typical of large, moderately nutrient-poor, oligo-humic Finnish lakes.

Stichogloea olivacea Chodat, *Synura* sp., *Mallomonas akrokomos* Ruttner, *M. tonsurata* Teiling, *Mallomonas* sp. and the greens *Botryococcus terribilis* Komárek and Marvan and *Monoraphidium dybowskii* (Woloszyńska) Hindák & Komárkova-Legenerová were typical of the phytoplankton communities after 1990. According to Brettum (1989), at least *Mallomonas akrokomos* and *Monoraphidium dybowskii* are typical of lakes with low phosphorus content, and together with *Synura* sp. are reported to be common in lakes with high N/P ratios (Brettum, 1989), which is consistent with the data for Lake Pyhäjärvi (mean $N/P = 58$).

Due to effective water protection measures, the phosphorus load from municipal wastewaters and also the airborne pollution affecting Lake Pyhäjärvi was reduced in 1997–1998 to only half of the load in 1990 (Kukkonen et al., 2005). In many Finnish lakes, a decreasing trend has been found in nutrient concentration during the past few decades (Räike et al., 2003). Phytoplankton is very sensitive to changes in the environment, and with no significant internal loading, a decrease in external loading can be seen quickly in phytoplankton communities. Our data agree with the long-term results on other eutrophicated large lakes, where phytoplankton communities have often responded in a pronounced way to changes in nutrient loading (Willén, 1992; Makarewicz, 1993; Lepistö et al., 1999).

At the end of this study period, the recovery of Lake Pyhäjärvi from the increased load situation in 1970–1980 is distinctly visible as increased water transparency, decreased phytoplankton biomass and changes in species composition. Owing to the special characteristics of this lake, it appears to be very sensitive even to a low nutrient load. In future, water pollution of this unique lake should be kept to a minimum by all possible protection measures in order to sustain its high ecological, economic and recreational value.

Acknowledgements Our thanks are due to the field and laboratory staff of the Ecological Research Institute of the University of Joensuu, and to the North Karelia Regional Environment Centre for collecting and analysing the water samples. We particularly thank Ismo J. Holopainen for his valuable comments. This study is a part of the joint Finnish-Russian research and development project "Assessment of the ecological state of the transboundary waters", financed by the Interreg IIIA Karelia programme.

References

Arvola, L., P. Eloranta, M. Järvinen, J. Keskitalo & A.-L. Holopainen, 1999. Phytoplankton. In Keskitalo J. & P.

Eloranta (eds), Limnology of Humic Waters. Backhus Publishers, Leiden, The Netherlands, 137–171.
Brettum, P., 1989. Alger som indicator på vannkvalitet. Planteplankton. Oslo. Niva-Rapport 0-86116, 111 pp (In Norwegian).
Brettum, P. & G. Halvorsen, 2004. The phytoplankton of Atnsjøen, Norway—a long-term investigation. Hydrobiologia 521: 141–147.
Forsius, M., J. Vuorenmaa, J. Mannio & S. Syri, 2003. Recovery from acidification of Finnish lakes: regional patterns and relations to emission reduction policy. The Science of the Total Environment 310: 121–132.
Heinonen, P., 1980. Quantity and composition of phytoplankton in Finnish inland waters. Publications of the Water Research Institute 37: 5–91.
Holopainen, A.-L., P. Huttunen, G. Letanskaya & E. Protopopova, 1996. The trophic state of Lake Ladoga as indicated by late summer phytoplankton. Hydrobiologia 366: 9–16.
Holopainen, A-L., L. Lepistö, R. Niinioja, A. Sharov & A. Rämö, 2004. Karjalan Pyhäjärven kasviplanktonin biomassa ja lajisto. Rajavesien ekologisen tilan arviointi - hanke. Pohjois-Karjalan ympäristökeskus. Alueelliset ympäristöjulkaisut 357: 7–36 (In Finnish).
Kukkonen, M., J. Kiiski, H. Luotonen & R. Niinioja, 2005. The plan of water and water ecosystem protection for Lake Karelian Pyhäjärvi. Regional Environmental publications 374, 53 pp.
Kukkonen, M., H. Simola, R. Niinioja & H. Luotonen, 2006. Diatoms as indicators of the present and past ecological state in Karelian Pyhäjärvi—an oligotrophic transboundary lake on the Finnish-Russian Border. Verhandlungen der Internationale Vereinigung für Limnologie 29: 1976–1980.
Lepistö, L. & U. Rosenström, 1998. The most typical phytoplankton taxa in four types of boreal lakes. Hydrobiologia 369: 89–97.
Lepistö, L., A. Räike & O.-P. Pietiläinen, 1999. Long-term changes of phytoplankton in a eutrophicated boreal lake during the past one hundred years (1893–1998). Algological Studies 94: 223–244.
Lepistö, L., A.-L. Holopainen & H. Vuoristo, 2004. Type specific and indicator taxa of phytoplankton as quality criterion for assessing the ecological status of Finnish boreal lakes. Limnologica 34: 236–248.
Lepistö, L., A.-L. Holopainen, H. Vuoristo & S. Rekolainen, 2006. Phytoplankton assemblages as a criterion in the ecological classification of lakes in Finland. Boreal Environment Research 11: 35–44.
Makarewicz, J. C., 1993. Phytoplankton biomass and species composition in Lake Erie 1970–1987. Journal of Great Lakes Research 19: 258–274.
Mitikka, S. & P. Ekholm, 2003. Lakes in the Finnish Eurowaternet: status and trends. The Science of the Total Environment 310: 37–45.
Niemi, J., P. Heinonen, S. Mitikka, H. Vuoristo, O.-P. Pietiläinen, M. Puupponen & E. Rönkä (eds), 2001. The Finnish Eurowaternet with information about Finnish water resources and monitoring strategies. Finnish environment 445: 1–62.
Niinioja, R., A.-L. Holopainen, T. Huttula, J. Sipura, A. Rämö & P. Mononen, 2000. Water chemistry, phytoplankton and hydrodynamics in Lake Pielinen, eastern Finland. Verhandlungen der internationale Vereinigung für Limnologie 27: 2212–2217.
Niinioja, R., A.-L. Holopainen, L. Lepistö, A. Rämö & J. Turkka, 2004. Public participation in monitoring programmes as a tool for lakeshore monitoring: the example of Lake Pyhäjärvi, Karelia, Eastern Finland. Limnologica 34: 154–159.
Niinioja, R., A.-L. Holopainen, L. Lepistö, M. Kukkonen, A. Rämö & P. Mononen, 2005. An overview of the state of two large lakes in North Karelia, Finland. Verhandlungen der internationale Vereinigung für Limnologie 29: 399–402.
Petrova, N., 1987. The phytoplankton of Ladoga and Onega lakes and its recent successional changes. Archiv für Hydrobiologie Beiheft Ergebnisse Limnologie 25: 11–18.
Rahkola-Sorsa, M., J. Tiainen & S. Staff, 2004. Eläinplanktonin ajallinen ja alueellinen vaihtelu Karjalan Pyhäjärvellä. Rajavesien ekologisen tilan arviointi -hanke. Alueelliset ympäristöjulkaisut 346: 5–46 (In Finnish).
Räike, A., O.-P. Pietiläinen, S. Rekolainen, P. Kauppila, H. Pitkänen, J. Niemi, A. Raateland & J. Vuorenmaa, 2003. Trends of phosphorus, nitrogen and chlorophyll *a* concentrations in Finnish rivers and lakes in 1975–2000. The Science of the Total Environment 310: 47–59.
Reynolds, C. S., 1984. The Ecology of Freshwater Phytoplankton. Cambridge University Press, Cambridge, 384 pp.
Rosén, G., 1981. Tusen sjöar. Växtplanktons miljökrav, Stockholm, LiberFörlag, 119 pp.
Talling, J. F., 1993. Comparative seasonal changes, and interannual variability and stability, in a 26-year record of total phytoplankton biomass in four English lake basins. Hydrobiologia 268: 65–98.
ter Braak, C. J. F., 1987. CANOCO—a FORTRAN program for canonical community ordinations by [partial] [detrended] [canonical] correspondence analysis and redundancy analysis (version 2.1). Wageningen, 95 pp.
ter Braak, C. J. F., 1990. Update notes: CANOCO version 3.10. Wageningen, 35 pp.
Utermöhl, H., 1958. Zur Vervollkommnung der quantitativen Phytoplankton Methodik. Mitteilungen der internationale Vereinigung für Limnologie 9: 1–38.
Willén, E., 1992. Long term changes in the phytoplankton of large lakes in response to changes in nutrient loading. Nordic Journal of Botany 12: 575–587.
Willén, E., 2002. Phytoplankton in water quality assessment—an indicator concept. In Heinonen P., G. Ziglio & A. Van der Beken (eds), Hydrological and Limnological Aspects of Lake Monitoring. John Wiley & Sons, Ltd, Chichester, 57–80.

Hydrobiologia (2008) 599:143–152
DOI 10.1007/s10750-007-9196-9

ELLS 2007

Plant-associated invertebrates and hydrological balance in the large volcanic Lake Bracciano (Central Italy) during two years with different water levels

Luciana Mastrantuono · Angelo G. Solimini · Peeter Nõges · Marcello Bazzanti

Abstract Lakes in the Mediterranean region experience an altered hydrological balance with large water level reductions and/or fluctuations within and between years. To date, little is known about the reaction of invertebrate assemblages inhabiting the littoral zone to water level changes in natural lakes. Here, we present the case of the volcanic Lake Bracciano, one of the largest and deepest Italian lakes. We compared the numerical composition and taxonomic richness of plant-associated invertebrate assemblages sampled from three sites and three depth ranges (0–4 m, 4–8 m and 8–15 m) between years with different water levels. Using historical data, we built a hydrological balance model to assess the role of different water budget components on the water levels. The hydrological balance showed that the volume of water currently abstracted for human use exceeds the potential changes caused by the projected warming of climate and decreasing amount of precipitation by nearly an order of magnitude. In a low water level year, littoral macrophytes and invertebrate assemblages exhibited qualitative and quantitative differences at all sampling sites and depths. The invertebrate assemblage showed a reduced taxonomic richness and lower numerical abundance of the more sessile forms (water mites, gastropods, nematodes, naidid oligochaetes), which feed directly on living plants or epiphytic algae, and an increase of more mobile and/or detritivore taxa.

Keywords Littoral zone · Macrophytes · Climatic change · Mediterranean region

Guest editors: T. Nõges, R. Eckmann, K. Kangur, P. Nõges, A. Reinart, G. Roll, H. Simola & M. Viljanen
European Large Lakes—Ecosystem changes and their ecological and socioeconomic impacts

L. Mastrantuono (✉) · M. Bazzanti
Department of Animal and Human Biology, University "La Sapienza", Viale dell'Università 32, Rome 00185, Italy
e-mail: luciana.mastrantuono@uniroma1.it

A. G. Solimini · P. Nõges
European Commission, Joint Research Centre, Institute for Environment and Sustainability, Ispra 21027, Italy

Introduction

Progressive increase of temperature and decrease in rainfall are the main projected features of climate change in the Mediterranean region (Dragoni, 1998; Christensen et al., 2007). The region is predicted to warm at a rate between 0.1 and 0.4°C per decade, twice that of northern Europe, to show a decline of mean annual precipitation south of 45°N (Haas, 2002) and an increase of extreme short-term precipitations. Consequently, lakes in the Mediterranean regions will experience large water level reductions and/or fluctuations within and between years. Superimposed on the variability of climate are anthropogenic water abstractions for agricultural, human or industrial needs that may become unsustainable, especially in years characterized by lower rainfall. Due to the relatively small hydrogeological basins, water levels of volcanic lakes in Central Italy are particularly sensitive to changing water balance, including human water abstraction.

Hydrological processes such as erosion, the retreat or advancement of shorelines and sediment resuspension affect the biota in the lake littoral zone. Submerged and emergent macrophytes respond sensitively to water level fluctuations, affecting the diversity and abundance of assemblages (Hellsten et al., 1996; Hill et al., 1998). As macrophytes and the associated periphyton are present only in this lake zone and provide food and physical habitat for invertebrates (Jeppesen et al., 1998), water level fluctuations probably also influence the littoral invertebrates by reducing the availability of the complex habitats provided by plants.

The volcanic Lake Bracciano is one of the largest and deepest Italian lakes, important as a source of drinking water for the city of Rome and for fishing and recreational activities. Over the last decade, low water levels have been observed in the lake, but the relative importance of climatic factors and human abstraction in determining those levels and the effects on the littoral biota are unclear. Using historical data, we built a hydrological balance model of the lake to assess the effects of climatic factors (precipitation, temperature) and anthropogenic water abstraction on the water level in the lake. To assess whether water level changes affected the littoral biota, we compared the invertebrate assemblages in the littoral zone between 2 years with different water levels. In particular, we considered the plant-associated invertebrates, which include mobile (sprawlers and climbers, excluding cladocerans and copepods because of their free-swimming habit), subsessile and sessile invertebrates that live within (miners) or attached to the plant substratum (Hutchinson, 1993; Merritt & Cummins, 1996).

Material and methods

Study area

Lake Bracciano (surface area 51 km^2, maximum depth 165 m; Fig. 1) is of volcanic origin, located at an altitude of 164 m a.s.l. in the "Monti Sabatini" area in the northern part of the Latium Region (Central Italy) and characterized by a long water residence time (137 years, Gaggino et al., 1985). Along most of its perimeter (31.5 km), the lake has a gently sloping bottom and large sandy beaches used for recreation.

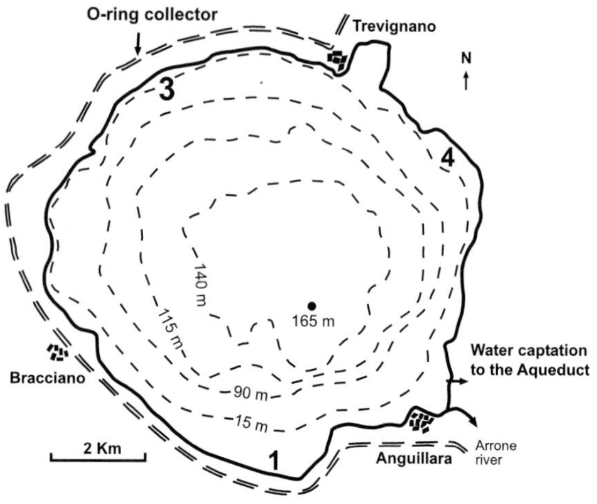

Fig. 1 Bathymetric map of Lake Bracciano with the three sampling sites

Water utilization by humans dates back to ancient times (Emperor Trajan's aqueduct was built in AD 110) and continues today. The pipeline of the aqueduct draws water from a depth of about 50 m in a locality near the Arrone river, the main effluent of the lake. In order to preserve the lake water from pollution, an O-ring collector connected with a sewage treatment plant was built around the lake in 1982–84 (Bruno, 1985; Martini, 1985). It discharges the treated waters directly into the Arrone river. The water quality parameters during the last decade showed good nutrient status (Pagnotta et al., 2002) and the lake was recently classified as oligo-mesotrophic (Margaritora et al., 2003).

Hydrological model

A hydrological catchment model based on annual values (1970–2004) of water inputs and losses, verified by measured lake levels, was used to assess the effects of different water budget components on the water level in Lake Bracciano. We used the following balance equation (all values annual in 10^6 m^3 y^{-1}):

$$P_L + P_C + G - E_L - E_C - A - Q_{out} - GW_{out} = f^*\Delta V_L + (1-f)^*\Delta V_G$$

where P_L and P_C are the amounts of precipitation over the lake area and the catchment area excluding the lake; G is the sub-surface inflow from the area by which the groundwater catchment exceeds the surface catchment; E_L is the evaporation from the lake surface; E_C is the evapotranspiration from the catchment area excluding the lake; A is the water abstraction from the basin; Q_{out} is the outflow; GW_{out} is the groundwater outflow; ΔV_L is the water accumulation in the lake; ΔV_G is the water accumulation in the ground and f is a dimensionless partitioning coefficient for water accumulation. Similar models have been used for other volcanic lakes (Dragoni et al., 2002, 2006), as they approximate the parameters useful for understanding the relative roles of hydrological components.

As input variables, we used the daily rainfall, air temperature and humidity, wind direction and speed from several stations in the basin and lake water levels. All relevant historical hydrological data were provided by the competent local authority. P_L and P_C were calculated for the areas of the lake and the catchment. We applied a constant inflow of 0.009 m^3 s^{-1} km^{-2} typical of the volcanic groundwater aquifers of the upper Latium region (Boni et al., 1986) to calculate the sub-surface inflow (G) from the area (24.8 km^2) by which the groundwater catchment exceeded the surface catchment. The evaporation from the lake surface (E_L) was calculated from monthly mean temperature and monthly mean air humidity using the equation of Visentini as described in Dragoni & Valigi (1994). To calculate the evapotranspiration from the rest of the catchment (E_C, mm y^{-1}) we used the equation of Turc (1963) based on annual mean temperature and precipitation values.

In order to balance the model, correction coefficients were applied to both evaporation components. The best fit (highest correlation between modelled and measured water levels) was achieved by applying the coefficients 0.978 and 0.985 to lake and basin evaporation, respectively. Data on water abstraction from the basin (A) were provided by the local water management company and used in the model without correction. Although it is known that Lake Bracciano is a practically closed lake where the riverine outflow (Q_{out}) is of very minor importance, a virtual outflow was created in the model to simulate various scenarios. Groundwater outflow (GW_{out}), not an easily measurable parameter, was given a constant value of 5^*10^6 m^3 y^{-1} that corresponded to the best model fit. Water accumulation in the basin (ΔV) was split into two compartments: accumulation in the lake (ΔV_L) and accumulation in the groundwater aquifer (ΔV_G). The partitioning coefficient (f), which divided the water accumulation between the two compartments, was calibrated against the measured water level amplitude in the lake. The amplitudes of the modelled and measured water levels were equal when 44.5% of the accumulated water was stored in the lake and 55.5% in the groundwater aquifer. When the model was calibrated and balanced we ran a regression analysis among the budget components to infer the relative contribution of each component to water level change.

Littoral invertebrates

Invertebrates were sampled every 2 months from three sampling stations (Fig. 1) in the same microhabitats (macrophyte substratum) during 2 years with different water levels: from May 1998 to February 1999 (hereafter: 1998 sampling campaign; for details see Mastrantuono & Mancinelli, 2005) and from September 2003 to July 2004 (hereafter: 2003 sampling campaign). The average year water level

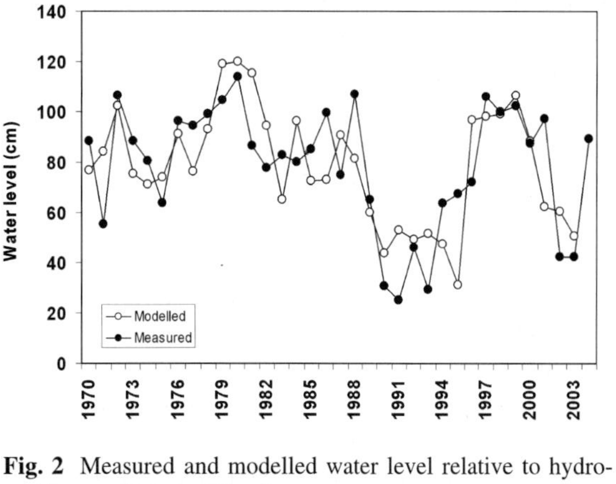

Fig. 2 Measured and modelled water level relative to hydrometric zero in Lake Bracciano

in 1998 was little higher than the long-term average (Fig. 2), while the water level of 2003–2004 was about 60 cm lower. Plant-associated invertebrates were sampled at each station and date (six dates per survey) with a sledge dredge (size: 35 × 20 cm, mesh size: 180 μm) along the entire plant-colonized belt at three depth intervals (0–4 m, 4–8 m, 8–15 m). At each depth range, the dredge was dragged parallel to the shoreline for about 50 m, following a sinusoidal path. In the 2003 sampling campaign, no aquatic vegetation occurred in the deepest zone investigated (8–15 m) at site 3, so no samples were taken. The dredge utilized is mounted on a steel framework (weight: 7 kg), which prevents sinking into the sediments. The mesh size adopted is suitable for capturing meiobenthic invertebrates such as micro-turbellarians, small nematodes and water mites. This instrument (Merrit & Cummins, 1996) appears to be suitable for sampling macrophyte-associated invertebrates when the bottom is not accessible by other devices, such as a pond net.

The coverage of each macrophyte species at the sampling sites was estimated approximately from its relative occurrence in the samples collected and assigned to one of three categories (<10%; 10–50%; >50%). In addition, the presence and distribution of plants were assessed visually when possible. The material collected was preserved in 5% formalin. In the laboratory, the macroinvertebrates were sorted from the bulk samples and separated from the meiobenthic organisms, which were subsampled and counted under a stereomicroscope at lower magnification. Specimens were identified at the lowest possible taxonomic level, in most cases the species level.

We used analysis of variance to assess whether the mean number of taxa per sample varied among years, sampling sites and depth ranges. We ran the analysis separately for each predominant taxonomic group (e.g. those with the higher number of species: Turbellaria, Nematoda, Oligochaeta, Diptera Chironomidae, Acarina and Gastropoda). Multiple comparisons were performed when necessary using the Bonferroni procedure. The mean number of taxa per sample was $\text{Log}_{10}(x + 1)$ transformed prior to statistical analysis to stabilize the variances.

To analyze the numerical counts of individuals per taxon, we generated an association matrix between samples using Bray–Curtis similarity and we ran a non-metric multidimensional scaling (N-MDS) to produce two-dimensional ordinal plots (Clarke & Warwick, 2001). To test the hypotheses that the abundance of taxa differed in years and depth zones, we computed a nonparametric two-way analysis of similarity (ANOSIM; Clarke & Warwick, 2001). Pair-wise permutation tests followed each ANOSIM to determine where the significant differences between the levels of the factors lay.

Similarity percentage (SIMPER) procedures were applied to identify the species that contributed most to the significant differences between years and among depth zones (Clarke & Warwick, 2001). The SIMPER algorithm determines the relative contribution of each species to the average similarity within a group and the average dissimilarity between groups. If a species consistently contributes to within-group similarity between pairs of samples, and to between-group dissimilarity between pairs of samples, then its percentage contribution to similarity is high and it can be considered a good discriminating species (Clarke & Warwick, 2001).

Results

The hydrological catchment model built for Lake Bracciano showed a good agreement between the modelled and the observed water levels in the lake (Fig. 2, $R^2 = 0.48$, $n = 33$, $P < 0.05$). If both series are smoothed with a 5-year moving average, R^2 increases to 0.85, which means that the model simulates the major dynamic of the water level well. Regarding the different components of the water budget, the amount of precipitation over the lake area (P_L) and the catchment area excluding the lake (P_C) together formed 94% of the budget input, which explained 88% of the water level changes ($n = 34$; $P < 0.001$). The sub-surface inflow from the extended part of the groundwater catchment (G) was $9.8{*}10^6\ m^3\ y^{-1}$ (6% of the water input).

Evaporation from the lake surface (E_L) constituted, on average, 43% of the total loss of water from the basin. Evapotranspiration from the catchment (E_C) formed 37% of the total loss. However, E_C had a statistically stronger relationship with the water level, explaining 53% ($n = 34$; $P < 0.001$) of the measured water level changes, while E_L accounted for 32% of it (n = 34; P = 0.001). During the whole period of investigation (1970–2000), human water abstraction accounted for between 4 and 21% (average 16%) of the total water loss from the basin. The modelled groundwater outflow (GW_{out}) made up 3.1% of the total water loss and also the river outflow, occurring only in rainy years, was a minor component (2%) of the budget.

The annual mean lake water levels differed by 60 cm between 1998 and 2003–2004. In the 2003 survey, aquatic plant species covered a reduced area and were more sparsely distributed at all sampling sites and depths (Table 1). This reduction was particularly apparent for the Charales, which constitute the dominant component of the aquatic plant assemblage, as they are present throughout the year at depths from 6 to 15–20 m. Moreover, the coverage of *Myriophyllum spicatum* and *Ceratophyllum demersum*, the most common species in central Italian lakes, was lower in the second survey.

Considering the 1998 and the 2003 surveys together, we collected a total of 113 invertebrate taxa: the largest number of species belonged to chironomids, oligochaetes, nematodes, acari, turbellarians and gastropods (Table 2). The overall number of taxa diminished drastically from 1998 to 2003 (from 106 to 70). The ANOVA results showed that the average number of taxa per sample (99 site, depth and date combinations) was significantly lower in the second survey for all taxa with the exception of chironomids (Table 3). The sampling site was not significant for most taxa, but the depth range was a significant factor for Turbellaria and Acarina (Bonferroni multiple comparisons: depth 0–4 m significantly different from the others). Chironomidae and Trichoptera showed a significant interaction between depth and site, because richness was lower at site 3 in the depth range 0–4 m (Bonferroni multiple comparisons: Site 3, depth 0–4 m different from the others). Interactions of all pairs of factors were significant for Gastropoda, as richness at site 3, depth 0–4 m and year 2003 was lower than the others (Bonferroni multiple comparisons: site 3, depth 0–4 m, year 2003 different from the others).

The non-metric multidimensional scaling (N-MDS) ordination of the samples, based on their similarity calculated from the numerical composition of the invertebrate assemblage, revealed a more detailed picture of the differences between years and depths (Fig. 3). The assemblages were significantly different between years (ANOSIM: R-statistic = 0.74, $P < 0.01$) and among depths (ANOSIM: R-statistic = 0.336, $P < 0.01$; all multiple comparisons among depths $P < 0.01$). The subsequent SIMPER analysis revealed substantial differences among years in the numerical composition of the assemblage at different depth ranges (Table 4). A few plant-associated taxa contributed to most of the dissimilarity between years and depths. Seven out of the 16 taxonomic groups and 12 out of the 113 species and/or genera represented the main contributors to dissimilarities between years at all three depth ranges (Table 4).

Table 1 Relative coverage of littoral macrophytes and Characeae in the littoral of Lake Bracciano

Year	1998									2003							
Site	1			3			4			1			3		4		
Depth zone (m.)	0–4	4–8	8–15	0–4	4–8	8–15	0–4	4–8	8–15	0–4	4–8	8–15	4–8	8–15	0–4	4–8	8–15
Characeae (sp. a)	2	1	3	2	2	3	1		2	2	3	2	1	2	2	2	1
Characeae (sp. b)	3	3	1	2	2	1	2	3	2		1	2	2	2		2	2
Elodea canadensis							1						1				
Myriophyllum spicatum	1						3						1				
Ceratophyllum demersum							2						1	2			
Potamogeton crispus	1						2										
Potamogeton perfoliatus							1			1			2				
Potamogeton pectinatus	1			1			1						1				
Potamogeton natans										1			1		2		
Potamogeton lucens													1				

Coverage codes are as follows: 1, less than 10% of bottom covered; 2, between 10 and 50%; 3, over 50%

Table 2 Number of taxa in each year, site and depth zone (all sampling dates cumulated)

Year	1998									2003							
Site	1			3			4			1			3		4		
Depth zone (m.)	0–4	4–8	8–15	0–4	4–8	8–15	0–4	4–8	8–15	0–4	4–8	8–15	4–8	8–15	0–4	4–8	8–15
Hydroida	1	1	1	1	1	1	1	1	1	1	1	1	1	1	1	1	1
Turbellaria	2	2	7	3	4	6	4	5	6	0	1	2	1	1	0	0	3
Nematoda	8	5	7	6	7	9	8	4	7	3	2	1	3	0	2	2	1
Oligochaeta	12	7	9	8	9	7	10	9	7	2	3	7	6	4	4	3	5
Hirudinea	0	0	0	0	0	0	0	0	0	0	0	0	1	0	0	0	0
Ostracoda	1	0	2	2	0	0	1	0	3	0	0	1	1	2	0	0	1
Amphipoda	1	1	1	1	1	1	1	1	1	1	1	1	1	1	1	1	1
Decapoda	1	1	1	1	1	1	1	1	1	1	1	1	1	1	1	1	1
Ephemeroptera	2	2	1	1	1	1	1	1	1	2	2	2	2	1	2	2	1
Odonata	2	0	0	1	0	2	1	1	0	1	0	1	1	2	2	0	1
Chironomidae	12	7	7	5	12	16	12	6	13	18	6	6	9	10	12	7	8
Lepidoptera	0	0	0	0	1	0	1	0	0	0	0	0	0	0	0	0	0
Trichoptera	4	2	2	4	3	3	4	2	2	2	1	1	3	3	3	1	2
Acarina	11	9	11	9	8	9	8	7	8	9	7	7	3	7	8	9	8
Gastropoda	5	1	7	5	2	5	5	1	5	4	1	4	4	6	4	1	5
Total	59	35	48	43	45	54	53	33	48	43	24	32	35	37	39	27	34

Table 3 Results of ANOVA on average taxonomic richness per sample (more abundant taxa only)

Taxon	Year	Depth	Site	Interactions
Turbellaria	$P < 0.01$	$P < 0.05$	ns	ns
Nematoda	$P < 0.01$	ns	ns	ns
Oligochaeta	$P < 0.01$	ns	ns	ns
Chironomidae	ns	–	–	Depth × site
Acarina	$P < 0.01$	$P < 0.05$	ns	ns
Gastropoda	–	–	–	Year × depth; Year × site; Depth × site
Trichoptera	$P < 0.01$	–	–	Depth × site

Factors were sampling sites (three sites), years (1998 and 2003) and depth zones (0–4 m; 4–8 m; 8–15 m). Significant interactions are also reported. Single effect factors were not tested (–), when interactions were significant

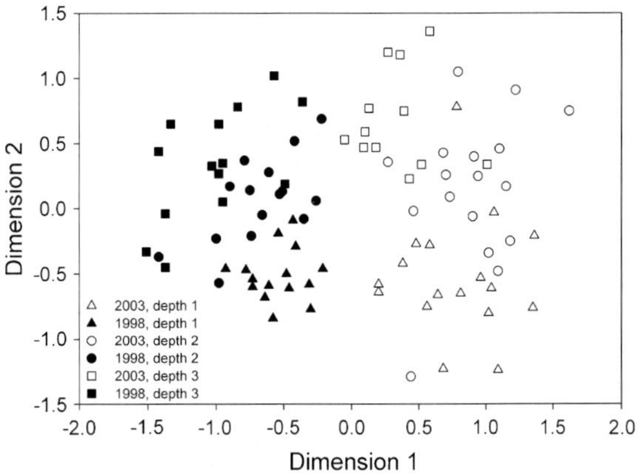

Fig. 3 Non-metric multidimensional scaling ordination of invertebrate samples (stress = 0.09) for each sampling date, depth and site combination. Upper triangles are samples from depth 1 (0–4 m), circles are samples from depth 2 (4–8 m) and boxes are samples from depth 3 (8–15 m). Closed symbols are samples from 1998 to 1999; open symbols are samples from 2003 to 2004

Discussion

In Lake Bracciano, the human water abstraction was constantly high in the 1970s and 1980s but was reduced in the middle of the 1990s when dry years occurred. During the whole period, water abstraction was one of the major components of water losses from the basin exceeding both the groundwater outflow and the river outflow. The 60 cm difference between annual mean water levels in 1998 and 2003–2004 was relatively small, but given the gentle slope of the littoral area, this decrease in water level resulted in a 15 m recession of the shoreline (varying

Table 4 Breakdown of average dissimilarity between 1998 and 2003 surveys at the three different depth intervals for selected taxa

Depth zone (m)	0–4			4–8			8–15		
	1998	2003	% C	1998	2003	% C	1998	2003	% C
Acarina	2039	820	16.4	5868	3168	21.8	14565	9489	20.1
*Unionicola**			2.4			3.2			3.2
*Limnesia**			2.3			4.6			3.1
*Arrenurus**			2.3			4.9			2.8
Halacaridae*			3.3			3.5			2.7
Acercus						4.1			3.1
Neumania						9.6			2.2
Chironomidae	3514	2847	23.5	1746	2473	18.2	3239	859	18.9
*Psectrocladius**			3.8			5.3			3.9
Cricotopus			3.4			2.2			
*Paratanytarsus**			3.4			2.9			2.6
Pentaneurini			2.9			2.6			
Nematoda	5648	195	13.9	6311	70	14.0	3330	55	11.8
Dorylaimina*			5.6			6.7			3.5
*Ethmolaimus pratensis**			3.6			4.3			2.5
Plectus sp.			2.5						
Monhystera sp.									2.2
Gastropoda	4722	2001	7.6	6554	2483	3.4	10567	2994	11.0
Theodoxus fluviatilis			2.5						3.2
*Bithynia tentaculata**			2.5			2.2			
Hydrobioidea									2.5
Valvata piscinalis									2.4
Oligochaeta	810	265	7.8	3145	216	13.5	13946	330	14.3
*Nais variabilis**			3.1			4.2			3.5
Chaetogaster diastrophus						2.9			
Stylaria lacustris									2.7
Dero sp.									2.4
Turbellaria	694	33	5.1	708	23	6.3	1338	206	7.1
*Macrostomum**			3.6			2.7			2.4
Dugesia tigrina						2.5			2.3
Trichoptera	953	376	4.4	592	635	4.2	750	92	2.5
*Leptocerus**			2.6			3.1			2.5
Ephemeroptera	168	642	5.5	43	880	5.3			
Caenis			3.2			3.6			
Cloeon gr. similis			2.3						
Amphipoda	222	647	2.4	33	166	2.1			
Echinogammarus			2.4			2.1			
Decapoda	1017	499	2.9						
Palaemonetes antennarius			2.9						

Taxa are ordered on the basis of their contribution to the dissimilarity among years. For each major taxonomic group, the total counts of individuals expressed as average over all sites and dates, total percentage contribution to dissimilarities among years (%C) and species breakdown are showed. Asterisks indicate taxa that contribute to among years dissimilarities at all depth ranges from 3 to 50 m in different parts of the lake), leaving 385,000 m^2 of littoral area dry over the whole perimeter of the lake. Therefore, the lake topography is likely to have a major effect on the response of communities to water level fluctuations, and both the rate of draw-down and the absolute amount are important.

The effects of water level changes on littoral assemblages of natural lakes have seldom been examined because most published studies refer to reservoirs or human-regulated lakes. The few studies dealing with littoral macrophytes and invertebrates in natural lakes have highlighted the complex interaction among different stressors in natural systems, making it difficult to isolate the effects on biota of water level changes from other pressures (Grimås, 1962; Palomäki, 1994; Savage & Beaumont, 1997; Hamabata & Kobayashi, 2002; Tronstad et al., 2005; Van Geest et al., 2005a, 2005b; Furey et al., 2006). As the water of Lake Bracciano does not receive major pollution from the watershed because of the O-ring collector of sewage, and as no other major stressors were reported during 2003, the observed changes in macrophyte and invertebrate communities can probably be ascribed to the water level reduction between the two study years.

The pattern of macrophyte abundance responds quickly to water level changes and is a reliable indicator of such stress (Hellsten & Dudley, 2006). The effects of water level changes on macrophytes are known from reservoirs (Wilcox & Meeker, 1991, 1992), floodplain and shallow lakes (Hamabata & Kobayashi, 2002; Van Geest et al., 2005a) and Scandinavian lakes (Hellsten, 2001), but less information is available from natural lakes in the Mediterranean area. Although a reduction in the number of species of submerged and floating leaved macrophytes has been reported by some authors (Van Geest et al., 2005b), the lowering of water level seems mainly to affect abundance patterns (Coops & van der Velde, 1995). This was also observed in the present study, where the lowering of water level was reflected by a reduction in macrophyte coverage, especially in the shallower zone.

Submerged macrophytes provide a complex multidimensional habitat and related food sources, such as detritus and periphytic algae (see Jeppesen et al., 1998), and greatly increase the ecological opportunities for invertebrates. Therefore, the observed change in macrophyte abundance pattern in the low water level year (2003) was probably the main cause of the diminution of invertebrate taxon richness in Lake Bracciano. This finding agrees with the results of other studies on a shallow lake (Savage & Beaumont, 1997) and on a large regulated lake in Central Italy (Lake Campotosto; Mastrantuono, 1987).

In the low water level year, the most evident structural change in the invertebrate assemblage was the increase of more mobile and/or feeding opportunistic taxa and the reduction of several sessile and/or herbivorous taxa. In particular, we observed a reduction of individuals between 1998 and 2003 in those taxa that use the aquatic vegetation as substrate or that feed directly on living plants or epiphythic algae (scrapers or grazers, according to Cummins & Wilzbach, 1985; Merritt and Cummins, 1996). Those include nematodes (Dorylaimina and *Ethmolaimus pratensis*), water mites (*Unionicola*, *Limnesia*, *Arrenurus*, Halacaridae), gastropods (*Bithynia tentaculata*) and naidid oligochaetes (*Nais variabilis*) (Table 4). In contrast, we observed a numerical increase in a few more mobile (climbers and sprawlers) and detritivore/omnivore invertebrates (shredders and collectors), such as the orthoclads *Psectrocladius* and *Cricotopus* at all depth ranges, the ephemeropteran *Caenis* and the macrocrustacean *Echinogammarus* at the shallowest (0–4 m) and intermediate depth ranges (4–8 m) (Table 4).

Our results suggest that the changes in macrophyte spatial pattern attributable to the lowering of water level could be tracked more efficiently by mobile invertebrates than by taxa that live attached to the plant substrata. From this point of view, more sessile forms that rely on macrophytes, for example, for obtaining food and for egg deposition (Merritt & Cummins, 1996), such as turbellarians, nematodes, water mites, gastropods and oligochaetes (see also Økland, 1983; Lodge & Kelly, 1985), may encounter unfavourable conditions in patchy and reduced macrophyte beds. In contrast, opportunistic forms (detritivore or omnivorous taxa) and mobile taxa (e.g. macrocrustaceans) may gain advantage from the availability of detritus coming from the bulk of decaying plants in lowered water level years. Interestingly, similar results were reported by Tronstad et al. (2005) for wetlands affected by decreasing water level, where mobile forms moved along with the receding water line.

In conclusion, the hydrological model for Lake Bracciano showed that the water volume that is currently abstracted for human use, exceeds the potential changes caused by the projected warming of climate and decreasing amount of precipitation during the first half of this century by nearly an order of magnitude. The decrease in water level will result in a loss of large areas of littoral habitat especially along the southern and eastern shores, where the bottom slopes less steeply. Our results indicate a marked difference in the structure of the invertebrate littoral assemblage in the low water level year. The numerical reduction in more sessile invertebrate taxa appears to be linked with the more patchy distribution of macrophytes observed in the low water level year. Overall, water level lowering seems to cause a significant loss of invertebrate diversity as indicated by a decline in both species richness and Shannon diversity (Shannon diversity range in 1998: 4.0–4.2; range in 2003: 3.2–3.9). The clear response of the littoral invertebrates to water level change underlines their potential usefulness as biological indicators for assessing current conditions and forecast future trends of the natural lakes in the Mediterranean.

Acknowledgements We thank G. Free and three anonymous referees for helpful comments on an early draft.

References

Boni, C., P. Bono & G. Capelli, 1986. Schema idrogeologico dell'Italia centrale. Memorie della Società Geologica Italiana 35: 991–1012.

Bruno, G., 1985. Bracciano sewerage system amd sewage treatment plant automatic control. Proceedings of the International Congress "Lakes pollution and recovery" (ANDIS), Roma: 194–198.

Christensen, J. H., B. Hewitson, A. Busuioc, A. Chen, X. Gao, I. Held, R. Jones, R. K. Kolli, W.-T. Kwon, R. Laprise, V. Magaña Rueda, L. Mearns, C. G. Menéndez, J. Räisänen, A. Rinke, A. Sarr & P. Whetton, 2007. Regional Climate Projections. In Solomon, S., D. Qin, M. Manning, Z. Chen, M. Marquis, K. B. Averyt, M. Tignor & H. L. Miller (eds), Climate Change 2007. The Physical Science Basis. Contribution of Working Group I to the Fourth Assessment Report of the Intergovernmental Panel on Climate Change. Cambridge University Press: 849–940.

Clarke, K. R. & R. M. Warwick, 2001. Change in marine communities: an approach to statistical analysis and interpretation. 2nd edn, Primer-E, Plymouth.

Coops, H. & G. van der Velde, 1995. Seed dispersal, germination, and seedling growth of six helophyte species in relation to water-level zonation. Freshwater Biology 34: 13–20.

Cummins, K. W. & M. A. Wilzbach, 1985. Field procedures for analysis of functional feeding groups of stream invertebrates. Appalachian Environmental Laboratory, University of Maryland, Frostburg, Contr. No. 1611.

Dragoni, W., 1998. Some considerations on Climatic Changes, Water Resources and Water Needs in the Italian Region South 43°N. In Brown, N. & A. Issar (eds), Water, environment and society in times of climatic change. Kluwer: 241–271.

Dragoni, W. & D. Valigi, 1994. Contributo alla stima dell'evaporazione dalle superfici liquide nell'Italia Centrale. Geologica Romana 30: 151–158.

Dragoni, W., F. Lotti, V. Piscopo & A. Sibi, 2002. Bilancio idrogeologico del lago di Vico (Lazio - Italia). Proceedings of the international conference Residence time in lakes: Science, Mangement, Education. Bolsena, Viterbo: 96–104.

Dragoni, W., V. Piscopo, L. Di Matteo, L. Gnucci, A. Leone, F. Lotti, M. Melillo & M. Petitta, 2006. Risultati del progetto di ricerca PRIN laghi 2003–2005. Giornale di Geologia Applicata 3: 39–46.

Furey, P. C., R. N. Nordin & A. Mazumder, 2006. Littoral benthic invertebrates under contrasting drawdown in a reservoirs and a natural lake. Journal of North American Benthological Society 25: 19–31.

Gaggino, G. F., R. Marchetti, E. Cappelletti & T. Calcagnini, 1985. La qualità delle acque dei laghi italiani negli anni '80. Atti Congresso Internazionale EWPCA, Andis ed.: 5–44.

Grimås, U., 1962. The effect of increased water level fluctuation upon the bottom fauna in Lake Blåsjön, Northern Sweden. Institute of Freshwater Research Drottningholm Report 44: 14–41.

Haas, L., 2002. Mediterranean water resources planning and climate change adaptation. Water, Wetlands and Climate Change: Building Linkages for their Integrated Management. Mediterranean Regional Roundtable, Athens, Greece, December 10–11, 2002. 62 pp.

Hamabata, E. & Y. Kobayashi, 2002. Present status of submerged macrophyte growth in Lake Biwa: recent recovery following a summer decline in the water level. Lakes & Reservoirs: Research and Managements 7: 331–338.

Hellsten, S., M. Marttunen, R. Palomäki, J. Riihimäki & E. Alasaarela, 1996. Towards an ecologically-based regulation practice in Finnish hydroelectric lakes. Regulated Rivers: Research & Management 12: 535–545.

Hellsten, S., 2001. Effects of lake water level regulation on aquatic macrophyte stands in northern Finland and options to predict these impacts under varying conditions. Acta Botanica Fennica 171: 1–47.

Hellsten, S. & B. Dudley, 2006. Hydromorphological pressures in lakes. In A. G. Solimini, A-S. Heiskanen & A. C. Cardoso (eds), Indicators and methods for the ecological status assessment under the Water Framework Directive. European Commission, EUR 22314 EN, Ispra: 135–140.

Hill, N. M., P. A. Keddy & I. C. Wisheu, 1998. A hydrological model for predicting the effects of dams on the shoreline

vegetation of lakes and reservoirs. Environmental Management 22: 723–736.
Hutchinson, G. E., 1993. A treatise on limnology. IV. The Zoobenthos. Wiley & Sons, Inc.
Jeppesen, E. M., M. Sondergaard, M. Sondergaard & K. Christoffersen, 1998. The structuring role of submerged macrophytes in lakes. Springer, New York: 452.
Lodge, D. M. & P. Kelly, 1985. Habitat disturbance and the stability of freshwater gastropod populations. Oecologia 68: 111–117.
Margaritora, F. G., M. Bazzanti, O. Ferrara, L. Mastrantuono, M. Seminara & D. Vagaggini, 2003. Classification of the ecological status of volcanic lakes in Central Italy. Journal of Limnology 62: 49–59.
Martini, P., 1985. The project CO.B.I.S. plants. Proceedings of the international congress, "Lakes pollution and recovery" (ANDIS), Roma: 194–198.
Mastrantuono, L., 1987. Invertebrate community in the littoral regulated area of a hydroelectric lake-reservoir (Lake Campotosto, Central Italy). Rivista di Idrobiologia 26: 17–32.
Mastrantuono, L. & T. Mancinelli, 2005. Littoral zoobenthos associated with aquatic plants and bioassessment of ecological status in Lake Bracciano (Central Italy). Journal of Limnology 64: 43–53.
Merritt R. W. & K. W. Cummins, (eds), 1996. An introduction to the aquatic insects of North America. Kendall/Hunt Publishing Company, Dubuque, USA: 862.
Økland, J., 1983. Factors regulating the distribution of freshwater snails (Gastropoda) in Norway. Malacologia 24: 277–288.
Pagnotta, R., D. Copetti & G. Tartari, 2002. Trophic classification of the main lakes of central Italy. Proceedings of the international conference Residence time in lakes: Science, Mangement, Education. Bolsena, Viterbo: 129–137.
Palomäki, R., 1994. Response by macrozoobenthos biomass to water level regulation in some Finnish lake littoral zones. Hydrobiologia 286: 17–26.
Savage, A. A. & D. L. Beaumont, 1997. A comparison of the benthic macroinvertebrate communities of a lowland lake, Oak Mere, in 1980 and 1994. Archiv für Hydrobiologie 139: 197–206.
Tronstad, L. M., P. Bryan, P. Tronstad & A. C. Benke, 2005. Invertebrate responses to decreasing water levels in a subtropical river floodplain wetland. Wetlands 25: 583–593.
Turc, L., 1963. Evaluation des besoins en eau d'irrigation, évapotranspiration potentielle, formulation simplifié et mise à jour. Annales Agronomiques 12: 13–49.
Van Geest, G. J., H. Coops, M. M. Roijackers, A. D. Buijse & M. Scheffer, 2005a. Succession of aquatic vegetation driven by reduced water-level fluctuations in floodplain lakes. Journal of Applied Ecology 42: 251–260.
Van Geest, G. J., H. Wolters, F. C. J. M. Roozen, H. Coops, M. M. Roijackers, A. D. Buijse & M. Scheffer, 2005b. Water-level fluctuation affect macrophyte richness in floodplain lakes. Hydrobiologia 539: 239–248.
Wilcox, D. A. & J. E. Meeker, 1991. Disturbance effects on aquatic vegetation on regulated and unregulated lakes in northern Minnesota. Canadian Journal of Botany 69: 1542–1551.
Wilcox, D. A. & J. E. Meeker, 1992. Implications for faunal habitat related to altered macrophyte structure in regulated lakes in northern Minnesota. Wetlands 12: 192–203.

Hydrobiologia (2008) 599:153–159
DOI 10.1007/s10750-007-9186-y

A comparison of zooplankton densities and biomass in Lakes Peipsi and Võrtsjärv (Estonia): rotifers and crustaceans versus ciliates

Priit Zingel · Juta Haberman

Abstract The abundance and biomass of ciliates, rotifers, cladocerans and copepods were studied in Lake Peipsi and Lake Võrtsjärv, both of which are shallow, turbid and large. Our hypothesis was that in a large shallow eutrophic lake, the ciliates could be the most important zooplankton group. The mean metazooplankton biomass was higher in Peipsi than in Võrtsjärv (mean values and SD, 1.8 ± 0.7 and 1.3 ± 0.6 mg WM l^{-1}). In Peipsi, the metazooplankton biomass was dominated by filtrators that feed on large-sized phytoplankton and are characteristic of oligo-mesotrophic waters. In Võrtsjärv, the metazooplankton was dominated by species characteristic of eutrophic waters. The planktonic ciliates in both lakes were dominated by oligotrichs. The biomass of ciliates was much greater in Võrtsjärv (mean 2.3 ± 1.4 mg WM l^{-1}) than in Peipsi (0.1 ± 0.08 mg WM l^{-1}). Ciliates formed about 60% of the total zooplankton biomass in Võrtsjärv but only 6% in Peipsi. Thus, the food chains in the two lakes differ: a grazing food chain in Peipsi and a detrital food-chain in Võrtsjärv. Consequently, top-down control of phytoplankton can be assumed to be much more important in Peipsi than in Võrtsjärv. When the detrital food chain prevails, the planktonic ciliates become the most important zooplankton group in shallow, eutrophic and large lake. Neglecting protozooplankton can result in serious underestimates of total zooplankton biomass since two-thirds of the zooplankton biomass in Võrtsjärv comprises ciliates.

Keywords Zooplankton · Ciliates · Detrital and planktonic food chain · Biomass · Densities

Guest editors: T. Nõges, R. Eckmann, K. Kangur, P. Nõges, A. Reinart, G. Roll, H. Simola and M. Viljanen
European Large Lakes—Ecosystem changes and their ecological and socioeconomic impacts

P. Zingel · J. Haberman
Võrtsjärv Centre for Limnology, Institute of Agricultural and Environmental Sciences, Estonian University of Life Sciences, Tartu, Estonia

P. Zingel (✉)
Department of Aquaculture, Institute of Veterinary Medicine and Animal Sciences, Estonian University of Life Sciences, Tartu, Estonia
e-mail: pzingel@zbi.ee

Introduction

Zooplankton forms a central link in the food chains of a freshwater lake. The capacity of zooplankton to utilize phytoplankton production and its suitability as food for fish determine the efficiency of the food chain. In some cases, phytoplankton is not efficiently grazed by zooplankton and enters the detrital food chain directly. In most studies, the term zooplankton is used to denote crustaceans and rotifers. However, planktonic ciliates also have an important role in aquatic food webs (Azam et al., 1983, Sherr & Sherr,

1984). Ciliates are important in controlling picoplankton and bacteria (Beaver & Crisman, 1982), but they also graze upon particles that can be ingested by crustacean and rotifer zooplankters (Sanders et al., 1989). There is clear evidence that planktonic ciliates are an important food resource for larger crustacean zooplankton (Gifford, 1991). It is now known that ciliates form an important trophic link between picoplankton and nanoplankton on one hand, and metazoans on the other.

There are large gaps in our knowledge of the feeding ecology of protozooplankton in many lakes. To our knowledge, few long-term studies have been published comparing the dynamics of ciliates and metazooplankton in lakes. The aim of current study was to analyse zooplankton data collected over 10 years from two large lakes in Estonia, Lake Peipsi and Lake Võrtsjärv. Our hypothesis was that in a large shallow eutrophic lake, the ciliates could be the most important zooplankton group.

Materials and methods

The Estonian lakes studied here are large (Võrtsjärv, 270 km^2; Peipsi, 3,555 km^2), shallow (mean depths 2.8 and 8.3 m, respectively) and eutrophic. Water temperature in these lakes reaches its maximum in July and ice cover lasts from November to April. In Võrtsjärv, the average total annual phosphorus concentration is 54 μg l^{-1}, total nitrogen concentration 1.6 mg l^{-1} and mean Secchi depth 1.1 m (Nõges et al., 1998). The shallowness of the lake and the wave-induced resuspension of bottom sediments contribute to the formation of a high seston concentration and high turbidity during summer. Most of the organic matter in the lake is autochthonous (Nõges et al., 1998).

Our study of Peipsi was carried out in the largest and deepest northern part, Peipsi *s.s.* (area 2,611 km^2), which is an unstratified eutrophic lake with mesotrophic features. The average annual total phosphorus concentration in Peipsi *s.s* is 35 μg P l^{-1}, total nitrogen concentration 0.7 mg N l^{-1} and mean Secchi depth 2.2 m (Starast et al., 2001).

The dataset used in the present article to analyse ciliates and metazooplankton (MZP) is based on data collected from Võrtsjärv and Peipsi *s.s.* from 1995 to 2005. Lake Peipsi was sampled from May to November and Võrtsjärv from January to December. The ciliate samples in Peipsi *s.s* were collected monthly from the central area of the lake and in Võrtsjärv from the deepest area of the lake. All MZP analyses were based on depth-integrated samples. The samples were preserved in 4% formaldehyde and counted by conventional quantitative methods (e.g. Haberman, 1998). Subsamples (2.5 or 5 ml) of each sample were counted under a binocular microscope in a chamber of dimensions 13 × 6 cm, capacity 8 ml, at 32–56× magnification. The masses of individual zooplankton taxa were derived from length/weight regressions according to Ruttner-Kolisko (1977) and Balushkina & Winberg (1979) and represented as wet mass (WM). The mean individual metazooplankter mass was calculated as the ratio between mean MZP biomass and abundance.

For analyses of ciliates, the entire water column was sampled. The integrated lake water was obtained by mixing the water collected by a 2 l Ruttner sampler from 1 m intervals through the entire water column. Integrated subsamples were preserved and fixed with acidified Lugol's iodine, and ciliates were counted after settling by the Utermöhl (1958) technique under an inverted microscope at 400–1000× magnification. The entire area of each Utermöhl chamber was surveyed. The first 20 specimens encountered for each taxon were measured. Biovolumes of each taxa were estimated by assuming simple geometric shapes. Specific gravity was assumed to be 1.0 g ml^{-1} (Finlay, 1982), so the biomass was expressed as wet mass (WM). Zooplankton taxa that made up 20% or more of the total zooplankton number or biomass were considered to be dominants. Nonparametric methods were used for statistical analysis since none of the data were either normally or log-normally distributed.

Results

Lake Peipsi

The zooplankton biomass was dominated by cladocerans and copepods (Fig. 1). Dominant MZP species were *Bosmina berolinensis*, *Daphnia galeata*, *D. cucullata* and *Eudiaptomus gracilis*. On some occasions, *Bosmina c. coregoni*, *B. gibbera*, *Daphnia cristata*, *Diaphanosoma brachyurum* and *Leptodora*

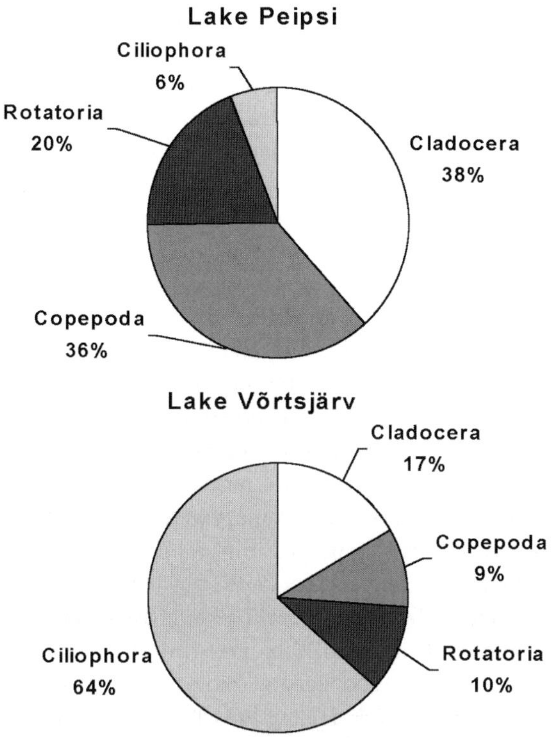

Fig. 1 Percentage of different groups in total mean zooplankton biomass in lakes Peipsi and Võrtsjärv

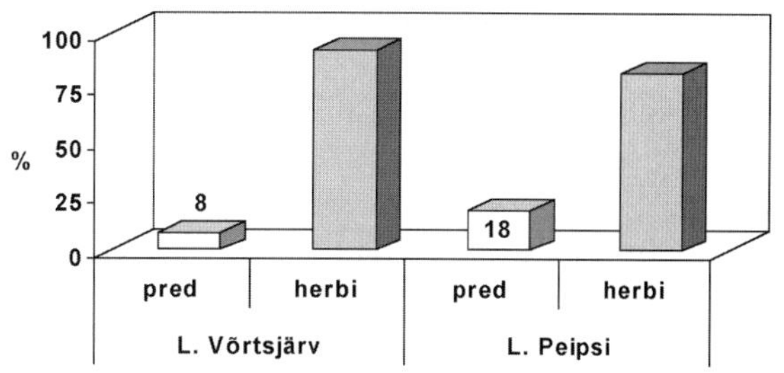

Fig. 2 Share of predators in metazooplankton biomass in lakes Peipsi and Võrtsjärv

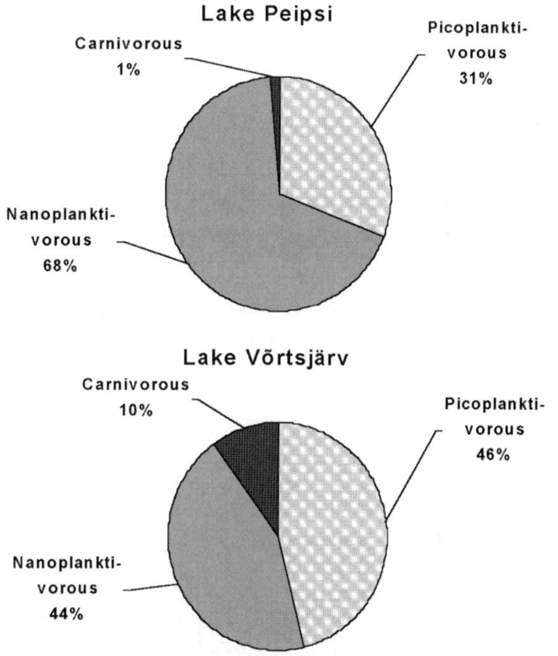

Fig. 3 Relative share of different trophic groups of ciliates in zooplankton biomass of lakes Peipsi and Võrtsjärv

kindti also dominated the biomass. Small rotifers (*Synchaeta verrucosa*, *Polyartha dolichoptera*, *Keratella cochlearis*) dominated with respect to number of zooplankton. The share of predatory MZP (e.g. *Leptodora kindti*, *Bythotrephes longimanus*) in the total MZP biomass was 18% (Fig. 2).

The population of ciliates was dominated by large oligotrichs: *Rimostrombidium* spp., *Pelagostrombidium* sp., *Limnostrombidium* sp., *Codonella cratera* and *Tintinnidium fluviatile*, most of which are nanoplanktivorous species (Fig. 3). Scuticociliates (*Uronema nigricans*, *Cyclidium claucoma*), prostomatids (*Urotricha* spp., *Balanion planktonicum*) and gymnostomes (*Askenasia volvox, Mesodinium pulex*) were less important. Altogether, 32 identifiable ciliate taxa were found.

The mean MZP numbers fluctuated between 138 and 1420 ind l^{-1} (mean 777 $\pm$ SD 534 ind l^{-1}). The highest values were recorded in May. The mean biomass was 1.8 $\pm$ 0.7 mg WM l^{-1}; values fluctuated between 1.1 and 3.0 mg WM l^{-1} (Fig. 4). The highest biomasses were recorded in July. Mean ciliate abundance was 4857 $\pm$ 3253 ind l^{-1} and mean biomass 0.1 $\pm$ 0.08 mg l^{-1}. The highest mean abundance and biomass were found in May, with a constant decrease thereafter. The biomass of MZP was higher than the ciliate biomass (*t*-test, $P = 0.0003$). Mean total zooplankton biomass (MZP + ciliates) was 1.9 $\pm$ 0.8 mg WM l^{-1}. Mean individual metazooplankter mass fluctuated between 1.4 and 7.7 µg WM ind^{-1}, with an average of 3.5 $\pm$ 2.2 µg WM ind^{-1}. Generally, mean metazooplankter mass increased from spring to autumn (Fig. 5). The biomass of ciliates was negatively correlated with the biomass of MZP ($R = -0.52$, $P < 0.01$; $n = 55$), although the degree of explanation was only 25%.

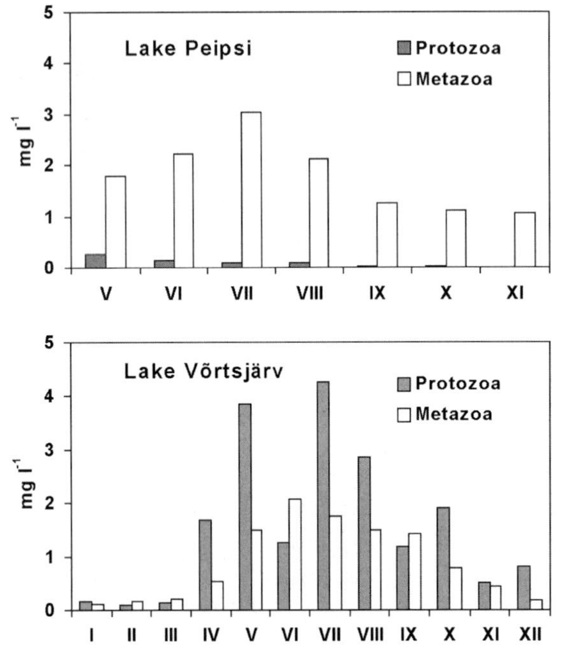

Fig. 4 Seasonal dynamics of zooplankton biomass (wet weight) in lakes Peipsi and Võrtsjärv

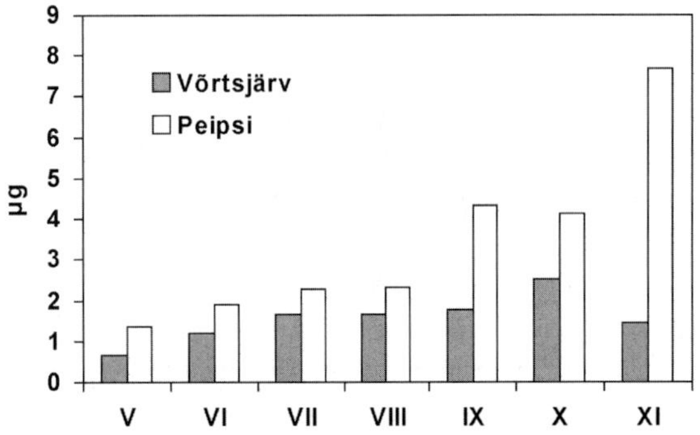

Fig. 5 Mean individual metazooplankton biomasses (wet weight) in lakes Peipsi and Võrtsjärv

Lake Võrtsjärv

In this lake the zooplankton was dominated by ciliates (Fig. 1). Oligotrichs were most abundant, especially small-sized picoplanktivorous species (*Rimostrombidium* spp., *Halteria grandinella*), followed by prostomatids (*Urotricha* spp., *Coleps* spp.), scuticociliates (*Uronema* sp., *Cyclidium* spp.) and gymnostomes (*Askenasia volvox, Mesodinium pulex, Dileptus* sp.). Altogether, 72 ciliate taxa were found. MZP biomass was dominated by cladocerans. The dominant species were *Chydorus sphaericus*, *Bosmina longirostris* and *Daphnia cucullata*. On some occasions, *Bosmina c. coregoni* also dominated. The rotifer species *Polyarthra dolichoptera, P. luminosa, Keratella cochlearis, Anuraeopsis fissa* and *Trichocerca rouseleti* were numerically dominant. The share of predatory MZP (e.g. *Leptodora kindti*, cyclopoids of IV juvenile stage) in the total MZP biomass was 8% (Fig. 2).

The mean MZP abundance was 1,040 ± 731 ind l^{-1} (ranging from 302 to 2,280 ind l^{-1}). The mean MZP biomass was 1.3 ± 0.6 mg WM l^{-1}. The highest mean biomass was recorded in June (Fig. 4). Mean ciliate abundance was 67,806 ± 45,434 ind l^{-1} and mean biomass 2.3 ± 1.4 mg WM l^{-1}. The highest mean abundance and biomass were found in July, but both values also peaked in May. Ciliate numbers declined after the summer peak, but there was also one minor peak in October. The biomass of ciliates was higher than the MZP biomass (*t*-test, $P = 0.02$). The mean total zooplankton biomass (MZP + ciliates) was 3.6 ± 1.7 mg WM l^{-1}. The mean individual metazooplankter mass fluctuated between 0.7 and 2.5 µg WM ind^{-1}, and the average was 1.6 ± 0.6 µg WM ind^{-1}. The mean metazooplankter weight in Võrtsjärv also increased from spring to autumn (Fig. 5). The biomass of ciliates was positively correlated with the biomass of MZP ($R = 0.43$, $P < 0.01$, $n = 70$), but again the degree of explanation (18%) was low.

The mean ciliate biomass was higher in Võrtsjärv (*t*-test, $P = 0.003$) compared to Peipsi. Conversely, the mean MZP biomass was higher in Peipsi (*t*-test, $P = 0.02$). Also mean individual metazooplankter mass was higher in Peipsi (*t*-test, $P = 0.03$).

Discussion

Species composition

Species characteristic of both eutrophic waters (*Daphnia cucullata*, *Bosmina c. coregoni*, *Anuraeopsis fissa* etc.) and oligo-mesotrophic waters (*Daphnia cristata*, *B. berolinensis*, *Bythotrephes longimanus* etc.) occur concurrently in Peipsi. Such coexistence is apparently facilitated by the meso-eutrophic to eutrophic character of the lake. Up to the early 1990s, the zooplankton of Peipsi was characterized by a steady decrease in the share of oligo-mesotrophic species and increase in the share of eutrophic species (Haberman, 2001). However, eutrophication was clearly stopped in the 1990s because of a decline in agricultural pollution, and the above-mentioned changes in the species composition of zooplankton also decreased. It is known that such changes in species composition occur during the meso-eutrophic stage of a lake (Karabin, 1985). Fundamental changes have occurred in the zooplankton composition of Võrtsjärv over the past 40 years: oligo-mesotrophic species have almost totally disappeared (Haberman, 1998). At present, the small MZP (rotifers, *Chydorus sphaericus*, *Bosmina longirostris*, etc.), known to be eutrophication indicators (Jeppesen et al., 2000), predominate in Võrtsjärv.

Impact of fish and phytoplankton

The mean individual metazooplankter mass in Võrtsjärv is less than half of that in Peipsi. The same trend is seen in the share of predators in MZP (e.g. *Leptodora kindti*): it is more than twice as high in Peipsi as in Võrtsjärv. The difference described could not be explained by a difference in impact of fish predation. The most abundant fish species in Võrtsjärv are bream, pikeperch and pike (Järvalt et al., 2004). The main zooplankton predators are the fish fry, which have the greatest impact in spring and early summer. The most abundant fish in Peipsi is smelt (Pihu & Kangur, 2001), which is known to be an efficient zooplankton predator. Recent studies have shown that smelt prefers larger zooplankters, especially large cladocerans (Vinni et al., 2004). Therefore, the predatory pressure of fish on MZP must be much stronger in Peipsi than in Võrtsjärv. Moreover, fish production is higher in Peipsi (30 kg ha^{-1}) than in Võrtsjärv (18 kg ha^{-1}); it is estimated that 3% of the phytoplankton production reaches fish in Võrtsjärv, while the corresponding value in Lake Peipsi is 6% (Haberman, 1998, 2001). The changes in the zooplankton must be partly linked to changes in the phytoplankton. In Võrtsjärv, the high concentrations of filamentous algae and detritus interfere with the feeding of large zooplankters (e.g. *B. berolinensis*, *E. gracilis*) (Nõges et al., 2004), conferring advantage on the smaller forms.

Importance of ciliates

Planktonic ciliates have an exceptionally important role in Võrtsjärv; they account for almost two-thirds of the total zooplankton biomass. The importance of ciliates is not a unique feature of Võrtsjärv but has also been recorded from other large lakes. Carrick & Fahnenstiel (1990), who studied the protozoan plankton of the Great Lakes Huron and Michigan, found that the ciliate biomass represented ca. 30% of the crustacean zooplankton biomass. Taylor & Heynen (1987) showed that ciliate biomass in Lake Ontario was of the same order of magnitude as crustacean zooplankton. The biomass of ciliates in Peipsi was negatively correlated with the biomass of MZP. It has been shown that ciliates are controlled top-down by large-sized cladocerans (e.g. Wickham & Gilbert, 1993), and this probably also applies in Peipsi. In Võrtsjärv, the above-mentioned correlation was positive, implying that the small-sized zooplankton dominating in this lake is clearly bottom-up or food-controlled. It has also been shown that the share of small edible algae in Võrtsjärv is low (Zingel et al., 2007). The ciliates peak in spring and late summer in Võrtsjärv. The spring peak is dominated by larger nanoplanktivorous species and the summer peak is formed mostly of smaller picoplanktivores. Bacteria seem to become the most important food item for zooplankton in summer. In most temperate lakes across the trophic spectrum, the maximum abundance of ciliates is achieved in late spring, as in Peipsi (e.g. Müller et al., 1991; Laybourn-Parry, 1992). Ciliates have the advantage in spring because their rapid division rates enable them to establish large populations relatively quickly. However, there are exceptions to this trend. In the large and shallow Neusiedler See, the maximum biomass of ciliates was recorded in late summer, not in spring (Schönberger, 1994). Gates & Lewg (1984) described late summer peaks in several lakes in Ontario, and Callieri & Heinimaa (1997) found the ciliate maximum in the large alpine Lake Maggiore in September.

Planktonic vs. benthic food chains

Earlier studies in Peipsi showed that both phytoplankton biomass and chlorophyll *a* concentration correlate negatively with the total biomass of MZP, which indicates that phytoplankton is most likely to be actively grazed by MZP. In earlier studies dealing with trophic relations in the plankton of Peipsi (Nõges et al., 1992, 1993), a strong direct relationship between zooplankton and phytoplankton was emphasized. It is assumed that filter-feeding MZP mainly ingest living algae, and detritus food is of little importance. In Võrtsjärv, large filamentous diatoms and cyanobacteria prevail, and these are not suitable food for small-bodied meta- and protozooplankton (Nõges et al., 1998). Therefore, the phytoplankton of Võrtsjärv cannot be extensively consumed in the planktonic grazing food chain. This implies that the benthic detrital food chain and the microbial loop are responsible for most of the transformation of organic matter (Nõges et al., 2004). This is reflected in the high bacterial production (70% of primary production) (Nõges et al., 2004) and the presence of an extremely abundant protist community (Zingel et al., 2007). Feeding experiments with fluorescent microspheres showed that ciliates in Võrtsjärv are potentially important grazers of nanoplankton: on average, ca. 20% of the standing stock of nanoplankton was grazed during the vegetation period (Zingel et al., 2007). Over the same time, MZP grazed only 0.43% of the standing stock of nanoalgae. This difference was even greater for picoplankton: ciliates grazed nearly 100% of the biomass production of suspended bacteria while MZP accounted for only 0.6%. The modest role of the detritus food chain and microbial loop in Peipsi compared with Võrtsjärv is confirmed by the very small biomass (5.5%) of ciliates compared with that of MZP and by the high ratio (0.24) of MZP to phytoplankton biomass. The respective values in Võrtsjärv were 175% and 0.05 (Nõges et al., 2004).

Conclusions

The lakes studied here represent opposite situations: in Peipsi the grazing food chain prevails, whereas in Võrtsjärv the detrital food chain is by far the most important. Consequently, the control of phytoplankton can be assumed to be largely top-down in Peipsi but not in Võrtsjärv. When the detrital food chain prevails, the planktonic ciliates become the most important zooplankton group in shallow, eutrophic and large lake. But as long as the grazing food chain remains uppermost, the major zooplankton groups are Crustacea and Rotifera. Neglecting protozooplankton can lead to severe underestimation of total zooplankton biomass. Two-thirds of the zooplankton biomass would be undetected in Võrtsjärv.

Acknowledgements We would like to thank the staff of the Võrtsjärv Centre for Limnology for valuable help in our study. Funding for this research was provided by Estonian target funding projects 0370208s98 and 0362483s03, by Estonian Science Foundation grants No. 2015, 2017, 4080, 5425 and by European Commission projects (contract EVK1-CT-1999-39 and contract EVK1-CT-2000-00076).

References

Azam, F., T. Fenchel, J. G. Field, J. S. Gray, L. A. Meyer-Reill & F. Thingstad, 1983. The ecological role of water-column microbes in the sea. Marine Ecology Progress Series 10: 257–263.

Balushkina, E. V. & G. G. Winberg, 1979. Relation between body mass and length in planktonic animals. In Winberg, G. G. (ed.), Obshchie Osnovy Izucheniya Vodnykh Ekosistem. Nauka, Leningrad: 169–172. (In Russian).

Beaver, J. R. & T. L. Crisman, 1982. The trophic response of ciliated protozoans in freshwater lakes. Limnology and Oceanography 27: 246–253.

Callieri, C. & S. Heinimaa, 1997. Microbial loop in the large subalpine lakes. Memorie dell'Istituto Italiano di Idrobiologia 56: 143–156.

Carrick, H. J. & G. L. Fahnenstiel, 1990. Planktonic protozoa in Lakes Huron and Michigan: seasonal abundance and composition of ciliates and dinoflagellates. Journal of Great Lakes Research 16: 319–329.

Finlay, B. J., 1982. Procedures for the isolation, cultivation and identification of protozoa. In Burns, R. G. & J. H. Slater (eds.), Experimental Microbial Ecology. Blackwell Scientific Publications, Oxford: 44–65.

Gates, M. A. & U. T. Lewg, 1984. Contribution of ciliated protozoa to the planktonic biomass in a series of Ontario lakes, quantitative estimates and dynamical relationships. Journal of Plankton Research 6: 443–456.

Gifford, D. J., 1991. The protozoan-metazoan trophic link in pelagic ecosystems. Journal of Protozoology 38: 81–86.

Haberman, J., 1998. Zooplankton of Lake Võrtsjärv. Limnologica 28: 49–65.

Haberman, J., 2001. Zooplankton. In Pihu, E. & J. Haberman (eds), Lake Peipsi: Flora and Fauna. Sulemees, Tartu: 50–68.

Jeppesen, E., J. P. Jensen, M. Sondergaard, T. Lauridsen & F. Landkildehus, 2000. Trophic structure, species richness and biodiversity in Danish lakes: changes along phosphorus gradient. Freshwater Biology 45: 201–218.

Järvalt, A., A. Kangur, K. Kangur, P. Kangur & E. Pihu, 2004. Fishes and fisheries management. In Haberman, J., E.

Pihu & A. Raukas (eds.), Lake Võrtsjärv. Estonian Encyclopaedia Publishers Ltd, Tallinn: 335–345.

Karabin, A., 1985. Pelagic zooplankton (Rotatoria, Crustacea) variation in the process of lake eutrophication. I. Structural and quantitative features. Ekologia Polska 33: 567–616.

Laybourn-Parry, J., 1992. Protozoan Plankton Ecology. Chapman & Hall, Ltd., London: 231.

Müller, H., A. Schöne, R. M. Pinto-Coelho, A. Schweizer & T. Weisse, 1991. Seasonal succession of ciliates in Lake Constance. Microbial Ecology 21: 119–138.

Nõges, T., J. Haberman, H. Tammert, M. Timm, I. Ott & P. Nõges, 1992. Ecological relations of main plankton components in the pelagial of Lake Peipsi (Peipus). Proceedings of the Estonian Academy of Sciences. Ecology 2: 137–155.

Nõges, T., J. Haberman, M. Timm & P. Nõges, 1993. The seasonal dynamics and trophic relations of the plankton components in lake Peipsi (Peipus). Internationale Revue der gesamten Hydrobiologie 78: 513–519.

Nõges, T., V. Kisand, P. Nõges, A. Põllumäe, L. Tuvikene, & P. Zingel, 1998. Plankton seasonal dynamics and its controlling factors in shallow polymictic eutrophic lake Võrtsjärv, Estonia. Internationale Revue der gesamten Hydrobiologie 83: 279–296.

Nõges, T., J. Haberman, A. Kangur, K. Kangur, P. Kangur, H. Künnap, H. Timm, P. Zingel & P. Nõges, 2004. Food webs in Lake Võrtsjärv. In Haberman, J., E. Pihu & A. Raukas (eds), Lake Võrtsjärv. Estonian Encyclopaedia Publishers Ltd, Tallinn: 335–345.

Pihu, E. & A. Kangur, 2001. Fishes and fisheries management. In Pihu, E. & J. Haberman (eds), Lake Peipsi: Flora and Fauna. Sulemees, Tartu: 100–111.

Ruttner-Kolisko, A., 1977. Suggestion for biomass calculation of planktonic rotifers. Archiv für Hydrobiologie 8: 71–76.

Sanders, R. W., K. G. Porter, S. J. Bennett & A. E. DeBiase, 1989. Seasonal patterns of bacterivory by flagellates, ciliates, rotifers, and cladocerans in a freshwater planktonic community. Limnology and Oceanography 34: 673–687.

Schönberger, M., 1994. Planktonic ciliated protozoa of Neusiedler See (Austria/Hungary)—a comparison between the turbid open lake and a reedless brown-water pond. Marine Microbial Food Webs 8: 251–263.

Sherr, E. B., & B. F. Sherr, 1984. Role of heterotrophic protozoa in carbon, energy flow in aquatic ecosystems. In Klug, M. J. & C. A. Reddy (eds), Current Perspectives in Microbial Ecology. American Society for Microbiology, Washington: 412–423.

Starast, H., A. Milius, T. Möls & A. Lindpere, 2001. Hydrochemistry of lake peipsi. In Nõges, T. (ed), Lake Peipsi: Meteorology, Hydrology, Hydrochemistry. Sulemees, Tartu: 97–131.

Taylor, W. D. & M. L. Heynen, 1987. Seasonal and vertical distribution of Ciliophora in Lake Ontario. Canadian Journal of Fisheries and Aquatic Sciences 44: 2185–2191.

Utermöhl, H., 1958. Zur Vervollkommung der quaatitativen Phytoplanktonmethodik. Mitteilung internationale Vereinigung für Limnologie 9: 1–38.

Vinni, M., J. Lappalainen, T. Malinen & H. Peltonen, 2004. Seasonal bottlenecks in diet shifts and growth of smelt in a large eutrophic lake. Journal of Fish Biology 64: 567–579.

Wickham, S. A. & J. J. Gilbert, 1993. The comparative importance of competition and predation by *Daphnia* on ciliated protists. Archiv für Hydrobiologie 126: 289–313.

Zingel, P., H. Agasild, T. Nõges & V. Kisand, 2007. Ciliates are the dominant grazers on pico- and nanoplankton in a shallow, naturally highly eutrophic lake. Microbial Ecology 53: 134–142.

Hydrobiologia (2008) 599:161–168
DOI 10.1007/s10750-007-9212-0

ELLS 2007

Validation of the MERIS products on large European lakes: Peipsi, Vänern and Vättern

Krista Alikas · Anu Reinart

Abstract The Medium Resolution Imaging Spectrometer (MERIS/Envisat) was launched in March 2002 for coastal zone monitoring. Preliminary data from MERIS show that its imagery of large lakes is superior to that of other common ocean colour sensors. The main objective of the present study is to evaluate the MERIS data on large European lakes, Vänern and Vättern in Sweden and Peipsi in Estonia/Russia. In these lakes, coloured dissolved organic matter (CDOM) can be a major contributor to the optical properties of the water. Another characteristic of the waters under investigation is the large temporal and spatial variability in the concentrations of chlorophyll (C_{Chl}) and suspended sediments (C_{TSS}). Potentially toxic cyanobacterial blooms occur in Lake Peipsi in late summer. We have compared the MERIS products from the latest reprocessing (finished in March 2006) with available in situ data. There is a reasonably good correlation between the MERIS algal_2 product and the measured C_{chl} over all three lakes, but no correlation was found for other optically active substances. A significant portion of the pixels (up to 90%) are flagged as invalid results after atmospheric correction.

Keywords Remote sensing · MERIS · Large lakes · Optical properties

Guest editors: T. Nõges, R. Eckmann, K. Kangur, P. Nõges, A. Reinart, G. Roll, H. Simola & M. Viljanen
European Large Lakes—Ecosystem changes and their ecological and socioeconomic impacts

K. Alikas · A. Reinart (✉)
Tartu Observatory, Toravere, Tartumaa 61602, Estonia
e-mail: anu.reinart@aai.ee

Introduction

Consistently implemented long-term programmes for monitoring large lakes are critical for detecting changes in water quality. They are required to assess the effects of (1) past and ongoing management programmes and (2) continuing anthropogenic influences, particularly the effects related to climate change; and (3) to meet the EU directives prescribing that water quality is adequately monitored and not deteriorating. However, conventional sampling methods cannot produce enough data about spatial and temporal heterogeneity in large lake systems.

Satellite-based remote sensing has become a useful tool for monitoring oceanic and marine waters (Case I) (Sathyendranath, 2000). Applications to coastal and inland waters (Case II), however, require more advanced methods and data analysis (Dekker et al., 2002).

MERIS (**Me**dium **R**esolution **I**maging **S**pectrometer) is an instrument on board the ENVISAT satellite put into orbit by the European Space Agency on 1 March 2002 (http://www.envisat.esa.int/instruments/

meris/). This instrument specifically addresses the needs of optically complex waters. Coastal and inland waters generally have a higher trophic status, which is associated with optically active substances (dissolved substances and suspended particles) originating from their drainage basins in amounts that exceed their concentrations in the open ocean many times (Arst, 2003). Until recently, two problems have reduced the utility of satellite monitoring of freshwaters in high latitudes: (1) interference of coloured dissolved organic matter (CDOM) in the estimation of chlorophyll concentration; and (2) infrequent satellite measurements and relatively frequent cloud cover.

The MERIS has spectral bands (centred at 650, 681 and 705 nm) that allow chlorophyll to be estimated in the presence of CDOM, and it uses a neural network (NN) technique instead of band ratio algorithms (Schiller & Doerffer, 1999). The MERIS atmospheric correction procedure has been especially adapted for turbid waters using a coupled hydrological-atmospheric model (Moore et al., 1999). MERIS (like other earth observation satellites) has high measurement frequencies, passing over the regions of interest each day and greatly increasing the chances of obtaining useful cloud-free images. The full resolution ($\sim$300 m) data provide the highest spatial resolution available from a satellite sensor, the radiometric sensitivity of which is optimised for measuring water, and could provide additional information to complement the systems for monitoring large lakes (Pierson & Strömbeck, 2000).

The present paper investigates the possibility of using the MERIS images to monitor water quality in European Union's three largest lakes (Table 1): Lake Peipsi in Estonia–Russia, and Lakes Vänern and Vättern in Sweden. We validate the MERIS products (chlorophyll *a*, suspended matter, absorption by dissolved organic matter) against in situ water quality parameters obtained by long-term monitoring programmes in these lakes (Pierson & Strömbeck, 2000; Strömbeck & Pierson, 2001; Reinart et al., 2004a). Because the MERIS images have not been widely available to researchers and the reprocessing has only recently been finished (March 2006), only a few publications have addressed the details and results. Therefore, the present paper gives valuable information for different users.

Table 1 Morphometric data (Kvarnäs, 2001; Nõges, 2001), concentrations of phytoplankton pigments (C_{chl}), total suspended matter (C_{TSS}) and absorption by CDOM (means ± 1 standard deviation) measured in water samples taken in the years 2003 and 2004, and ranges of Secchi depth in the three study lakes

Parameter	Peipsi	Vänern	Vättern
Area, km^2	3,555	5,648	1,856
Mean depth, m	7	27	40
Maximum depth, m	15.3	106	128
C_{chl}, mg m^{-3}	20.7 ± 17.5	3.6 ± 1.9	1.1 ± 0.3
C_{TSS}, g m^{-3}	5.9 ± 4.5	0.6 ± 0.1	0.5 ± 0.1
$a_{CDOM}(443)$, m^{-1}	2.6 ± 1.0	1.2 ± 0.5	0.3 ± 0.1
Secchi depth, m	0.4–4.8	2.0–7.9	6.5–15.0

Materials and methods

MERIS products

MERIS Level 2 processing includes atmospheric correction and the application of algorithms for retrieving concentrations of phytoplankton pigment, total suspended matter and CDOM from the radiation from the water (http://www.envisat.esa.int/handbooks/meris/). The optical properties of phytoplankton are described in Schiller & Doerffer (1999) and ATBD_2_12.

Case II water processing uses the inverse radiative transfer modelling technique with a NN parameterisation. The NN uses all visible bands (nine, excluding the fluorescence band at 681 nm) and outputs the scattering coefficient (b), absorption by pigments ($a_{pigment}$) and absorption by CDOM and detritus (a_{CDOM}). The phytoplankton pigment index is expressed as the chlorophyll concentration in mg m^{-3} (product name "algal_2"), and the non-absorbing suspended particle concentration in g m^{-3} ("total_susp"). CDOM is expressed by absorption at 443 nm in units of m^{-1} ("yellow_subs").

Concentrations are calculated from the scattering and absorption coefficients according to empirical relationships (Doerffer, 2006; Krasemann, 2006):

$$\text{total\_susp} = 1.73 * b(443)$$

$$\text{algal\_2} = 21 * a_{pigment}(443)^{1.04}$$

As the NN gives optical properties rather than concentrations, it makes these algorithms regionally adjustable, as successfully shown by Peters (2006). In order to allow comparison with other ocean colour sensors, there is also another special product for Case I waters—algal_1. Algal_1 is calculated by a single polynomial based on the highest reflectance ratio (bands 442, 490, 510 and 590), clipping within 0.01–30 mg m^{-3} (ATBD_2_12).

In addition to water quality parameters, several flags are incorporated in the Level 2 files. Some of them are set during processing to indicate whether anything may be wrong (invalid reflectance, algal_1 or algal_2 algorithm application), or for information such as Case II sediment-rich and CDOM-rich waters ("case2_s", "case2_y").

Description of lakes

All three lakes studied are located on almost the same latitude (57.8–59.4 N). The two Swedish lakes are west of the Baltic Sea and are more affected by the Atlantic Ocean and the Gulf Stream while Peipsi, on the border between Estonia and Russia, is more affected by the continental climate (Fig. 1).

Lake Peipsi is a large shallow lake (Table 1). The nutrient load from the Estonian River Suur-Emajõgi and the Russian River Velikaya has caused strong eutrophication of the lake. The clearest water is observed in the north-western corner of Peipsi and the most turbid water in its southernmost part (Lake Pihkva). The River Suur-Emajõgi brings large amounts of humic matter from the surrounding bog areas into the lake, which makes the water close to the river inflow brownish and usually less transparent than the water in the deeper parts of Peipsi (Reinart et al., 2004b). Water transparency in Peipsi depends primarily on the development of phytoplankton in its northern part, while in other parts it also depends on the amount of humic substances. Toxin-producing cyanobacteria have been recorded in Peipsi at concentrations up to 100 mg m^{-3} by the end of summer.

Lake Vänern in central Sweden is divided into two basins by a shallow archipelago. Its water quality is classified as moderately nutrient-rich, and measurements of algal biomass indicate oligotrophic conditions (Willén, 2001). The southern basin is slightly more turbid than the northern, but river inflow mainly into the northern basin makes it more

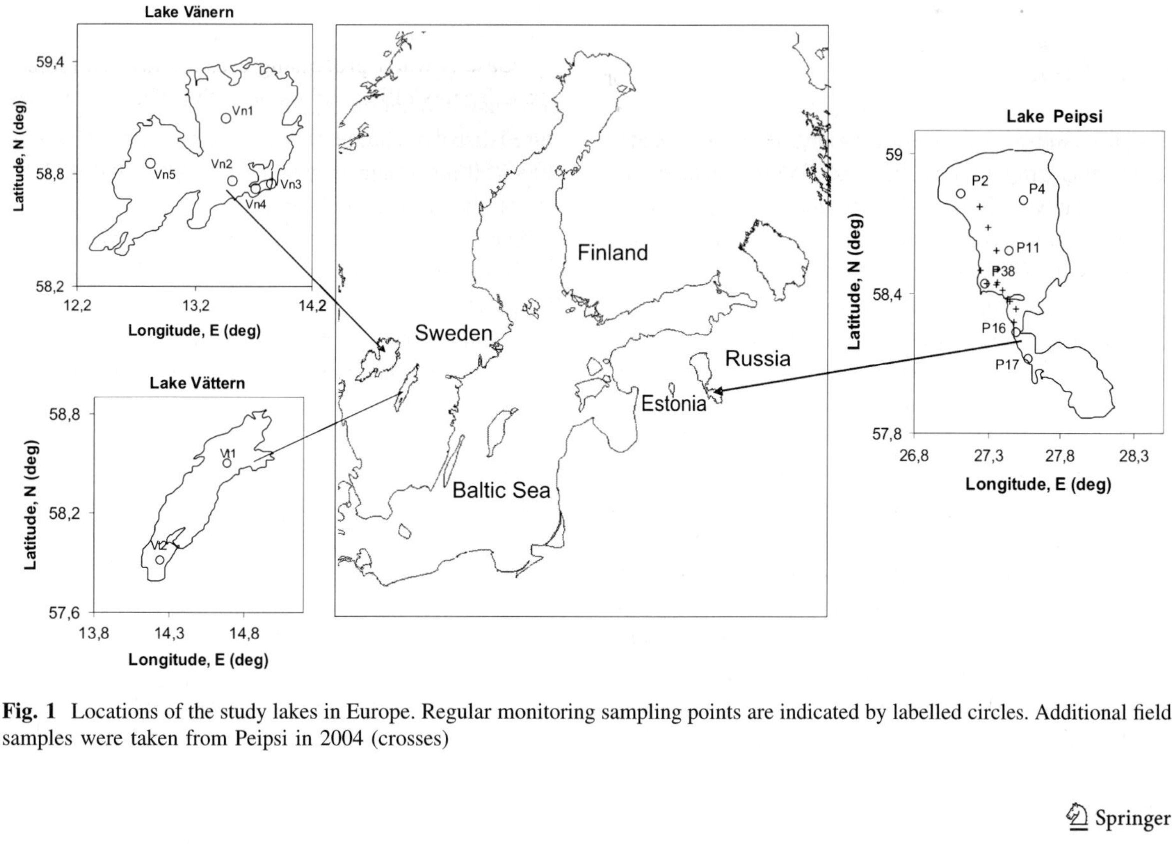

Fig. 1 Locations of the study lakes in Europe. Regular monitoring sampling points are indicated by labelled circles. Additional field samples were taken from Peipsi in 2004 (crosses)

humic. From long-term monitoring of this lake it is known that the suspended sediment concentration may vary between 0.2 and 4.5 g m^{-3}, and the chlorophyll concentration between 0.3 and 7.4 mg m^{-3} (Strömbeck & Pierson, 2001). In this lake, the bloom of phytoplankton in spring (May–early June) is often more pronounced than the second bloom in August.

Lake Vättern is about 20 km east of Vänern and comprises only one rather narrow basin (width less than 15 km). It has a relatively small watershed, and seasonal variations in discharge from the tributaries have very little impact on the lake's water quality. Ultra-oligotrophic conditions prevail in this lake (Willén, 2001).

The water transparency measured by Secchi depth is highest in Vättern, 6.5–15.0 m; it is lower in Vänern (2.0–7.9 m) and lowest in Peipsi (0.4–4.8 m). Overall, these three lakes represent a wide range of optical variability in water bodies. Compared to the two Swedish lakes, Peipsi has higher concentrations of OAS (optically active substances: phytoplankton pigments and other suspended particles or dissolved organic substances). Therefore, the lake complements the continuum with Vänern and Vättern both spatially and in terms of the optical properties of the water.

In situ data

The first sources of in situ data are the regular state monitoring programmes. In Lakes Peipsi (state monitoring database: http://www.seiremonitor.ee), Vänern and Vättern (Swedish University of Agricultural Sciences, SLU database: http://www.info1.ma.slu.se/), regular monitoring includes the measurement of chlorophyll *a* concentration. Data about suspended sediments and absorption by CDOM, however, are available only occasionally (Table 2). Regular sampling in Peipsi in the years 2003 and 2004 was performed 2–3 times per month (May–November) at 6 points, but only 5 times per year in Lakes Vänern (5 points) and Vättern (2 points). The sampling points in each lake are shown in Fig. 1. We have used data for the period of May–September only, because in October and November the sun is too low for correct satellite validation and the sky is mainly cloudy.

As regular monitoring in Peipsi does not include the estimation of CDOM absorption, additional sampling was performed at the monitoring points during four field campaigns in 2003 and 2004 (Reinart et al., 2004a, b). The concentrations of chlorophyll, suspended sediments and CDOM were measured from these samples; thus, many more in situ data are shown in Table 2 for Peipsi (159) than for Vänern (41) or Vättern (11).

Table 2 Numbers of satellite images (2003–2004) and measured data used in the present study

	Satellite images	In situ measurements		
		Chl	CDOM	TSS
Peipsi	86 (18)	104 (52)	58 (39)	161 (68)
Vänern	110 (17)	45 (20)	35 (19)	6 (2)
Vättern	101 (3)	8 (4)	6 (4)	4 (3)
Sum	297 (38)	157 (76)	99 (62)	171 (73)

Numbers in brackets indicate how many image/data pairs were used for comparison between satellite and measured data

Lindell et al. (1999) have suggested methods for remote sensing of lakes. For chlorophyll concentration, water was filtered through Whatman GF/F-filters, and the chlorophyll *a* + pheophytin *a* concentration (C_{chl}) was measured spectrophotometrically in ethanol extracts of the filters. The concentration of total suspended matter (C_{TSS}) was measured gravimetrically (precision of 0.01 mg) after a measured volume of water was filtered through pre-weighed and precombusted Whatman GF/F-filters. Absorption by coloured dissolved matter (a_{CDOM}) was measured over the 400–750 nm range in water filtered through 0.2 μm filters; the value at 443 nm is reported here.

The Swedish database contains absorbance at 420 nm (abs_420_F) measured with a 5 cm cuvette. These values were recalculated to obtain the absorption values at 443 nm, assuming an exponential decrease with wavelength as described by Strömbeck & Pierson (2001):

$$a_{CDOM}(443) = \frac{2.303}{0.05} * \text{abs\_420\_F} \exp[-0.014(443 - 420)].$$

Satellite images

The MERIS images were reprocessed twice in accordance with input from scientific teams. Products with the latest official version have been available since 8 May 2006 (http://www.earth.esa.int/pub/ESA_DOC/ENVISAT/MERIS). The MERIS L2 RR (**r**educed **r**esolution) images processed by this method are available to the user community via the MERCI catalogues at http://www.brockmann-consult.de/merci. Access to these images was possible through a Category 1 user project (ID 3180, PI A. Reinart) "Testing the applicability of MERIS L2 products for monitoring humic coastal and lake waters in the Baltic Sea region".

We collected a total of 297 images from Lakes Peipsi, Vänern and Vättern (Table 2) during the ice-free period (May–September) in 2003 and 2004. Because of the frequent cloud cover and sporadic in situ measurements, only 38 images could be used to compare satellite products with direct measurements. We have used images that were acquired on the day of in situ measurement or on a neighbouring day (Table 2).

Images were visualised and analysed using the software tool BEAM 3.5 (Brockmann Consult/ESA). In situ measurement points were marked with a pin in the MERIS RR L2 images and a 3 × 3 pixel area around the pin was examined. Average concentrations and standard deviations were calculated at every measurement point for every date on the basis of nine pixel values for the algal_1, algal_2, total suspended matter and yellow substance products.

Results

Chlorophyll

Satellite data allow both the temporal and spatial variations of phytoplankton development to be studied in more detail than point measurements. In Fig. 2, the temporal variation of the product algal_2 is shown for two points in each lake.

In Peipsi, point P4 is located in the centre of the lake and represents the deepest and clearest region, while P17 is in a narrow region between the southern and northern basins (Fig. 1). Satellite data confirm a notable spring bloom (C_{Chl} up to 45 mg m^{-3}) over the whole of Peipsi. The bloom is slightly lower in open waters than close to the coast, where the water heats up more quickly. In June–July, C_{Chl} decreases to 5 mg m^{-3} at P4, but at P17 the midsummer bloom (mainly cyanobacteria) starts again. The MERIS data also show an increase in C_{Chl} in August–September. At P17, no such increase is notable and the values are even lower than in spring. However, it is known from field observations that cyanobacteria may form a dense layer on the surface and concentrations may rise to several hundred mg m^{-3}.

In Vänern (Fig. 2b) point Vn3 is located in a small bay and Vn2 in an open area between the northern and southern basins (Fig. 1). At both points, a spring bloom is notable in May. Thereafter the C_{Chl} values remain rather low and do not vary much in the open lake, but at Vn3 the MERIS data indicate rather high values in June and again later in August–September (occasionally up to 30 mg m^{-3}). Such high values are not supported by measurements. The in situ dataset gives an average C_{Chl} for Vänern of 3.6 ± 1.9 mg m^{-3}. One or two samplings during summer, however, may easily overlook rapid changes in water quality.

In Vättern there is basically no difference between the northern (Vt1) and southern points (Vt2), and algal_2 values stay very low with no clear seasonal trend. Some peaks shown in Fig. 2c are associated with a large standard deviation among the pixels examined. This may be caused by processing errors or small clouds rather than representing a real variation in chlorophyll concentration.

In general, the MERIS algal_2 is lowest and has the smallest variation in Vättern (1.9 ± 0.9 mg m^{-3}), and is highest in Peipsi (15.1 ± 11.6 mg m^{-3}) (Table 3). This order agrees with the field measurements. However, the average values from the MERIS results are higher than the values measured in Lakes Vättern and Vänern, though they are lower than the values measured in Peipsi (Tables 1 and 3).

The MERIS derived algal_2 and the in situ measured C_{Chl} are compared directly in Fig. 3. Although all points are located around the 1:1 line, the scatter is rather wide. A good correlation can be obtained only when all the lakes are included ($y_{MERIS} = 1.16x_{in\ situ}$, $R^2 = 0.52$, $N = 76$). In Peipsi, it was observed that when the measured C_{Chl} was extremely high (>45 mg m^{-3}) the MERIS results were particularly low (~1 mg m^{-3}). These values are not shown in Fig. 3, and they are not included in the correlation as they exceed the limit of the algorithm.

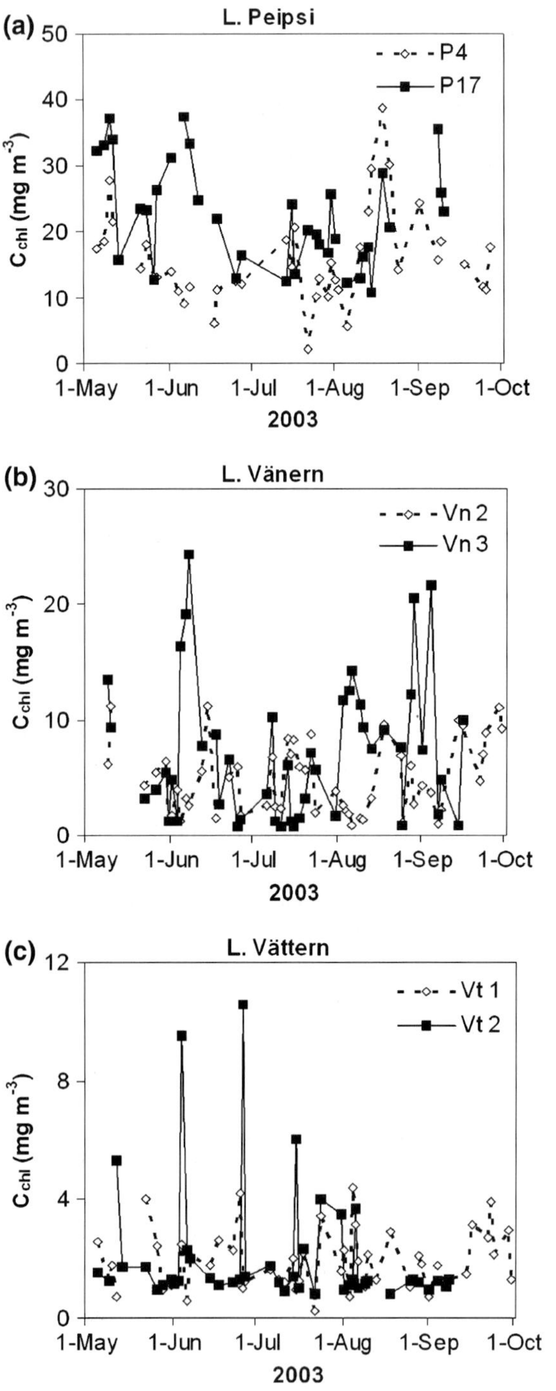

Fig. 2 Temporal variation of C_{chl} in lakes Peipsi (**a**), Vänern (**b**) and Vättern (**c**) in 2003 as estimated by the MERIS product algal_2. Two sampling points (cf. Fig. 1) from each lake were selected for line plots

Table 3 Concentrations of optically active substances and absorption by CDOM (means ± 1 standard deviation) in the three study lakes in 2003–2004, estimated from MERIS images

Parameter	Peipsi	Vänern	Vättern
C_{chl}, mg m^{-3}	15.1 ± 11.6	6.4 ± 5.8	1.9 ± 0.9
C_{TSS}, g m^{-3}	3.9 ± 4.5	1.6 ± 1.1	0.6 ± 1.2
$a_{CDOM}(443)$, m^{-1}	1.5 ± 1.9	0.4 ± 0.5	0.4 ± 0.3

Total suspended matter

The in situ C_{TSS} in Peipsi varied between 1 and 26 g m^{-3} with the higher values in the southern shallow regions towards autumn. The overall mean was 5.9 ± 4.5 g m^{-3}, which is markedly higher than in the two deep Swedish lakes (Table 1). The MERIS results shown in Table 3 are lower than the measured values for Peipsi, but higher than the measured values for Vänern and Vättern.

In Vänern, measurements are in the range of 0.4–0.8 g m^{-3}, but MERIS gives values up to 6.4 g m^{-3} for the same points. Unfortunately, C_{TSS} in the SLU database values shown in Table 1 do not include the most turbid coastal points. In other studies (Strömbeck & Pierson, 2001; Reinart et al., 2004a), the average C_{TSS} of 1.2 ± 0.4 g m^{-3} was close to the MERIS results in Table 3. We have very few in situ data for Vättern (0.4–0.5 g m^{-3}) but they actually match the MERIS results (mean: 0.6 g m^{-3}). However, MERIS often gives much higher values, which have never been found in samples from Vättern. No

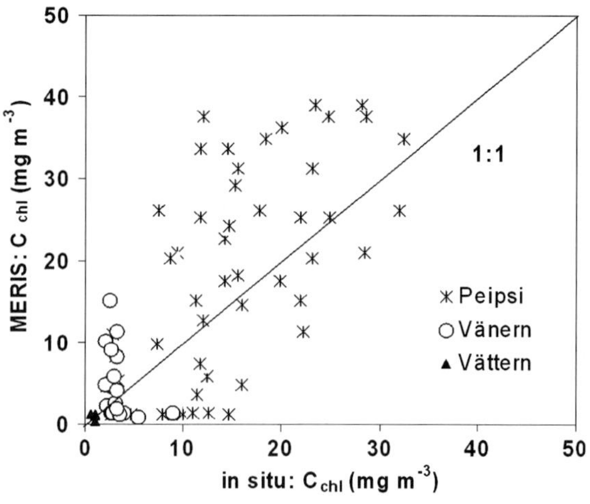

Fig. 3 Comparison of the MERIS-estimated C_{chl} (product algal_2) and in situ measured C_{chl} (2003–2004) for the three lakes

correlation was found between the measured and satellite-based estimates of C_{TSS} in the lakes studied.

Coloured dissolved organic matter

Yellow substance may be produced in a lake or transported into a lake from its drainage area. It was previously shown (Strömbeck & Pierson, 2001; Reinart et al., 2004b) that CDOM in the lakes studied is clearly of allochthonous origin. The values of $a_{CDOM}(443)$ in Lakes Peipsi and Vänern are higher in samples from river mouths or shallower areas close to the shore. Therefore, the standard deviations shown in Table 1 are rather large. The minimal values measured in Peipsi are comparable with the maximal ones in Vänern, while $a_{CDOM}(443)$ was much lower in Vättern (Table 1) and less variable than in the other two lakes.

Comparison of the measured values with MERIS estimates showed no correlation. The MERIS data mostly underestimated $a_{CDOM}(443)$ in Lakes Peipsi and Vänern, and there was also high pixel-to-pixel variation. For Vättern, the satellite estimates were close to the in situ data (0.4–0.5 m^{-1}), but this result was based on only four data points. There is no difference in $a_{CDOM}(443)$ between Lakes Vänern and Vättern on the basis of the satellite data.

Discussion

There is a special flag to indicate CDOM-loaded waters in the MERIS processing (case2_y), but this was not activated over the lakes. The same is true for the flag showing anomalously high scattering (case2_anom), though in some cases, especially in Peipsi, scattering should have been really high. A flag indicating sediment-dominated turbid waters (case2_s) was active over the whole of Peipsi and the southern bays of Vänern. It was also sometimes active in the central area of Vänern, where scattering cannot be high.

Flag PCD_1_13, which indicates invalid results after atmospheric correction at any of the spectral bands, was active almost everywhere over the lakes. This is the probable reason why the algal_1 product was always invalid over the lakes. Flag "invalid algal2_tsm_ys" (any of the case_2 water products is invalid) was most commonly activated over Lakes Vänern and Vättern. In Peipsi, 0–3 pixels were typically flagged for the coast and, more often, for the southern lake regions. Overall, MERIS Case II processing resulted in invalid reflectance over the lakes and invalid concentrations of chlorophyll and suspended sediments or absorption by CDOM in Lakes Vänern (70–90% of pixels), Vättern (90–100% of pixels) and Peipsi (5–47% of pixels).

On the basis of our data we can conclude that the MERIS Case II chlorophyll concentration product (algal_2) and the in situ measured C_{Chl} are correlated ($R^2 = 0.52$). The MERIS results make clear distinctions among the three lakes studied, with Peipsi having the highest and Vättern the lowest values, and seasonal variations (blooms) can be detected. MERIS strongly underestimates the very high chlorophyll values during the autumn bloom in Peipsi (>45 mg m^{-3}), and it overestimates the values in Lakes Vänern and Vättern almost twofold. For suspended sediment and CDOM there were no correlations between the MERIS and the measured data. Total_susp still provides a reasonable spatial distribution for most of the lake areas (excluding the coastal region of Vänern). Absorption by CDOM is strongly underestimated in Lakes Vänern and Peipsi.

The MERIS products are still under validation. Preliminary results have only been presented at meetings and none have been published yet. Generally good results have been obtained for the North Sea (Krasemann, 2006; Peters, 2006) because the inherent optical dataset was also derived on the basis of the North Sea data (Doerffer, 2006). In this case, there was a notable underestimation of $a_{CDOM}(443)$ when the values were above $\sim$1 m^{-1} and an overestimation when $a_{CDOM}(443)$ was less than $\sim$1 m^{-1}, as we also found in the three lakes we studied.

The present MERIS Case II algorithm was developed for the ranges $a_{CDOM}(443)$: 0.005–5.0 m^{-1}, algal_2: 0.02–43 mg m^{-3}, and total_susp: 0.01–51 g m^{-3} (Doerffer, 2006). Therefore, in Peipsi, some very high values during a heavy bloom and in river inflow may really exceed the processing limits. However, these lakes are mostly within the range for which the MERIS algorithms were developed. The lack of correlation between measured and satellite-derived C_{TSS} over the entire dataset may be due to temporal and spatial changes in the composition of suspended matter (organic and inorganic). The scattering coefficient of mineral particles is markedly different from that of phytoplankton. There are even large differences among different phytoplankton species. The set of optical properties selected for the NN probably cannot be applied to intense plankton blooms.

From the limited amount of data presented here, we see that MERIS has already shown its capability and potential for detecting changes in the optical properties of lakes. Before MERIS images can be used in long-term programmes for monitoring water quality in large lakes, validation of the MERIS products should continue, including validation of the relationships between OAS and the inherent optical properties to specify local algorithms. This will allow estimates to be improved, especially for the parameter most often included in monitoring programmes and also directly related to ecological conditions, C_{Chl}.

Acknowledgements We thank ESA/ENVISAT, EO Help Desk and Brockmann Consult for the production and distribution of images. This study is supported by the Estonian Science Foundation Grant's 6814 and 5738 and the Swedish National Space Board grant 132/03. We thank the Võrtsjärve Limnological Station for water sample analyses.

References

Arst, H., 2003. Optical properties and remote sensing of multicomponental water bodies. Praxis Publishing, Springer, Chichester, UK.

ATBD_2_12. Pigment index, sediment and gelbstoff retrieval from directional water-leaving reflectances using inverse modelling technique. http://www.envisat.esa.int/instruments/meris/pdf/atbd_2_12.pdf.

Dekker, A. G., R. J. Vos & S. W. M. Peters, 2002. Analytical algorithms for lake water TSM estimation for retrospective analyses of TM and SPOT sensor data. International Journal of Remote Sensing 23: 15–35.

Doerffer, R., 2006. MERIS Case I validation. Proceedings of the Second Working Meeting on MERIS and AATSR Calibration and Geophysical Validation. CD SP-615.

Krasemann, H., 2006. Case II Water Validation. Proceedings of the Second Working Meeting on MERIS and AATSR Calibration and Geophysical Validation. CD SP-615.

Kvarnäs, H., 2001. Morphometry and hydrology of the four large lakes of Sweden. Ambio 30: 467–474.

Lindell, T., D. Pierson, G. Premazzi & E. Zilioli (eds), 1999. Manual for Monitoring European Lakes Using Remote Sensing Techniques. Office for Official Publications of the European Communities, Luxembourg.

Moore, G. F., J. Aiken & J. Lavender, 1999. The atmospheric correction of water colour and the quantitative retrieval of suspended particulate matter in Case II waters: application to MERIS. International Journal of Remote Sensing 20: 1713–1733.

Nõges T. (ed.), 2001. Lake Peipsi. Meteorology, Hydrology, Hydrochemistry. Sulemees Publishers, Tartu, Estonia.

Peters, S., 2006. MERIS Water Product Validation Case II waters. Validation at the North Sea. Proceedings of the Second Working Meeting on MERIS and AATSR Calibration and Geophysical Validation. CD SP-615.

Pierson, D. C. & N. Strömbeck, 2000. A modelling approach to evaluate preliminary remote sensing algorithms: Use of water quality data from Swedish Great Lakes. Geophysica 36: 177–202.

Reinart, A., B. Paavel, D. Pierson & N. Strömbeck, 2004a. Inherent and apparent optical properties of Lake Peipsi. Estonian Boreal Environment Research 9: 429–445.

Reinart, A., B. Paavel & L. Tuvikene, 2004b. Effect of coloured dissolved organic matter on the attenuation of photosynthetically active radiation in Lake Peipsi. Proceedings of the Estonian Academy of Sciences. Biology. Ecology 53: 88–105.

Sathyendranath, S. (ed.), 2000. Remote sensing of ocean colour in coastal and other optically complex waters. Report of the International Ocean Colour Coordinating Group, No. 3. IOCCG, Dartmouth, Canada.

Schiller, H. & R. Doerffer, 1999. Neural network for emulation of an inverse model operational derivation of Case II water properties from MERIS data. International Journal of Remote Sensing 20: 1735–1746.

Strömbeck, N. & D. Pierson, 2001. The effects of variability in the inherent optical properties on the estimations of the chlorophyll a by remote sensing in Swedish freshwaters. The Science of the Total Environment 268: 123–137.

Willén, E., 2001. Phytoplankton and water quality characterization: Experiences from the Swedish large lakes Mälaren, Hjälmaren, Vättern and Vänern. Ambio 30: 529–537.

Hydrobiologia (2008) 599:169–176
DOI 10.1007/s10750-007-9213-z

ELLS 2007

Relations of phytoplankton in situ primary production, chlorophyll concentration and underwater irradiance in turbid lakes

Helgi Arst · Tiina Nõges · Peeter Nõges · Birgot Paavel

Abstract In order to study the relationships among primary productivity, chlorophyll content and light conditions (underwater quantum irradiance, diffuse attenuation coefficient) in turbid lakes, we performed a series of measurements in three Estonian lakes (Peipsi, Võrtsjärv and Harku) during 2003–2005. We focused on (1) comparison of in situ vertical profiles of planar and scalar quantum irradiances in the water, (2) extremely high values of chlorophyll content and primary production in some turbid lakes and (3) analysis of the formation of the vertical profiles of primary production (shape and the depth of its maximum value) as a function of incident irradiance. The following parameters were measured: (a) primary production at different depths, (b) incident planar quantum irradiance in the region of 400–700 nm, (c) underwater downwelling planar and scalar quantum irradiances at different depths in the same waveband, (d) relative transparency of water (Secchi depth), (e) concentrations of chlorophyll *a* and (f) spectra of beam attenuation coefficient for unfiltered and filtered water in the wavelength range of 350–700 nm (from water samples). During 36 measurement days (14 in Peipsi, 14 in Võrtsjärv and 8 in Harku), and we collected reliable dataset for 53 measurement series. Our data showed the relative difference between underwater planar and scalar quantum irradiance in a range of about 25–65% implying big differences in primary production models if instead of the preferable scalar quantum irradiance planar irradiance data are used. The results showed a large variation of vertical profiles of primary production in relation to illumination conditions. The depth of maximum primary production depended on water turbidity and incident irradiance, and was located at the water surface in case of low illumination. The relationship between vertically integrated primary production and chlorophyll concentration could be described by a power function with an exponent of 0.64 ($R^2 = 0.81$).

Guest editors: T. Nõges, R. Eckmann, K. Kangur, P. Nõges, A. Reinart, G. Roll, H. Simola & M. Viljanen
European large lakes—Ecosystem changes and their ecological and socioeconomic impacts

H. Arst (✉) · B. Paavel
Estonian Marine Institute, University of Tartu, Mäealuse St. 10A, 12618 Tallinn, Estonia
e-mail: Helgi.Arst@sea.ee

T. Nõges · P. Nõges
Centre for Limnology, Institute of Agricultural and Environmental Sciences, Estonian University of Life Sciences, 61101 Rannu, Tartumaa, Estonia

P. Nõges
European Commission, Joint Research Centre, Institute for Environment and Sustainability, Ispra, Italy

Keywords Primary production · Underwater irradiance · Optically active substances

Introduction

Although a large proportion of lakes over the world are turbid and highly eutrophic, their bio-optical modelling (e.g. Talling et al., 1973; Kirk, 1981, 1996; Dekker & Peters, 1993) seems to have gained less attention compared to moderate and clear-water lakes (e.g. Bukata et al., 1985, 1988; Shanz, 1985; Chekhin, 1987; Koenings & Edmunson, 1991; Gallie & Murtha, 1992; Gitelson et al., 1994; Dekker et al., 2001; Giardino et al., 2001; Härma et al., 2001; Strömbeck & Pierson, 2001, etc.).

During last years our group of marine optics performed a number of studies on turbid lakes in Estonia (Arst et al., 1999, 2006; Herlevi, 1999, 2002; Erm et al., 2001, 2002; Arst, 2003). Turbid lakes deserve more attention because several water parameters like concentrations of optically active substances, plankton biomass and primary production can gain extreme values in such lakes, which have to be taken into account when calibrating of different functional models describing water ecosystems.

The main aim of the productivity measurements of aquatic ecosystems, which started in the second half of the twentieth century (Steeman Nielsen, 1952; Ryther, 1956; Vinberg, 1961; Peterson, 1980), was to get a better understanding of the food chain relationships and of the functioning of the ecosystems. Traditional methods for primary production measurements are based on the gas exchange part of the process of photosynthesis and measure either the release rate of oxygen from plants (Gaarder & Gran, 1927) or the uptake of carbon dioxide (Steeman Nielsen, 1952). In the latter case the radioactive ^{14}C method is used. Both methods measure in fact the instantaneous rate of the photosynthesis during short (1–2 h) exposures. Using of the longer exposures biases the results: in the case of the oxygen method, part of the oxygen produced in the experimental bottles may form gas bubbles when a supersaturation is generated, in case of ^{14}C method, cells start releasing the isotope label when the exposure exceeds the turnover time. The methodological differences of the methods have been discussed, e.g. by Strickland (1960) and Patten & Chabot (1966).

Lake types differ by the ability to bind carbon and the differences can be attributed mainly to differences in their productivity. Due to changing light conditions the planktonic photosynthesis has a pronounced daily dynamics. In order to get integrated results over longer time periods (days, months, years), a large number of consecutive measurements of instantaneous photosynthesis rate should be carried out and integrated. However, this time-consuming method cannot be used for routine measurements and therefore the involving of bio-optical model calculations could solve the problem. Modelling is exceptionally important in turbid lakes of high productivity where the sharp light gradient may cause big errors if traditional field methods are applied.

Our final purpose (beyond the analyses included in this paper) is to build a model for computing primary production in the water based on the values of photosynthetically absorbed radiation and quantum yield of carbon fixation. Based on theoretical considerations from previous bio-optical models and some empirical relationships (Bannister, 1974; Morel, 1978; Kiefer & Mitchell, 1983; Morel & Berthon, 1989; Smith et al., 1989; Kiefer & Gullen, 1990; Kyewalyanga et al., 1992; Berthon & Morel, 1992; Kirk, 1996; Sosik, 1996; Nõges & Nõges, 1998) we made an hypothesis that the model can be quantified on the basis of the spectral/integral incident quantum irradiance and diffuse attenuation coefficient of light in the water, and the respective chlorophyll *a* content.

The purpose of the first stage of our project was to collect a database, containing a large number of primary production profiles and corresponding solar irradiances (incident and underwater) as well as chlorophyll *a* values. In the analysis, we focused on the following aspects scarcely discussed in the literature: (1) comparison of in situ measured vertical profiles of planar and scalar quantum irradiances measured in the water, (2) extremely high values of chlorophyll *a* and primary production (*P*) in some turbid lakes, (3) analysis of the formation of the vertical profiles of *P* (and the depth of its maximum value) as a function of incident irradiance and (4) collecting a rather impressive database describing the values of bio-optical characteristics and their variations in turbid lakes. In the present article the main results of this sub-programme are discussed.

Material and methods

In 2003–2005 we collected the following in situ and laboratory data:

- Primary production at different depths ($P(z)$, mg C m^{-3} h^{-1}) was estimated in situ at six depths in the lake using $^{14}CO_2$ assimilation technique (Steeman Nielsen, 1952) and 2-h incubation around midday. After incubation the water was acidified (pH < 2) by 0.5 N HCl to remove the rest of inorganic ^{14}C (Niemi et al., 1983; Hilmer & Bate, 1989; Lignell, 1992). Then radioactivity of the samples was measured by a LSC RackBeta 1211 counter (Wallac, Finland) using external standardization for DPM calculations and scintillation coctail Optiphase 'HiSafe 3' (Wallac, Finland). $P(z)$ was calculated by the standard formula (Nielsen & Bresta, 1984). Non-photosynthetic carbon fixation was measured in dark vials and subtracted from light assimilation. Integral values of primary production, P_{int}, were calculated by integrating $P(z)$ over depth.

- Incident planar quantum irradiance (q_{PAR}(inc) μmol m^{-2} s^{-1}; using quantum sensor LI-192 SA, LI-COR Corporation SA, Lincoln, Nebraska, USA) in the photosynthetically active region (PAR, 400–700 nm) every two minutes above the water during the 2-h incubation period of $P(z)$ bottles.
- Underwater downwelling planar and scalar quantum PAR irradiances, $q_{PAR}(z)$ (quantum sensor LI-192 SA) and $q_{0,PAR}(z)$ (quantum sensor LI-193 SA) at different depths (z), 2–8 depth vertical profiles during 2-h period.
- Relative transparency of water by Secchi disk (z_{SD}, m).
- Concentrations of chlorophyll *a* (C_{chl}, in mg m^{-3}) according to the Lorenzen (1967) method, total suspended matter (C_s, in mg l^{-1}) by dry weight.
- Spectra of the light attenuation coefficient in filtered and unfiltered water samples in the wavelength range of 350–700 nm using the laboratory spectrophotometer *Hitachi U*1000. On the basis of these results we derived the values of the diffuse attenuation coefficient (Arst et al., 2002; Arst, 2003) to check the in situ downwelling planar irradiance (could be biased by waves).

The measurements were carried out in three Estonian lakes, Peipsi, Võrtsjärv and Harku (Table 1). Altogether there were 36 measurement days: 14 in both Peipsi and Võrtsjärv, and 8 in Harku. Two consecutive measurement series were planned for each day, the first before or at midday (marked as a.m. in Figs. 4–7) and the second in the afternoon (p.m.), but sometimes due to the weather conditions or technical problems some data were missing or not acceptable. Altogether we got reliable results from 53 field and 68 laboratory series.

Results

Some examples of the vertical profiles of planar and scalar irradiances measured simultaneously are shown in Figs. 1 and 2.

The turbidity of water is often described using the depth-averaged diffuse attenuation coefficient of PAR, $K_{d,PAR}(z)$. To determine its value we plotted the logarithm of $q_{PAR}(z)$ *vs.* depth and measured $K_{d,PAR}(z)$ as the slope of the least-square regression line through these points. The seasonal variations of z_{SD}, C_{chl} and $K_{d,PAR}$ are shown in Fig. 3a–c. The corresponding temporal change in the primary production integrated over depth (P_{int}) is presented in Fig. 4.

Some vertical profiles of primary production are shown in Figs. 5–7. We also investigated the relationships between the depth of maximum primary production ($z(P_{max})$) and the optical parameters. The following multiple regression formula was obtained:

$$z(P_{max}) = 0.000544 z_{SD} \times q_{PAR}(\text{inc}) + 0.0398 \quad R^2(\text{adj}) = 0.794, \quad n = 53, \quad p < 0.0001 \tag{1}$$

where z and z_{SD} are in m and q_{PAR}(inc) in μmol m^{-2} s^{-1}.

The minimum and maximum values of C_{chl}, $K_{d,PAR}$, P_{int} and P_{max} (the maximum of a primary production profile) measured during 2003–2005 are shown in Table 2. We also calculated the values of hourly chlorophyll-specific primary production (assimilation number, AN, mgC m^{-3} mg Chla^{-1}) as ratio of C_{chl} and $P(z)$, and the depth-averaged values of $P(z)$ and AN for each measurement series. In Arst et al. (2006) the regressions C_{chl} vs. P_{int} and AN(av) vs. P(av) were studied using the first ten measurement series from 2003. In the present study we repeated these calculations for all 53-measurement series (Figs. 8 and 9).

Table 1 Trophic type, main morphometric data and typical variation of Secchi depth (z_{SD}) in summer in the lakes under investigation

Lake	Trophic type	Area (km^2)	Average depth (m)	Maximum depth (m)	z_{SD} (m)
Peipsi	Meso/eutrophic	2,611	8.3	12.9	0.9–3.0
Võrtsjärv	Eutrophic	270	2.8	6.0	0.3–1.6
Harku	Hypertrophic	1.64	2	2.5	0.1–1.0

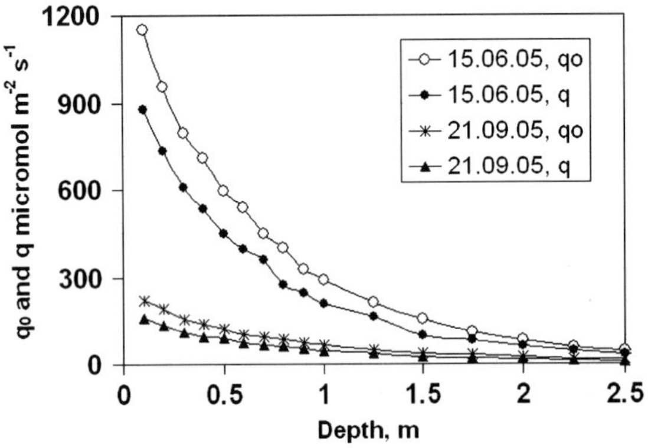

Fig. 1 Vertical profiles of planar (q) and scalar (q_0) irradiances measured in Peipsi (the values were scaled so as to correspond to the mean value of incoming surface irradiance during the measurement procedure)

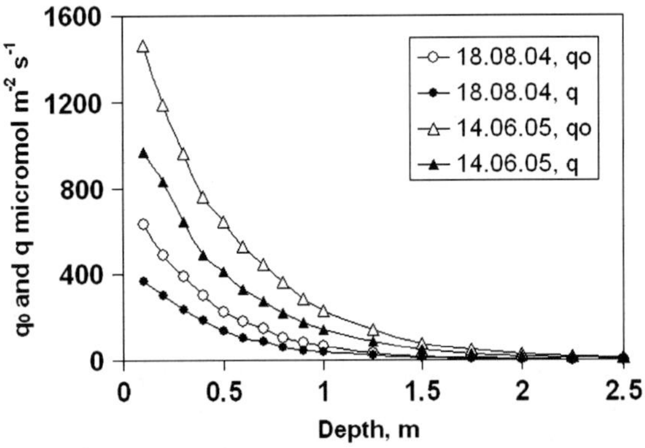

Fig. 2 Vertical profiles of planar (q) and scalar (q_0) irradiances measured in Võrtsjärv (the values were scaled so as to correspond to the mean value of incoming surface irradiance during the measurement procedure)

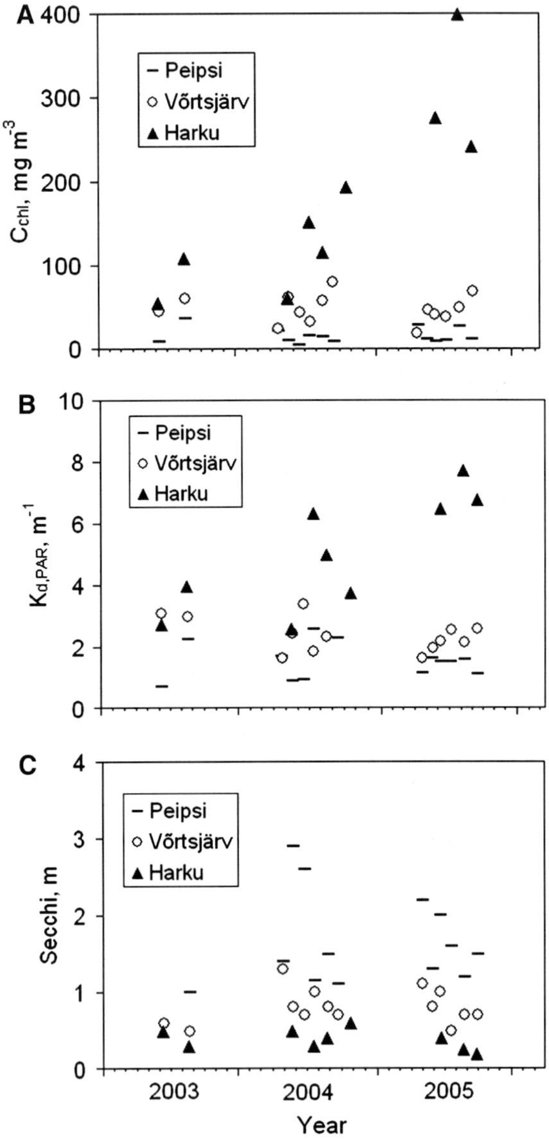

Fig. 3 Temporal variation of chlorophyll *a* concentration (C_{chl}), diffuse attenuation coefficient in PAR band ($K_{d,PAR}$), and Secchi depth in lakes Peipsi, Võrtsjärv and Harku in 2003–2005 (averaged values for each measurement day are presented)

Discussion

In Figs. 1 and 2 all cases show that $q_0 > q$, while the absolute differences decreased with increasing depth. However, relative differences between these two parameters practically did not depend on depth: according to all our data they were 25–45% for

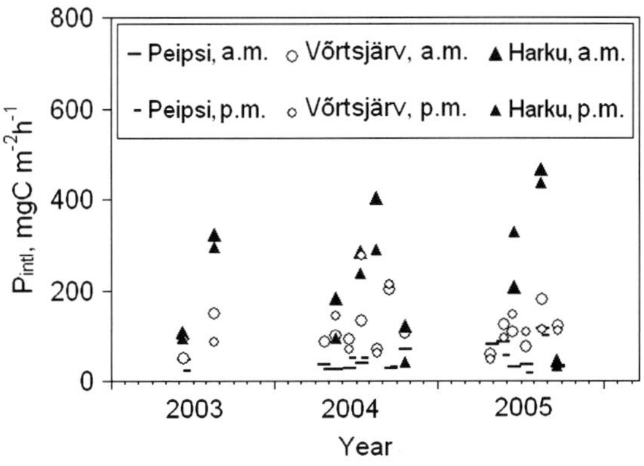

Fig. 4 Temporal variation of integral primary production (P_{int}) in lakes Peipsi, Võrtsjärv and Harku in 2003–2005. a.m., measurement series starting before or at noon; p.m., measurement series made in the afternoon

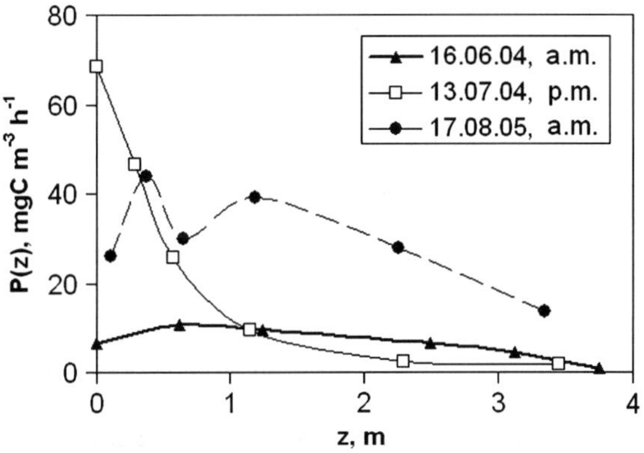

Fig. 5 Some vertical profiles of primary production $P(z)$ measured in Peipsi: (1) 16.06.04, C_{chl} = 5.7 mg m^{-3}, q_{PAR}(inc) = 620 μmol m^{-2} s^{-1}, (2) 13.07.04, C_{chl} = 14.5 mg m^{-3}, q_{PAR}(inc) = 350 μmol m^{-2} s^{-1}, (3) 17.08.05, C_{chl} = 33.1 mg m^{-3}, q_{PAR}(inc) = 1,060 μmol m^{-2} s^{-1}. C_{chl} is chlorophyll *a* concentration in the water, q_{PAR}(inc) is incident planar quantum irradiance; a.m., measurement series starting before or at noon; p.m., measurement series made in the afternoon

Peipsi, 30–55% for Võrtsjärv and about 65% in Harku (in case of algal bloom these numbers could be even bigger). It is important to note the substantial variation of these differences both within and between the lakes. Due to the fact that the algal cells are illuminated from all directions primary production calculations should be performed using the data of underwater scalar quantum irradiance (Kirk, 1996; Sosik, 1996). Because of the large variability of the differences between q and q_0, the use of q without a correction coefficient in bio-optical models instead of q_0 should be avoided.

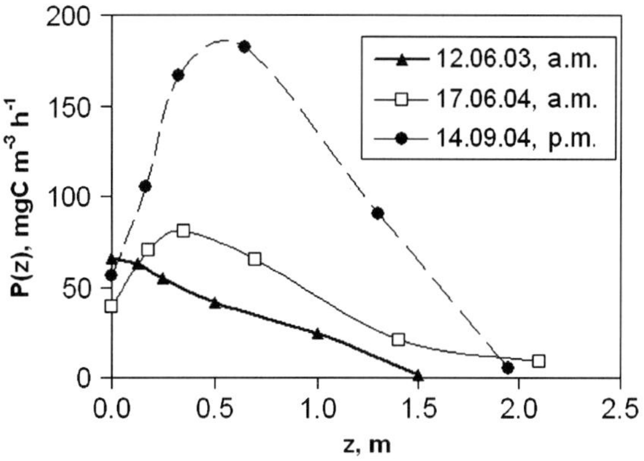

Fig. 6 Some vertical profiles of primary production $P(z)$ measured in Võrtsjärv: (1) 12.06.03, C_{chl} = 44 mg m^{-3}, q_{PAR}(inc) = 425 μmol m^{-2} s^{-1}, (2) 17.06.04, C_{chl} = 43 mg m^{-3}, q_{PAR}(inc) = 560 μmol m^{-2} s^{-1}, (3) 14.09.04, C_{chl} = 79 mg m^{-3}, q_{PAR}(inc) = 1,300 μmol m^{-2} s^{-1}. C_{chl} is chlorophyll *a* concentration in the water, q_{PAR}(inc) is incident planar quantum irradiance; a.m., measurement series starting before or at noon; p.m., measurement series made in the afternoon

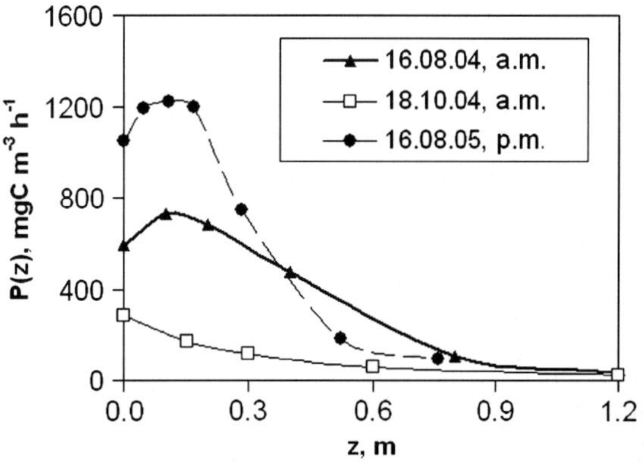

Fig. 7 Some vertical profiles of primary production $P(z)$ measured in Harku: (1) 16.08.04, C_{chl} = 123 mg m^{-3}, q_{PAR}(inc) = 1,300 μmol m^{-2} s^{-1}, (2) 18.10.04, C_{chl} = 160 mg m^{-3}, q_{PAR}(inc) = 80 μmol m^{-2} s^{-1}, (3) 16.08.05, C_{chl} = 389 mg m^{-3}, q_{PAR}(inc) = 990 μmol m^{-2} s^{-1}. C_{chl} is chlorophyll *a* concentration in the water, q_{PAR}(inc) is incident planar quantum irradiance; a.m., measurement series starting before or at noon; p.m., measurement series made in the afternoon

The seasonal behaviour of C_{chl} and $K_{d,PAR}$ was similar (Fig. 3a, b)—their values depended on the biological activity (maxima in August–September). A general increase of these variables from 2003 to 2005 was observed only in Harku. Also the variability of suspended matter was in good accordance with the corresponding C_{chl} values (not shown). The variation of the Secchi depth between the different lakes (Fig. 3c) was much lower than

Table 2 The minimum and maximum values of chlorophyll *a* concentration (C_{chl}), chlorophyll *a* content in the euphotic layer (C_{chl}(euf)), diffuse attenuation coefficient determined from q_{PAR} vertical profile ($K_{d,PAR}$), primary production integrated over depth (P_{int}), and the vertical maximum value of primary production (P_{max}) in lakes Peipsi, Võrtsjärv and Harku in 2003–2005 (taken from separate measurement series)

Lake	C_{chl} (mg m^{-3})	C_{chl}(euf) (mg m^{-2})	$K_{d,PAR}$ (m^{-1})	P_{int} (mg C m^{-2} h^{-1})	P_{max} (mg C m^{-3} h^{-1})
Peipsi	4.2–37	24–117	0.74–2.58	16.6–107	7.8–116
Võrtsjärv	28–79	61–142	1.65–3.40	42.8–273	23.4–182
Harku	54–389	87–239	2.50–7.7	39–435	96–1,532

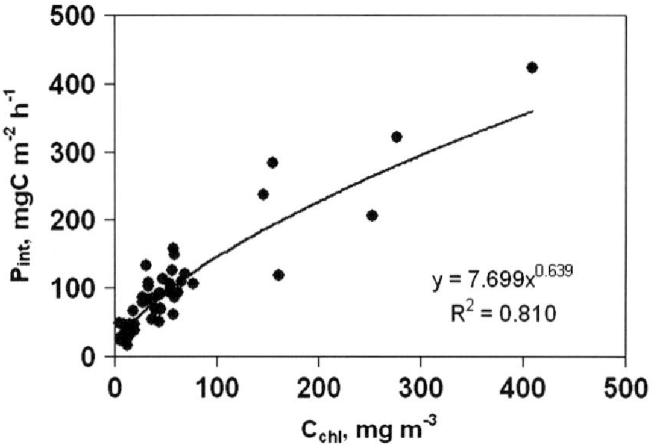

Fig. 8 Regression C_{chl} vs. P_{int} obtained from the data of 2003–2005 ($N = 53$, $p < 0.001$). C_{chl} is chlorophyll *a* concentration in the water, P_{int} is the integral primary production

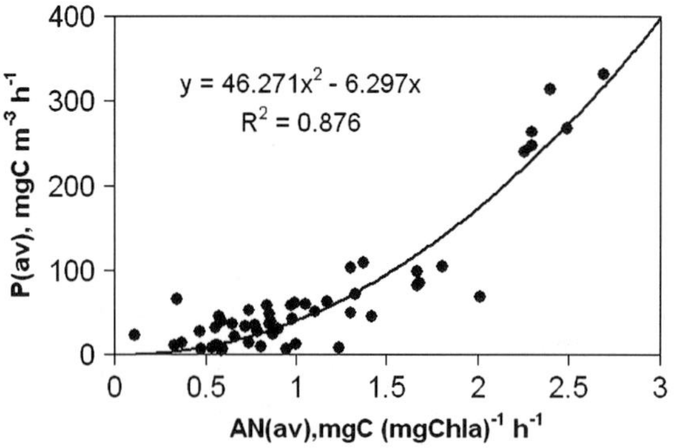

Fig. 9 Regression AN(av) vs. *P*(av) obtained from the data of 2003–2005 ($N = 53$, $p < 0.001$). AN(av) and *P*(av) are correspondingly vertically averaged assimilation number and primary production for each measurement series

that of C_{chl} and $K_{d,PAR}$. The low sensitivity of the Secchi depth to changes of optically active substances in very turbid waters is more thoroughly discussed by Arst (2003). Similarly to C_{chl}, P_{int} had the maximum values in August (Figs. 3a and 4), but dropped in September as a result of the decrease of the incoming irradiance.

The vertical profiles of primary production and the value of maximum primary production (P_{max}) demonstrate the dependence of the *P*(*z*) on plankton abundance and on incoming and underwater irradiance (Figs. 5–7). The depth of maximum primary production ($z(P_{max})$) depends mainly on illumination conditions. In cases of low illumination $z(P_{max})$ is close to zero (the maximum primary production was observed at the surface, e.g. on 12.06.03, 13.07.04 and 18.10.04 in Figs. 5–7). In the multiple regression formula (Eq. 1) the product of Secchi depth and incident irradiance, $z_{SD}q_{PAR}$(inc), explained nearly eighty percent of the variability of $z(P_{max})$.

Our results (Table 2) show that the investigated three turbid lakes (Secchi depth below 3 m) can be very different by the values of optically active substances, $K_{d,PAR}$ and z_{SD}, and also by biological productivity. A really exceptional, extremely turbid lake is Harku, showing maximum values of C_{chl}, C_{chl}(euf), $K_{d,PAR}$ and P_{int} respectively 398 mg m^{-3}, 239 mg m^{-2}, 7.7 m^{-1} and 435 mg C m^{-2} h^{-1}. The value of P_{int} was rather close to the record highest rates of photosynthesis, 530–900 mg C m^{-2} h^{-1} measured in the Ethiopian soda lakes (Talling, 1973, oxygen production converted to carbon using a coefficient of 0.375). The chlorophyll content per square metre in the euphotic layer (C_{chl}(euf)) in Harku ranged from 87 to 239 mg m^{-2}, with the highest values reaching the maximum holding capacity of the euphotic layer (~200–300 mg m^{-2}) expected on theoretical grounds (Steeman Nielsen, 1962; Nõges, 2000).

Comparing the regressions C_{chl} vs. P_{int} obtained by us and presented in Arst et al. (2006) it turned out that the general trend was the same, but instead of polynomial regression formula (in Arst et al., 2006) we got an exponent (Fig. 8). For the AN(av) *vs.* *P*(av) relationship the general form of the regression formula remained the same:

$$P(\mathrm{av}) = T_1(\mathrm{AN}(\mathrm{av}))^2 + T_2\mathrm{AN}(\mathrm{av}), \quad (2)$$

but the constants T_1 and T_2 changed, respectively, from 41.5 to 46.3 and from 5.1 to −6.3 (Fig. 9).

Conclusions

- According to our study turbid lakes (Secchi depth below 3 m) can be very different by the values of optically active substances, $K_{d,PAR}$ and z_{SD}, and also by biological productivity. Lake Harku, showed exceptionally high values of C_{chl}, C_{chl}(euf), $K_{d,PAR}$ and P_{int} respectively 398 mg m^{-3}, 239 mg m^{-2}, 7.7 m^{-1} and 435 mg C m^{-2} h^{-1}, which are close to the theoretical maximum values.
- We observed large variability (both within and between lakes) of the relative differences between planar and scalar quantum irradiance in PAR region ranging from 25% to 65% and practically not depending on depth. Because of this variability, the underwater scalar quantum irradiance data required in primary production models cannot be correctly derived from planar quantum irradiance data.
- Depth at which $P(z)$ has its maximum value depends on the turbidity of the water and on the incident irradiance at the water surface. The product of Secchi depth and incident irradiance, $z_{SD} \times q_{PAR}(\mathrm{inc})$, explained nearly eighty percent of the variability of the depth of maximum production. In case of low illumination P_{max} was located at the surface.
- We can compare the lakes by their productivity only when using the long-term integral productivity (e.g. annual sums), but not by separate profiles of primary production, measured during one-two hours. Each of these profiles is formed under the simultaneous influence of two factors: (1) the amount of phytoplankton and (2) the light field in the air and in the water.

Acknowledgements The authors are indebted to Estonian Ministry of Education (target funding project 03962480s03 and 0712699s05) and to Estonian Science Foundation (Grant 5594) for financial support to this investigation, and to Ants Erm, Tõnu Feldmann, Medhat Hussainov, Kersti Kangro, Evi Lill and Ilmar Tõnno for their help at field measurements and laboratory analyses.

References

Arst, H., A. Erm, M. Hussainov, T. Kutser, S. Mäekivi, A. Reinart & A. Herlevi, 1999. Investigation of Estonian and Finnish lakes by optical measurements in 1992–97. Proceedings of Estonian Academy of Sciences. Biology. Ecology 48: 5–24.

Arst H., A. Erm, A. Reinart, L. Sipelgas & A. Herlevi, 2002. Calculating irradiance penetration into water bodies from the measured beam attenuation coefficient II: application of improved model to different types of lakes. Nordic Hydrology 33(2): 207–226.

Arst, H., 2003. Optical Properties and Remote Sensing of Multicomponental Water Bodies. Springer, Praxis-Publishing, Chichester.

Arst, H., A. Erm, K. Kangro, T. Nõges & P. Nõges, 2006. Comparison of spectral and broad-band models for computing photosynthetically absorbed radiation in turbid waters. Boreal Environmental Research 11: 55–65.

Bannister, T. T., 1974. Production equations in terms of chlorophyll concentration, quantum yield, and upper limit to production. Limnology & Oceanography 19: 1–12.

Berthon, J. -F. & A. Morel, 1992. Validation of spectral light-photosynthesis model and use of the model in conjunction with remotely sensed pigment observations. Limnology & Oceanography 37: 781–796.

Bukata R. P., J. E. Bruton, & J. H. Jerome, 1985. Application of direct measurements of optical parameters to the estimation of lake water quality indicators. Environment Canada, Inland Waters Directorate, National Water Research Institute, Canada Centre for Inland Waters, Burlington, Ontario, Scientific Series 140: 1–35.

Bukata, R. P., J. H. Jerome & J. E. Bruton, 1988. Particulate concentrations in Lake St. Clair as recorded by shipborne multispectral optical monitoring system. Remote Sensing Environment 25: 201–229.

Chekhin, L. P., 1987. Light regime in the water bodies. Karelia Section of the Academy of Sciences USSR, Petrozavodsk: 128 pp. (in Russian).

Dekker, A. G. & S. W. M. Peters, 1993. The use of Thematic Mapper for the analysis of eutrophic lakes: a case study in the Netherlands. International Journal of Remote Sensing 14: 799–821.

Dekker, A. G., R. J. Vos & S. W. M. Peters, 2001. Comparison of remote sensing data, model results and in situ data for total suspended matter (TSM) in the southern Frisian lakes. Science of the Total Environment 268: 197–214.

Erm, A., H. Arst, T. Trei, A. Reinart & M. Hussainov, 2001. Optical and biological properties of Lake Ülemiste, a water reservoir of the city of Tallinn I: water transparency and optically active substances in the water. Lakes and Reservoirs: Research and Management 6: 63–74.

Erm, A., H. Arst, P. Nõges, T. Nõges, A. Reinart & L. Sipelgas, 2002. Temporal variations in bio-optical properties of four North Estonian lakes in 1999–2000. Geophysica 38: 89–111.

Gaarder, T. & H. H. Gran, 1927. Investigation of the production of plankton in Oslo Fjord. Rapport du Conseil Internationale pour l'Exploration de la Mer 42: 1–48.

Gallie, E. A. & P. A. Murtha, 1992. Specific absorption and scattering spectra for suspended minerals and chlorophyll

a in Chilko Lake, British Columbia. Remote Sensing Environment 39: 103–118.
Giardino, C., M. Pepe, P. A. Brivio, P. Ghezzi & E. Zilioli, 2001. Detecting chlorophyll, Secchi disk depth and surface temperature in sub-alpine lake using Landsat imagery. Science of the Total Environment 268: 19–29.
Gitelson, A., M. Mayo, Y. C. Yacobi, A. Paparov & T. Berman, 1994. The use of high-spectral-resolution radiometer data for detection of low chlorophyll concentrations in Lake Kinnert. Journal of Plankton Research 16: 993–1002.
Herlevi, A., H. Virta, H. Arst & A. Erm, 1999. Results of light absorption measurements in Finnish and Estonian lakes in summer 1997. Proceedings of Estonian Academy of Sciences. Biology Ecology 48/1: 46–62.
Herlevi, A., 2002. Inherent and apparent optical properties in relation to water quality in Nordic waters. Academic dissertation in Geophysics. Report Series in Geophysics, No 45, University of Helsinki, Finland.
Hilmer, T. & G. C. Bate, 1989. Filter types, filtration and post-filtration treatment in phytoplankton production studies. Journal of Plankton Research 11: 49–63.
Härmä, P., J. Vepsäläinen, T. Hannonen, T. Pyhälahti, J. Kämäri, K. Kallio, K. Eloheimo & S. Koponen, 2001. Detection of water quality using simulated satellite data and semi-empirical algorithms in Finland. Science of the Total Environment 268: 107–121.
Kiefer, D. A. & B. G. Mitchell, 1983. A simple steady state description of phytoplankton growth based on absorption cross section and quantum efficiency. Limnology & Oceanography 28: 770–776.
Kiefer, D. A. & J. J. Cullen, 1990. Phytoplankton growth and light absorption as regulated by light, temperature and nutrients. Polar Research 10: 163–172.
Kirk, J. T. O., 1981. Estimation of the scattering coefficient of natural waters using underwater irradiance measurements. Australian Journal Marine and Freshwater Research 32: 533–539.
Kirk, J. T. O., 1996. Light and Photosynthesis in Aquatic Ecosystems. Cambridge University Press, Cambridge.
Koenings, J. P. & J. A. Edmundson, 1991. Secchi disk and photometer estimates of light regimes in Alaskan lakes: effects of yellow colour and turbidity. Limnology & Oceanography 36: 91–105.
Kyewalyanga, M., T. Platt & S. Sathyendranath, 1992. Ocean primary production calculated by spectral and broad-band models. Marine Ecology Progress Series 85: 171–185.
Lignell, R., 1992. Problems in filtration fractionation of ^{14}C primary productivity samples. Limnology & Oceanography 37: 172–178.
Lorenzen, C. J., 1967. Determination of chlorophyll and phaeopigments; spectrophotometric equations. Limnology & Oceanography 12: 343–346.
Morel, A., 1978. Available, usable and stored radiant energy in relation to marine photosynthesis. Deep-Sea Research 25: 673–688.
Morel, A. & J.-F. Berthon, 1989. Surface pigments, algal biomass, and potential production of the euphotic layer: relationships reinvestigated in view of remote-sensing applications. Limnology & Oceanography 34: 1545–1562.
Nielsen G. A. & A. M. Bresta (eds), 1984. Guidelines for the measurement of phytoplankton primary production. The Baltic Marine Biologists (BMB) Publ. 1: 2nd edn., 23 pp.
Niemi, M., J. Kuparinen, A. Uusi-Rauva & K. Korhonen, 1983. Preparation of algal samples for liquid scintillation counting. Hydrobiologia 106: 149–159.
Nõges, P. & T. Nõges, 1998. Advantages and limitations of optical methods in the estimation of the primary production of lakes. In Pulkkinen, K. (ed.), Report Series in Geophysics 38, Department of Geophysics, University of Helsinki: 59–67.
Nõges, P., 2000. Euphotic holding capacity for optically active substances. Geophysica 36: 169–176.
Patten, B. C. & B. F. Chabot, 1966. Factorial productivity experiments in a shallow estuary: characteristics of response surfaces. Chesapeake Science 7: 117–136.
Ryther, J. H., 1956. The measurement of primary production. Limnology and Oceanography 1: 72–84.
Peterson, B. J., 1980. Aquatic primary productivity and the 14C–CO_2 method: A history of the productivity problem. Annual Review of Ecology and Systematics 11: 359–385.
Schanz, F., 1985. Vertical light attenuation and phytoplankton development in Lake Zürich. Limnology & Oceanography 30: 299–310.
Smith, R. C., B. B. Prezelin, R. R. Bidigare & K. S. Baker, 1989. Bio-optical modeling of photosynthetic production in coastal waters. Limnology & Oceanography 34: 1524–1544.
Sosik, H. M., 1996. Bio-optical modelling of primary production: consequences of variability in quantum yield and specific absorption. Marine Ecology Progress Series 143: 225–238.
Steeman Nielsen, E., 1952. The use of radioactive carbon (14C) for measuring primary production in the sea. Journal du Conseil. Conseil Permanent Internationalpour l'Exploration de la Mer 18: 117–140.
Steemann Nielsen, E., 1962. On the maximum quantity of plankton chlorophyll per surface unit of a lake or the sea. Internationale Revue der gesamten Hydrobiologie und Hydrographie 47: 333–338.
Strickland, J. D. H., 1960. Measuring the production of marine phytoplankton. Fisheries Research Board of Canada Bulletin 122: 1–172.
Strömbeck, N. & D. S. Pierson, 2001. The effects of variability in the inherent optical properties on estimations of chlorophyll *a* by remote sensing in Swedish freshwaters. Science of the Total Environment 268: 123–137.
Talling, J. F., R. B. Wood, M. V. Prosser & R. M. Baxter, 1973. The upper limit of photosynthetic productivity by phytoplankton: evidence from Ethiopian soda lakes. Freshwater Biology 3: 53–76.
Vinberg, G. G. (ed.), 1961. Primary production of seas and continental waters Minsk, 407 pp. (In Russian, contents in English).

Hydrobiologia (2008) 599:177–182
DOI 10.1007/s10750-007-9210-2

Models as tools for understanding past, recent and future changes in large lakes

Thorsten Blenckner

Abstract Large lakes currently exhibit ecosystem responses to environmental changes such as climate and land use changes, nutrient loading, toxic contaminants, hydrological modifications and invasive species. These sources have impacted lake ecosystems over a number of years in various combinations and often in a spatially heterogeneous pattern. At the same time, many different kinds of mathematical models have been developed to help to understand ecosystem processes and improve cost-effective management. Here, the advantages and limitations of models and sources of uncertainty will be discussed. From these considerations and in view of the multiple environmental pressures, the following emerging issues still have to be met in order to improve the understanding of ecosystem function and management of large lakes: (1) the inclusion of thresholds and points-of-no-return; (2) construction of general models to simulate biogeochemical processes for a large number of lakes rather than for individual systems; (3) improvement of the understanding of spatio-temporal variability to quantify biogeochemical fluxes accurately; and (4) inclusion of biogeochemical linkages between terrestrial and aquatic ecosystems in model approaches to assess the effects of external environmental pressures such as land-use changes. The inclusion of the above-mentioned issues would substantially improve models as tools for the scientific understanding and cost-effective management of large lakes that are subject to multiple environmental pressures in a changing future.

Keywords Ecology · Large lakes · Modelling · Uncertainty · Emergent issues · Management

Guest editors: T. Nõges, R. Eckmann, K. Kangur, P. Nõges, A. Reinart, G. Roll, H. Simola & M. Viljanen
European Large Lakes—Ecosystem changes and their ecological and socioeconomic impacts

T. Blenckner (✉)
Erken Laboratory, Department of Ecology and Evolution, Evolutionary Biology Centre, Uppsala University, Norr Malma 4200, 761 73 Norrtälje, Sweden
e-mail: Thorsten.Blenckner@ebc.uu.se

Introduction

Large lakes are increasingly exhibiting ecosystem responses owing to high pressures from a combination of sources such as climate change and variability, land use changes, hydrological modifications, toxic contaminants and invasive species. The effects of climate change and variability have been analysed in a number of lakes with significant changes in lake temperature, period of ice cover, stratification strength, bottom oxygen concentration, spring phytoplankton bloom, clear water phase, fish abundance and match–mismatch of predator–prey interactions (see for example Magnuson et al., 1990; Schindler et al., 1990, 1996; Scheffer et al., 2001a; Winder & Schindler, 2004; Blenckner, 2005; Mooij et al., 2005). Land use changes also have an important effect on lakes by changing nutrient concentrations.

Today, non-point sources are especially highly relevant (Schindler, 2006). Furthermore, hydrological modifications, i.e. alterations of stream flow by human activities, may increase non-point source pollution to lakes (see for example Schelske et al., 2005). In dry climatic regions, these modifications can also have a serious impact on water quantity if water withdrawal exceeds inflow rates, as for instance in the Aral Sea region or Lake Kinneret.

Toxic contaminants, in particular heavy metals such as mercury and lead and Persistant Organic Pollutants (POPs) such as dioxin, PCBs and DDT, are chemical substances that persist in the environment. As they bio-accumulate within the food web (Binelli et al., 2004; Fernandez et al., 2005), they represent a growing risk to human health (Dewailly et al., 1989). Finally, invasive species cause considerable problems in aquatic systems as they can contribute to massive extinctions of native fauna and alter food webs severely. For example, the zebra mussel alone has resulted in millions of dollars of damage to local water users, i.e. municipalities and industries in the Great Lakes area, USA, since its invasion in 1985/1986 (http://www.great-lakes.net/envt/flora-fauna/invasive/zebra.html). Despite strong management efforts and research, all the above-mentioned environmental pressures affect lakes markedly and often have combined or cumulative effects.

During recent decades, many different mathematical models have been developed to understand ecosystem responses to single environmental pressures and to improve cost-effective management options. Nevertheless, a predominant focus on small scales and a natural time delay between the induction of an environmental pressure and the consequent impact have led to indecisive management recommendations. Furthermore, no profound scientific understanding of ecosystem responses to multiple environmental pressures has been achieved so far.

Various characteristics make research on large lakes, and management of such lakes, particularly challenging. Because of their size, many biotic processes are largely wind-driven, and this often results in complex hydrodynamic behaviour (e.g. seiche, Kelvin waves; see for example Kvarnäs, 2001). This becomes increasingly complex if the morphometry is diversified (sub-basins). Some large lakes are very shallow and eutrophic with substantial water level changes (Võrtsjärv, Estonia) while others are very deep and oligotrophic (Vättern, Sweden) and considerable spatial variability is likely, e.g. a horizontal gradient in trophic status (Boegman et al., 2001; Kvarnäs, 2001). Generally, large lakes often have a longer hydrological water residence time, leading to a slower response to external pressures, combined with an often longer recovery time to the external influence of undesirable substances (Tilzer & Bossard, 1992).

Understanding combinations of these specific characteristics of large lakes with the multiple environmental factors affecting them requires new approaches with specially constructed tools (models).

The aim of this paper is not to review current models in aquatic science (see, for example, Jørgensen, 1995) with their pitfalls and domains, but to assess the requirements of models that specifically address large lakes, to present their advantages and limitations and to offer a realistic outlook for handling multiple environmental pressures and developing a better understanding of ecosystem processes and management options for large lakes.

Models

Dynamic process-based models (hereafter referred to as models, which include the dynamics (time) of interactions between defined entities (Minns, 1992)) are used as tools since they predict ecosystem responses to new or changing conditions (e.g. nutrient load). In contrast, statistical models are invalid for conditions outside the current range (Klepper, 1997).

In general, models represent the qualitative understanding gained from our experience, training, observations, quantitative experiments and monitoring and the application of that understanding in a conceptual model, using a set of equations (Beven, 2002). Many water quality models have been developed during recent decades (e.g. Jørgensen, 1995; Janse, 1997; Krivtsov et al., 2001; Håkanson & Boulion, 2002). Those for small and shallow lakes mostly focus on food-web interactions or simple mass-balance calculations where the need for detailed hydrodynamic simulation is negligible. In contrast, to model the water quality of large lakes, a detailed hydrodynamic module in the model is necessary, as large lakes experience a highly wind-driven environment associated with often complex hydrodynamic structures, which influence chemical and biological variables markedly (Padisák & Koncsos, 2002; Beletsky et al., 2003; Gerstner et al., 2003).

One of the main advantages of applying models lies in predicting future changes (e.g. climate change) or testing management options (e.g. nutrient reduction). For the latter, it is possible to calculate the costs necessary to improve water quality to a certain level (e.g. total annual phosphorus load). As restoration management of large lakes has been very cost-intensive, the advantages of calculating the most cost-effective option are obvious.

However, models will always be only simplifications of reality and the uncertainty of model results needs to be included in their presentation and is a valuable component for further improvements of the model. Several sources of uncertainty in models are elaborated below.

Uncertainty

Overall, uncertainty in model applications is seldom addressed and if it is, then only a brief estimate based on sensitivity analysis is taken into account (Beven, 2002). However, three main sources of uncertainty ought to be considered.

Firstly, the uncertainty in the input data needed to drive the model. These are mostly monitoring data with particular uncertainties arising from measurement techniques and from temporal and spatial resolution. For example, a mass-balance nutrient model is driven by single-point inflow measurements, which are known to have large uncertainties because of the high variability of the discharge.

Secondly, the uncertainty in model structure (design of the model including the interactions of different model components) and state parameters, which describe or assume the underlying mechanisms. These parameters are hardly measurable in the field and values must therefore either be estimated or obtained from experiments under controlled conditions. Interactions between the different parameters in the model are often complex and may compensate each other within different processes.

Thirdly, we have to consider the uncertainty in the observations against which the model is calibrated and validated. Here again, observational data are necessary and their uncertainty is similar to that described for the input data.

To present and calculate the uncertainty in models, several methods for uncertainty and sensitivity analysis have been applied (see for example Aldenberg et al., 1995; Janse, 1997; Klepper, 1997; Omlin et al., 2001). To determine the number of samples (n) needed for instance to obtain a given level of uncertainty in the driving or calibration data, the following formula from Håkanson (1984) can be used:

$$n = \left(\frac{t \times CV}{L}\right)^2 + 1$$

where t is the Student's t value, CV is the coefficient of variation and L is the expected error. The value $L = 0.1$ implies that the user accepts an error of 10% (see for more details Håkanson, 1984).

However, it is difficult to separate the observational error from the error in the model output. This is usually done by residual statistics of the comparison between model predictions and observations (Beven, 2002). One way to overcome this problem is the Generalised Likelihood Uncertainty Estimation (GLUE) method (Beven & Binley, 1992; Ratto et al., 2001), which addresses the uncertainty in a particular model more specifically and may help to reject models scientifically if their predictions fall outside the observational range.

Emerging issues

Today, I see the following emerging issues that need to be addressed more thoroughly to improve the application of models as tools for research and management of large lake ecosystems.

Thresholds and resilience

Thresholds and points-of-no-return are essential concepts in ecology in general and in aquatic ecology in particular (Scheffer et al., 2001a, b; Folke et al., 2004). Nonlinear interactions and feedbacks across scales and their associated thresholds are common features of ecosystems. These nonlinear interactions challenge scientists and engineers to understand and model the behaviour of the system, as cross-scale effects often result in a "point-of-no-return" (passing thresholds) or unexpected events with considerable consequences for the environment and human welfare. Good examples are floods, initiated by heavy rainfall, which spread nonlinearly over extensive areas, in this case a result of positive feedback between weather, river basin management and urbanization; only by considering all three components together can the area affected by the flood be determined. Nonlinear processes can thus have marked impacts on ecosystem function, local and regional economy and human health (Schimel & Baker, 2002).

Several examples from shallow lakes have shown that abrupt changes in, for instance, the dominance of phytoplankton or macrophytes could be related to relatively small changes in nutrient loading (Scheffer et al., 2001a), a hypothesis that has been successfully tested by models (Janse, 1997; Van Nes et al., 2003). It is consequently very important to identify methods that avoid trespassing over ecological thresholds, which can lead to a collapse of ecosystem structure and function, in particular in large lakes with a long response time but a slow recovery time (Tilzer & Bossard, 1992).

Wide-range models

Focus should be to construct models of biogeochemical fluxes and ecosystem function for a large set of lakes, streams and wetlands in landscapes rather than for individual bodies of water. A recent climate impact model study exemplifies the wider range of sites and generalized processes that were examined. Here, a physical lake model and a mechanistic phosphorus model were combined with two future emission scenarios generated by a regional climate model (RCM) for three lake sites in central Sweden (for details see Malmaeus et al., 2005). The general phosphorus model (LEEDS) required only small, lake-specific adjustments before it was applied to all lake sites (for more information on the LEEDS model see Malmaeus & Håkanson, 2004). In the simulations, Lake Erken was much more sensitive to climate warming, with epilimnetic dissolved phosphorus concentrations being almost doubled in spring, than the two basins of Lake Mälaren (each in Sweden). The result stresses the importance of internal processes. As the phytoplankton biomass in Lake Erken is mainly phosphorus-limited, the potential for phytoplankton production is almost twice as high, which implies that in Lake Erken, and in other eutrophic lakes with long water residence times, eutrophication may become a serious problem in future. In conclusion, this study exemplifies how general models, after relatively small adjustments to several sites, allow comparative studies to be made and thus constitute a good opportunity to reveal new processes of key importance. Further examples of more general models are LakeWeb (Håkanson & Boulion, 2002), PCLake (Janse, 1997) and the Rostherme model (Krivtsov et al., 2001).

However, it should be noted that the model results may be difficult to interpret in applications with future climate conditions. This is important because several model state parameters and their interactions are often valid only under particular calibration conditions, which will probably change in the future or during extreme years. In addition, the factor climate needs to be elaborated more in the model calibration and, as the model example above shows, also in water management guidelines, e.g. the European Water Framework Directive.

Scaling

Ecosystems are characterized by high variability in space and time, which is not fully understood (Hughes et al., 2005; Humbert & Dorigo, 2005). Improving this understanding forces aquatic scientists to quantify biogeochemical fluxes accurately and to make better assessments of the responses of aquatic ecosystems to climate and human-induced changes in external forcing.

The main problem in this respect is the availability of data (Porter et al., 2005). Many national environmental monitoring programmes are declining in both the quantity and quality of sampling. However, as the importance of diminishing the uncertainty of models is increasingly recognized, data monitoring on different scales is highly necessary. In addition, continual advances in affordable sensors and wireless communication make the development of automated sensing systems with remote communication affordable for many ecological research programmes (Porter et al., 2005). These in situ instruments provide high-frequency data on key variables that previously were measured intermittently and by hand. The use of these automatic stations will increase the volume of ecological data generated at different scales, which are particularly needed for modelling large lakes.

Further, it is important to choose the right scale (or the right model with the appropriate model schematization) for a particular question. For example, the question of how the nutrient concentration in the lake changes with a certain nutrient reduction in the inflow may not be answered appropriately by a complex, fine-scale model with a high associated uncertainty. Here, a simpler mass-balance model might be more adequate.

Watershed perspective

New, innovative approaches are needed to understand the processes controlling the biogeochemical linkages between terrestrial and aquatic ecosystems. Although this is an old issue (see for example Likens et al., 1970) there is still a need to quantify these linkages. There have been significant recent advances in modelling the biogeochemical processes in the watershed (see for example Zhang & Jørgensen, 2005; Grunwald & Qi, 2006; Kavvas et al., 2006; Markel et al., 2006); nevertheless, innovative approaches are needed to relate terrestrial, hydrological and biogeochemical processes to aquatic ecosystem function. GIS-based and remote sensing expertise with tools for transforming geographical data into export models are being used increasingly in aquatic sciences (Kim et al., 2006), but full development and implementation of this approach will require comprehensive datasets and knowledge of the transformation of GIS classification sets into model parameters.

Outlook

The inclusion of what, in my opinion, are the four most emerging issues (thresholds, wide-range models, scaling and watershed perspective) in new state-of-the-art model designs will improve the modelling of large lakes and also diminish the remaining uncertainty. As scientists and engineers make progress on these issues, the approach will be to perform adaptive management and build high tolerance to environmental pressures into the system, i.e. increase the resilience of the system (Janssen & Carpenter, 1999; Folke et al., 2004), which will reduce the impact of possible environmental factors in future. Until models reach greater certainty or, in other words, until a lower risk of failure becomes feasible, management aimed at high stress tolerance remains necessary. Therefore, ecosystems should be managed in an adaptive and precautionary way on the basis of the evaluation of the ongoing monitoring programme so that their natural resilience is maintained, and thus be prepared for future changes.

References

Aldenberg, T., J. H. Janse & P. R. G. Kramer, 1995. Fitting the dynamic model PCLake to a multi-lake survey through Bayesian statistics. Ecological Modelling 78: 83–90.

Beletsky, D., D. J. Schwab, P. J. Roebber, M. J. McCormick, G. S. Miller & J. H. Saylor, 2003. Modeling wind-driven circulation during the March 1998 sediment resuspension event in Lake Michigan. Journal of Geophysical Research 108: 20-1.

Beven, K., 2002. Towards a coherent philosophy for environmental modelling. Proceedings of the Royal Society of London Series A 458: 2465–2484.

Beven, K. & A. Binley, 1992. The future of distributed models: Model calibration and uncertainty prediction. Hydrological Processes 6: 279–298.

Binelli, A., F. Ricciardi & A. Provini, 2004. Present status of POP contamination in Lake Maggiore (Italy). Chemosphere 57: 27–34.

Blenckner, T., 2005. A conceptual model of climate related effects on lake ecosystems. Hydrobiologia 533: 1–14.

Boegman, L., M. R. Loewen, P. F. Hamblin & D. A. Culver, 2001. Application of a two-dimensional hydrodynamic reservoir model to Lake Erie. Canadian Journal of Fisheries and Aquatic Sciences 58: 853–869.

Dewailly, D. E., A. A. Nantel, J.-P. Weber & F. Meyer, 1989. High levels of PCBs in breast milk of Inuit women from Arctic Quebec. Bulletin Environmental Contamination Toxicology 43: 641–646.

Fernandez, P., G. Carrera & J. O. Grimalt, 2005. Persistent organic pollutants in remote freshwater ecosystems. Aquatic Sciences 67: 263–274.

Folke, C., S. R. Carpenter, B. Walker, M. Scheffer, T. Elmqvist, L. Gunderson & C. S. Holling, 2004. Regime shifts, resilience, and biodiversity in ecosystem management. Annual Review of Ecology and Systematics 35: 557–581.

Gerstner, G., W. Melendez, D. E. Endicott & K. R. Rygwelski, 2003. High resolution modeling of PCB congeners in lake Michigan using the lake Michigan level 3 (LM3) contaminant model. Global Threats to Large Lakes 46: 14.

Grunwald, S. & C. Qi, 2006. GIS-based water quality modeling in the Sandusky watershed, Ohio, USA. Journal of the American Water Resources Association 42: 957–973.

Håkanson, L., 1984. Aquatic contamination and ecological risk – An attempt to a conceptual framework. Water Resources 18: 1107–1118.

Håkanson, L. & V. V. Boulion, 2002. The Lake Foodweb. Backhuys Publishers, Leiden, 344 pp.

Hughes, T. P., D. R. Bellwood, C. Folke, R. S. Steneck & J. Wilson, 2005. New paradigms for supporting the resilience of marine ecosystems. Trends in Ecology & Evolution 20: 380–386.

Humbert, J. E. & U. Dorigo, 2005. Biodiversity and aquatic ecosystem functioning: A mini-review. Aquatic Ecosystem Health & Management 8: 367–374.

Janse, J. H., 1997. A model of nutrient dynamics in shallow lakes in relation to multiple stable states. Hydrobiologia 342/343: 1–8.

Janssen, M. A. & S. R. Carpenter, 1999. Managing the resilience of lakes: A multi-agent modeling approach. Conservation Ecology 3: 15.

Jørgensen, S. E., 1995. Handbook of environmental and ecological modelling. CRC Press Inc. 688 pp.

Kavvas, M. L., J. Yoon, Z. Q. Chen, L. Liang, E. C. Dogrul, N. Ohara, H. Aksoy, M. L. Anderson, J. Reuter & S. Hackley, 2006. Watershed environmental hydrology model: Environmental module and its application to a California watershed. Journal of Hydrologic Engineering 11: 261–272.

Kim, S.-C., C. F. Cerco & B. H. Johnson, 2006. Three-dimensional management model for Lake Washington, Part I: Introduction and hydrodynamic modeling. Lake and Reservoir Management 22: 103–114.

Klepper, O., 1997. Multivariate aspects of model uncertainty analysis: Tools for sensitivity analysis and calibration. Ecological Modelling 101: 1–13.

Krivtsov, V., C. Goldspink, D. C. Sigee & E. G. Bellinger, 2001. Expansion of the model 'Rostherne' for fish and zooplankton: Role of top-down effects in modifying the prevailing pattern of ecosystem functioning. Ecological Modelling 138: 153–171.

Kvarnäs, H., 2001. Morphometry and hydrology of the four large lakes of Sweden. Ambio 30: 467–474.

Likens, G. E., F. H. Bormann, N. M. Johnson, D. W. Fisher & R. S. Pierce, 1970. Effects of forest cutting and herbicide treatment on nutrient budgets in the Hubbard Brook watershed-ecosystem. Ecological Monographs 40: 23–47.

Magnuson, J. J., J. D. Meissner & D. K. Hill, 1990. Potential changes in thermal habitat of great lakes fish after global climate warming. Transactions American Fisheries Society 119: 254–264.

Malmaeus, J. M. & L. Håkanson, 2004. Development of a lake eutrophication model. Ecological Modelling 171: 35–63.

Malmaeus, J. M., T. Blenckner, H. Markensten & I. Persson, 2005. Lake phosphorus dynamics and climate warming: A mechanistic model approach. Ecological Modelling 190: 1–14.

Markel, D., F. Somma & B. M. Evans, 2006. Using a GIS transfer model to evaluate pollutant loads in the Lake Kinneret watershed, Israel. Water Science & Technology 53: 75–83.

Minns, C. K., 1992. Use of models for integrated assessment of ecosystem health. Journal of Aquatic Ecosystem Health 1: 109–118.

Mooij, W. M., S. Huelsmann, L. N. De Senerpont Domis, B. A. Nolet, P. L. E. Bodelier, P. C. M. Boers, L. M. Dionisio Pires, H. J. Gons, B. W. Ibelings, R. Noordhuis, R. Portielje, K. Wolfstein & E. H. R. R. Lammens, 2005. The impact of climate change on lakes in the Netherlands: A review. Aquatic Ecology 39: 381–400.

Omlin, M., R. Brun & P. Reichert, 2001. Biogeochemical model of Lake Zuerich: Sensitivity, identifiability and uncertainty analysis. Ecological Modelling 141: 105–123.

Padisák, J. & L. Koncsos, 2002. Trend and noise: Long-term changes of phytoplankton in the Keszthely-basin of Lake Balaton, Hungary. Verhandlungen der internationale Vereinigung für theoretische und angewandte Limnologie 28: 194–203.

Porter, J., P. Arzberger, H.-W. Braun, P. Bryant, S. Gage, T. Hansen, P. Hanson, C.-C. Lin, F.-P. Lin, T. Kratz, W. Michener, S. Shapiro & T. Williams, 2005. Wireless sensor networks for ecology. Bioscience 55: 561–572.

Ratto, M., S. Tarantola & A. Saltelli, 2001. Sensitivity analysis in model calibration: GSA-GLUE approach. Computer Physics Communications 136: 212–224.

Scheffer, M., D. Straile, E. H. Van Nes & H. Hosper, 2001a. Climatic warming causes regime shifts in lake food webs. Limnology and Oceanography 46: 1780–1783.

Scheffer, M., S. Carpenter, J. A. Foley, C. Folke & B. Walker, 2001b. Catastrophic shifts in ecosystems. Nature 413: 591–596.

Schelske, C. L., E. F. Lowe, L. E. Battoe, M. Brenner, M. F. Coveney & W. F. Kenney, 2005. Abrupt biological response to hydrologic and land-use changes in Lake Apopka, Florida, USA. Ambio 34: 192–198.

Schimel, D. S. & D. Baker, 2002. Carbon cycle: The wildfire factor. Nature 420: 29–30.

Schindler, D. W., 2006. Recent advances in the understanding and management of eutrophication. Limnology and Oceanography 51: 356–363.

Schindler, D. W., K. G. Beaty, E. J. Fee, D. R. Cruikshank, E. D. DeBruyn, D. L. Findlay, G. A. Linsay, J. A. Shearer, M. P. Stainton & M. A. Turner, 1990. Effects of climate warming on lakes of the central boreal forest. Science 250: 967–970.

Schindler, D. W., S. E. Bayley, B. R. Parker, K. G. Beaty, D. R. Cruikshank, E. J. Fee, E. U. Schindler & M. P. Stainton, 1996. The effects of climate warming on the properties of boreal lakes and streams at the experimental lake area, northwestern Ontario. Limnology and Oceanography 41: 1004–1017.

Tilzer, M. M. & P. Bossard, 1992. Large lakes and their sustainable development. Aquatic Sciences 54: 91–103.

Van Nes, E. H., M. Scheffer, M. S. Van den Berg & H. Coops, 2003. Charisma: A spatial explicit simulation model of submerged macrophytes. Ecological Modelling 159: 103–116.

Winder, M. & D. E. Schindler, 2004. Climate change uncouples trophic interactions in an aquatic ecosystem. Ecology 85: 2100–2106.

Zhang, J. & S. E. Jørgensen, 2005. Modelling of point and non-point nutrient loadings from a watershed. Environmental Modelling & Software 20: 561–574.

Hydrobiologia (2008) 599:183–189
DOI 10.1007/s10750-007-9201-3

The ice cover on small and large lakes: scaling analysis and mathematical modelling

Matti Leppäranta · Keguang Wang

Abstract Lake ice cover is described by its thickness, temperature, stratigraphy and overlying snow layer. When the ratio of ice thickness to lake size is above $\sim 10^{-5}$, the ice cover is stable; otherwise, mechanical forcing breaks the ice cover, and ice drifting takes place with lead-opening and ridging. This transition enables a convenient distinction to be made between small and large lakes. The evolution of the ice cover on small lakes is solved by a wholly thermodynamic model, but a coupled mechanical–thermodynamic model is needed for large lakes. The latter indicates a wide distribution of ice thickness, and frazil ice may be formed in openings. Ecological conditions in large lakes differ markedly from those in small lakes because vertical mixing and oxygen renewal may take place during the ice season, and the euphotic zone penetrates well into the water column in thin ice regions. Mesoscale sea ice models are applicable to large lakes with only minor tuning of the key parameters. These model systems are presented and analysed using Lake Peipsi as an example. As the climate changes, the transition size between small and large lake ice cover will change.

Keywords Large lakes · Small lakes · Ice · Physics · Scaling · Mathematical modelling · Mechanics · Thermodynamics

Guest editors: T. Nõges, R. Eckmann, K. Kangur, P. Nõges, A. Reinart, G. Roll, H. Simola & M. Viljanen
European Large Lakes—Ecosystem changes and their ecological and socioeconomic impacts

M. Leppäranta (✉) · K. Wang
Division of Geophysics, University of Helsinki, P.O. Box 64, Gustaf Hällströmin katu 2, 00014 Helsinki, Finland
e-mail: matti.lepparanta@helsinki.fi

Introduction

The length of the lake ice season in northern Europe is up to 7 months, and the maximum annual thickness of ice is 10–100 cm. The ice is a solid sheet with high mechanical strength and high thermal insulating properties. It is normally covered with snow, which is an even better insulator and also has a small optical depth compared with water and ice. The winter water body beneath the ice is quiet and dark with weak circulation (Bengtsson, 1996). The presence of ice stabilizes the lake temperature conditions. The surface water is kept at 0°C, and only weak upward heat fluxes exist from the lake bottom to the water body and out through the ice into the atmosphere. The ice captures impurities from the lake water, lake bottom and atmospheric fallout. If ice coverage is complete, oxygen renewal is not possible (e.g. Golosov et al., 2006). The reduction of liquid water volume by ice formation is of concern in shallow lakes. The physical conditions created by the ice cover further influence the water chemistry and plankton populations (Abgeti & Smol, 1995; Järvinen et al., 2002).

In large lakes, wind forcing is sometimes able to break the ice and drive displacements, creating leads and pressure ridges (e.g. Wake & Rumer, 1983; Burda, 1999; Wang et al., 2006a). Leads make ventilation of the water body possible, and ice growth is re-initiated when they freeze, resulting in a wide distribution of ice thicknesses across the lake. Thin ice areas allow light to be transmitted into the water body and primary production will be possible below the ice all winter. The formation of frazil ice in leads introduces frazil layers into the ice sheet; this process is effective in capturing suspended particles from the water.

There is no rational definition of a 'large lake' but quite often a surface area of 100 km^2 is taken as the criterion. From the point of view of the ice season, 'large lakes' are rather those in which mechanical displacements are common. There are four such lakes in northern Europe: Ladoga, Peipsi, Vänern and Onega. As an example, Fig. 1 shows the ice conditions in Lake Peipsi for March 19, 2002. The dark elongated zones on the western and central coasts indicate major leads, and small leads and fractures can also be seen far from the coast. In spite of the major role of ice mechanics in large lakes, very little work has been done in respect of full thermal–mechanical modelling. The only major effort was made for the North American Great Lakes around the year 1980 (e.g. Wake & Rumer, 1983).

This paper presents results from recent research on differences in physical scaling between small and large lakes and the revision of mathematical models for large lakes. During the past 25 years there have been several developments in sea ice modelling (see Leppäranta, 2004). The models have shown good results from polar oceans down to 100 km size basins (Wang et al., 1994, 2003, 2006b), so it is natural to utilize this knowledge of sea ice for modelling large lakes. Here, a modern viscous-plastic lake ice model is presented. Finally, the basis and methods for full three-dimensional lake ice models are discussed.

Fig. 1 Lake Peipsi as seen in MODIS image of the Terra/Aqua satellite, 19 March 2002

Growth and melting of lake ice

The fundamental processes and their implications

A lake ice sheet may contain three types of ice: congelation ice, snow-ice and frazil ice (e.g. Leppäranta & Kosloff, 2000). In lakes with stable ice cover, only congelation and snow ice are found. The growth of congelation ice is governed by the heat conduction equation

$$\frac{\partial \rho c T}{\partial t} = \frac{\partial}{\partial z}\left(\kappa \frac{\partial T}{\partial z} - I\right) \qquad (1)$$

where ρ is the ice density, c is the specific heat of ice, T is the temperature, t is the time, κ is the thermal conductivity of ice, I is the solar radiation penetrating into the ice and z is the vertical coordinate. The boundary conditions describe the continuity of heat fluxes with the bottom temperature kept at freezing point:

$$\text{Top}: \ \kappa\frac{\partial T}{\partial z} = Q_o, T < 0°\text{C}; \quad \kappa\frac{\partial T}{\partial z} = \rho L\frac{dh}{dt} + Q_o, T = 0°\text{C} \tag{2a}$$

$$\text{Bottom}: \ T = 0°\text{C}, \ \rho L\frac{dh}{dt} = \kappa\frac{\partial T}{\partial z} - Q_w \tag{2b}$$

where Q_o is the heat loss from the surface to the atmosphere, L is latent heat of freezing and Q_w is the heat flux from the water to the ice. Snow on top of the ice forms slush when mixed with liquid water from flooding, rain or meltwater, and formation of snow-ice is possible. This is modelled in the same way as the growth of congelation ice, but the release of latent heat is reduced to $\nu\rho L$, where ν is the liquid water content of the slush layer. This process may fuse the slush layers into the ice sheet. The first author has observed algae growth in such slush pockets in Lake Pääjärvi, southern Finland. The compaction of snow on ice is an important metamorphic process because it influences the thermal conductivity (Yen, 1981).

In a lead in a large lake, ice may grow in much the same way as in small lakes. However, if the surface conditions are disturbed, ice formation may possibly result in free frazil ice crystals, given by

$$\frac{dh_F}{dt} = \frac{Q_o}{\rho L} \tag{3}$$

where h_F is the volume of frazil per unit surface area of the water column. Frazil ice generation is quite efficient because the heat losses from open water surfaces are large (Gerard, 1990), and the frazil ice may subsequently attach to the bottom of an existing ice sheet or to the lake bottom.

The melting season begins when the surface radiation balance turns positive. The ice then melts at the boundaries and also inside the ice sheet because of the solar radiation; the mechanical strength of the ice decreases. The solar radiation penetrates the ice efficiently and triggers heating of the water and primary production. According to measurements by Arst et al. (2006) in several Finnish and Estonian lakes, the light level beneath snow-covered ice is sufficient for primary production when the ice thickness is less than about 40 cm.

Mathematical modelling

In simple analytical models, the ice thickness is a function of the sum of freezing-degree-days (Stefan, 1891; Ashton, 1986). The main problem is the snow cover. The snow-ice and the congelation ice system grow interactively, and a solution is possible only by numerical methods (Leppäranta, 1983; Saloranta, 2000; Leppäranta & Uusikivi, 2002). Models in this situation are formulated on the basis of the heat conduction law (Eqs. 1 and 2), and snow accumulation is added to the surface boundary conditions. Lead openings continuously create new starts for ice thickness histories in large lakes, so several thickness categories exist. Figure 2 shows an idealized case in which new ice starts to grow every 20 days; thus (for example) there are ice fields of about 25, 45 and 60 cm thickness at day 50. Frazil ice models have been constructed for rivers (Ashton, 1986), but no modelling attempts for large lakes are known to the present authors.

The length scale of thermal conduction in the ice cover is $L \sim [t\kappa/(\rho c)]^{1/2}$. For $t \sim 1$ month, we have $L \sim 1$ m, so thermal conduction processes are very local compared to lake size. When the horizontal grid size of a lake model is 5 m or more, thermal changes in the ice sheet can be taken to be purely local at each grid point.

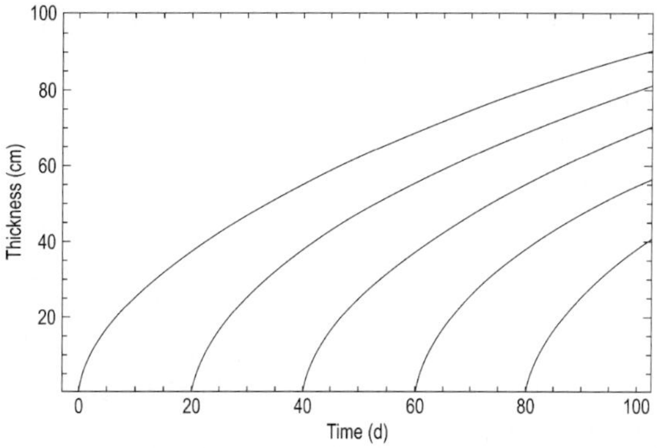

Fig. 2 Idealized ice growth calculations with new ice formation starting every 20 days. The air temperature is fixed at −10°C

Lake ice mechanics

Stability of ice cover

Drift ice is considered as a plastic medium driven by winds and currents and thermal forcing, and the ice cover medium is characterized by the mean ice thickness H and compactness A. The plasticity allows us first to examine whether the ice cover is stable or movable. The yield strength can be expressed as (Hibler, 1979)

$$\sigma_Y = P * H \exp[-C(1 - A)] \tag{4}$$

where $P*$ is the ice strength constant, equal to the compressive strength of compact ice of unit thickness, and C is the strength reduction factor for lead opening. The magnitudes of these parameters are $P* \sim$ 10–100 kPa and $C \gg 1$, normally taken as $C = 20$. When a compact ice cover ($A = 1$) is forced by a wind stress τ_a, the critical requirement for a mechanical breakage event becomes (Leppäranta, 2004)

$$h/\ell < \tau_a/P* \tag{5}$$

where ℓ is the fetch. The wind stress is given by $\tau_a = \rho_a C_a U_a^2$, where ρ_a is the density of air, $C_a \approx 1.5 \times 10^{-3}$ is the air–ice drag coefficient, and U_a is the wind speed. The lake size L represents the maximum fetch. Taking $P* = 30$ kPa, for $U_a = 10$ m/s we have $\tau_a/P* \sim 10^{-5}$. In Lake Peipsi, $L \sim 50$ km and $h \sim 30$ cm, so $h/L \sim 0.6 \times 10^{-5} < 10^{-5}$; and for a typical lake in Finland, $L \sim 10$ km and $h \sim 40$ cm, so $h/L \sim 4 \times 10^{-5} > 10^{-5}$. Thus the ice cover is movable in Lake Peipsi but stationary in typical Finnish lakes. However, larger Finnish lakes may become mobile if the ice becomes thinner and/or wind stresses become higher.

The drag force on the ice bottom is given by an analogous quadratic law, $\tau_w = \rho_w C_w |\boldsymbol{U}_w - \boldsymbol{u}|^2$, where ρ_w is water density, $C_w \sim 10^{-3}$ is the ice–water drag coefficient, and $\boldsymbol{u}$ and $\boldsymbol{U}_w$ are the ice and current velocities. Under stationary ice, for $U_w < 10^{-1}$ m/s we have $\tau_w < 10^{-2}$ Pa, much less than normal wind stresses; and since fetches beneath the ice are also less than those above the ice, current forcing is much less likely to break the ice cover than the wind.

Ice drift modelling

When yield stress is achieved, the ice breaks and starts to drift. The motion is governed by the momentum equation and conservation of ice (Leppäranta, 2004):

$$\rho H\left(\frac{\partial \boldsymbol{u}}{\partial t} + \boldsymbol{u} \cdot \nabla \boldsymbol{u} + f\boldsymbol{k} \times \boldsymbol{u}\right) = \tau_a + \tau_w + \nabla \cdot \boldsymbol{\sigma} \tag{6a}$$

$$\frac{\partial H}{\partial t} = -\nabla \cdot (H\boldsymbol{u}) + \phi_h, \quad \frac{\partial A}{\partial t} = -\nabla \cdot (A\boldsymbol{u}) + \phi_A \tag{6b}$$

where $\boldsymbol{\sigma}$ is the internal ice stress, f is the Coriolis parameter, $\boldsymbol{k}$ is the unit vector vertically upward, and φ_h and φ_A describe the influence of thermal changes on the thickness and compactness, respectively. Mechanical displacements may be of magnitude 1–10 km, depending on the wind stress and ice thickness. The motion of ice results in opening of leads in divergent zones and formation of pressure ridges in compression zones, and the movement continues until the internal stress has fallen below the yield strength everywhere. Under compression, the ice thickness increases in ridging, so for a fixed wind stress the ice motion stops when the thickness has reached the critical level (see Eq. 4).

For moderate or strong winds, the wind stress is of the order 0.1 Pa and much greater than the ice–water stress. The inertial time scale then becomes $t* = \rho H u/\tau_a \sim 10^2$–$10^3$ s for $u \sim 10$ cm/s and $H \sim$ 10–100 cm. With this velocity scale, a 1-km lead opens in 3 h. The ratio of advective acceleration to local acceleration is $ut*/L \ll 1$, so it can be ignored.

Consider the one-dimensional case with compact ice of constant initial thickness H_o (Leppäranta, 2004). If the yield stress is attained, the ice drift results in lead-opening on the lee side and ridging on the windward side (Fig. 3). Since the inertial time scale is very short, a quasi-steady-state approximation can be made:

$$\tau_a - \rho_w C_w \mathrm{sgn}(u) u^2 - P * \frac{\partial H}{\partial x} = 0 \tag{7a}$$

$$\frac{\partial H}{\partial t} = -\frac{\partial}{\partial x}(uH) + \phi_h \tag{7b}$$

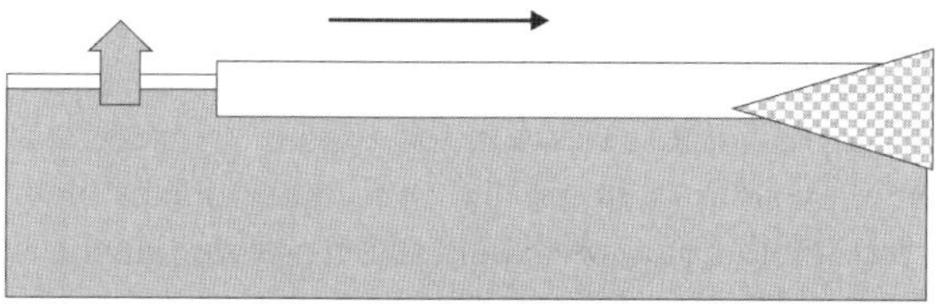

Fig. 3 One-dimensional thermal–mechanical lake ice model. Mechanics opens a lead, ridging takes place on the windward side, and new ice forms in the lead

We may choose $u > 0$. It is then clear that $\tau_a > 0$ and $\partial H/\partial x > 0$, and finally

$$u = \sqrt{\frac{1}{\rho_w C_w}\left(\tau_a - P * \frac{\partial H}{\partial x}\right)} \quad (8)$$

During the ridging process the ice is piled up as long as $\tau_a > P^* \partial H/\partial x$ (the solution for u is real). However, the gradient of thickness increases because of the ridging, and finally $u \equiv 0$. In this case, $\partial H/\partial t \equiv 0$, and $H = H_o + \beta x$ where $\beta = \tau_a/P^*$. The width of the lead, b, is obtained from

$$H_o L = [H_o + \frac{1}{2}\beta(L - b)](L - b) \quad (9)$$

where L is the length of the lake. Consequently, during the course of the ice season, chains of the processes lead opening—frazil ice formation—ice covered lead—breakage event take place, and a wide ice thickness distribution results.

Next, idealized model simulations are run for two-dimensional cases for Lake Peipsi (Fig. 1). This lake is the fourth largest in Europe, bordering Estonia and Russia. It covers an area of 3,555 km^2, with an average depth of 8 m and maximum depth of 15.3 m. The lake is covered with ice in every winter. In an average winter, the ice thickness increases to 50–60 cm until March. The present model is based on our sea ice model for the Gulf of Riga (Wang et al., 2003), which has dimensions and ice thickness similar to Lake Peipsi. Results from idealized simulations are shown in Fig. 4. The wind speed is 10 m/s, constant across the lake and over time, and the nominal ice thickness is 10 cm. The key ice property in the model is the compressive strength of the compact ice, $P = P^*H$; in these simulations the constant P^* was taken as 27.5 kPa. By the scaling analysis, the ice in this case is movable up to the order of 50 cm; this case is shown here in order to provide a clearer illustration. Simulations with

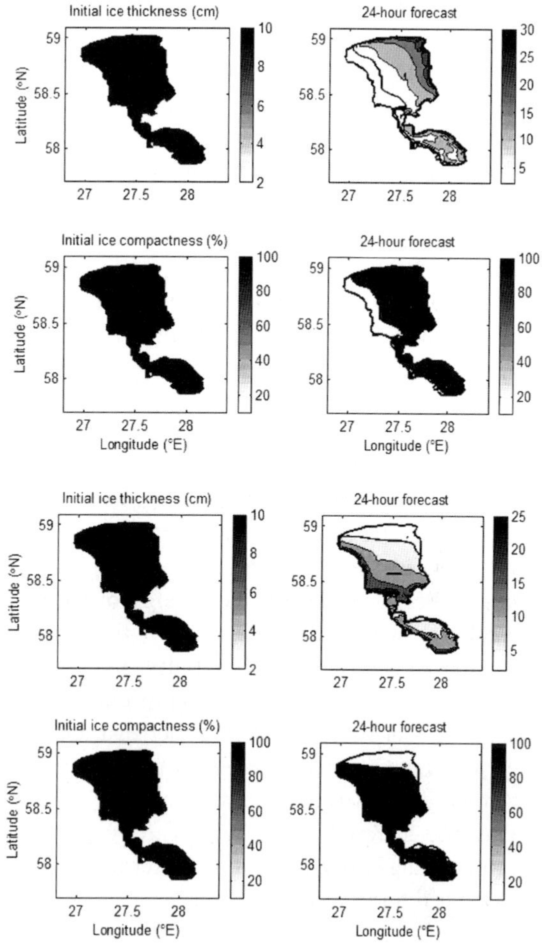

Fig. 4 Model simulations of the mechanical displacement of the ice cover in Lake Peipsi. Ice thickness is 10 cm, wind speed 10 m/s, and wind direction southwest (upper case) or north (lower case). The white area indicates open water

greater ice thicknesses showed that large displacements were also obtained with 30 cm thickness but with no more for 50 cm.

In the case of a southwest wind (upper part of Fig. 4), a lead about 25 km wide opens at the western side in 1 day. The ice edge is very sharp, the compactness jumping immediately to 90% at the ice edge and approaching 100% at 10 km distance, since the strength of the ice is very sensitive to compactness (see Eq. 4). The thickness map shows an increase at the eastern side up to double the original level. In the southern basin of Lake Peipsi (Lake Pihkva), a very weak shift can be observed in the compactness chart. Similar features appear in the case of a north wind (Fig. 4, lower part). Detailed inspection shows that the ridging zone in the northern basin has three peaks forced by the basin geometry, and the displacement is greater in the southern basin than in the case of the southwest wind.

These idealized calculations show that sea ice models can be used to develop dynamic–thermodynamic models for large lakes. The model parameters are the same as those used in the Gulf of Riga (Wang et al., 2003). With further fine-tuning of the model against real data the parameters can be optimized, but it is anticipated that the change will not be very large (see Leppäranta, 2004). The main tuning parameters are the air–ice and ice–water drag coefficients and the strength constant P^*. The drag coefficients are likely to be smaller by 20–30% in Lake Peipsi because the ice surfaces are smoother, but the net influence will be quite small. The strength constant is governed by friction between ice blocks in ridging, mechanical breakage of ice, and the potential energy of the ice. For breakage, more stress is needed for fresh water ice than for the brackish ice of the Baltic Sea, but the overall impact on the strength constant is expected to be small. This is supported by the fact that approximately the same constant is used for brackish ice as for the considerably weaker normal first-year sea ice. Full dynamic–thermodynamic models are constructed by incorporating a one-dimensional thermodynamic model at each horizontal grid point.

Conclusions

A treatise on mathematical modelling of the lake ice season has been presented, with discussion of ecological aspects. Physically, the evolution of the ice season is a coupled thermal–mechanical process in which the ice grows and decays as forced by the atmosphere and the liquid water body below. In addition, the wind forcing may exceed the yield strength of the ice cover in large lakes, creating mechanical displacements with leads and pressure ridges. In 'medium size' lakes the ice cover is thick and stable in cold winters and may be mobile in warm winters (e.g. Lakes Saimaa and Oulujärvi in Finland). Mathematical models are one-dimensional, vertical for thermodynamics and light transfer, and two-dimensional, horizontal for mechanics. The most recent developments concern snow metamorphosis, slush and snow-ice formation and plastic rheologies for the mechanics.

Lake ice models provide crucial information for investigating the ecology of ice-covered lakes. Beneath a solid ice cover, the motion of waters is forced by solar radiation and bottom heat fluxes only, so the circulation is weak and stable and turbulence is damped. The temperature conditions are stable with permanent stratification until the snow melts and sunlight consequently penetrates through the ice. The oxygen balance is negative in winter, which leads to oxygen deficit problems with fish kills in long ice seasons. In shallow lakes, freezing stores a large portion of the water mass in the solid state, and the oxygen problem becomes then even more severe. In large lakes, the mobility of ice introduces forcing for circulation, turbulence and—through leads—oxygen renewal.

Ice cover protects the lake water from atmospheric fallout and in general the water chemistry is changed (Järvinen et al., 2002). In the melting season the accumulated fallout is rapidly released into the surface waters, which introduces risks from harmful substances. Owing to its high albedo and low transparency, the snow cover is the main limiting factor for primary production in winter. On the other hand, snow and liquid water may form semi-persistent slush layers on top of the ice, constituting a biological habitat. Over time, a slush layer may also become interlocked with the ice sheet. Different plankton populations have been observed in the lake water body in ice-covered lakes (Abgeti & Smol, 1995; Adrian et al., 1999). Leads in large lakes that receive much sunlight have potential for primary production throughout the winter.

In the winter ecology of freezing lakes, the critical ice cover characteristics are the snow and the mobility of the ice (large lakes). The former largely determines the light conditions in the lake, while the latter allows mechanical force to be transferred through the ice so that leads and pressure ridges form, both of which are particular biological habitats. The present modelling approach allows a physically realistic solution of these two characteristics to be obtained and consequently a proper basis for evaluating the influence of climate changes on freezing lakes to be established. Snow on ice may be more sensitive to climate warming than the ice itself, and its future depends on the temperature change and the amount and phase of precipitation. With regard to mechanical stability, even small changes in the thickness of the ice may influence its mobility. Thus the ice seasons will not only become shorter with climate warming but also their quality may be essentially different.

Acknowledgements Financial support has been provided by the Academy of Finland, project "Hydrodynamic focusing of sediment oxygen consumption to the deepest parts of large lakes—a multidisciplinary approach" (Kalevi Salonen).

References

Abgeti, M. D. & J. P. Smol, 1995. Winter limnology: A comparison of physical, chemical and biological characteristics in two temperate lakes during ice cover. Hydrobiologia 304: 221–234.

Adrian, R., N. Walz, T. Hintze, S. Hoeg & R. Rusche, 1999. Effects of ice duration on plankton succession during spring in shallow polymictic lake. Freshwater Biology 41: 621–632.

Arst, H., A. Erm, M. Leppäranta & A. Reinart, 2006. Radiative characteristics of ice-covered fresh- and brackish-water bodies. Proceedings of the Estonian Academy of Science. Geology 55: 3–23.

Ashton, G. (ed.), 1986. River and Lake Ice Engineering. Water Resources Publications, Littleton, Colorado.

Bengtsson, L., 1996. Mixing in ice-covered lakes. Hydrobiologia 322: 91–97.

Burda, N. Yu., 1999. Computer simulation for Ladoga Lake ice dynamics based on remotely sensed data. International Geoscience and Remote Sensing Symposium 2: 937–939. Institute of Electrical and Electronics Engineers, New York.

Gerard, R., 1990. Hydrology of floating ice. In Prowse, T. D. & C. S. C. Ommanney (eds), Northern Hydrology. Canadian Perspectives. National Hydrology Research Institute, Saskatoon, Saskatchewan, 103–134.

Golosov, S., O. A. Maher, E. Schipunova, A. Terzhevik, G. Zdorovennova & G. Kirillin, 2006. Physical background of the development of oxygen depletion in ice-covered lakes. Oecologia (online). DOI: 10.1007/s00442-006-0543-8.

Hibler, W. D., III, 1979. A dynamic and thermodynamic sea ice model. Journal of Physical Oceanography 9: 815–846.

Järvinen, M., M. Rask, R. Ruuhijärvi & L. Arvola, 2002. Temporal coherence in water temperature and chemistry under the ice of boreal lakes (Finland). Water Resources 36: 3949–3956.

Leppäranta, M., 1983. A growth model for black ice, snow ice and snow thickness in subarctic basins. Nordic Hydrology 14: 59–70.

Leppäranta, M., 2004. The Drift of Sea Ice. Springer-Praxis, Heidelberg, Germany.

Leppäranta, M. & P. Kosloff, 2000. The thickness and structure of Lake Pääjärvi ice. Geophysica 36: 233–248.

Leppäranta, M & J. Uusikivi, 2002. The annual cycle of the Lake Pääjärvi ice. Lammi Notes 29: 4–9.

Saloranta, T., 2000. Modeling the evolution of snow, snow ice and ice in the Baltic Sea. Tellus 52A: 93–108.

Stefan, J., 1891. Über die Theorie der Eisbildung, insbesondere über Eisbildung im Polarmeere. Annales Physik, 3rd Series 42: 269–286.

Wake, A. & R. R. Rumer, 1983. Great lakes ice dynamics simulation. Journal of Waterway, Port, Coastal and Ocean Engineering 109: 86–102.

Wang, J., L. A. Mysak & R. G. Ingram, 1994. A numerical simulation of sea ice cover in Hudson Bay. Journal of Physical Oceanography 24: 2515–2533.

Wang, K., M. Leppäranta & T. Kõuts, 2003. A sea ice dynamics model for the Gulf of Riga. Proceedings of the Estonian Academy of Science. Engineering 9: 107–125.

Wang, K., M. Leppäranta & A. Reinart, 2006a. Modeling ice dynamics in Lake Peipsi. Internationale Vereinigung für theoretische und angewandte Limnologie 29: 1443–1446.

Wang, K., M. Leppäranta & T. Kõuts, 2006b. A study of sea ice dynamic events in a small bay. Cold Regions Science and Technology 45: 83–94.

Yen, Y.-C., 1981. Review of thermal properties of snow, ice and sea ice. CRREL Report 81–10: 1–27.

Hydrobiologia (2008) 599:191–199
DOI 10.1007/s10750-007-9197-8

ELLS 2007

Effects of warmer world scenarios on hydrologic inputs to Lake Mälaren, Sweden and implications for nutrient loads

Karen Moore · Donald Pierson · Kurt Pettersson · Elliot Schneiderman · Patrick Samuelsson

Abstract A simple, rapid, and flexible modelling approach was applied to explore the impacts of climate change on hydrologic inputs and consequent implications for nutrient loading to Lake Mälaren, Sweden using a loading function model (GWLF). The first step in the process was to adapt the model for use in a large and complex Swedish catchment. We focused on the Galten basin with four rivers draining into the western region of Mälaren. The catchment model was calibrated and tested using long-term historical data for river discharge and dissolved nutrients (N, P). Then multiple regional climate model simulation results were downscaled to the local catchment level, and used to simulate possible hydrological and nutrient loading responses to warmer world scenarios. Climate change projections for the rivers of Galten basin show profound changes in the timing of discharge and nutrient delivery due to increased winter precipitation and earlier snow melt. Impacts on total annual discharge and load are minimal, but the alteration in river flow regime and the timing of nutrient delivery for future climate scenarios is strikingly different from historical conditions.

Keywords Catchment modeling · Hydrologic transport · Nutrient loading · Climate impact assessment

Introduction

One challenge for climate change impact assessment of large lakes is to consider changes in hydrologic and nutrient inputs from large catchment areas with diverse land cover and land use. Changes in the magnitude and timing of river flows are possible consequences of climate warming (Schindler, 2001). Increased nitrogen and phosphorus loadings from rivers are among the effects projected for areas where precipitation increases are anticipated (Murdoch et al., 2000). For Sweden, regional increases in air temperature and precipitation are expected, with the largest increases occurring in the winter months (Räisänen et al., 2003). What response in river flow and nutrient transport might accompany such

Guest editors: T. Nõges, R. Eckmann, K. Kangur, P. Nõges, A. Reinart, G. Roll, H. Simola & M. Viljanen
European Large Lakes—Ecosystem changes and their ecological and socioeconomic impacts

K. Moore (✉) · K. Pettersson
Department of Ecology and Evolution, EBC, Norr Malma Field Station, Uppsala University, Uppsala, Sweden
e-mail: karen.moore@ebc.uu.se

D. Pierson · E. Schneiderman
New York City Department of Environmental Protection, New York, USA

P. Samuelsson
Swedish Meteorological and Hydrological Institute, Norrköping, Sweden

changes? How would projected changes be compared with the levels of enrichment from cultural eutrophication observed in the past? We explored these questions for Lake Mälaren, the third largest lake in Sweden. The objective of this article is to use existing data and simulation models to estimate the hydrologic inputs and associated dissolved inorganic nitrogen and phosphorus loads (assuming simple loading functions) in rivers flowing into one basin of Lake Mälaren, Sweden for historical conditions and under future climate scenarios. Many model choices with varying degrees of complexity are currently available (Silgram & Schoumans, 2004). We used GWLF, the Generalized Watershed Loading Functions model (Haith et al., 1992), a simple hydrologic and nutrient transport model, to first estimate stream flow and nutrient export from large catchments with mixed land uses under present day conditions. We then use state-of-the-art climate models for future scenarios to assess the pattern and magnitude of change in river water quantity and quality for major inflows to this vitally important large lake.

The study area

Lake Mälaren is the water supply for the City of Stockholm and is the dominant freshwater source to the Stockholm Archipelago, with a mean outflow averaging about 165 $m^3 s^{-1}$ (Boesch et al., 2006). Our study focused on the western basin (Galten) and its four river inflows: Arbogaån, Hedströmmen, Köpingsån and Kolbäcksån. Galten is a rapidly flushed basin with a short water retention time of 0.07 $year^{-1}$ on average (Willén, 2001) with a large catchment to lake area ratio (8,508 km^2: 61.2 km^2), and consequently, external nutrient loading exerts a strong influence on lake water quality. Historically, inputs of nitrogen and phosphorus from wastewater treatment plants have been dramatically reduced, but nutrient inputs to the Stockholm Archipelago from land-based sources are still considerable (Boesch et al., 2006). Land cover characteristics and stream gauge units for hydrological modelling of the Galten inflows are shown in Table 1. Ungauged areas were modelled using coefficients and parameters developed for areas with river discharge gauges.

Materials and methods

Overall approach

The general procedure for evaluating impacts of hypothetical climate change on hydrological behaviour and nutrient loads was to:

(1) Parameterise the GWLF hydrologic model and nutrient loading functions for the historical period using observed data for river discharge and nutrient concentration for model calibration and validation.
(2) Apply a simple statistical downscaling method to transfer the climate signal from regional climate model output to hydrological and nutrient transport models.
(3) Run GWLF simulations for multiple realisations of observed and future scenario weather sequences to get a picture of seasonal pattern and range of variability.

GWLF was chosen because it can be applied to catchments with mixed land uses ranging several orders of magnitude in size, and the minimal data requirements make it relatively simple and rapid to use.

GWLF model description

The Generalized Watershed Loading Functions (GWLF) model dynamically simulates the hydrologic components of runoff from different land uses and baseflow; nutrient loads associated with these flow paths are estimated by empirically derived, source-specific nutrient concentrations that are input to the model (Haith & Shoemaker, 1987; Schneiderman, 2006). Although chemical transformation processes are not simulated, the implications of hydrological changes due to climate change on nutrient loading are of major importance given the dominant role that hydrology plays as the primary mechanism for nutrient transport. To evaluate the effects of climate change on hydrology and consequent implications for nutrient loading, it was assumed that land use and management remained constant. We used a version of GWLF that was written in the Vensim (Ventana Systems, Inc.) visual modelling software (Schneiderman et al., 2002; Schneiderman, 2006).

Table 1 Drainage area description

Drainage basin characteristics				Land cover characteristics (km^2)							
River sub-catchment*	Gauged (km^2)	Ungauged (km^2)	Total area (km^2)	Conifer forest	Other forest	Water bodies	Transition woodland shrub	Pasture	Arable land	Peat bog	Other
Arbogaån			3808	1744	386	264	665	68	446	141	94
Hammarby	891	171	1062								
Dalkarlshyttan	1183	128	1311								
Kåfalla	413	80	494								
Fellingsbro	298	1	299								
Direct drainage		642	642								
Hedströmmen			1050	528	72	85	191	19	85	58	12
Dömsta	998	52									
Köpingsån			287	137	19	14	48	6	43	9	11
Odensvibron	110	177									
Kolbäcksån			3119	1623	277	272	552	40	129	116	110
Ramnäs Krv	2849	156	3005								
Berg	36	17	54								
Direct drainage		60	60								

* Sub-catchment name corresponds to river gauge name in SMHI (1994)

GWLF is driven by daily temperature and precipitation data, and water balances are calculated on a daily interval. Streamflow consists of runoff from different land uses and baseflow. Runoff is calculated using the SCS curve number method (Ogrosky & Mockus, 1964). Swedish curve numbers were estimated from soil and landuse data available for the Mälaren catchment. Dissolved nutrient loads are derived by multiplying runoff by a land use-specific nutrient concentration. There are also watershed-wide nutrient concentrations associated with baseflow. The contributions of nutrients from septic systems were estimated from literature values. Point source loadings from treated sewage and industrial effluents are also input and accounted for in the overall estimates of nutrient flux. Dissolved nutrient loads can be summed by month, season or year. Sediment and particulate nutrient loss were not included in our analysis, so only dissolved nutrient estimates were modelled.

Data requirements and data sources

Land use data from the Swedish Land Survey (Lantmäteriet) were based on the CORINE (European Commission Coordination of Information on the Environment) level 3 classification scheme. Soil data from the Swedish Geological Survey (SGU) map (1: 1,000,000) were used for runoff curve number calculation. Meteorological data (daily precipitation and air temperature) from a 4 × 4 km grid interpolation of local stations (Johansson, 2002) were used for calibration of the hydrological model for the period of 1980–1991. The fit of the model was optimised using measured stream discharge. Both meteorological and hydrological data were from the Swedish Meteorological and Hydrological Institute (SMHI). Simulations for the historical period were based on a 30-year period (1961–1990).

Point sources of nutrients from wastewater treatment plants and industry were based on published data from the TRK (Transport–Retention–Källfördelning) study (Brandt & Ejhed, 2002). Nutrients from septic systems were reported as an annual estimate for the TRK study, and these loadings were converted to fixed daily rates when used as inputs to the model.

In GWLF, nutrient loads from diffuse sources are partitioned by land use, and information about land use-specific nutrient concentrations were drawn from average published values from the TRK study

(Brandt & Ejhed, 2002). Information for agricultural land uses was used for selected areas in the JRK (Typområden på jordbruksmark) study (Kyllmar & Johnsson, 1998). Concentrations in groundwater were taken from long-term water quality data (SLU database, Swedish Agricultural University, http://www.info1.ma.slu.se). Groundwater concentrations were calculated as the mean of water quality samples collected at or below the 20th percentile of historical streamflow. Long-term monthly water quality data from four stations near each inflow to Galten basin were used for calibration of the nutrient model (SLU database). The model was optimised to minimise the residuals between long-term average measured and modelled N and P concentrations.

Climate change scenario application and downscaling to the catchment level

Climate forcings for this study were taken from the Rossby Centre Atmosphere-Ocean (RCAO) regional climate model and the Hadley Centre regional climate model (HadRM3p) using boundary conditions from two general circulation models: the Max Planck Institute ECHAM4/OPYC3 and the Hadley Centre HadAM3H for two IPCC emission scenarios (A2, B2) (Samuelsson, 2004; Table 2). A 30-year "time slice" was used for a control period (1961–1990) and future period (2071–2100). For both A2 and B2 scenarios, increases in population growth are projected, but the rate of increase is greater for scenario A2 (Houghton et al., 2001). The use of multiple models allows for some consideration of the range of uncertainty in model representation of the climate system.

The method used to downscale climate forcings from the regional to the local catchment scale is based on monthly average changes in precipitation and temperature, known as the "delta change" approach in its simplest form, as described by Hay et al. (2000).

The difference in average monthly temperature between control and future RCM model scenarios is added or subtracted from the observed daily temperature (Table 3). For precipitation, the ratio of the monthly total precipitation for control and future RCM scenarios is used as a multiplier that is applied to the daily record of observed precipitation (Table 3).

One limitation in this method is that the frequency of precipitation events is not varied (Andreásson et al., 2004a). To avoid this pitfall, and examine the widest range of variability in climate, we used a resampling approach to create multiple weather sequences from the control and future climate scenarios. Using this method, monthly weather records were randomly recombined to create multiple 30-year data series that could be used to drive the GWLF model. For example, a random choice of one of the 30 January records was made followed by a random choice of one of the 30 February records and so on to make up 100 30-year synthetic weather timeseries for the purpose of hydrological modelling. This approach does not account for any interdependence of monthly weather between successive months. Hydrological and nutrient yield results are presented as medians and variance of multiple model realisations.

Results

Hydrological model performance

To derive the best set of hydrological model coefficients, eight sub-catchments, where streamflow measurements were available, were modelled. Hydrologic model coefficients were optimised to obtain the

Table 2 Climate models and scenarios used for future projections of climate

Abbreviation	General Circulation Model (GCM)	Regional Climate Model (RCM)	SRES Scenario (IPCC, 2001)
E A2	ECHAM4/OPYC3	RCAO	A2
E B2	ECHAM4/OPYC3	RCAO	B2
H A2	HadAM3H	RCAO	A2
H B2	HadAM3H	RCAO	B2
Had A2	HadAM3P	Had RM3p	A2
Had B2	HadAM3P	Had RM3p	B2

Table 3 Monthly change factors applied to observed weather timeseries for future scenario (2071–2100) simulations

Month	Temperature change factor (°C)[a]						Precipitation change factor (proportion)[b]					
	E A2	E B2	H A2	H B2	Had A2	Had B2	E A2	E B2	H A2	H B2	Had A2	Had B2
Jan	5.76	4.23	4.18	2.85	5.13	4.52	1.58	1.32	1.50	1.16	1.53	1.27
Feb	6.45	5.17	3.26	2.46	2.66	2.69	1.43	1.50	1.40	1.30	1.37	1.16
Mar	6.51	5.05	3.36	2.17	4.24	2.92	1.39	1.23	1.18	1.17	1.16	1.14
Apr	5.00	3.74	3.96	2.71	3.98	3.31	1.38	1.12	1.02	1.06	1.15	1.24
May	3.59	2.73	3.98	2.74	3.64	2.98	1.04	1.18	1.35	1.04	1.11	1.23
Jun	3.35	2.67	3.01	1.24	4.04	3.06	0.93	0.92	0.86	1.05	0.99	0.90
Jul	3.77	2.81	3.21	1.45	4.79	4.21	0.88	1.02	0.86	0.78	0.86	0.74
Aug	4.54	3.42	3.49	2.24	5.03	4.44	0.67	0.74	0.70	0.79	0.83	0.74
Sep	4.15	3.11	3.80	2.55	4.37	3.13	0.90	1.02	0.73	1.00	0.76	0.79
Oct	4.49	3.24	3.73	2.64	3.85	3.04	1.38	1.26	1.19	1.06	0.80	0.96
Nov	4.96	4.01	4.18	3.18	4.36	3.50	1.40	1.29	1.22	1.24	0.95	1.05
Dec	4.65	3.75	4.65	2.94	4.85	3.59	1.44	1.34	1.30	1.17	1.20	1.16

[a] Model (scenario—control) mean monthly temperature

[b] Proportion model (scenario: control) total monthly precipitation

best match between simulated and measured river discharge. The coefficients were then used to simulate river discharge for the total river drainage areas for each of the four inflows. Summaries and statistics for evaluating model efficiency are given in Table 4. The hydrological model calibration period was 1981–1990. The same model coefficients were tested for a 10-year validation period (1991–2000) and the Nash-Sutcliffe statistics (Nash & Sutcliffe, 1970) remained high, indicating a close correspondence between measured and modelled streamflow (Table 4).

Patterns in climate forcings

In Figs. 1–4, only results for Arbogaån, the largest river drainage in Galten basin, are shown, since the patterns in results for all other sub-basins are consistent with those for Arbogaån. The temperature factors applied in Table 2 result in a change in mean daily air temperature at the catchment scale, as shown in Fig. 1. Scenario E A2 shows the greatest overall increase in mean daily air temperature, and all scenarios show the greatest temperature increases during the winter months (Table 3).

Changes in precipitation are variable at a monthly interval between different scenario projections (Fig. 2). Overall, there is an increase in annual precipitation for all scenarios except for the Hadley Centre Had B2 results. In general, the future scenarios suggested that the increase would occur between the fall and spring, and the summers would be somewhat drier.

Impacts on modelled snowpack

In GWLF, a simple algorithm is used to estimate the partitioning of precipitation into rain and snow, and to estimate the accumulation and melting of snow. On any day when mean daily temperature drops below 0°C, precipitation is accumulated as snow. To account for snowmelt, losses from the snowpack are modelled as a function of mean daily air temperature using a simple degree-day melt factor (Haith et al., 1992). A great reduction in snowpack occurs in all future scenario simulations, due to higher winter temperatures.

Catchment response: seasonal changes in river discharge and nutrient loads

The greatest change in streamflow resulted from increased winter precipitation in the form of rain rather than snow (Fig. 2), lower snow accumulation and decreased spring snow melt. This translated to a change in the timing of peak river discharge, with a shift or disappearance of the springmelt peak (Fig. 3). A similar shift is seen in nutrient loadings as a

Table 4 Performance of GWLF hydrological sub-model for gauged sub-catchments

River	Calibration period	Validation period	Summary for calibration period (1981–1990) (mm)		
Sub-catchment*	Nash-Sutcliffe R^2 streamflow (monthly) 1981–1990	Nash-Sutcliffe R^2 streamflow (monthly) 1991–2000	Mean annual streamflow (observed)	Mean annual evapotranspiration	Mean annual precipitation
Arbogaån					
Hammarby	0.76	0.80	387 (387)	471	853
Dalkarlshyttan	0.70	0.78	436 (434)	473	905
Kåfalla	0.64	0.78	454 (453)	493	943
Fellingsbro	0.69	0.70	391 (390)	496	882
Hedströmmen					
Dömsta	0.74	0.79	395 (394)	491	880
Köpingsån					
Odensvibron	0.75	0.75	293 (293)	467	756
Kolbäcksån					
Hallstahammar	0.56	0.61	328 (326)	468	756
Berg	0.69	0.59	276 (275)	426	701

* Sub-catchment name corresponds to river gauge name (SMHI, 1994)

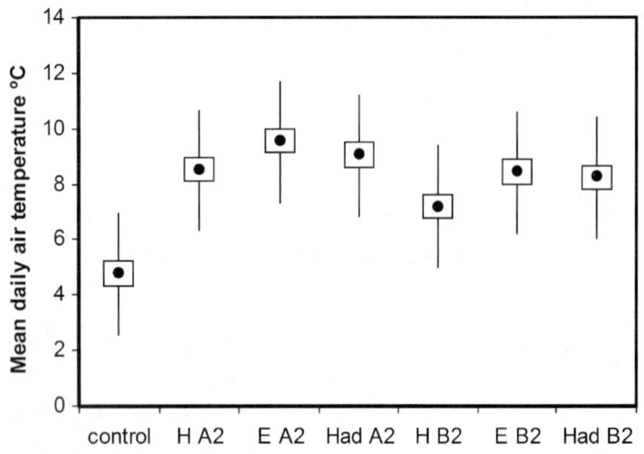

Fig. 1 Mean daily air temperature—100 30-year simulations obtained by resampling. Boxplots indicate medians and interquartile ranges, and whiskers show minimum and maximum values for 100 30-year sequences of mean daily air temperature for a control (based on observed weather) and future scenarios (based on observed weather perturbed by factors given in Table 2)

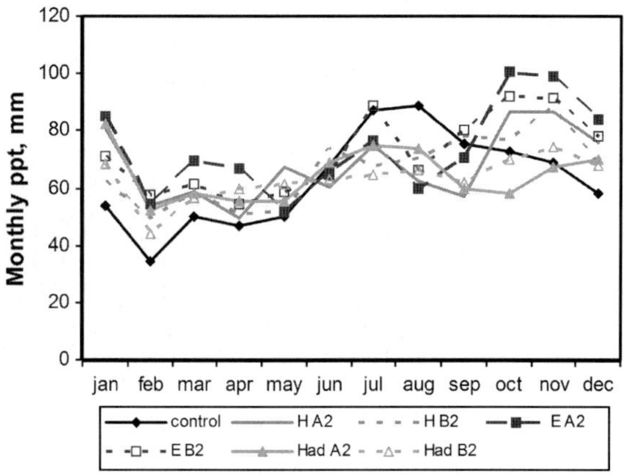

Fig. 2 Monthly precipitation (ppt) based on 100 30-year simulations obtained by resampling

consequence, with decreased loading during the period from spring to autumn and increased loading in the winter months. The pattern for dissolved inorganic nitrogen (DIN) load is shown in Fig. 4. The pattern for dissolved phosphorus load is similar, with a maximum increase of 12% in the annual load for the most extreme future scenario (E A2), with the majority of this increase occurring during the winter months. DIN and DIP loads decreased by 12–14% and 4–6%, respectively for the HadRM3p A2 and B2 scenarios due to a decrease in annual river discharge. The impact on total annual loads is shown in the range of median values for streamflow, dissolved phosphorus and dissolved inorganic nitrogen in Table 5.

Discussion

The net effect of future climate projections on changes in river flow is variable in terms of total

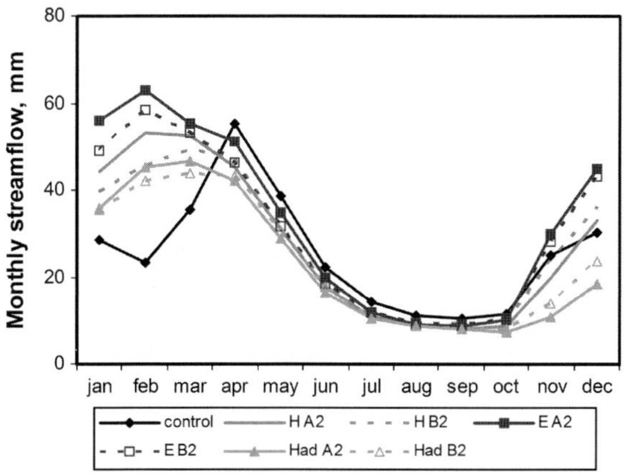

Fig. 3 Monthly streamflow based on 100 30-year simulations obtained by resampling

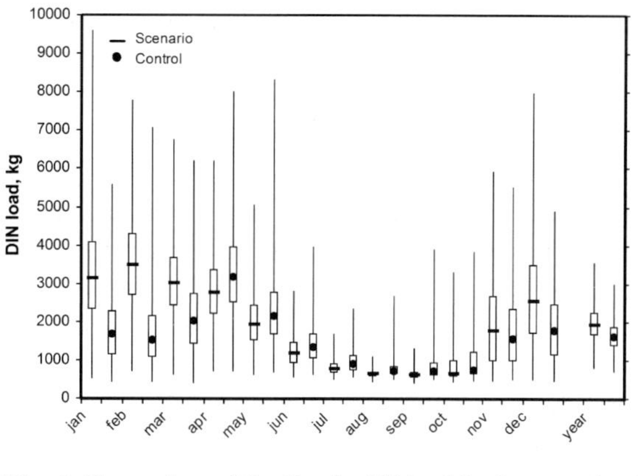

Fig. 4 Comparison of the dissolved N load for the control and E A2 climate scenario based on 100 30-year simulations obtained by resampling. Boxplots indicate medians and interquartile ranges, and whiskers show minimum and maximum values

increase (3–22%), but the change in flow regime from snowmelt-dominated spring runoff to one dominated by maximum winter discharge is the prevailing pattern for all future climate scenarios. The Hadley Center RCM results are less extreme, but show a similar pattern. Each of the climate models has its own biases, but the outcome is not markedly different when the impacts on hydrology and nutrient transport are considered with this simple modelling approach. Hydrological modelling results for the HBV model applied to a comprehensive study for the whole of Sweden showed a change in runoff ranging from +2 to −35% for the Norrström basin (Andréasson et al., 2004a, b). Our results for the same scenarios range from +3 to +22%. Andréasson et al. (2004a, b) presented results on a regional scale, and their approach differed from ours, particularly in how the climate signal was transferred to the hydrological model. However, a similar seasonal pattern in runoff was shown, for an example from southern Sweden for four future climate scenarios that matched our discharge results. The impacts of shifts in the timing of peak flows and nutrient pulses from spring to winter

Table 5 Median streamflow and dissolved nutrient loads for principal rivers of Galten basin

Parameter	Model Scenario						
	CTL	E A2	E B2	H A2	H B2	Had A2	Had B2
Streamflow (mm year^{-1})							
Arbogaån	328	398	376	339	336	285	292
Hedströmmen	339	409	387	350	347	296	299
Köpingsån	266	336	310	281	277	237	241
Kolbäcksån	270	318	303	270	270	219	226
DIP (kg km^2 year^{-1})							
Arbogaån	6.43	7.16	6.82	6.57	6.43	6.10	6.04
Hedströmmen	3.47	4.12	3.85	3.62	3.52	3.24	3.20
Köpingsån	5.61	7.07	6.48	5.88	5.70	5.05	5.46
Kolbäcksån	2.75	2.75	2.75	2.75	2.75	2.75	2.75
DIN (kg km^2 year^{-1})							
Arbogaån	148	177	166	153	150	131	132
Hedströmmen	168	205	193	175	172	147	149
Köpingsån	148	186	174	157	153	131	133
Kolbäcksån	86.0	92.8	90.7	86.3	86.3	79.8	80.7

have important implications for receiving waters. The magnitude and variability in winter discharge may increase when precipitation from individual winter storms is not stored in a snowpack (Gibson et al., 2005). This accounts for the winter increases in DIN, as shown in Fig. 4. Galten basin, with a short retention time, would experience earlier flushing at a time when light and temperature limit phytoplankton growth. So increases in dissolved nutrient loads may not result in a growth response in phytoplankton. Other basins of Lake Mälaren have much longer residence times than Galten (Willén, 2001), and as a result impacts in these areas could be greater.

Conclusions

The GWLF model was used to estimate the effects of hypothetical climate change scenarios on hydrology and the consequent implications for nutrient loading, for four rivers spanning three orders of magnitude in drainage area size. The most profound change for future scenarios was in the seasonal distribution of river flow and nutrient loads. Increases in export occur earlier in the winter, with decreases during the spring snow melt period and summer months for all future scenarios. Seasonal changes in water quantity and quality from diffuse sources projected from climate impacts are unparalleled in the historical record. A logical next step to our modelling work would be to evaluate management strategies and land use change in combination with climate projections to get a better picture of their combined effects.

Acknowledgements We thank Pamela Naden and Eleanor Jennings for their contributions to statistical summary methods and programmes, Hampus Markensten for assistance with data acquisition and management and Mark Zion for helpful comments on the manuscript. The European Commission Environment and Sustainable Development Programme provided funding support under contract EVK1-CT-2002-00121 (CLIME).

References

Andréasson, J., S. Bergström, B. Carlsson, L. P. Graham & G. Lindström, 2004a. Hydrological Change – Climate change impact simulations for Sweden. Ambio 33: 228–234.

Andréasson, J., G. Lindström, G. Grahn & B. Johansson, 2004b. Runoff in Sweden – Mapping of climate change impacts on hydrology. Nordic Hydrological Programme, NHP Report 48: 625–632.

Boesch, D., R. Hecky, C. O'Melia, D. Schindler & S. Seitzinger, 2006. Eutrophication of Swedish Seas. Swedish Environmental Protection Agency (Naturvårdsverket) Report 5509, Stockholm.

Brandt, M. & H. Ejhed, 2002. TRK Transport–Retention–Källfördelning. Belastning på havet. Naturvårdsverket Rapport 5247, Stockholm. (In Swedish.)

Gibson, C. A., J. L. Meyer, N. L. Poff, L. E. Hay & A. Georgakakos, 2005. Flow regime alterations under changing climate in two river basins: Implications for freshwater ecosystems. River Research and Applications 21: 849–864.

Haith, D. A., R. Mandel & R. S. Wu, 1992. GWLF, Generalized Watershed Loading Functions Version 2.0 User's Manual. December 1992 (corrected and reprinted, January 1996). Department of Agricultural and Biological Engineering, Cornell University, Ithaca, New York.

Haith, D. A. & L. L. Shoemaker, 1987. Generalized watershed loading functions for stream flow nutrients. Water Resources Bulletin 23: 471–478.

Hay, L. E., R. L. Wilby & G. H. Leavesley, 2000. A comparison of delta change and downscaled GCM scenarios for three mountainous basins in the United States. Journal of the American Water Resources Association 36: 387–397.

Houghton, J. T., Y. Ding, D. J. Griggs, M. Noguer, P. J. van der Linden, X. Dai, K. Maskell & C. A. Johnson (eds), 2001. Climate Change 2001: The Scientific Basis. Contribution of Working Group I to the Third Assessment Report of the Intergovernmental Panel on Climate Change. Cambridge University Press, Cambridge.

Johansson, B., 2002. Estimation of areal precipitation for hydrological modelling in Sweden. Doctoral Thesis, Department of Physical Geography, Göteborg University, Göteborg.

Kyllmar, K. & H. Johnsson, 1998. Nutrient losses from arable land within the period 1984–1995. Results from the water quality monitoring programme Typområden på jordbruksmark. Swedish University of Agricultural Sciences, Division of Water Quality Management, Ekohydrologi 44.

Murdoch, P. S., J. S. Baron & T. L. Miller, 2000. Potential effects of climate change on surface-water quality in North America. Journal of the American Water Resources Association 36: 347–366.

Nash, J. E. & I. V. Sutcliffe, 1970. River flow forecasting through conceptual models. Part 1: A discussion of principles. Journal of Hydrology 10: 282–290.

Ogrosky, H. O. & V. Mockus, 1964. Hydrology of agricultural lands. In Chow, V. T. (ed.), Handbook of Applied Hydrology. McGraw-Hill, New York.

Räisänen, J., U. Hansson, A. Ullerstig, R. Döscher, L. P. Graham, C. Jones, M. Meier, P. Samuelsson & U. Willén, 2003. GCM driven simulations of recent and future climate with the Rossby Centre coupled atmosphere Baltic Sea regional climate model RCAO. Reports Meteorology and Climatology 101, SMHI, Norrköping, Sweden.

Samuelsson, P., 2004. RCAO and HadRM3p simulation results with focus on CLIME sites. CLIME Deliverable 5.

Schindler, D. W., 2001. The cumulative effects of climate warming and other human stresses on Canadian freshwaters in the new millennium. Canadian Journal of Fisheries and Aquatic Sciences 58: 18–29.

Schneiderman, E. M., D. C. Pierson, D. G. Lounsbury & M. S. Zion, 2002. Modeling the hydrochemistry of the Cannonsville watershed with GWLF. Journal of the American Water Resources Association 38, 1323–1347.

Schneiderman, E. M., 2006. Modeling Phosphorus with the Generalized Watershed Loading Functions (GWLF) Model. In Radcliffe, D. E. & M. L. Cabrera (eds.), Modeling Phosphorus in the Environment. CRC Press, Boca Raton, FL.

Silgram, M. & O. F. Schoumans (eds), 2004. Modelling approaches: Model parameterisation, calibration and performance assessment methods in the EUROHARP project. EUROHARP report 8-2004. NIVA report SNO 4740-2003, Oslo. 18 pp.

SMHI, 1994. Svenskt Vattenarkiv: Avrinningsområden I Sverige Del 3. Vattendrag till Egentliga Östersjön och Öresund. (In Swedish.).

Willén, E., 2001. Four decades of research on the Swedish large lakes Mälaren, Hjälmaren, Vättern and Vänern: The significance of monitoring and remedial measures for a sustainable society. Ambio 30: 458–466.

Hydrobiologia (2008) 599:201–211
DOI 10.1007/s10750-007-9200-4

ELLS 2007

Variability of bio-optical parameters in two North-European large lakes

Birgot Paavel · Helgi Arst · Anu Reinart

Abstract The bio-optical properties of some North-European large lakes were examined during 1995–2005 using field data and laboratory measurements. The key variables were optically active substances (OAS: chlorophyll, total suspended matter and dissolved organic matter), Secchi depth, and the "spectrometric" and diffuse light attenuation coefficients. Our main study sites were Lake Peipsi and Lake Võrtsjärv in Estonia, both eutrophic with mean Secchi depth below 3 m. The measured water parameters were compared with those obtained from two clear-water Swedish lakes, Lake Vänern and Lake Vättern. This comparison describes the bio-optical differences of the water in eutrophic and oligotrophic lakes. The variability of water parameters in the turbid Estonian lakes was rather high, e.g. the chlorophyll content varied from 1.8 to 102 mg m^{-3} and the diffuse light attenuation coefficient from 0.92 to 6.5 m^{-1}. The change in water properties depends on the season and the biological activity of phytoplankton. We found no apparent long-time trend in water properties. Regression analysis showed that in the turbid Estonian lakes the optical properties were well correlated with chlorophyll and suspended matter, but not with dissolved organic matter. The highest determination coefficients (between 0.73 and 0.89) were obtained when the optical parameters were correlated with all three OAS together (multiple regressions). Our results concerning the variability and interconnections among bio-optical parameters in two Estonian large lakes illustrate the effect of OAS and light field on the ecological conditions of lakes in general.

Keywords Underwater light field · Optically active substances · European large lakes

Introduction

Among the different types of water bodies, lakes deserve special attention because of the high variability of their water properties. Inspection of published data on lakes throughout the world reveals that the concentrations of optically active substances (OAS) and diffuse light attenuation coefficient can differ by more than one hundred fold (Kirk, 1994; Lindell et al., 1999; Arst, 2003). The relative values of OAS can also differ: in one lake the dominating substance may be coloured dissolved organic matter,

Guest editors: T. Nõges, R. Eckmann, K. Kangur, P. Nõges, A. Reinart, G. Roll, H. Simola & M. Viljanen
European Large Lakes—Ecosystem changes and their ecological and socioeconomic impacts

B. Paavel (✉) · H. Arst
Estonian Marine Institute, University of Tartu, Mäealuse St. 10A, 12618 Tallinn, Estonia
e-mail: birgot.paavel@ut.ee

A. Reinart
Tartu Observatory, Tõravere, Tartumaa 60602, Estonia

in another it may be phytoplankton. To obtain a full picture on the global scale a huge amount of data is necessary. For oceans and seas, optical remote sensing has become a useful tool for monitoring the spatial and temporal properties of the aquatic environment, and the development of new sensors has also made satellite monitoring of lake and coastal waters possible (Schiller & Doerffer, 1999; Kutser et al., 2005). However, there are difficulties concerning the quantitative interpretation of satellite data, so in situ measurements retain their value. Reliable ground data on water properties are needed in case satellite measurements are missing (for which there can be several reasons, including cloudiness) and also to aid in the interpretation of remote-sensing data. Each database (even for 1–2 lakes) helps build a mosaic for describing global variability in the optical properties of lakes and elucidates the connections between types of lakes and environmental/climatic conditions. The present study, describing the variability and interconnections among bio-optical parameters in two large Estonian lakes, is a fragment in this mosaic.

Optical properties also play major role in the ecological structure and function of lakes, which are likely to be highly sensitive to climate changes (Schindler et al., 1990; Vincent et al., 1998) as well as local changes in land use (Goldman, 1988; Rae et al., 2001). Typical problems include increases in lake productivity, decreases in water transparency and increased frequency and severity of algal blooms (Dokulil & Teubner, 2003). Optical methods are used to monitor the properties of water bodies. Information on the types and concentrations of OAS—suspended matter, coloured dissolved organic matter and phytoplankton pigments—is needed to build remote-sensing models (Kutser et al., 2001; Dall'Olmo & Gitelson, 2005) of optically complex water, which have to be parameterized and validated using in situ data (Pierson & Strömbeck, 2000; Darecki & Stramski, 2004; Kutser et al., 2006; Metsamaa et al., 2006).

In the present study, the spatial-temporal variations in the bio-optical parameters of some North-European large lakes are described using field data and laboratory measurements. Our main study sites were Lake Peipsi and Lake Võrtsjärv. The first is located on the border of Estonia and Russia, the other is the largest entirely Estonian lake; in situ measurements were carried out in the years 2001–2005 and 1995–2005, respectively. Both lakes are rather turbid (mean Secchi depth below 3 m). The measured water parameters were also compared with those obtained from two clear-water Swedish lakes, Lake Vänern and Lake Vättern, and published previously (Pierson & Strömbeck, 2000; Reinart et al., 2004). Our purpose was not only to describe the spatial-temporal variations of the water properties, but also to identify general relationships between the OAS and optical parameters.

Description of lakes

Estonian lakes

Both Lake Peipsi and Lake Võrtsjärv are large unstratified eutrophic lakes, which are interconnected by the River Emajõgi discharging from Võrtsjärv and running into Peipsi. Lake Peipsi on the Estonian-Russian border is the fourth largest lake in Europe and is unique in character: shallow, eutrophic, biologically very productive and bordered by many wetland areas along its coast. Peipsi (3,555 km^2) consists of three parts (Fig. 1): (1) the northernmost, largest and deepest (average depth: 8.3 m), Lake Peipsi sensu stricto (Lake Peipsi s.s.), is unstratified and eutrophic; (2) the southernmost Lake Pihkva is shallower (5.3 m) and hypertrophic; and (3) the narrow Lake Lämmijärv (2.3 m) connects the first two and is close to hypertrophic (Nõges, 2001). Lake Võrtsjärv (270 km^2, mean depth: 2.8 m) is the largest wholly Estonian lake. Võrtsjärv water is optically very turbid (Table 1) and its underwater light climate is very strongly affected by the water level and ice conditions (Reinart & Nõges, 2004).

Swedish lakes

Sweden has about 100 000 lakes with a total area of 41,000 km^2, unevenly spread over 450,000 km^2 in a glacially formed landscape. The largest of these, Lake Vänern, has area of 5,648 km^2 (Table 1) and is the third largest lake in Europe. The second largest lake in Sweden, Vättern (Fig. 1), is the fifth largest in Europe (Willén, 2001). Vättern and Vänern are clear-water lakes with relatively large Secchi depth; Vänern has more intense colour but lower turbidity

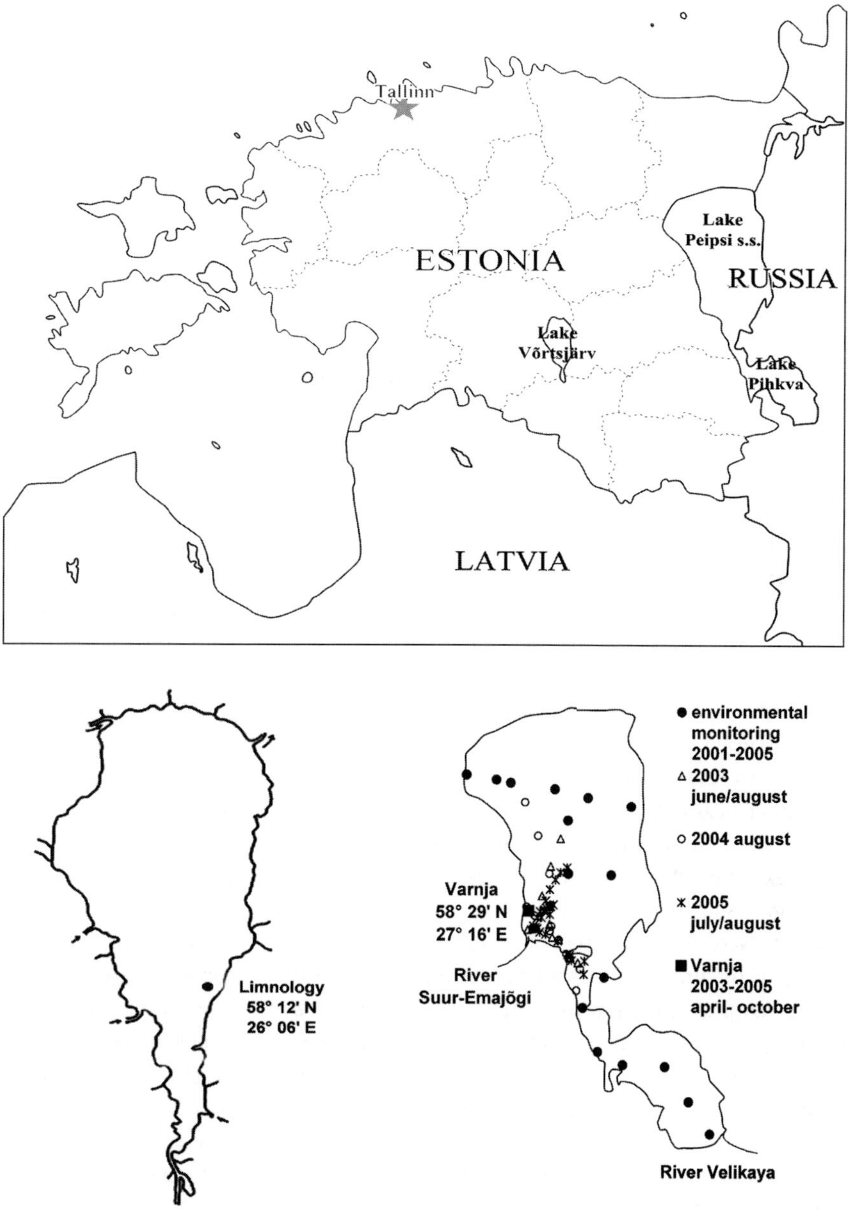

Fig. 1 Location map of Lake Peipsi and Lake Võrtsjärv. Points show sampling stations for optical measurements and regular state monitoring

Table 1 Morphometric data and some parameters of Lakes Peipsi and Võrtsjärv (Nõges, 2001; Arst, 2003) compared with similar data for large Swedish lakes (Willén, 2001)

Parameter	L. Peipsi	L. Võrtsjärv	L. Vänern	L. Vättern
Area (km^2)	3,555	270	5,648	1,856
Mean depth (m)	7	2.8	27	40
Maximum depth (m)	15	6.0	106	128
Chlorophyll *a* (mg m^{-3})[a]	1.8–95	20–102	0.2–10.6	<0.1–5.2
Secchi depth (m)[a]	0.4–4.8	0.15–1.0	2.0–7.9	6.5–15.0

[a] Data from long-time monitoring (http://www.seiremonitor.ee/, http://www.info1.ma.slu.se/)

of the two (Pierson & Strömbeck, 2000; Reinart et al., 2004).

Materials and methods

Both in situ and laboratory measurements were made. The programme of our study included: (1) water transparency measurements by Secchi disk, (2) concentrations of chlorophyll *a* and suspended matter in water samples, (3) "spectrometric" light attenuation coefficient spectra from filtered and unfiltered water samples, and (4) underwater downwelling planar quantum irradiance measurements in the photosynthetically active region (400–700 nm).

Water samples were collected from Lake Peipsi at several stations, but from only one station in Lake Võrtsjärv (Fig. 1). We have long-term (but intermittent) datasets from one sampling station in Peipsi (Varnja, 58°29′ N, 27°16′ E) and from one in Võrtsjärv (Limnology, 58°12′ N, 26°06′ E).

At the Varnja (Peipsi) and Limnology (Võrtsjärv) Stations, we collected water samples from the 0.2 to 2 m layer using a standard water sampler. At the other stations in Peipsi (Fig. 1) the samples were taken from different depths (79 individual samples altogether). All samples were stored in the dark and cold for less than 7 h before filtering. The "spectrometric" attenuation coefficients, $c^*(\lambda)$ and $c_f^*(\lambda)$, were determined from unfiltered and filtered water samples only for the years 2003–2005. The term 'spectrometric attenuation coefficient' requires explanation. As is known (Zaneveld et al., 1992; Bricaud et al., 1995), experimental determination of the true values of the beam attenuation coefficient (c) is complicated. Theoretically, the beam transmittance should contain no contribution from scattered light, but in reality small-angle forward scattering does reach the detector. The measured transmittance therefore exceeds the theoretical value and the attenuation coefficient determined from the measured transmittance is lower than the true value. The direct spectrophotometer reading, $c^*(\lambda)$, is actually:

$$c^*(\lambda) = c(\lambda) - Fb(\lambda) - c_d(\lambda), \quad (1)$$

where $c(\lambda)$ is the beam attenuation coefficient in the water at wavelength λ, $b(\lambda)$ is the light-scattering coefficient, $c_d(\lambda)$ is the beam attenuation coefficient for distilled water and F is a coefficient showing the contribution of small-angle forward scattering to the radiation measured by the spectrophotometer. For filtered water, Eq. 1 is also valid, but the term $Fb(\lambda)$ is considerably smaller owing to the decrease in scattering after filtration. We performed our laboratory measurements with a commercial Hitachi U1000 spectrophotometer. Unfortunately, the value of F for this instrument is not known and therefore we describe the measurement result not as a 'beam attenuation coefficient', but as a 'spectrometric' attenuation coefficient.

The amount of coloured dissolved organic matter was characterized by its absorption coefficient at 380 nm ($a_y(380)$). However, with spectrometers such as the Hitachi U1000 we cannot directly measure $a_y(380)$, but the attenuation coefficient $c_f^*(380)$ (Kirk, 1994). For filtered water, $c_f^*(380)$ is not identical to $a_y(380)$ because some very small mineral particles can also pass through the filter. However, the differences are small, about 2–8% (Sipelgas et al., 2003).

For chlorophyll concentrations, C_{Chl}, 0.5–1 l of water was filtered through Whatman GF/F-filters (0.7 μm pore size) and the chlorophyll *a* and phaephytin *a* concentrations were measured spectrophotometrically in ethanol extracts of the filters according to the ISO standard method (ISO 10260, 1992 (E)). The concentration of total suspended matter, C_{TSS}, was determined by dry weight (precision of weights 0.01 mg) after filtration of 0.5–1 l of water through pre-weighed and pre-combusted (103–105°C for 1 h) Whatman GF/F-filters (ESS method 340.2, 1993).

On the basis of $c^*(\lambda)$ spectra measured between 400 and 700 nm (the photosynthetically active region, PAR), the mean value for this waveband, c^*_{PAR}, was determined. It has been shown (Arst, 2003) that in addition to the Secchi depth and diffuse attenuation coefficient, c^*_{PAR} is a rather good indicator of the optical quality of the water. Using underwater quantum irradiance $q_{PAR}(z)$ measurements, obtained using an LI-192 SA quantum sensor (LI-COR Corporation, Lincoln, Nebraska, USA), we also calculated the depth-averaged diffuse attenuation coefficient of PAR ($K_{d,PAR}$). For this we used a semi-logarithmic plot of $q_{PAR}(z)$ versus depth and calculated $K_{d,PAR}$ as the slope of the least-square regression line through these points.

Results

Temporal and spatial variations of optical parameters

The minimum and maximum values of two important water parameters for Peipsi and Võrtsjärv, chlorophyll *a* concentration and Secchi depth, are presented in Table 1; for comparison, these parameters are shown also for the two Swedish clear-water lakes. The temporal variations of all three OAS measured at the two sampling stations in Peipsi and in Võrtsjärv are shown in Fig. 2. The corresponding temporal variations of Secchi depth, diffuse and spectrometric attenuation coefficients ($K_{d,PAR}$ and c^*_{PAR}) are presented in Fig. 3. From these figures we can see that the variability of all six parameters is irregular, and with no apparent long-term trend.

The range of variation and corresponding mean values of water parameters for the Varnja and Limnology Stations are shown in Table 2. The variability of the water properties is high in both lakes, except that the Secchi depth was low and less variable. At Varnja Station it ranged from 1.1 to 2.75 and at Limnology Station it was consistently below 1.3 m.

During our measurement campaigns we also examined the spatial variability of OAS in the north-south direction in Lake Peipsi. The results are shown in Table 3.

Correlation analysis

Lakes are multicomponent systems in which all three OAS influence the optical parameters of the

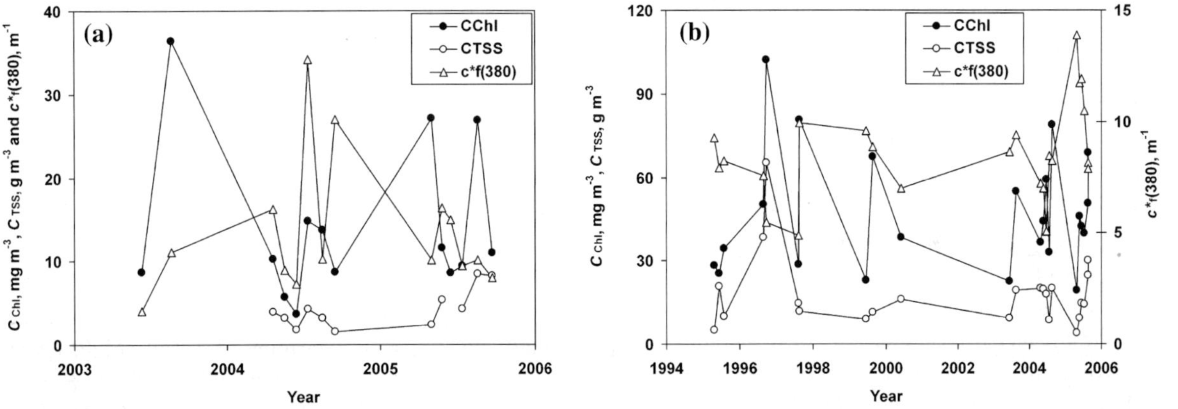

Fig. 2 Temporal distribution of OAS during 1995–2005 in Estonian lakes: (**a**) Lake Peipsi (Varnja Station) and (**b**) Lake Võrtsjärv (Limnology Station)

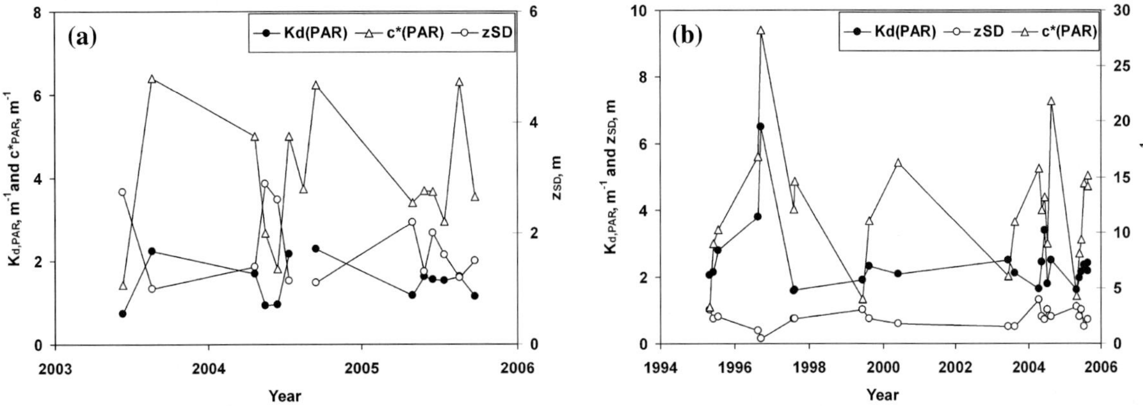

Fig. 3 Temporal distribution of Secchi depth and diffuse and "spectrometric" light attenuation coefficients during 1995–2005 in Estonian lakes: (**a**) Lake Peipsi (Varnja Station) and (**b**) Lake Võrtsjärv (Limnology Station)

Table 2 Variation range of the optical properties of the water for Peipsi (Varnja) and Võrtsjärv (Limnology)

Variable	Station Varnja	N	Station Limnology	N
C_{Chl}, mg m^{-3}	3.8–36.4 (14)	60	19–102 (47)	23
C_{TSS}, g m^{-3}	1.6–8.5 (4.2)	60	3.9–65 (18)	23
$c_f^*(380)$, m^{-1}	4–34 (13)	60	4.9–14 (8.6)	23
z_{SD}, m	1.1–2.8 (1.8)	12	0.2–1.3 (0.8)	23
K_d, m^{-1}	0.7–2.3 (1.5)	12	1.6–6.5 (2.4)	23
c_{PAR}^*, m^{-1}	1.4–6.4 (4)	60	3.3–28 (12)	23

The mean values of the parameters are shown in parentheses

Table 3 Variation of "spectrometric" light attenuation coefficient, Secchi depth and concentrations of OAS (mean and standard deviation) in north-south direction in Lake Peipsi

Latitude N	C_{Chl}, mg m^{-3}	C_{TSS}, g m^{-3}	$c_f^*(380)$, m^{-1}	c_{PAR}^*, m^{-1}	z_{SD}, m^{-1}
58.8–59.0	17.1 ± 11.6	4.7 ± 3.1	5.5 ± 2.5	3.5 ± 1.9	–
58.6–58.8	18.7 ± 14.0	7.0 ± 5.1	5.1 ± 2.5	2.6 ± 0.8	2.0 ± 0.3
58.4–58.6	16.6 ± 11.6	5.7 ± 4.0	9.5 ± 6.5	3.5 ± 1.8	1.8 ± 0.7
58.2–58.4[a]	33.3 ± 23.4	13.3 ± 6.0	8.1 ± 3.4	5.7 ± 2.6	0.9 ± 0.4
58.0–58.2[b,c]	57.1 ± 15.4	22.8 ± 9.1	9.7 ± 4.9	12.9 ± 2.2	–
58.0–57.8[c]	48.8 ± 24.6	15.4 ± 12.9	7.8 ± 6.7	12.5 ± 2.5	–

[a] Mouth of River Emajõgi

[b] Mouth of River Velikaya

[c] Measurements were made only in August and October

water. We can usually obtain reliable regression formulae using databases from different types of lakes, where the overall variation of parameters is large. However, it is also of interest to examine lake-specific regressions using databases from the two turbid lakes separately. We determined the regressions $K_{d,PAR}$ versus each OAS separately, as well as c_{PAR}^* versus each OAS separately. In addition, the regressions $K_{d,PAR}$ versus z_{SD} and C_{Chl} versus C_{TSS} were studied. The determination coefficients of the regression predicting $K_{d,PAR}$ were in most cases only 0.25–0.35. The remaining regression formulas predicting c_{PAR}^*as a function of each OAS separately are shown in Table 4. The optical property c_{PAR}^* correlated well with chlorophyll and total suspended matter. The determination coefficients were relatively high for both lakes (from 0.67 and 0.72), but the correlation C_{Chl} versus C_{TSS} was good only for Peipsi (Table 4). However c_{PAR}^* was not strongly correlated with $c_f^*(380)$ in either lake; the correlation coefficients in Peipsi and Võrtsjärv were 0.24 and 0.29.

Since all three OAS simultaneously influence water transparency, we also determined multiple

Table 4 Relationships between OAS and $c_{PAR}^*(380)$ in two large Estonian lakes

y	x	Relationship	R^2	SE	N
Lake Peipsi (all stations)					
C_{Chl}	C_{TSS}	$y = 2.21x + 5.06$	0.64	8.89	238
c_{PAR}^*	C_{Chl}	$y = 0.14x + 1.32$	0.72	1.32	135
c_{PAR}^*	C_{TSS}	$y = 0.38x + 1.54$	0.70	1.40	120
c_{PAR}^*	$c_f*(380)$	$y = 0.28x + 0.98$	0.24	2.20	134
Lake Võrtsjärv (Station Limnology)					
C_{Chl}	C_{TSS}	$y = 1.09x + 27.10$	0.44	16.40	23
c_{PAR}^*	C_{Chl}	$y = 0.21x + 2.24$	0.63	3.57	23
c_{PAR}^*	C_{TSS}	$y = 0.36x + 5.7$	0.67	3.37	23
c_{PAR}^*	$c_f^*(380)$	$y = -1.42x + 24.3$	0.29	4.95	23

SE is standard error, N is the number of water samples

regressions predicting c^*_{PAR} or $K_{d,PAR}$ from all three OAS together. The regression formulas obtained and the corresponding statistical characteristics are shown below:

parameters are known. The relationships c^*_{PAR}(meas) versus c^*_{PAR}(regr)and $K_{d,PAR}$(meas) versus $K_{d,PAR}$(regr) (both lakes together) are presented in Figs. 4 and 5.

Lake Peipsi (all stations)
$$c^*_{PAR} = 0.212C_{TSS} + 0.066C_{Chl} + 0.208c^*_f(380) - 0.718 \quad R^2 = 0.89, \text{Std. err.} = 0.84\ m^{-1}, N = 115, P < 0.0001 \tag{2}$$

Lake Võrtsjäv (Limnology Station)
$$c^*_{PAR} = 0.18C_{TSS} + 0.125C_{Chl} - 0.512c^*_f(380) + 7.45 \quad R^2 = 0.80, \text{Std. err.} = 2.70\,m^{-1}, N = 23, P < 0.00001 \tag{3}$$

Lake Peipsi (all stations)
$$K_{d,PAR} = 0.078C_{TSS} + 0.010C_{Chl} + 0.051c^*_f(380) + 0.404 \quad R^2 = 0.82, \text{Std. err.} = 0.37\,m^{-1}, N = 18, P < 0.00001 \tag{4}$$

Lake Võrtsjäv (Limnology Station)
$$K_{d,PAR} = 0.065C_{TSS} + 0.0023C_{Chl} - 0.0038c^*_f(380) + 1.17 \quad R^2 = 0.73, \text{Std. err.} = 0.58\ m^{-1}, N = 23, P < 0.00001 \tag{5}$$

Here C_{TSS} is measured in g m^{-3}, C_{Chl} in mg m^{-3} and $c^*_f(380)$in m^{-1}. When other units are used, the coefficients in the regression formulas will change. These equations provide a practical means for determining any parameter if the other three

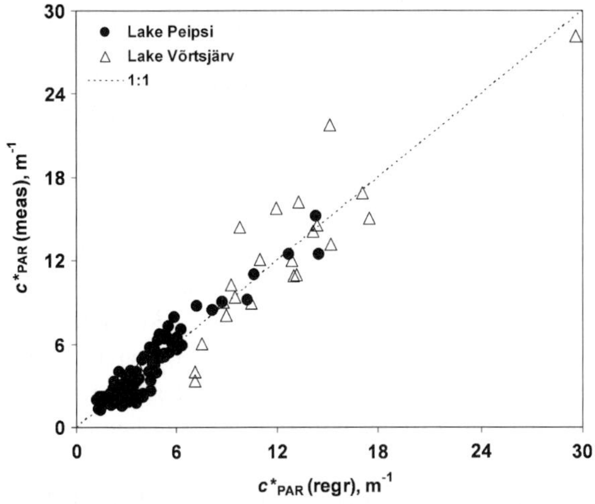

Fig. 4 Relationship between measured and calculated (from multiple regression analysis) spectrometric attenuation coefficient in Peipsi (all stations) and in Võrtsjärv (Limnology Station)

The relationship z_{SD} versus $K_{d,PAR}$ found for two Estonian lakes and two Swedish lakes on the basis of our data (Reinart et al., 2004) is shown in Fig. 6. The results show that this relationship can be described by a power function with an exponent of 1.38. This

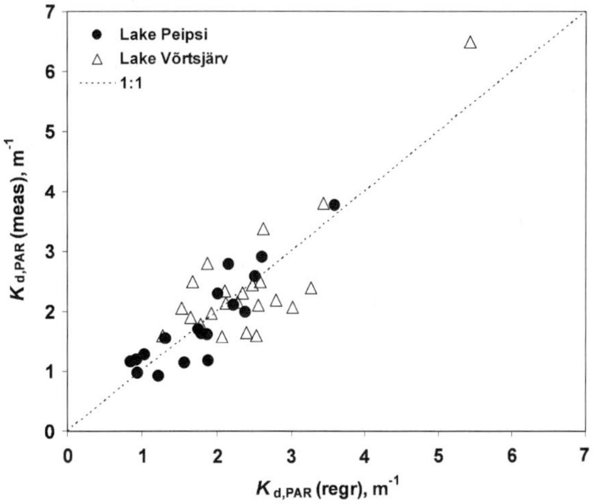

Fig. 5 Relationship between measured and calculated (from multiple regression analysis) diffuse attenuation coefficient values in Peipsi and Võrtsjärv

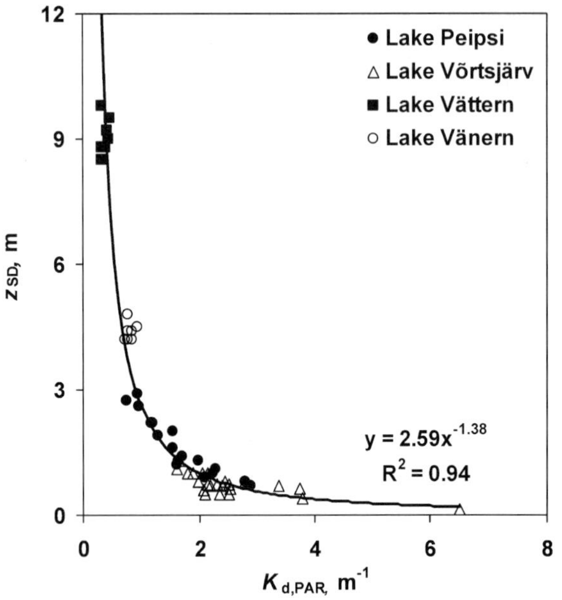

Fig. 6 Relationship between the values of Secchi disk depth and diffuse attenuation coefficient in four large North-European lakes (for Estonian lakes the data for all stations are included)

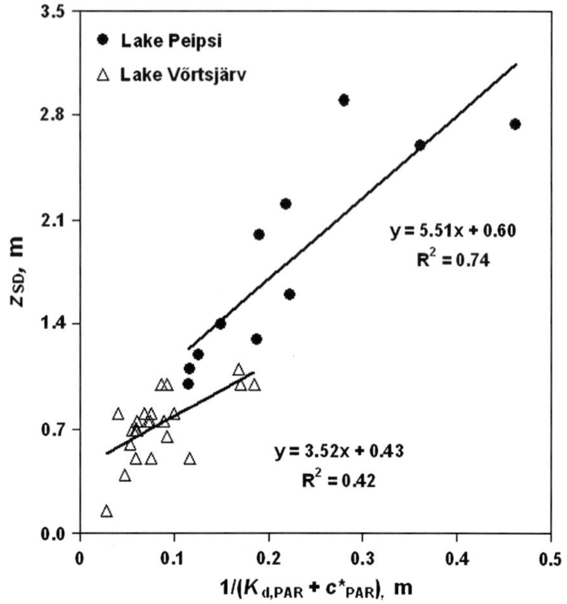

Fig. 7 Relationship z_{SD} versus $1/(K_{d,PAR} + c^*_{PAR})$, separately for Võrtsjärv and Peipsi

value is rather close to those obtained by Davies-Colley & Vant (1988) and Koenings & Edmundson (1991) (the exponent is 1) and also by Reinart & Nõges (2004) for various Estonian lakes (1.35).

In Kirk (1994) the connection between z_{SD}, $K_{d,PAR}$ and is described in the following way:

$$z_{SD} = \frac{C}{K_{d,PAR} + c_{PAR}} \quad (6)$$

where C is a constant ranging from 8.7 to 9.4 (estimated for coastal waters and New Zealand lakes). We tried to determine the value of C for Peipsi and Võrtsjärv (Varnja and Limnology Stations), using not c_{PAR}, but c^*_{PAR}(i.e. the attenuation caused by pure water was not taken into account). This hardly influences the result, because the contribution of pure water is very small, 1–4%, in turbid waters. Taking the two Estonian lakes together, the statistical characteristics of the regression z_{SD} versus $1/(K_{d,PAR} + c^*_{PAR})$ were: $C = 6.44$, intercept = 0.25, $R^2 = 0.82$, Std. error = 0.29 m and $P > 0.0001$. The corresponding regressions obtained separately for both lakes are presented in Fig. 7.

Discussion

In general, any change in the type and concentrations of OAS changes the optical properties of water. This includes temporal changes due to phytoplankton blooms as well as the springtime increase of dissolved organic matter associated with snowmelt runoff or suspended particles associated with river discharge or resuspensions. From Table 1 we can see that the water properties of the two large Estonian lakes differ markedly from those of the large Swedish lakes. For instance, the maximum values of C_{Chl} in Estonian lakes exceeded those in Lakes Vänern and Vättern by about 10–20 times. The concentrations of all OAS in Lake Peipsi and Lake Võrtsjärv were noticeably higher than those in Vänern and Vättern. The data collected at Varnja (Peipsi) and Limnology (Võrtsjärv) Stations showed that chlorophyll *a* concentrations in both lakes were highest during the intensive algal blooms that occurred in late summer (Fig. 2), which is typical of turbid lakes situated in agricultural and rural areas (Marshall & Peters, 1989). At Limnology Station the maximum level of chlorophyll *a* developed in the second half of the growing period, with C_{Chl} maximum values reaching as high as 102 mg m^{-3} (Fig. 2b). Vernal chlorophyll peaks were also observed at the end of April (in Peipsi) or in May (in Võrtsjärv).

The considerable increases of humic substances in Lake Peipsi in July and September 2004 were due to summer rainfall, which resulted in intensive inflow from rivers to the lake. The low $a_y(380)$ values in April 2005 corresponded to a very severe winter, when snowmelt runoff peaks were shifted from April to early May (Fig. 2). According to these parameters, Peipsi and Võrtsjärv fall in the same range as the Estonian lakes investigated previously (Arst, 2003).

Temporal variations in the diffuse attenuation coefficient ($K_{d,PAR}$) were closely related to the variations in OAS. The lowest $K_{d,PAR}$ values (from 0.74 to 1.70 m^{-1}) in Peipsi were observed in spring and summer, but by the end of August (in some years even in September) $K_{d,PAR}$ had increased to 2.28 m^{-1} because of phytoplankton blooms (Fig. 3a). In spring $K_{d,PAR}$ values in Võrtsjärv varied from 1.60 to 2.44 m^{-1} but increased during late summer and autumn up to 3.80 m^{-1}. Note that in September 1996, when the Secchi depth was only 0.15 m, there was one extremely high value of $K_{d,PAR}$, equal to 6.5 m^{-1} (Fig. 3b). In general, the values of c^*_{PAR} showed a seasonal dynamics similar to $K_{d,PAR}$, but sometimes their behaviours conflicted because $K_{d,PAR}$, unlike c^*_{PAR}, depends on the angular distribution of irradiance (Fig. 3). For instance, in August 2005 the increase in C_{Chl} values in Peipsi was clearly reflected in c^*_{PAR}, but not in $K_{d,PAR}$. The variation range of Secchi depth (z_{SD}) was 0.15–1.3 m at the Limnology Station (Võrtsjärv) and 1.1–2.8 m in Varnja (Peipsi).

We can compare the vernal diffuse attenuation coefficient in the eutrophic Estonian lakes with that measured in the Swedish lakes Vänern and Vättern during spring 2002. $K_{d,PAR}$ ranged between 0.74 and 1.70 m^{-1} in Peipsi and from 1.60 to 2.44 m^{-1} in Võrtsjärv, while in Vänern and Vättern the $K_{d,PAR}$ values were respectively 0.71 and 0.33 m^{-1}. Water in the deep Swedish lakes is more transparent than in the shallow Estonian lakes; it contains fewer nutrients because the rivers bring less organic matter from the rocky upland catchments into these lakes (Willén, 2001) than to the Estonian lakes, which are surrounded by wetlands.

The variations in Secchi depth between Peipsi and Võrtsjärv were much lower than the corresponding variations in $K_{d,PAR}$. The low sensitivity of Secchi depth for describing changes in OAS in turbid waters is more thoroughly discussed by Arst (2003).

We observed noticeable spatial variability of OAS in the north-south direction in Lake Peipsi. The Velikaya River (entering Lake Pihkva) accounts for about 50% and the Emajõgi River (entering Lake Peipsi s.s.) for 39% of the inflowing water to Lake Peipsi as a whole (Laugaste & Yastremskij, 2000). Since these large, strongly eutrophied rivers are major inflows in the southern coast of Lake Peipsi, the concentrations of coloured dissolved organic matter as well total suspended matter and chlorophyll *a* concentrations are higher in the south and especially near the mouths of the Rivers Emajõgi and Velikaya (Table 3). The relatively high concentrations of C_{Chl} and C_{TSS} in the southern area of Lake Peipsi were also at least partly caused by the fact that measurements were made primarily to monitor the conditions associated with algal blooms.

From Table 4 we can see that in Peipsi the correlation between c^*_{PAR} and each OAS separately was highest for chlorophyll concentration (R^2 = 0.72), while in Võrtsjärv the strongest relationship was observed for suspended matter (R^2 = 0.67). The depth of the lakes is probably the most important factor here: in Võrtsjärv, the large wind-exposed area causes extensive resuspensions of sediments, while Peipsi is somewhat deeper and algal blooms affect the optical properties more strongly. The correlation c^*_{PAR}versus $c^*_f(380)$was weak in both lakes (in Võrtsjärv it was actually negative), which is explained by fact that the relative temporal change of $c^*_f(380)$is noticeably smaller than the relative variations in phytoplankton concentration in both lakes (Erm et al., 2002). Note that the contribution of $c^*_f(380)$to the scattering coefficient is extremely small.

The determination coefficients obtained between $c^*_{PAR}$and each OAS separately (Table 4) and between $K_{d,PAR}$ and OAS are smaller than those for the multiple regression (Eqs. 2–5). The values of R^2 for c^*_{PAR}(Eqs. 2, 4) are higher than those for $K_{d,PAR}$ (Eqs. 3, 5). The reason is that $K_{d,PAR}$ depends not only on the concentrations of substances, but also on the angular distribution of irradiance.

The regression formula $z_{SD} = 6.44/(K_{d,PAR} + c^*_{PAR}) + 0.25$ obtained from the regression for the two lakes together (Varnja and Limnology Stations) showed that the constant C was lower than that presented in Kirk (1994) and there was also an intercept in our algorithm. The determination coefficient R^2 was 0.82, but for the regressions obtained separately for Peipsi and Võrtsjärv, the values of R^2 were correspondingly 0.74 and 0.42 (Fig. 7). The reason for the low value of R^2 in Võrtsjärv may be the small range of variation in z_{SD} and $K_{d,PAR}$; also, the relative errors of these parameters are noticeable in very turbid water (Secchi depth mostly below 1 m). Comparing the values of C obtained by Kirk (8.7–9.4) with those for Peipsi and Võrtsjärv (together 6.6, separately 5.6 and 3.5), we can propose the hypothesis that the constant C in Eq. 6 decreases with increasing turbidity of water.

Conclusions

We have demonstrated large differences between the water properties and optical characteristics of eutrophic and oligotrophic lakes. Variability in the water parameters was also observed in the two turbid Estonian lakes where the chlorophyll concentration varied between 1.8 and 102 mg m^{-3} and the diffuse light attenuation coefficient varied between 0.92 and 6.5 m^{-1}.

In Peipsi and Võrtsjärv we found no apparent long-term trend in the temporal changes of the water parameters. Rather, it seems that their variability is connected with the season and the biological activity of the water. There was some spatial variation in the OAS and light attenuation coefficient from north to south in Peipsi.

In turbid lakes, all three OAS contribute significantly to the bio-optical parameters of the water. For this reason, correlations between optical variables with the OAS separately are often moderately good for C_{TSS} and C_{Chl}, but weak for c_f*(380). However, the determination coefficients of multiple regressions between all three OAS and the spectrometric attenuation coefficient were remarkably good, ranging between 0.80 and 0.89 for Võrtsjärv and Peipsi, respectively. For $K_{d,PAR}$ versus OAS, the determination coefficients were 0.73 (Võrtsjärv) and 0.82 (Peipsi). This illustrates the additive effects of all OAS on the optical properties of the water.

Acknowledgements The authors are indebted to the Estonian Science Foundation (Grants 5594, 6814, 5738) and to the Estonian Marine Institute for financial support of this investigation. We also are grateful to Ants Erm, Liis Sipelgas and Medhat Hussainov for assistance in fieldwork and to Dr. Tiina Nõges (Centre for Limnology, Estonia) for Lake Peipsi data.

References

Arst, H., 2003. Optical Properties and Remote Sensing of Multicomponental Water Bodies. Springer, Praxis Publishing, Chichester, UK: 231.

Bricaud, A., C. Roesler & J. R. V. Zanevald, 1995. In situ methods for measuring the inherent optical properties of ocean waters. Limnology and Oceanography 40: 393–410.

Dall'Olmo, G. & A. A. Gitelson, 2005. Effect of bio-optical parameter variability on the remote estimation of chlorophyll-a concentration in turbid productive waters: experimental results. Applied Optics 44: 412–422.

Darecki, M. & D. Stamski, 2004. An evaluation of MODIS and SeaWiFS bio-optical algorithms in the Baltic Sea. Remote Sensing of Environment 89: 326–350.

Davies-Colley, R. J. & W. N. Vant, 1988. Estimates of optical properties of water from Secchi disk depths. Water Resources Bulletin 24: 1329–1335.

Dokulil, M. & K. Teubner, 2003. Klimaeinfüsse auf Seen in Europe (CLIME). Österreichs Fischerei 56: 176–180.

Erm, A., H. Arst, P. Nõges, T. Nõges, A. Reinart & L. Sipelgas, 2002. Temporal variations in bio-optical properties of four North Estonian lakes in 1999–2000. Geophysica 38: 89–111.

ESS method 340.2, 1993. Total suspended solids, mass balance (dried at 103–105° C), volatile suspended solids (Ignited at 550° C). Environmental Sciences Section, ESS 3: 189–192.

Goldman, C. R., 1988. Primary productivity, nutrients, and transparency during the early onset of eutrophication in ultra-oligotrophic Lake Tahoe, California-Nevada. Limnology and Oceanography 33: 1321–1333.

ISO 10260, 1992 (E). Water Quality – Measurement of Biochemical Parameters – Spectrophotometric Determination of Chlorophyll-a Concentration. Geneva, Switzerland: ISO: 1–6.

Kirk, J. T. O., 1994. Light and Photosynthesis in Aquatic Ecosystems. University Press, Cambridge: 509.

Koenings, J. P. & J. A. Edmundson, 1991. Secchi disk and photometers estimates of light regimes in Alaskan lakes: Effects of yellow color and turbidity. Limnology and Oceanography 36: 91–105.

Kutser, T., A. Herlevi, K. Kallio & H. Arst, 2001. A hyperspectral model for interpretation of passive optical remote sensing data from turbid lakes. The Science of the Total Environment 268: 47–58.

Kutser, T., D. Pierson, K. Kallio, A. Reinart & S. Sobek, 2005. Mapping lake CDOM by satellite remote sensing. Remote Sensing of Environment 94: 534–540.

Kutser, T., E. Vahtmäe & G. Martin, 2006. Assessing suitability of multispectral satellites for mapping benthic macroalgal cover in turbid coastal waters by means of model simulations. Estuarine Coastal and Shelf Science 67: 521–529.

Laugaste, R. & V. Yastremskij, 2000. Role of inflows in the phytoplankton compositions of lake Peipsi. Proceedings of the Estonian Academy of Sciences. Biology, Ecology 49: 19–33.

Lindell T., D. Pierson, G. Premazzi & E. Ziliolli (eds), 1999. Manual for Monitoring European Lakes Using Remote Sensing Techniques. Office for Official Publications of the European Communities, Luxembourg: 161.

Marshall, C. T. & R. H. Peters, 1989. General patterns in the seasonal development of chlorophyll a for temperate lakes. Limnology and Oceanography 34: 856–867.

Metsamaa, L., T. Kutser & N. Strömbeck, 2006. Recognising cyanobacterial blooms based on their optical signature: a modelling study. Boreal Environment Research 11: 493–506.

Nõges, T. (ed.), 2001. Lake Peipsi. Meteorology, Hydrology, Hydrochemistry. Sulemees Publishers, Tartu, Estonia: 163.

Pierson, D. C. & N. Strömbeck, 2000. A Modelling approach to evaluate preliminary remote sensing algorithms: Use of water quality data from Swedish Great Lakes. Geophysica 36: 177–202.

Rae, R., C. Howard-Williams, I. Hawes, A.-M. Schwarz & W. F. Vincent, 2001. Penetration of solar ultraviolet radiation into New Zealand lakes: Influence of dissolved organic carbon and catchment vegetation. Limnology 2: 79–89.

Reinart, A. & P. Nõges, 2004. Light conditions in lake Võrtsjärv. In Haberman J., E. Pihu & A. Raukas (eds), Lake Võrtsjärv. Estonian Encyclopedia Publishers, Tallinn: 141–149.

Reinart, A., B. Paavel, D. Pierson & N. Strombeck, 2004. Inherent and apparent optical properties of Lake Peipsi, Estonia. Boreal Environment Research 9: 429–445.

Schiller, H. & R. Doerffer, 1999. Neural network for emulation of an inverse model operational derivation of Case II water properties from MERIS data. International Journal Remote Sensing 20: 1735–1746.

Schindler, D. W., K. W. Beaty, E. J. Fee, D. R. Cruikshank, E. R. DeBruyn, D. L. Findlay, G. A. Linsey, J. A. Shearer, M. P. Stainton & M. A. Turner, 1990. Effect of climatic warming on lakes of central boreal forest. Science 250: 967–970.

Sipelgas, L., H. Arst, K. Kallio., A. Erm, P. Oja & T. Soomere, 2003. Optical properties of dissolved organic matter in Finnish and Estonian lakes. Nordic Hydrology 34: 361–386.

Vincent, W. F., I. Laurion & R. Pienitz, 1998. Artic and Antarctic lakes as optical indicators of global change. Antarctic Glaciology 27: 691–696.

Willén, E., 2001. Four decades of research on the Swedish large lakes Mälaren, Hjälmaren, Vättern and Vänern: the significance of monitoring and remedial measures for a sustainable society. Ambio 30: 458–466.

Zaneweld, J. R. V., J. C. Kitchen, A. Bricaud & C. Moore, 1992. Analysis of in situ spectral absorption meter data. In Ocean Optics II. Proceedings SPIE 1750: 187–200 (In Russian).

Hydrobiologia (2008) 599:213–220
DOI 10.1007/s10750-007-9189-8

ELLS 2007

Contributions of DOC from surface and groundflow into Lake Võrtsjärv (Estonia)

Toomas Tamm · Tiina Nõges · Arvo Järvet · Faycal Bouraoui

Abstract Hydrological changes have the greatest impact on shallow lakes where they alter the water volume and lake depth noticeably. Dissolved organic carbon (DOC), which is markedly affected by hydrological factors, has an important role in many biogeochemical processes. The DOC load supplied to Lake Võrtsjärv, the second largest lake in Estonia, was studied on the scale of the subcatchments discharging into the lake. Daily discharges and biweekly or monthly DOC concentrations were measured close to the river outlets over the years 1990–2002. The stream flow data were separated into groundflow and surface flow by applying local minimum and recursive digital filtering methods. Constituent load estimation software, LOADEST, was used to estimate DOC concentrations and load. LOADEST performed well for three of the four rivers. The total estimated DOC load to Võrtsjärv from all four main rivers varied from 1,320 to 4,934 t year^{-1}. The average annual load over the 13-year period was 3,265 t year^{-1} or 1.18 g C m^{-2} year^{-1}. Baseflow separation analysis indicated that the DOC load originating from groundflow contributed 79% and 69% of the total load according to the digital filter and local minimum methods, respectively. The results of our study demonstrate the utility of linking the rating-curve method and baseflow separation to assess the allochthonous DOC load to Võrtsjärv both currently and under changing climatic conditions.

Keywords Dissolved organic carbon (DOC) · DOC load · Baseflow separation

Guest editors: T. Nõges, R. Eckmann, K. Kangur, P. Nõges, A. Reinart, G. Roll, H. Simola & M. Viljanen
European large lakes—Ecosystem changes and their ecological and socioeconomic impacts

T. Tamm (✉)
Department of Water Management, Estonian University of Life Sciences, Kreutzwaldi 5, Tartu 51014, Estonia
e-mail: toomas.tamm@emu.ee

T. Nõges
Institute of Agricultural and Environmental Sciences, Estonian University of Life Sciences, Tartu, Estonia

A. Järvet
Institute of Geography, Tartu University, Tartu, Estonia

T. Tamm · F. Bouraoui
European Commission, DG Joint Research Centre, Institute for Environment and Sustainability, Ispra, Italy

Introduction

Lakes function as sinks for different chemical substances, among which major nutrients such as nitrogen (N) and phosphorus (P) in soluble or particulate forms have been most extensively investigated. However, apart from N and P, attention should also be paid to other substances, among which carbon is important. Dissolved organic carbon (DOC) produced in the surrounding catchment area and washed into the lake has received scientific attention because it is important in terrestrial and aquatic biogeochemistry. For example, DOC is the primary substrate for bacterial growth in ecosystems (De Lange et al., 2003) and controls the pelagic food chain (Jansson et al., 2000). It supports denitrification when leached below the rooting zone in the watershed (Brye et al., 2001) and in streams (Seitzinger et al., 2006). Humic or tannin DOC affects the color of streams and lakes (e.g., Watts et al., 2001). Absorption of solar ultraviolet radiation (UVR) by aquatic ecosystems is primarily controlled by DOC (Morris et al., 1995; De Lange et al., 2003). The factors affecting the DOC load to the lake can be divided into (1) climatic effects on allochthonous DOC production (temperature, radiation, CO_2, and moisture; e.g., Freeman et al., 2004, Clark et al., 2005), (2) climatic effects on discharge (Arvola et al., 2004), and (3) landscape type, often paired with anthropogenic impact (e.g., clear-cut forest; Nieminen, 2004). Thus, changes in DOC input to lakes may exert significant control over productivity and biogeochemical cycles, and either accurate measurement or reliable estimation of the mass of allochthonous DOC transported in streams is needed. Export of DOC in transport models is often calculated using regressions of stream discharge and stream DOC concentrations (e.g., Moore, 1989; Hinton et al., 1997; Boyer et al., 2000; Carey, 2003).

Võrtsjärv, the second largest lake in Estonia, is highly eutrophic, receiving nutrients and DOC from the catchment area mostly via four main rivers. It has been found that the seasonal dynamics of the nutrient load from the catchment area is mostly determined by the dynamics of the hydrological load rather than by changes in the nutrient concentrations of inflowing water (Järvet, 2004). Recent findings on lake hydrology (Nõges & Nõges, 1999), light conditions (Reinart & Nõges, 2004), nitrogen and phosphorus budget (Nõges, et al., 1998a), primary production (Nõges & Nõges, 1998), phytoplankton dynamics (Frisk, et al., 1999), and food web interactions (Nõges et al., 1998b) have provided a comprehensive overview, showing that water-level fluctuations have a leading role in the functioning of this lake. However, the role of DOC in the functioning of the ecosystem of Võrtsjärv has received less attention to date; though for a light-limited and microbial loop-dominated ecosystem, DOC has to be considered a particularly important driver, controlling light conditions and providing substrate for bacteria. The present study represents a first step toward filling these gaps in knowledge and its aim is to understand the factors determining DOC inflow into Võrtsjärv.

The objectives of this study were: (1) to examine the relationship between DOC concentration and discharge using sporadic DOC data to estimate the total load and (2) to apply baseflow separation analysis to evaluate the surface and groundflow contributions of DOC to Võrtsjärv.

Material and methods

Site description

Võrtsjärv is located in the southern part of Estonia in a shallow depression of preglacial origin. The submeridionally elongated drainage basin of the lake covers 3,374 km^2 including the lake itself. Its drainage basin through the outflowing River Emajõgi belongs to the drainage basin of Lake Peipsi, the fourth largest lake in Europe. Peatlands and meadows occupy approximately 20% and forest 30–50% of total watershed area (Table 1). The lake is natural and unregulated, and water level-discharge measurements date back to 1876. Among the 18 inflows to Võrtsjärv, the rivers Väike-Emajõgi,

Table 1 Contributing area and land use upstream of each sampling point

Watershed (river-sampling point)	Total (km^2)	Acerland (%)	Forest (%)	Meadows (%)	Bogs (%)	Other (%)
Väike-Emajõgi-Pikasilla	1,270	25.2	46.9	18.9	3.5	5.5
Õhne-Suislepa	573	24.4	47.1	10.5	13.1	4.9
Tänassilma-Oiu	453	30.9	41.9	11.0	11.3	4.9
Tarvastu-Linnaveski	108	41.7	32.4	13.9	3.7	8.3

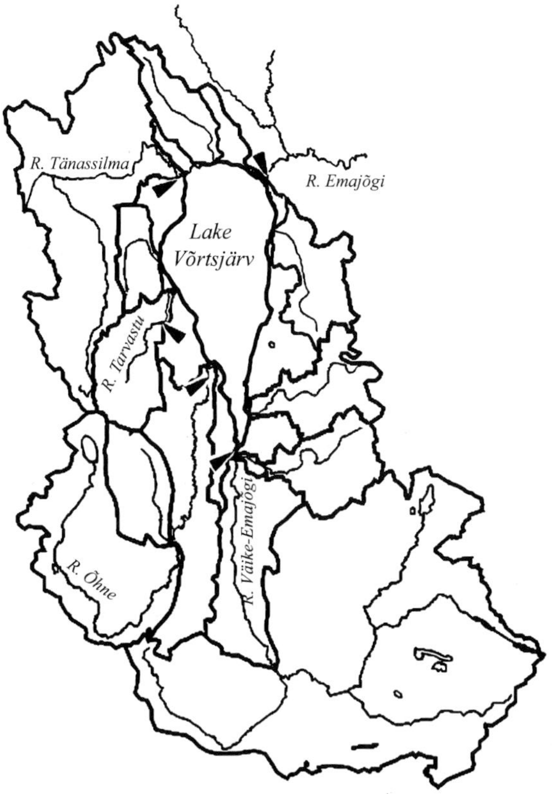

Fig. 1 Main watersheds and drainage basins of Lake Võrtsjärv. Black triangles denote sampling points

Õhne, Tänassilma, and Tarvastu (Fig. 1) constitute 70–75% of the total discharge from the catchment area (Järvet, 2004.). The mean depth of the lake is 2.8 m, with water-level fluctuations averaging 1.4 m amplitude with a maximum range of 3.2 m. Changes in discharge and load may be critical for the overall state of Võrtsjärv.

Data and statistical analysis

The period 1990–2002 is analyzed in this article. Daily riverine discharges were estimated by the Estonian Institute of Hydrology and Meteorology. Chemical oxygen demand in the rivers was measured at least monthly by the State Monitoring Programme at Tartu Environmental Research Ltd. using permanganate (COD_{Mn}, mg O l^{-1}). We used COD_{Mn} as proxy for DOC, since mg C l^{-1} was calculated according to the molar composition of CO_2:2 atoms, or 32 mg of O consumed in wet burning by permanganate was considered to be equivalent to 1 atom or 12 mg of C in DOC. Non-parametric seasonal Mann-Kendall statistics were used to analyze trends.

Load estimates

The basic approach to estimating loads in a stream is by a simple calculation, daily flow multiplied by concentration, which summed over the study period provides an estimate of the load transported. However, measured flow data are more numerous than chemical concentration data, and missing values can often be interpolated using a nearest neighbor routine (e.g., Carey, 2003) or calculated from a site-specific regression or rating-curve method based on the relationship between concentration and river discharge along with a continuous record of river discharge (e.g., Hinton et al., 1997; Carey, 2003).

In the present study, the USGS software LOADEST was used to estimate daily concentrations and constituent loads using daily log-streamflow data and event constituent log-concentration data by the rating-curve method (Runkel et al., 2004). Given a time series of streamflow and constituent concentration, LOADEST helps the user to develop a regression model for estimating explanatory variables as functions of data variables (e.g., streamflow and time in Eq. 1). LOADEST suggests the implementation of an adjusted maximum likelihood estimate (AMLE), maximum likelihood (MLE), when the model residuals are normally distributed, or a least absolute deviation (LAD) method when this condition is not fulfilled.

Hydrological flowpaths—hydrograph separation

Changes in temperature, precipitation, and evapotranspiration, due either to normal climatic variability or to climatic change, may influence hydrological flowpaths for DOC transport (Hinton et al., 1998; Boyer et al., 2000; Carey, 2003), e.g., the decline of surface runoff attributable to warmer winters.

Among the several methods for disaggregating a stream hydrograph into base and direct flow, the following two were selected: *recursive digital filter* (Nathan & McMahon, 1990; Arnold & Allen, 1999)

and *local minimum technique* (HYSEP hydrograph separation program, Sloto & Crouse, 1996). The local-minimum technique is less conservative than the recursive digital filter, i.e., it yields higher percentages of baseflow.

Method for calculating baseflow-separated DOC load

For baseflow-separated load, the following workflow was used (based on Schilling & Zhang, 2004). Firstly, concentrations and total DOC load were estimated using LOADEST. Among the predefined empirical equations in LOADEST, the following model was the best and AMLE was used, since all residuals in the model were normally distributed:

$$\begin{aligned}\text{Ln}(\text{parameter}) = {} & a_0 + a_1\text{Ln}Q + a_2\text{Ln}Q^2 \\ & + a_3\text{Sin}(2pi\,\text{dtime}) \\ & + a_4\text{Cos}(2pi\,\text{dtime}) \qquad (1)\end{aligned}$$

where parameter is load, kg day^{-1}, or concentration, mg l^{-1}; Q is daily discharge, m^3 s^{-1}; dtime is adjusted decimal time (=decimal time − center of decimal); a_0, a_1, a_2, a_4 are model coefficients.

Secondly, the two selected hydrograph separation techniques, i.e., the recursive digital filter and local minimum technique, were used to distinguish groundflow and surface flow. Thirdly, the DOC data record was screened according to flow character, i.e., (a) DOC originating from groundflow when groundflow formed 90–100% of the total streamflow and (b) the rest originating from mixed flow including both groundflow and surface flow. This procedure was applied for both hydrograph separation techniques. Fourthly, LOADEST was used a second time by applying it to the screened DOC data, i.e., estimating the DOC load only from groundflow. Fifthly, the latter load was subtracted from the previously estimated total load to yield the DOC load originating from surface flow.

Results

Concentration

Because DOC concentrations have increased in some northern European rivers in recent decades (Freeman et al., 2004; Worrall & Burt, 2004), non-parametric seasonal Mann-Kendall statistics were applied to analyze trends over the period 1990–2002. However, no significant trend was found at $P < 0.05$. Estimated DOC concentrations in the rivers were quite evenly distributed over the year with relatively small amplitudes.

Regressions between LOADEST-modeled and measured DOC concentrations yielded r^2 values of 0.530 ($P < 0.001$, $n = 175$), 0.585 ($P < 0.001$, $n = 170$), 0.161 (not significant), and 0.545 ($P < 0.001$, $n = 172$) for Väike-Emajõgi, Õhne, Tänassilma, and Tarvastu, respectively. Visual inspection of the same data displayed as a time series showed that the dynamics of DOC concentrations was best captured for the River Väike-Emajõgi (Fig. 2a). For the River Õhne (Fig. 2b), two DOC concentrations emerged as outliers (11.3 mg l^{-1} in May 1991 and 1.2 mg l^{-1} in January 2002). The r^2 was negligible for the River Tänassilma (Fig. 2c), where the average concentration of the entire dataset yielded a total load very close to that estimated by regression. The predictive power of the model was second highest for the River Tarvastu (Fig. 2d), but LOADEST could not model a few high concentration values that occurred at the beginning of May in 1993 and 2000, and at the beginning of July in 1991 and 1998. In fact, the selected model (Eq. 1) inherently entails adjustment by calibration with a certain 'average' oscillation, so a few very high or low values cannot be captured.

The log-transformed regression between measured DOC concentration and discharge (Q) was significant ($P < 0.001$) for three of the four rivers, with r^2 values of 0.360 ($n = 175$), 0.470 ($n = 170$), 0.062 (not significant, $n = 179$), and 0.502 ($n = 172$) for Väike-Emajõgi, Õhne, Tänassilma, and Tarvastu, respectively.

Baseflow-separated DOC concentration datasets were analyzed to assess the effect of the hydrological flowpath on DOC concentration and load. The dataset was screened as: (a) groundflow (90–100% of total flow was groundflow) and (b) mixed discharge (groundflow was less than 90%). Using a more rigid definition of groundflow (e.g., up to 100%) could result in only a few DOC samples being calibrated in the regression model (as in Schilling & Zhang, 2004); e.g., using the local minimum method for the River Väike-Emajõgi, the number of days with DOC data samples during groundflow could be decreased from 89 to 25.

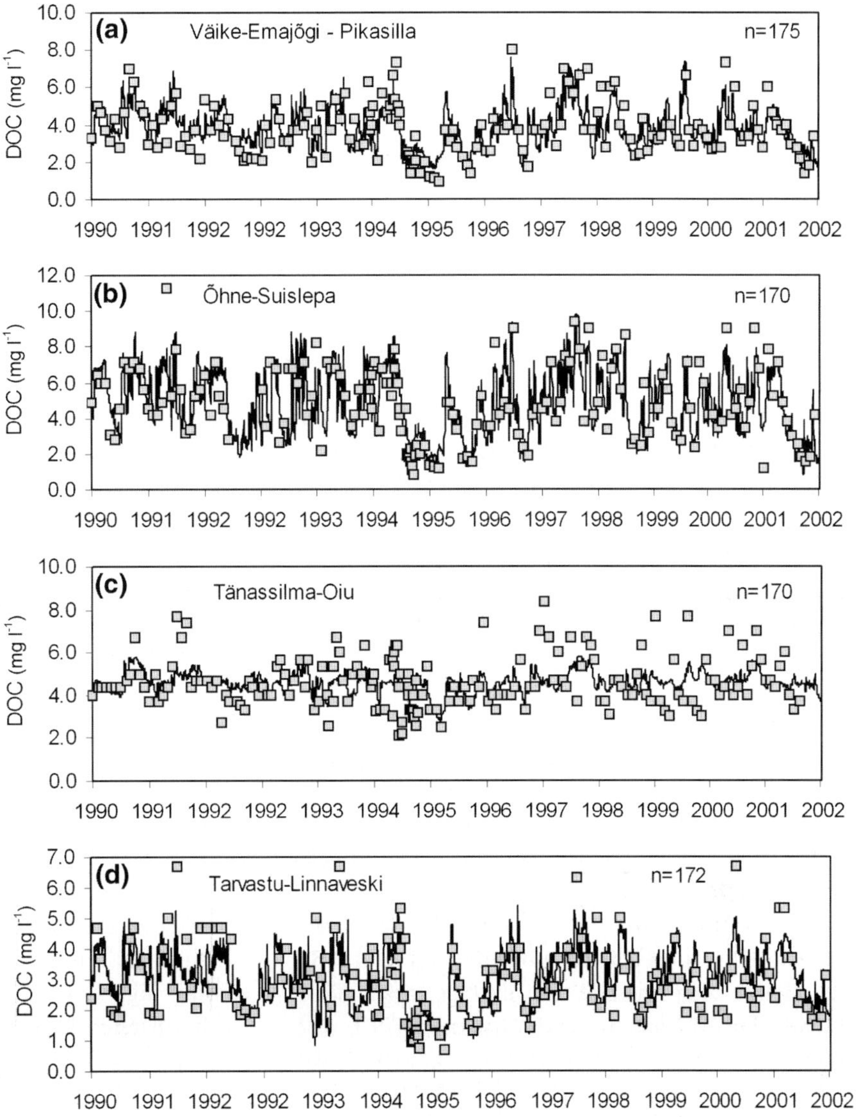

Fig. 2 Time-series of measured (boxes) and modeled (line) concentrations of dissolved organic carbon (DOC) in the main rivers discharging into Lake Võrtsjärv during 1990–2002

Data screening revealed that both groundflow and mixed discharge were log-correlated with DOC concentrations, and that on days with mixed discharge the concentration tended to be lower than on days when groundflow was dominant (Fig. 3). A common explanation for such patterns is that runoff is diluted either with rainwater or with shallow soil water during periods of storm flow (Anderson et al., 1997) or else with snowmelt water. However, this finding applies only to the Rivers Väike-Emajõgi and Õhne (Fig 3), because the groundflow and mixed runoff datasets for the River Tarvastu did not differ, and the data for the River Tänassilma were of too poor quality to allow baseflow separation to be analyzed.

Load

The flow-weighted DOC loads estimated by LOADEST were calculated for total and baseflow-separated discharge. The DOC load was mainly controlled by hydrological conditions (Fig. 4), e.g., it was lowest for Väike-Emajõgi in 1996 (annual precipitation, P, 545 mm; DOC load 651 t year^{-1} or 0.51 g C m^{-2} year^{-1}) and highest in the very wet years 1990 ($P = 987$ mm, 2,301 t year^{-1} or 1.81 g C m^{2} year^{-1}) and 1998 ($P = 911$ mm, 1,936 t year^{-1} or 1.52 g C m^{2} year^{-1}). The annual pattern of the cumulative load (Fig. 4) was consistent with the discharge pattern. The total estimated load to Võrtsjärv from all four rivers

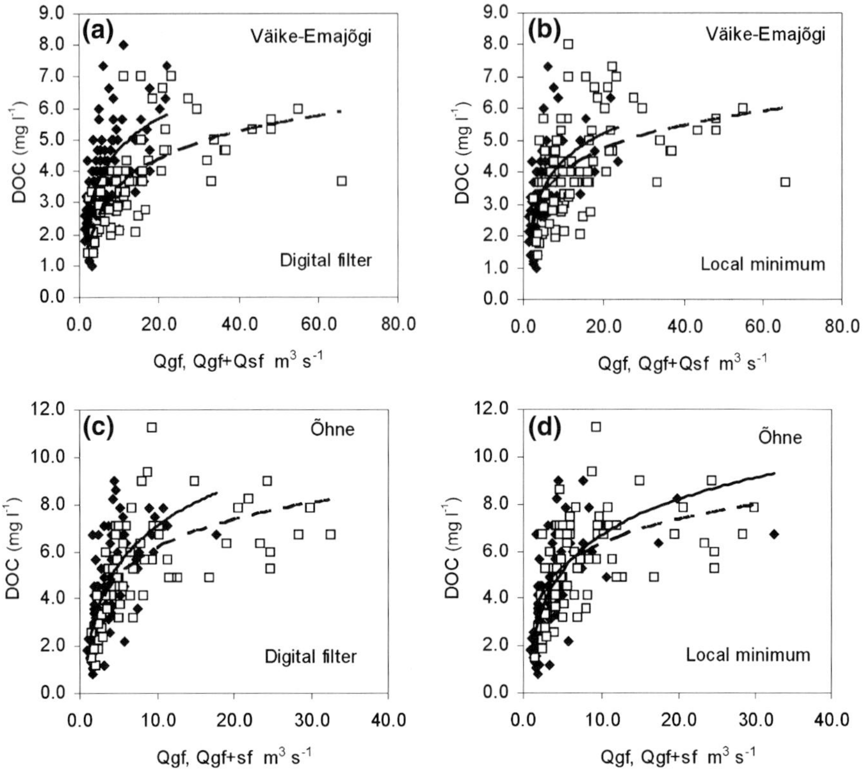

Fig. 3 Relationship ($y = a \cdot \mathrm{Ln}(x) + b$) of baseflow-separated concentrations of dissolved organic carbon (DOC) and discharges (groundflow Q_{gf}: solid diamonds and solid line; mixed flow as groundflow + surface flow Q_{gf+sf}: open boxes and dashed line) in the main rivers discharging into Lake Võrtsjärv during 1990–2002. 'Digital filter' and 'local minimum' represent two different approaches to flow calculation. The coefficients of determination for groundflow and mixed flow are (**a**) 0.473, 0.463; (**b**) 0.404, 0.277; (**c**) 0.494, 0.433; (**d**) 0.501, 0.295, respectively

varied from 1,320 t year^{-1} in 1996 to 4,934 t year^{-1} in 1990. The annual average load over the 13-year period was 3,265 t year^{-1} or 1.18 g C m^{2} year^{-1}.

Baseflow separation revealed that most of the DOC load originated from groundflow (Fig. 4). By the digital filter method the average groundflow fraction was 77% and the load fraction was 79%, and by the more conservative local minimum method the corresponding values were 71% and 69%. However, these results may be questioned because they tend to overestimate the role of groundflow: the model does not include runoff conditions peculiar to Northern latitudes, where frozen soil may effectively increase surface runoff during the spring flood (Arvola et al., 2004). Thus, additional studies with more frequent sampling during the snowmelt period might reveal an increase in the surface flow fraction.

A slight dilution effect on DOC concentrations during high flows (Fig. 3) was confirmed by flow-path-separated DOC load analysis. In general, a 5% increase in the fraction of groundflow volume estimated by the digital filter method increased the fraction of groundflow DOC load by 6.7% ($r^2 = 0.708$, $P < 0.001$, figure not shown). By the local minimum method the increase was 8.2% ($r^2 = 0.844$, $P < 0.001$, figure not shown), i.e.,

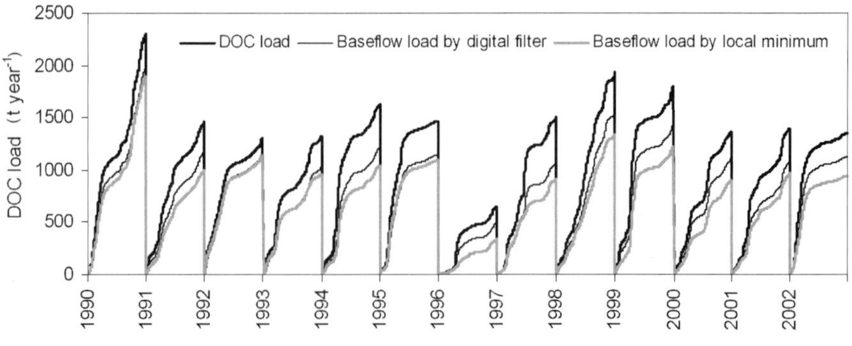

Fig. 4 Yearly cumulative load of dissolved organic carbon (DOC) into Lake Võrtsjärv by River Väike-Emajõgi

DOC concentrations must have been higher in the groundflow to produce that result.

Discussion

Assessing the hydrological consequences of climate change for the DOC load to Võrtsjärv, we may conclude that if winters become warmer then spring floods will be less developed, implying a relative increase in concentrations as the dilution effect of the floods decreases. However, the total load will be still mainly dependent on total discharge. Processes that are favored by increased temperature and higher precipitation, such as organic matter production and decomposition, may further increase the availability of DOC. However, many factors are involved in DOC production and fluxes, so it is nearly impossible to predict changes with real confidence. For example, Laudon et al. (2004) found, for total organic carbon (TOC), that significant variations in runoff patterns occurred during snowmelt-driven runoff episodes. Much of the snowmelt runoff from forested areas reaches the stream via a subsurface flowpath, increasing TOC concentration; in contrast, a larger component of the snowmelt from overland flow over the frozen wetland surface diluted the TOC concentrations. Rapid changes in DOC concentration that have been reported within one rainstorm or snowmelt-dominated runoff event (Boyer et al., 2000; Inamdar et al., 2004) reveal a need to establish a more frequent sampling program at the main rivers discharging to Võrtsjärv, and more generally for other lakes.

Conclusions

The DOC load from the four main rivers discharging into Võrtsjärv was estimated by combining the rating-curve method for load estimation with two different baseflow-separation methods for analyzing flowpath effects. The annual average load over the 13-year period was 3,265 t year^{-1}. LOADEST performed well for three of the four rivers, estimating the modeled DOC concentration ($r^2 > 0.5$) and load on the basis of biweekly or monthly DOC measurements and daily discharge data. Both the baseflow-separation methods showed an unexpectedly high groundflow contribution to the total DOC load, around 70%. We conclude that hydrological flowpath analysis with more frequent and detailed DOC sampling may give useful insights into allochthonous DOC production and load to Võrtsjärv both currently and under changing climatic conditions.

Acknowledgments Funding for this research was provided by Estonian target funding project 03962480s03, by Estonian Science Foundation grant 5738 and by EC project CLIME (EVK1-CT-2002-00121). We used data obtained in frames of the State Monitoring Program of the Estonian Ministry of Environment.

References

Anderson, P. S., W. E. Dietrich, R. Torres, D. R. Montgomery & K. Loague, 1997. Concentration-discharge relationships in runoff from a steep, unchanneled catchment. Water Resources Research 33: 211–225.

Arnold, J. G. & P. M. Allen, 1999. Automated methods for estimating baseflow and ground water recharge from streamflow records. Journal of American Water Resources Association 35: 411–424.

Arvola, L., A. Räike, P. Kortelainen & M. Järvinen, 2004. The effect of climate and landuse on TOC concentrations in Finnish rivers. Boreal Environment Research 9: 381–387.

Boyer, E. W., G. M. Hornberger, K. E. Bencala & D. M. McKnight, 2000. Effects of asynchronous snowmelt on flushing of dissolved organic carbon: a mixing model approach. Hydrological Processes 14: 3291–3308.

Brye, K. R., J. M. Norman, L. G. Bundy & S. T. Gower, 2001. Nitrogen and carbon leaching in agroecosystems and their role in denitrification potential. Journal of Environmental Quality 30: 58–70.

Carey, S. K., 2003. Dissolved organic carbon fluxes in a discontinuous permafrost subartic alpine catchment. Permafrost and Periglacial Processes 14: 161–171.

Clark, J. M., P. J. Chapman, J. K. Adamson & S. N. Lane, 2005. Influence of drought-induced acidification on the mobility of dissolved organic carbon in peat soils. Global Change Biology 11: 791–809.

De Lange, H. J., D. P. Morris & C. E. Williamson, 2003. Solar ultraviolet photodegradation of DOC may stimulate freshwater food webs. Journal of Plankton Research 25: 111–117.

Freeman, C., N. Fenner, N. J. Ostle, H. Kang, D. J. Dowrick, B. Reynolds, M. A. Lock, D. Sleep, S. Hughes & J. Hudson, 2004. Export of dissolved organic carbon from peatlands under elevated carbon dioxide levels. Nature 430: 195–198.

Frisk, T., Ä. Bilaletdin, H. Kaipainen, O. Malve & M. Möls, 1999. Modelling phytoplankton dynamics of the eutrophic Lake Võrtsjärv, Estonia. Hydrobiologia 414: 59–68.

Hinton, M. J., S. L. Schiff & M. C. English, 1997. The significance of storms for the concentration and export of dissolved organic carbon from two Precambrian Shield catchments. Biogeochemistry 36: 67–88.

Hinton, M. J., S. L. Schiff & M. C. English, 1998. Sources and flowpaths of dissolved organic carbon during storms in two forested watersheds of the Precambrian Shield. Biogeochemistry 41: 175–197.

Inamdar, S. P., S. F. Christopher & M. J. Mitchell, 2004. Export mechanisms for dissolved organic carbon and nitrate during summer storm events in a glaciated forested catchment in New York, USA. Hydrological Processes 18: 2651–2661.

Jansson, M., A.-K. Bergstrom, P. Blomqvist & S. Drakare, 2000. Allochthonous organic carbon and phytoplankton/bacterioplankton production relationships in lakes. Ecology 81: 3250–3255.

Järvet, A., 2004. Pollution load to Lake Võrtsjärv. In Haberman, J., E. Pihu & A. Raukas (eds), Lake Võrtsjärv. Estonian Encyclopaedia publishers, Tallinn: 151–171.

Laudon, H., S. Köhler & I. Buffam, 2004. Seasonal TOC export from seven boreal catchments in northern Sweden. Aquatic Sciences 66(2): 223–230.

Moore, T. R., 1989. Dynamics of dissolved organic carbon in forested and disturbed catchments, Westland, New Zealand. 1. Maimai. Water Resources Research 25: 1321–1330.

Morris, D. P., H. Zagarese, C. E. Williamson, E. G. Balseiro, B. R. Hargreaves, B. Modenutti, R. Moeller & C. Queimalinos, 1995. The attenuation of solar UV radiation in lakes and the role of dissolved organic carbon. Limnology and Oceanography 40: 1381–1391.

Nathan, R. J. & T. A. McMahon, 1990. Evaluation of automated techniques for base flow and recession analyses. Water Resources Research 26: 1465–1473.

Nieminen, M., 2004. Export of dissolved organic carbon, nitrogen and phosphorus following clear-cutting of three Norway spruce forests growing on drained peatland in southern Finland. Silva Fennica 38: 123–132.

Nõges, P., L. Tuvikene, A. Järvet & T. Nõges, 1998a. The budgets of nitrogen and phosphorus in shallow eutrophic Lake Võrtsjärv (Estonia). Hydrobiologia 363: 219–227.

Nõges, T. & P. Nõges, 1998. Primary production of Lake Võrtsjärv. Limnologica 28: 29–40.

Nõges, T., P. Nõges, J. Haberman, V. Kisand, K. Kangur, A. Kangur & A. Järvalt, 1998b. Food web structure in shallow eutrophic Lake Võrtsjärv (Estonia). Limnologica 28: 115–128.

Nõges, T. & P. Nõges, 1999. The effect of extreme water level decrease on hydrochemistry and phytoplankton in a shallow eutrophic lake. Hydrobiologia 408/409: 277–283.

Reinart, A. & P. Nõges, 2004. Light conditions in Lake Võrtsjärv. In Haberman, J., E. Pihu & A. Raukas (eds), Lake Võrtsjärv. Estonian Encyclopaedia publishers, Tallinn: 141–149.

Runkel, R. L., C. G. Crawford & T. A. Cohn, 2004. Load Estimator (LOADEST): A FORTRAN Program for Estimating Constituent Loads in Streams and Rivers: U.S. Geological Survey Techniques and Methods Book 4, Chapter A5, 69 pp.

Schilling, K. & Y.-K. Zhang, 2004. Baseflow contribution to nitrate–nitrogen export from a large, agricultural watershed, USA. Journal of Hydrology 295: 303–316.

Seitzinger, S., J. A. Harrison, J. K. Böhlke, A. F. Bouwman, R. Lowrance, B. Peterson, C. Tobias & G. Van Drecht, 2006. Denitrification across landscapes and waterscapes: a synthesis. Ecological Applications 16: 2064–2090.

Sloto, R. A. & M. Y. Crouse, 1996. HYSEP: a computer program for streamflow hydrograph separation and analysis. Water-Resources Investigations. Report 96–4040, US Geological Survey, Reston, VA.

Watts, C. D., P. S. Naden, J. Machell & J. Banks, 2001. Long term variation in water colour from Yorkshire catchments. The Science of the Total Environment 278: 57–72.

Worrall, F., & T. Burt, 2004. Time series analysis of long-term river dissolved organic carbon records. Hydrological Processes 18: 893–912.

Hydrobiologia (2008) 599:221–226
DOI 10.1007/s10750-007-9214-y

ELLS 2007

Implications of flexibility in European Community environmental law: exemptions from environmental objectives in the Water Framework Directive

Elisabeth Grönlund · Tapio Määttä

Abstract The present article is a brief review of the legal characteristics of water quality objectives and legally permissible exemptions from these objectives, as enacted in the EC Water Framework Directive. Six different types of exemptions have been identified in total. These vary markedly in the legal premises of their feasibility, ranging from the set deadlines for the environmental objectives to application of less stringent environmental objectives for certain water bodies.

Keywords Water Framework Directive · Environmental objectives · Environmental law

Introduction

The European Community environmental policy has two principal objectives: to enhance environmental protection in the Community region and on the global scale, and to contribute to the integration of environmental policies within and among the present 27 member states of the Community. The policy aims at a high level of environmental protection and attempts to take into account the diversity of situations in the member states. However, from the environmental standpoint, the European environmental law and the institutions involved have a dual task: to find an acceptable balance between the requirements of the internal market economy and environmental protection in the national and community legal contexts. The environmental policy of the European Community is implemented in the member states, predominantly through directives. At the moment, one of the most important environmental directives is the Water Framework Directive (WFD; Directive, 2000), which will shape the water policies and legislative measures in the EC and member states for several decades to come.

The legal position of the WFD, as well as other directives, is based on article 249 para. 3 of the Treaty of the European Community (Treaty of Nice, 2001). The article states that "a directive shall be binding as to the result to be achieved, upon each Member State to which it is addressed, but shall leave to the national authorities the choice of form and methods". In practice this means that the member states have a certain degree of flexibility and discretion in implementing the objectives of the directive, e.g. in amending existing national

Guest editors: T. Nõges, R. Eckmann, K. Kangur, P. Nõges, A. Reinart, G. Roll, H. Simola & M. Viljanen.
European Large Lakes—Ecosystem changes and their ecological and socioeconomic impacts.

E. Grönlund (✉)
Ecological Research Institute, University of Joensuu, P.O. Box 111, 80101 Joensuu, Finland
e-mail: Elisabeth.Gronlund@joensuu.fi

T. Määttä
Department of Law, University of Joensuu, P.O. Box 111, 80101 Joensuu, Finland
e-mail: Tapio.Maatta@joensuu.fi

legislation or in enacting new legislation. However, as is characteristic of environmental regulations in general, the WFD comprises formal procedural, organisational and also relatively detailed technical provisions through which the environmental objectives are to be achieved.

The WFD was enacted by the co-decision and majority-voting procedures of the Council and the Parliament, and its legal base is the article 175 para. 1 of the Treaty. This means that the objective of the WFD is in the field of water protection and not the field of water economy. The chosen legal base enables article 176 of the Treaty to be applied, which allows the member states to apply more stringent environmental objectives in their own legislation than are actually required by the directive itself.

The Water Framework Directive was implemented in the Finnish legal system by the enactment of the Act on the Organisation of River Basin Management in 2004 (1229/2004), a related Decree on River Basin Districts (1303/2004) and simultaneous amendments to the Environmental Protection (86/2000) and the Water (264/1961) Acts. Two recent Decrees have been enacted on the contents and drafting of the River Basin Management Plans (1040/2006) and on hazardous substances and pollutants in fresh surface waters (1022/2006).

The principal environmental objective of the WFD for surface waters is that the ecological and chemical status of all water bodies will be good or excellent by 2015, and that in none of them does the status deteriorate below its present condition (WFD art. 1, 4.1a). The environmental objectives set out in the WFD are legally binding on the member states. However, the directive includes several exemptions pertaining to specified conditions and requirements (art. 4.1c, 4.3, 4.4–7) that allow the member states to evade the achievement of good status. This might be considered a reflection of the general flexibility common to the present EC environmental law. In addition, the permitted exemptions enable the member states to alleviate the eventual technical, economical or developmental burdens of complying with the water quality requirements of the WFD (e.g. Grimeaud, 2004).

The implementation of the environmental objectives and corresponding administrative measures set out in the WFD are realised in the form of River Basin Management Plans (art. 13), which are to be reported to the Commission in 6-year cycles (art. 15). Each state makes autonomous decisions about the legal procedures in relation to national decision making and official approval of the river basin management plans. In Finland, for instance, the plans are prepared and drafted by the environmental authorities and ultimately approved by the government.

The aim of the present study is to identify the particular circumstances in which the specified exemptions are enforceable, to discuss some issues concerning the characteristics of the provisions and to add some remarks comparing the WFD with the corresponding Finnish legislation as an example of WFD implementation in a member state.

Provisions of exemptions and their prerequisites in the WFD

We have identified a total of six different situations (1–6 below) in which exemption from the environmental objectives is feasible, thus giving a member state liberty to breach the directive. The exemptions differ in respect of their legal bases and prerequisites of enforcement and their legal effects. The common denominator of all the exemptions is that they must be included and reported in the water management plans, with annotations of the grounds of choice in each individual case according to the legal requirements specified in the respective provision.

The allowed circumstances or corresponding situations for exemptions may be systematised as follows:

1. Artificial and heavily modified surface waters (art. 4.3, 4a para. iii). Water bodies of this type constitute a category separate from other surface waters, and they are specifically identified and designated in the water management plans. The water quality requirement for this category is formulated as "good ecological potential" and "good chemical status". A water body may be included in this category only if the changes to the hydro-morphological characteristics needed to achieve good ecological status would have significant negative effects on e.g. the wider environment, navigation, recreation, water storage, power generation, flood protection or "other

equally important sustainable human development activities". In other words, designation of artificial and heavily modified waters implies that a lower ecological quality is accepted to promote a certain significant human purpose, or if the pursuit of good ecological status would be contrary to acceptable environmental, technical or economical causes. The artificial or heavily modified water bodies and the grounds for their designation have to be explicated in the river basin management plans, and to be reviewed every sixth year. All the other exemptions mentioned below (2–6) are also applicable to designated artificial and heavily modified surface water bodies.

2. Waters in areas protected under specific Community law, the protection level of which shall be at the required level of the appropriate regulation by 2015 (art. 4.1c). According to articles 6, 7 and Annex IV, these include: (a) areas designated for the abstraction of drinking water and for the protection of economically significant aquatic species; (b) areas regulated by the Bathing Water Directive (76/160/EEC), the nitrate-vulnerable zones of the Nitrate Directive (91/676/EEC), and areas sensitive to eutrophication by nitrogen and phosphorus under Urban Waste Water Directive (91/271/EEC); (c) designated protection areas under the Habitat and Wild Bird Directives (92/43/EEC, 79/409/EEC, respectively). For example, if a water body is designated as a special protection area (SPA), its conservation value may depend directly on the maintenance of a certain water quality. Thus, its water quality status is defined by the Bird Directive obligations, irrespective of the water quality status demanded for the given lake basin according to the WFD. One example of this type is the largest lake in the British Isles, Lough Neagh (Northern Ireland), a shallow hypertrophic SPA-lake hosting a rich avian fauna, the management of which is quite complicated because of somewhat conflicting anthropogenic pressures on the lake water quality by, e.g. agriculture, recreation and its importance as a drinking water source for Belfast.
3. Postponing the achievement of deadlines for environmental objectives by a phased achievement (art. 4.4). The prerequisites for prolonging the period are that it would not be feasible to enhance the ecological status of a water body for technical reasons and/or because of disproportionate economic costs. Member states must ensure that the delay does not lead to further deterioration of the water body below its current quality status. The grounds for the delay have to be explained and an anticipated timetable for implementing the required status must be set out in the river basin management plan. An extension is permissible for two subsequent updates (12 years) of the river basin management plan. An exemption may also be made if the natural conditions are such that the objectives are unattainable within the designated 12-year period. This latter exemption appears rather vague, as nothing definite is provided in the article for further procedures or time limitations of designation for water bodies of this kind. This exemption might be especially applicable to lake basins in which a long-term period of recovery may be anticipated, for instance lakes with a significant internal input of phosphorus from anoxic sediments, as shown for Lake Varese in Northern Italy by Premazzi et al. (2003).
4. Application of less stringent environmental objectives is permissible if the water body is affected by human activities to such a degree, or its natural condition is such, that the objectives cannot be achieved without disproportionate economic costs, or the water quality requirements are technically unfeasible (art. 4.5). The human activities referred to are ones that cause strain on the particular water body, e.g. installations leading to point source and diffuse pollution (Art. 5.1, App. II, 1.4.1–2). A definite prerequisite for this exemption is that the environmental and socio-economic needs served by the particular human activities cannot be fulfilled by other means. Even though these other means would present a significantly better environmental option, they would also result in disproportionate costs if applied to meet the environmental and socio-economic needs. However, it is not permissible to lower the quality status of the affected water body. Also, the member state has to ensure the highest possible ecological and chemical status of the water body, given the impacts that could not reasonably have been

avoided in view of the nature of the human activity or pollution. This provision appears quite unclear or even obscure, leaving plenty of room for interpretation and the possibility of application to almost any kind of water body.

5. Temporary exemptions concerning the status of water bodies are allowed in circumstances of exceptional and reasonably unforeseeable natural causes or *force majeure* situations, e.g. extreme floods, long-lasting droughts or accidents (art. 4.6). By this provision, in contrast to the preceding ones, a deterioration of water quality is allowed. The conditions to apply the provision are *inter alia* that all practicable steps have been taken to prevent any further decline of the status of the affected water body, or other unaffected water bodies. Also, the conditions of the exceptional or reasonably unforeseen situations have to be accounted for in the management plans and the effects of the circumstances have to be reviewed annually.
6. New modifications to the physical characteristics of a water body allow for an exemption from the objectives of good ecological status or good potential as well as from actions to prevent deterioration in its status. However, deterioration of the water's chemical status is not permissible in this case. Also, it is permissible to decrease the water quality status from high to good if the reason is a new sustainable human development. The provision requires that all practical steps have been taken to mitigate the adverse effects of the modifications or new development activities. In order to apply this exemption, the member state has to demonstrate that the actions are justified by "overriding public interest". It must also be shown that the improvement in regard to public health, safety or sustainable development would outweigh the societal and environmental benefits that would otherwise be realised by the achievement of the environmental objectives. It is also required that the benefits of the actions cannot be achieved by other means because of technical unfeasibility or disproportionate costs. In addition, all practicable steps have to be taken to reduce negative effects on the water body status and to explain the reasons and environmental consequences for such new modifications in the river management plans. The provision appears in particular to address large future projects that are considered very important for the society. However, projects subject to this provision might also face more stringent permission authorisation procedures at national level.

In addition to the above exemptions, transboundary lake basins form a certain exception in their own category. Member state(s) shall endeavour to coordinate with the relevant non-member state with the aim of achieving the WFD water quality objectives, and to ensure proper application of the WFD in their own territory (art. 3 para. 5). There are a number of transboundary lakes in Europe: for example, Lake Constance (shared by Germany and Switzerland), Lake Peipsi (Estonia and Russia), Lake Geneva (France and Switzerland), Lake Mikri Prespa (Greece, Albania and Macedonia) and Lake Pyhäjärvi (Finland and Russia). These transboundary lakes are managed through bi- or multilateral cooperation between the member and the non-member states. It should also be noted that use of these lakes is legally covered by the Convention on the Protection and Use of Transboundary Watercourses and International Lakes (Convention, 1992), which was adopted by the United Nations Economic Commission for Europe (UNECE) and is presently ratified by all the EU member states along with e.g. Russia, Albania and Switzerland.

Discussion

Several recent authors in environmental law publications have expressed somewhat apprehensive and even sceptical views about the realisation of the environmental objectives of the WFD (see e.g. Krämer, 2000; Howarth & McGillivray, 2001; Fairley et al., 2002; Grimeaud, 2004; Howarth 2006). The discussion has ranged from the difficulty or practical unattainability of a legal definition of ecological status for aquatic communities because of the natural variability and scientifically recognised instability of ecosystems (Howarth, 2006), to more general concern that fixing a legal objective of good water quality might lead to enhanced differentiation and little integration of environmental law in the Community, as the individual member states are left to define good status (Krämer, 2000). Also, the risks and uncertainties connected with climate change must be considered seriously in respect of applying the WFD, along with allowed exemptions, in the member states. As discussed by Wilby et al. (2006), the WFD offers a means for connecting policies for adaptation to climate change on the national and also on the Community scale.

Another assessment is that at present member states tend to favour the application of derogations provided by many environmental directives, which may suggest a political preference for measures that allow less strict environmental standards. In addition, the wide spectrum of exemptions allowed in the WFD might actually lead to quite an extensive number of water bodies in which the good status requirement is not fully realised (Howarth & McGillivray, 2001). The latter possibility will naturally emerge only if many member states apply the provisions for exemptions liberally in the future.

Seppälä (2004) has commented that the use of the concept of sustainable development as legal grounds for derogation from the environmental objectives in WFD and the corresponding Finnish act appear somewhat problematic and contradictory. Although sustainable development is an extremely flexible concept per se, the provisions of the WFD seem to reflect a novel tendency that justifies the social and economic dimension of sustainable development to be utilised as grounds for such human activities, which are actually harmful to the environment (Seppälä, 2004).

Most of the above-mentioned provisions for exemption contain a fair number of flexible or open expressions, e.g. "significant", "reasonable" "disproportionate costs", "overriding public interest" and "benefit to environment and society", the exact meaning of which seem to be unclear, leaving a wealth of possibilities for interpretation. In the future, such expressions will require due and diligent scrutiny of each individual case in the decision-making processes by member states, which would greatly benefit from further and detailed guidance by the European Community for interpreting the provisions for exemptions (Fairley et al., 2002; Grimeaud, 2004; on dimensions of flexibility in environmental law in general, see e.g. Scott, 2000; Määttä, 2005).

In the Finnish Water Resources Management Act, the provisions for permissible exemptive situations have been integrated into the overall planning and decision-making procedure as the exemptions are realised in the national river basin management plans, which are approved by the government. A judicially interesting issue in the future will be the integration of the possible exemption situations and the legal system for environmental permit authorisation under the Finnish Water Act and the Environmental Protection Act. Ultimately, the national courts of the member states and/or the European Court of Justice will be the forums in which the legality of the implementation of the Water Framework Directive and application of its specific provisions in the member states are assessed.

References

Convention, 1992. Convention on the Protection and Use of Transboundary Watercourses and International Lakes, Helsinki 17 March 1992. International Legal Material, 330 pp.

Directive, 2000. Directive 2000/60/EC of the European Parliament and of the Council of 23 October 2000 establishing a framework for Community action in the field of water policy. Official Journal of the European Communities L 327/1: 1–73.

Fairley, R., C. Smith, H. van Geen, M. Hoedl-Adick, G. van Thyne & D. De Roover, 2002. Riding the new wave on European water law: How member states are tackling the Water Framework Directive. European Environmental Law Review August/September 2002: 232–239.

Grimeaud, D., 2004. The EC Water Framework Directive—an instrument for integrating water policy. Review of European Community and International Environmental Law 1: 27–39.

Howarth, W., 2006. The progression towards ecological quality standards. Journal of Environmental Law 1: 3–35.

Howarth, W. & D. Gillivray, 2001. Water Pollution and Water Quality Law. Shaw and Sons. Crayford.

Krämer, L., 2000. Differentiation in EU environmental policy. European Environmental Law Review May 2000: 133–140.

Määttä, T., 2005. Joustavien normien kiinteytys-, täsmentämis- ja konkretisointimekanismit ympäristöoikeudessa. In Lohi, T. (ed.), Kaavoitus, rakentaminen, varallisuus. Juhlajulkaisu Vesa Majamaa. Edita, Helsinki, 265–299, (In Finnish).

Premazzi, G., A. Dalmiglio, A. C. Cardoso & G. Chiaudani, 2003. Lake management in Italy: the implications of the Water Framework Directive. Lakes & Reservoires: Research and Management 8: 41–59.

Scott, J., 2000. Flexibility in the implementation of EC environmental law. Oxford University Press, Yearbook of European Environmental Law 1: 37–60.

Seppälä, M., 2004. Vesienhoitosuunnitelmien huomioon ottaminen ympäristönsuojelulain ja vesilain mukaisessa lupaharkinnassa. Ympäristöjuridiikka 3–4: 91–103, (In Finnish).

Treaty of Nice, 2001. Treaty of Nice amending the Treaty on European Union, the Treaties establishing European Communities and certain related Acts. Official Journal of the European Communities C 80/1: 1–87.

Wilby, R. L., H. G. Orr, M. Hedger, D. Forrow & M. Blackmore, 2006. Risks posed by climate change to the delivery of Water Framework Directive objectives in the UK. Environment International 23: 1043–1055.

Hydrobiologia (2008) 599:227–237
DOI 10.1007/s10750-007-9207-x

ELLS 2007

Analysis of long-term ecological status of Lake Balaton based on the ALMOBAL phytoplankton database

Éva Hajnal · Judit Padisák

Abstract Lake Balaton (Hungary), one of the largest lakes in Europe, has undergone eutrophication and restoration during the last two decades. The first quantitative phytoplankton records date back to the 1930s, and since that time thousands of data have been published or accumulated in counting protocols or computer sheets. These data provide material for both scientific analyses (e.g. effects of global change) and applications (e.g. estimation of reference state for the Water Framework Directive). The ALMOBAL phytoplankton database was developed to provide computing support for these applications. It stores data in standardized forms, handles synonyms and allows analyses to be conducted on the basis of floral records, numbers or biomass. The analysis includes records of about 3000 phytoplankton samples taken during the past 60 years from two representative basins in Lake Balaton. This article represents the first attempt at historical reconstruction of the ecological status and compares it with changes in trophic state and current water quality. The results indicate that the phytoplankton biomass and community structure found in the early 1960s could be regarded as reference conditions. Statistical analyses support the view that late summer phytoplankton assemblages are the most sensitive indicators of trophic change, and clearly show the eutrophication of the lake that occurred from the mid-1970s to the mid-1990s and the restoration during the last decade. An additional advantage is that, since quality estimation is based on relative biomass, the method can be used to reconstruct water quality in cases when counting protocols are available, but some basic data for quantitative estimates are missing.

Keywords WFD · Phytoplankton · Database · Functional assemblages · Monitoring · Long-term changes · Eutrophication · Restoration · *Q*-index

Guest editors: T. Nõges, R. Eckmann, K. Kangur, P. Nõges, A. Reinart, G. Roll, H. Simola & M. Viljanen.
European Large Lakes—Ecosystem changes and their ecological and socioeconomic impacts.

É. Hajnal (✉) · J. Padisák
Department of Limnology, University of Pannonia, Egyetem u. 10, 8200 Veszprem, Hungary
e-mail: hajnal@limnologia.hu

J. Padisák
e-mail: padisak@almos.uni-pannon.hu

Introduction

Human activity impacts lake water quality in many different ways, and on various temporal and spatial scales. Enrichment with plant nutrients originating from the catchment leads to eutrophication. Conversely, reduction of the load results sooner or later in a decrease in the trophic state of the lake. Since it is the primary acceptor of the load, the phytoplankton reflects the changes very rapidly both quantitatively and qualitatively. In addition to the increase in biomass, the community structure may also change, sometimes dramatically. Moreover, it has been increasingly recognized that global climate change can also be traced through changes in the structural and functional parameters of phytoplankton; however, few analyses have been conducted worldwide, primarily because coherent long-term phytoplankton datasets are lacking (Padisák, 1998).

Some long-term studies have been published for lakes with long histories of phytoplankton research, or at least measurements such as Secchi transparency that can be used as proxies for phytoplankton changes (Lund, 1979; Makarewicz & Baybutt, 1981; Findlay & Kasian, 1987; Kümmerlin, 1991; Neale et al., 1991; Reynolds & Bellinger, 1992; Willén, 1992; Talling, 1993; Mineeva & Litvinov, 1998; Padisák, 1998; Padisák & Reynolds, 1998; Lepistö et al., 1999). However, cross-lake comparisons are complicated by inconsistency in the variables used for long-term assessments. For example, the following variables were used in the above-cited studies: average numbers of cells of dominant species, contribution of dominant species to total biomass, epilimnetic biomass of dominant phytoplankton groups, minimum and maximum values of the spring and late summer biomass, phytoplankton biomass and chlorophyll-a.

As agreed by a group of phytoplankton ecologists (Kristiansen, 2003), the relevant basis for ecological studies of phytoplankton is the volumetrically estimated biomass. This approach led to a much deeper understanding of spatial and temporal patterns than any other kind of analysis (Padisák et al., 1993; Descy et al., 1994; Alvarez-Cobelas et al., 1998; Reynolds et al., 2000; Naselli-Flores et al., 2003; Naselli-Flores et al., 2007). However, the need for species-specific data in long-term analyses constitutes an additional difficulty. The assemblage concept for analysing community patterns appeared quite recently (Reynolds et al., 2002) and an increasing number of analyses have supported its usefulness (Crosetti & Bicudo, 2005; da Silva et al., 2005; Findlay et al., 2005; Lopes et al., 2005; Naselli-Flores & Barone, 2005; Townsend, 2005; Devercelli, 2006, Morabito et al., 2007).

On the basis of the assemblage concept and to satisfy the needs of the EC Water Framework Directive (hereafter WFD; EC Parliament and Council, 2000), a phytoplankton assemblage index (Q) was developed that includes the relative share of each functional group in the total biomass (p_i, N: number of species) multiplied by a factor (F_i) established for the ith functional group in the lake (Padisák et al., 2006):

$$Q = \sum_{i=1}^{N} p_i F_i$$

The objective of this article is to use the phytoplankton biomass and assemblage structure expressed by the Q-index to reconstruct the history of water quality in Lake Balaton and to quantify the changes in water quality during the eutrophication and restoration phases. In this study, we were interested primarily in the applicability of this water quality index. Our article investigates the seasonal changes in water quality and outlines the major long-term trends in two characteristically different basins of Lake Balaton.

Site description, materials and methods

Lake Balaton is the largest shallow lake of Central Europe with a surface area of 593 km^2, average depth of 3.14 m, catchment area of 5182 km^2 and theoretical retention time of 3–8 years. The lake consists of four basins. The River Zala enters the lake at the western basin and provides 45% of the discharge and most of the nutrient load. There are 51 minor inflows, 20 of which are permanent. The water level of the lake is regulated by the single artificial outflow, the Sió Canal (eastern basin), which came into operation at the end of the 19th century. As a result of the increased phosphorus load, the originally mesotrophic lake underwent rapid eutrophication from the 1960s onwards. In the early 1980s, a large-scale eutrophication management programme was initiated to restore the water quality. A large reservoir to protect water quality was constructed at the throat of the River Zala (Pomogyi, 1993), a sewage diversion canal was built around the eastern basins of the lake and sewage treatment plants with P-removal steps were installed. This management resulted in a more than 50% reduction of the nutrient load (Istvánovics et al., 2002; Istvánovics et al., 2007).

The earliest quantitative phytoplankton data were limited to cell counts, but these were soon expanded to include biovolume estimation as well. The quantitative data used in this article were obtained from two areas in Lake Balaton with sampling frequencies varying from daily to monthly, including protocols from 27 of the last 40 years in the western basin and from 30 of the last 60 years in the eastern basin. The large quantity of data enabled us to analyse the features of the phytoplankton during eutrophication and restoration, as well as to explore methods for assessing water quality as required by the WFD.

The first quantitative records of phytoplankton were obtained in the 1930s (Entz et al., 1937) and published as extended printed sheets until the end of the 1970s (Tamás, 1955, 1965, 1967, 1969, 1972, 1974, 1975; Sebestyén, 1960; Herodek & Tamás, 1973, 1975; G.-Tóth & Padisák 1978). The earliest samples were pre-sedimented and counted in a newly developed chamber (similar to what is known as a Kolkwitz-chamber). From the 1940s onwards, Utermöhl's (1936, 1958) method and inverted microscopy were applied. Some of the phytoplankton data used in this article were obtained from the archives of the Middle Transdanubian Water Authority (1985–1990, 1998–2003) containing biomass datasets. Our own archives were used for the years between 1987 and 2005.

A database, ALMOBAL 1.0, was developed for data analysis. It enables all types of phytoplankton datasets to be stored and queried, storing basic information about species and their taxonomic classifications, functional group codes, synonyms, mean cell volumes etc. To compare the datasets, the biomass of each species in every sample was estimated on the basis of biovolume data. Some of these volume data have been published (Tamás, 1955), while others were estimated from the database by extracting mean volumes. The volumes of infrequently occurring species (especially Chlorococcales) were estimated by calculating their volumes from sizes given in the identification manuals (Komárek & Fott, 1983). According to our calculations, mean biovolumes have not changed during past 10 years. However, the biovolumes of some species had quite large standard deviations, which necessitated checking our results. Our estimates were controlled against the total biomasses published in the 1970s (Tamás, 1974; Vörös, 1980) and were found to match approximately.

ALMOBAL 1.0 is a local client-server database created in the Interbase 5.0 server program package. A user interface program was developed for data load and queries using Borland Delphi 5.0 Professional Edition (Hajnal & Padisák, 2006a, b). The ALMOBAL now contains more than 60,000 records derived from about 3000 individual samples, and can import them automatically from Excel sheets. Forms and Structured Query Language (SQL) queries can be applied, and results can be exported into Excel sheets for subsequent calculations and analyses.

For this study, the species-specific biomasses in each sample were pooled into functional groups (Reynolds et al., 2002; Padisák et al., 2003, 2006). Two different regions were analysed, one (Keszthely) located in the middle of the westernmost basin of Lake Balaton, the other (Tihany) in the eastern basin. The western basin received the bulk of the nutrient load during the eutrophication period, and at that time became hypertrophic. The eastern basin did not become hypertrophic, but remained in a eutrophic state except in the years 1982, 1992 and 1994, with heavy *Cylindrospermopsis raciborskii* blooms (Padisák & Reynolds, 1998).

The monthly average biomasses of the functional groups at each location were calculated to reduce the variability and/or temporal and spatial heterogeneity in individual samples. The results were organized in pivot tables. For cluster analysis, the August and September average biomasses of functional groups were calculated; Euclidean distances and the UPGMA algorithm were applied using the SynTax III program package (Podani, 1988).

Results

Seasonal changes of community structure

In the present analysis, the data from the pre- and post-management periods were separated.

The monthly average Q indices for Lake Balaton show that the water quality in the eastern basin was consistently better by 0.5–1 units (Fig. 1). The differences between the two basins were lowest in the cold months. Detailed analyses (not shown here) indicated that structural differences existed, and the winter phytoplankton was almost dominated by small flagellated forms, as is common in many lakes (Salmaso & Padisák, 2007). There are relatively high differences between the two locations in April. June represents the clear water phase after the spring

Fig. 1 Monthly averages of Q index ($\pm$ SD) over the past 40 (western basin Keszthely: (**a**) and 60 (eastern basin Tihany: (**b**) years in Lake Balaton

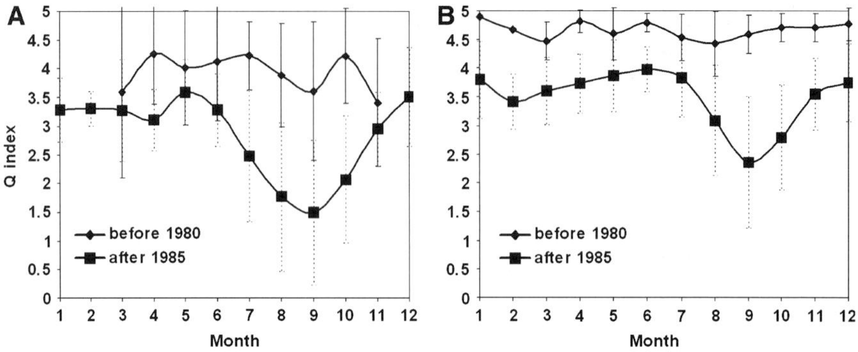

diatom maxima, usually dominated by small flagellates and partly by green algae (Padisák & G.-Tóth, 1991).

The lowest Q index values and the largest standard deviations were observed in late summer at both locations. The water quality minima were reached in August-September in the western basin and September-October in the eastern basin. In other words, the periods of critical water quality were delayed by about 2 weeks in the eastern compared to the western basin.

The monthly average contribution of diatom species with different sizes and features belonging to the **A**, **C** and **D** assemblages (Fig. 2, Table 1) exhibited a winter and a more defined spring maximum in both the western (Fig. 2a) and eastern (Fig. 2b) basins. Functional group **A** (comprising centric diatoms in general) is typical of the less eutrophic eastern basin (30%). In the more eutrophic western basin, species of group **D** were more dominant.

The $\mathbf{L_0}$ functional group mainly comprises "original" species (*Ceratium, Snowella*, etc.) in Lake Balaton. Species belonging to this group are present in every month but their dominance increases after the spring diatom phase. The $\mathbf{L_0}$ maximum (average 30%) was observed in August in the eastern basin (Fig. 2c) and in July (<20%) in the western basin (Fig. 2d).

In the summer communities, the contribution of Cyanobacteria (**S1**, $\mathbf{S_N}$) to the biomass is important. The biomass maximum of the **S1** (Fig. 2) group in the western basin (Fig. 2c) typically occurred in October and amounted to ca. 15%, whereas in the eastern basin its share was only 5%. The biomass contribution of the $\mathbf{S_N}$ group in the western basin peaked in August or September at about 30%, while the average biomass contribution of this group in the eastern basin was ~20% and peaked in September.

The monthly average proportions of the functional groups including different flagellates (Fig. 2e, f) showed the following features. In the winter months, especially in the western basin, **Y** (comprising potentially mixotrophic flagellates) was a very characteristic functional group; its average content sometimes exceeded 20%. **X1** and **X2** groups were more dominant in the western (**X1** ~10%, **X2** ~30%) than the eastern (**X1** ~5%, **X2** ~15%) basin.

Long-term trends in biomass and community structure

These analyses concentrated on the data from August to September. The long-term dynamics of total biomass in August (Fig. 3) and September (not shown here) were similar in the two basins.

Periods with high biomass had low Q indices. As Fig. 3 clearly shows, there was only a slight increase in biomass at the beginning of eutrophication (1966), but serious decrease in the Q index that fell into the weak-medium range. From the 2000s onwards, the total biomass was not high, but the low Q indices indicated rather high percentages of alien species of Cyanobacteria (*Cylindrospermopsis raciborskii*, many but not each *Anabaena* and *Aphanizomenon* spp.) in the summer communities (Padisák et al., 2004, 2005).

Examining the dendrogram that reflects the changes in abundance of the main functional groups and the Q index in August/September in the western basin, four period groups could be distinguished at a 0.3 dissimilarity level (Fig. 4). The Q index of the first group varied between 0.8 and 2.9 (poor-medium), and its dominant assemblages were **H1** and $\mathbf{L_0}$. The second group had Q index values between 0.6 and 1.8 (poor-weak). In this group, **S1**

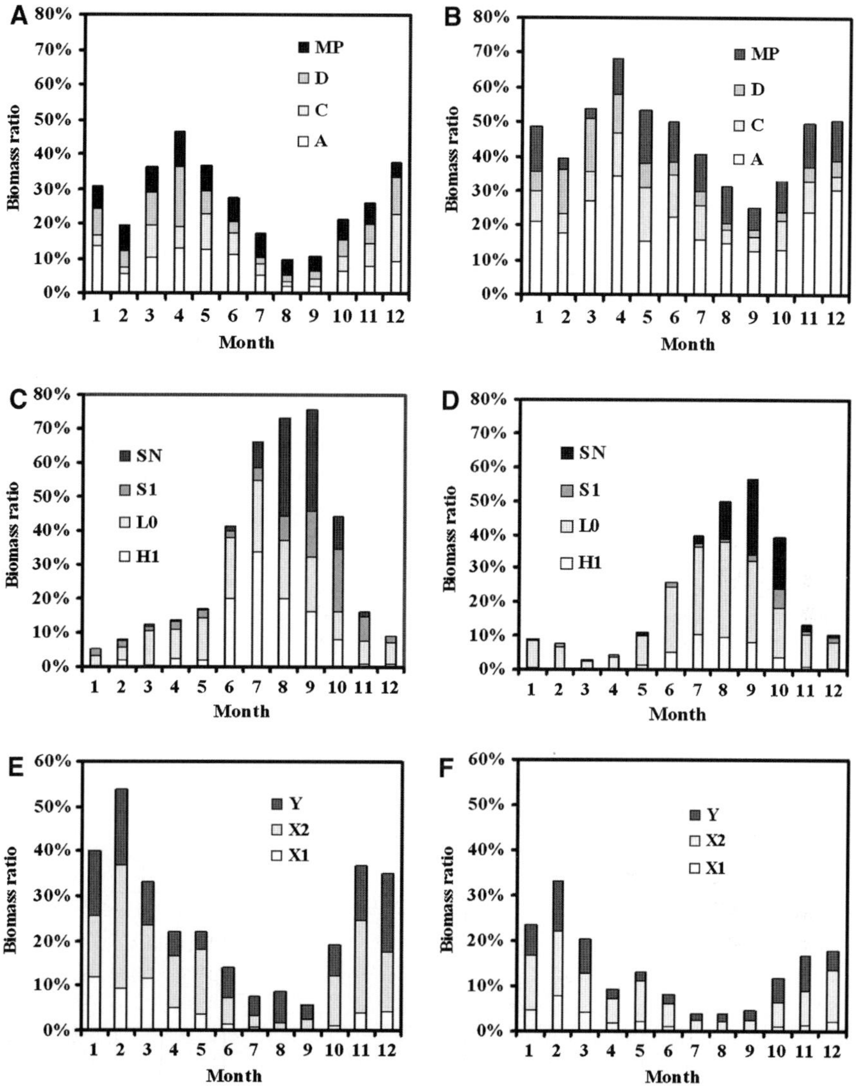

Fig. 2 Average contributions of the main functional assemblages to total biomass in the western (Keszthely: **a**, **c**, **e**) and eastern (Tihany: **b**, **d**, **f**) basins of Lake Balaton. See Table 1 and also the nomenclature of assemblages in Reynolds et al. (2002) and Padisák et al. (2006)

species (representing more than 10% of the total biomass) occurred among the $\mathbf{S_N}$ assemblage. The water quality associated with the third group was poor, with Q values between 0.3 and 0.8. Its characteristic functional group was $\mathbf{S_N}$ (*Cylindrospermopsis raciborskii*), which contributed more than 50% to the total biomass. The fourth group in the dendrogram indicates good to excellent water quality (Q between 3 and 5) with $\mathbf{L_0}$, **A** and **H1** species domination (>40%). The years 1980 and 1983 did not match any group.

In the eastern basin, the years clustered into three groups (not shown here). However, grouping by water quality index was not clear in this basin.

Long-term and seasonal variation of ecological status

The interannual and seasonal variations in Q index values in both basins are shown in Fig. 5. The light-coloured areas represent the lowest water quality, and the dark areas show better water quality. In the western basin, water quality was excellent during the 1960s throughout the whole year. The Q index first decayed in 1966 with a patch in August. From 1975 onwards, the deterioration of water quality was apparent throughout most seasons. From the 1980s onwards, a characteristic late summer drop in water quality developed. This period of poor water quality

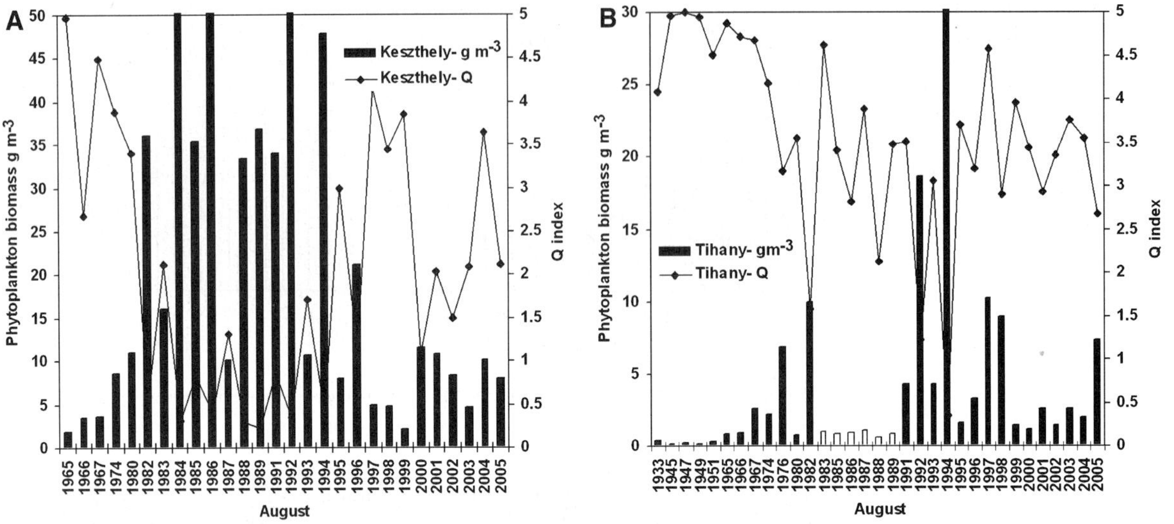

Fig. 3 Long-term dynamics of the average total biomass, and water quality index (Q) in August, in the western (Keszthely: **a**) and eastern (Tihany: **b**) basins of Lake Balaton. Cell density data in the eastern basin during the period 1980–1990 were unrealistically low in the available protocols: the absolute biomasses calculated from these data are indicated as empty bars in (**b**)

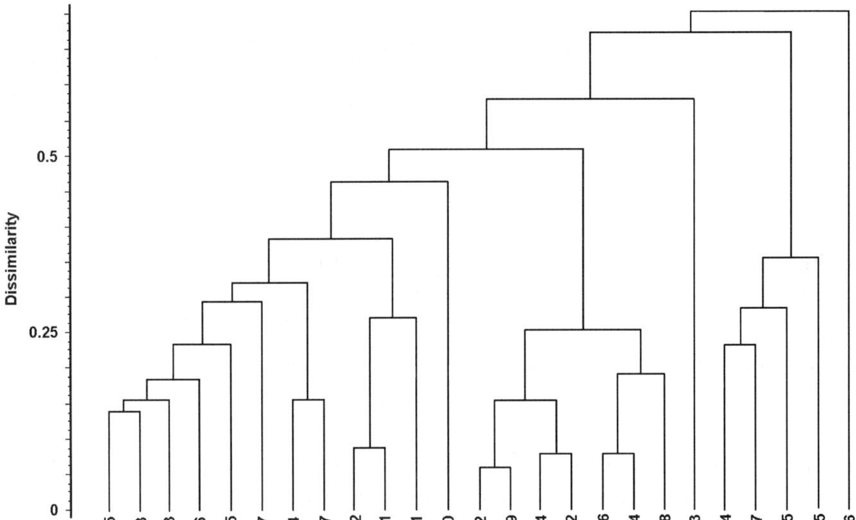

Fig. 4 Hierarchical cluster analysis of dissimilarity among the assemblages of late summer phytoplankton in the western basin of Lake Balaton

was longer during the 1990s. Some years later, the water quality management programme was initiated, and there was a noticeable improvement in water quality. After that, the summer periods of poor water quality became shorter and after 2000 more fragmented, when excellent water quality patches occurred in the springs and autumns. We observed a similar pattern in the eastern basin (Fig. 5b).

Discussion

The differences between Q indices of the eastern and western basins of Lake Balaton reflect that the western basin receives much higher loadings from the tributary (Fig. 1). Structural differences exist in winter but remain hidden because the functional assemblages of both locations have similar factor

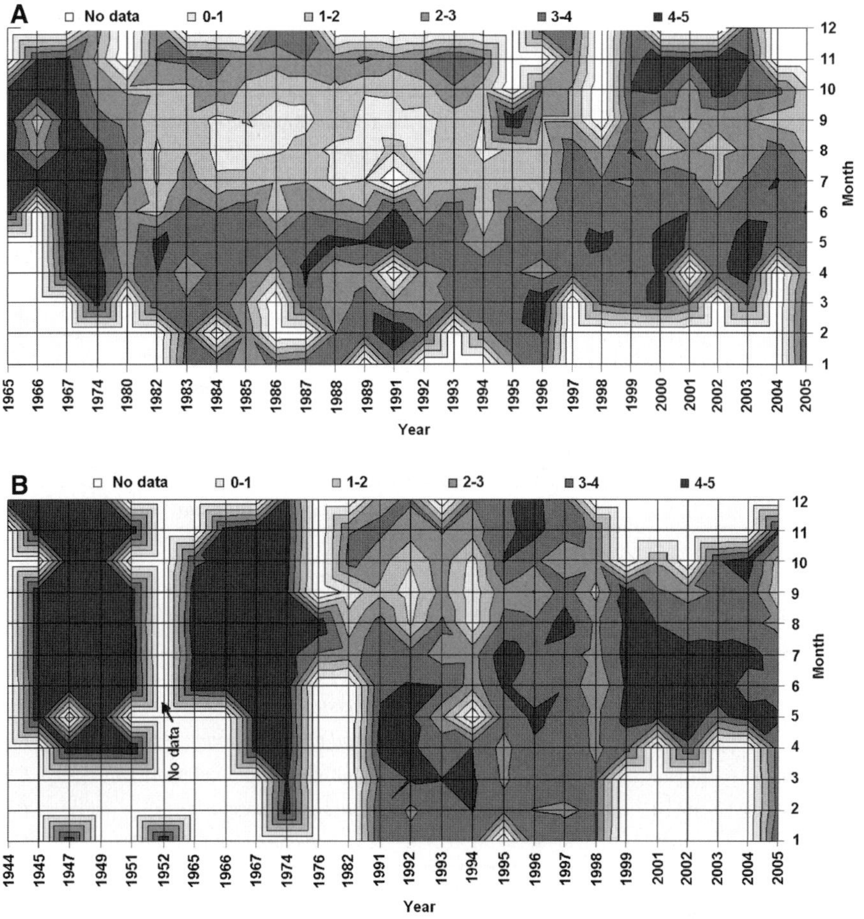

Fig. 5 Contour-map of water quality index (*Q*) by years and months in the western (Keszthely: **a**) and eastern (Tihany: **b**) basins of Lake Balaton. Light areas represent poor water quality whereas dark areas represent good water quality

values. In addition, winter sampling is sporadic owing to the often unstable ice, reducing the reliability of the winter dataset.

The differences between the two locations in April are caused by the development of the spring diatom peak (*Cyclotella* spp. in low trophic state periods and *Synedra* or exceptionally *Stephanodiscus* in higher ones). However, the timing of the short spring diatom maximum depends strongly on the timing of ice-melt (Padisák et al., 2006) and therefore, no clear temporal pattern can be discerned in such a rough comparison.

The water quality minimum in the late summer corresponds to the general experience that the symptoms of eutrophication are most severe in late summer, and that this is the most crucial period for observing changes in trophic state (Padisák et al., 2006). In this study, periods with high standard deviations (considerable differences) of *Q* indicated months when the assemblages were most strongly affected by eutrophication/restoration. Here we must note that while a narrow standard deviation around the mean indicates reliable results in most scientific studies, this study is different: large differences between the averages with high and overlapping SDs indicate that the given period was strongly influenced by external driving forces (eutrophication in this case) and therefore have the highest indicative value. A possible reason of the delay of the eastern basin is that the slower temperature rise in the deeper eastern basin may delay the germination of the akinetes of heterocytic blue-green algae.

According to both expert judgement and statistical analyses of the water quality indices (Fig. 1) and indices of dissimilarity between assemblage structures, summer communities are the best indicators of trophic change.

Since the *Q* index is based on the relative biomass contributions of the assemblages, high biomass appears as "poor quality" only if the factor numbers of the corresponding assemblages are low. However,

Table 1 Most abundant genera in the main functional assemblages in the western (Keszthely) and eastern (Tihany) basins of Lake Balaton

Functional group	Most important taxa	
	At Keszthely	At Tihany
A	*Cyclotella comta*	*Cyclotella comta*
B	Centrales spp. ∅ > 5–20 μm	Centrales spp. ∅ > 5–20 μm
C	Centrales spp. ∅5–20 μm	Centrales spp. ∅5–20 μm
D	*Synedra acus*	*Nitzschia sigmoidea*
E	*Dinobryon sociale* small sized chrysoflagellates	*Dinobryon sociale* (unique event), small sized chrysoflagellates
F	*Coelastrum microporum*	*Botryococcus braunii*
$\mathbf{H_1}$	*Aphanizomenon flos-aquae*	*Aphanizomenon flos-aquae*
$\mathbf{L_0}$	*Planktolyngbya limnetica, Ceratium hirundinella, Peridinium* spp.	*Aphanizomenon klebahnii, Snowella lacustris, Ceratium hirundinella*
M		*Microcystis aeruginosa*
MP	*Aulacoseira granulata*	*Aulacoseira granulata*
P	*Navicula* spp.	*Navicula* spp.
$\mathbf{S_1}$	*Planktothrix agardhii*	*Planktothrix agardhii*
$\mathbf{S_N}$	*Cylindrospermopsis raciborskii*	*Cylindrospermopsis raciborskii*
$\mathbf{W_1}$	*Euglena oxyuris*	*Euglena* spp.
$\mathbf{X_1}$	*Ankistrodesmus* spp.	*Ankistrodesmus* spp.
$\mathbf{X_2}$	*Rhodomonas minuta*	*Rhodomonas minuta*
$\mathbf{X_3}$	small chlorelloid cells	*Chrysococcus* spp.
$\mathbf{X_{PH}}$	*Phacotus lenticularis*	*Phacotus lenticularis*
Y	*Cryptomonas erosa, C. reflexa*	*Gymnodinium* spp., *Cryptomonas erosa*

cell density data in the eastern basin in the period of 1980–1990 appear unrealistically low in the protocols available. This discrepancy indicates that changes in the flora preceded those in the biomass, as predicted by Sas's (1989) model of eutrophication and also found by Padisák & Kovács (1997) and Padisák & Reynolds (1998) during the early restoration period. From the 2000s onwards, the total biomass was not high, but the low Q indices indicated rather high percentages of alien species of Cyanobacteria (*Cylindrospermopsis raciborskii*, many but not each *Anabaena* and *Aphanizomenon* spp.) in the summer communities (Padisák et al., 2004, 2005).

Another excellent feature of the *Q* index is that it enables water quality to be reconstructed from historical archives in which identifications are correct, but some essential data (such as volume of the sedimentation chamber, number of fields/transects counted or the magnification) are missing from the counting protocols, as observed in some cases (Padisák et al., 1998).

Examining the dendrogram, the fourth group was prominent before eutrophication in 1966–1974 and again in 1995, reflecting the recent recovery from eutrophication. However, in the past several years the mean *Q* index has remained below three because the proportion of alien species is still high. In 1980 and 1983, there was a serious change in community structure preceding the bloom of *Cylindrospermopsis raciborskii*, which could be understood as a signal of system instability or, in other terminology, as a point-of-no-return (Blenckner, this volume).

In the eastern basin, grouping by water quality index was not clear because phytoplankton associations typically consisted of only a few functional groups that contributed almost equally to the biomass.

The changes in phytoplankton assemblages in the eastern and western basins of Lake Balaton during the past 60 years show that both the community structure and the biomass responded some years after management measures were introduced to reduce the nutrient load. In agreement with Padisák et al. (2006), the most representative month for eutrophication monitoring is August (or September). The *Q* index, which integrated the biomass and the indicative values of the different phytoplankton groups, showed that water quality was poorest in spring and late summer. However, this feature was not clearly evident in the pre-eutrophication period. In view of the *Q* index values, the 1960s were considered to represent the reference status of phytoplankton assemblages according to the WFD.

Acknowledgement We acknowledge financial support from the Hungarian Programme for Research and Development under contract 3B/022/2004 (BALÖKO).

References

Alvarez-Cobelas, M., C. S. Reynolds, P. Sanchez-Castillo & J. Kristiansen, 1998. Phytoplankton and Trophic Gradients. Developments in Hydrobiology, Vol. 129. Kluwer Academic Publishers, Dordrecht, Boston, London, pp. 372.

Blenckner, T., This volume. Models as tools to understand past, recent and future changes of large lakes. Hydrobiologia.

Crosetti, L. O. & C. E. D. Bicudo, 2005. Structural and functional phytoplankton responses to nutrient impoverishment in mesocosms placed in a shallow eutrophic reservoir (Garças Pond), São Paulo, Brazil. Hydrobiologia 541: 71–85.

da Silva, C. A., S. Train & L. C. Rodrigues, 2005. Phytoplankton assemblages in a Brazilian subtropical cascading reservoir system. Hydrobiologia 537: 99–109.

Descy, J.-P., C. S. Reynolds & J. Padisák, 1994. Phytoplankton in Turbid Environments. Rivers and Shallow Lakes. Developments in Hydrobiology, Vol. 100. Kluwer Academic Publishers, Dordrecht, Boston, London, 214.

Devercelli, M., 2006. Phytoplankton of the Middle Paraná River during an anomalous hydrological period: a morphological and functional approach. Hydrobiologia 563: 465–478.

EC Parliament and Council, 2000. Directive of the European Parliament and of the Council 2000/60/EC establishing a framework for community action in the field of water policy. European Commission PE-CONS 3639/1/100 Rev 1, Luxembourg.

Entz, G., J. Kottász & O. Sebestyén, 1937. Quantitatív tanulmányok a Balaton bioszesztonján. Annales Instituti Biologici Tihany 9: 1–144. (Quantitative studies on the bioseston of Lake Balaton).

Findlay, D. L. & S. E. M. Kasian, 1987. Phytoplankton community responses to nutrient addition in Lake 226, Experimental Lakes Area, Northwestern Ontario. Canadian Journal of Fisheries and Aquatic Sciences 44: 35–46.

Findlay D. L., J. J. Paterson, L. L. Hendzel & H. J. Kling, 2005. Factors influencing *Gonyostomum semen* blooms in a small boreal reservoir lake. Hydrobiologia 533: 243–252.

G.-Tóth, L. & J. Padisák, 1978. Short term investigations on the phytoplankton of Lake Balaton at Tihany. Acta Botanica Academiae Scientiarum Hungaricae 24: 187–204.

Hajnal, É. & J. Padisák, 2006a. Balatoni fitoplankton adatbázis (ALMOBAL) létrehozása és alkalmazhatósága vízminőségi monitorozásra. Hidrológiai Közlöny 86: 149–150. (The ALMOBAL DataBase for water quality monitoring of Lake Balaton. In Hungarian with English summary).

Hajnal, É. & J. Padisák, 2006b. Elaboration phytoplankton database (ALMOBAL) of Lake Balaton for monitoring water quality. Kandó Conference, Budapest ISBN 9637154426, 11.

Herodek, S. & G. Tamás, 1973. The primary production of phytoplankton in Lake Balaton April-September 1972. Annales Instituti Biologici Tihany 40: 207–218.

Herodek, S. & G. Tamás, 1975. The primary production of phytoplankton in the Keszthely Basin of Lake Balaton in 1973–74. Annales Instituti Biologici Tihany 42: 175–190.

Istvánovics, V., L. Somlyódy & A. Clement, 2002. Cyanobacteria-mediated internal eutrophication in shallow Lake Balaton after load reduction. Water Research 36: 3314–3322.

Istvánovics, V., A. Clement, L. Somlyódy, A. Specziár, L. G.-Tóth & J. Padisák, 2007. Updating water quality targets for shallow Lake Balaton (Hungary), recovering from eutrophication. Hydrobiologia 581: 305–318.

Komárek, J. & B. Fott, 1983. Chlorophyceae (Grünalgen). Ordnung: Chlorococcales. In Huber-Pestalozzi, G. (ed.), Das Phytoplankton des Süsswassers. Systematik und Biologie, Vol. 7/1. Schweitzerbart'sche Verlagbuchhandlung, Stuttgart: 1044.

Kristiansen, J., 2003. From IAP to IAP. Hydrobiologia 502: 1–2.

Kümmerlin, R. E., 1991. Long-term development of phytoplankton in Lake Constance. Verhandlungen der internationle Vereinigung für throretische und angewandte Limnologie 24: 826–830.

Lepistö, L., A. Räike & O.-P. Pietiläinen, 1999. Long-term changes of phytoplankton in a eutrophicated boreal lake during the past one hundred years (1893–1998). Algological Studies 94: 223–244.

Lopes, M. R. M., C. E. de M. Bicudo & M. C. Ferragut, 2005. Short term spatial and temporal variation of phytoplankton in a shallow tropical oligotrophic reservoir, southeast Brazil. Hydrobiologia 542: 235–247.

Lund, J. W. G., 1979. Changes in the phytoplankton of an English lake, 1945–1977. Freshwater Biological Association Publications 1306.

Makarewicz, J. C. & R. I. Baybutt, 1981. Long-term (1927–1978) changes in the phytoplankton community of Lake Michigan at Chicago. Bulletin of the Torrey Botanical Club 108: 240–254.

Mineeva, N. M. & A. S. Litvinov, 1998. Long-term variation of chlorophyll content in Rybinsk Reservoir (Russia) in relation to its hydrological regime. In George, D. G., J. G. Jones, P. Puncochar, C. S. Reynolds & D. W. Sutcliffe (eds), Management of Lakes and Reservoirs During Global Change. Kluwer Academic Publisher, Dordrecht, Boston, London: 111–125.

Morabito, G., A. Oggioni, E. Caravati & P. Panzani, 2007. Seasonal morphological plasticity of phytoplankton in Lago Maggiore (N. Italy). Hydrobiologia 578: 47–57.

Naselli-Flores, L., J. Padisák & M. Dokulil, 2003. Phytoplankton and Equilibrium Concept: The Ecology of Steady-State Assemblages. Developments in Hydrobiology, Vol. 172. Kluwer Academic Publishers, Dordrecht, Boston, London, 403.

Naselli-Flores, L. & R. Barone, 2005. Water-level fluctuations in Mediterranean reservoirs: setting a dewatering threshold as a management tool to improve water quality. Hydrobiolgia 548: 85–99.

Naselli-Flores, L., J. Padisák & M. Albay, 2007. Shape and size in phytoplankton ecology: Do they matter? Hydrobiologia 578: 157–161.

Neale, P., J. F. Talling, S. I. Heaney, C. S. Reynolds & J. W. G. Lund, 1991. Long time series from the English Lake District: Irradiance-dependent phytoplankton dynamics during spring maximum. Limnology Oceanography 36: 751–760.

Padisák, J., 1998. Sudden and gradual responses of phytoplankton to global climate change: case studies from two large, shallow lakes (Balaton, Hungary and the Neusiedlersee Austria/Hungary). In George, D. G., J. G. Jones, P. Puncochar, C. S. Reynolds & D. W. Sutcliffe (eds), Management of Lakes and Reservoirs During Global Change. Kluwer Academic Publisher, Dordrecht, Boston, London: 111–125.

Padisák, J. & L. G.-Tóth, 1991. Some aspects of the ecology of the subdominant green algae in a large nutrient limited shallow lake (Balaton, Hungary). Archiv für Protistenkunde 139: 225–242.

Padisák, J., C. S. Reynolds & U. Sommer, 1993. Intermediate Disturbance Hypothesis in Phytoplankton Ecology. Developments in Hydrobiology, Vol. 81. Kluwer Academic Publishers, Dordrecht, Boston, London, 200.

Padisák, J. & A. Kovács, 1997. *Anabaena compacta* (Nygaard) Hickel—új kékalga faj a Balaton üledékében és planktonjában. Hidrológiai Közlöny 77: 29–32. (*Anabaena compacta* (Nygaard) Hickel—a new blue-green algal species in the sediments and plankton of lake Balaton. In Hungarian with English summary).

Padisák, J., L. Krienitz, W. Scheffler, R. Koschel, J. Kristiansen & I. Grigorszky, 1998. Phytoplankton succession in the oligotrophic Lake Stechlin (Germany) in 1994 and 1995. Hydrobiologia 369/370: 179–197.

Padisák, J. & C. S. Reynolds, 1998. Selection of phytoplankton associations in Lake Balaton, Hungary, in response to eutrophicatioon and restoration measures, with special reference to cyanoprokaryotes. Hydrobiologia 384: 41–53.

Padisák, J., G. Borics, G. Fehér, I. Grigorszky, I. Oldal, A. Schmidt & Z. Zámbóné-Doma, 2003. Dominant species and frequency of equilibrium phases in late summer phytoplankton assemblages in Hungarian small shallow lakes. Hydrobiologia 502: 157–168.

Padisák, J., É. Soróczki-Pintér, É. Hajnal & Zs. Zámbóné-Doma, 2004. A balatoni fitoplankton tér- és időbeli mintázata 2003-ban. In Mahunka, S. & J. Banczerowski (eds), A Balaton kutatásának 2003. évi eredményei. MTA-Amulett'98, Budapest, 16–26. (Spatial and temporal patterns of Lake Balaton phytoplankton in 2003. Results of Lake Balaton Research in 2003. In Hungarian).

Padisák, J., É. Soróczki-Pintér, É. Hajnal & Zs. Zámbóné-Doma, 2005. A balatoni fitoplankton tér- és időbeli mintázata 2004-ben. In Mahunka, S. & J. Banczerowski (eds), A Balaton kutatásának 2004. évi eredményei. MTA-Amulett'98, Budapest, 16–26. (Spatial and temporal patterns of Lake Balaton phytoplankton in 2004. Results of Lake Balaton Research in 2003. In Hungarian).

Padisák, J., G. Borics, I. Grigorszky & É. Soróczki-Pintér, 2006. Use of phytoplankton assemblages for monitoring ecological status of lakes within the water framework directive: the assemblage index. Hydrobiologia 553: 1–14.

Podani, J., 1988. Syn-Tax III. User's manual. Abstracta Botanica 12: 1–183.

Pomogyi, P., 1993. Nutrient retention by the Kis-Balaton Water Protection System. Hydrobiologia 251: 309–320.

Reynolds, C. S. & E. G. Bellinger, 1992. Patterns of abundance and dominance of the phytoplankton on Rostherne Mere, England: evidence from a 18-year data set. Aquatic Sciences 54: 10–36.

Reynolds, C. S., M. T. Dokulil & J. Padisák, 2000. The Trophic Spectrum Revisited. Developments in Hydrobiology, Vol. 150. Kluwer Academic Publishers, Dordrecht, Boston, London, pp. 152.

Reynolds, C. S., V. Huszar, C. Kruk, L. Naselli-Flores & S. Melo, 2002. Towards a functional classification of the freshwater phytoplankton. Journal of Plankton Research 24: 417–428.

Salmaso, N. & J. Padisák, 2007. Morpho-functional groups and phytoplankton development in two deep lakes (Lake Garda, Italy and Lake Stechlin, Germany). Hydrobiologia 578: 97–112.

Sas, H., 1989. Lake Restoration and Reduction of Nutrient Loading: Expectations, Experiences, Extrapolations. Academia Verlag Richarz, St Augustin.

Sebestyén, O., 1960. Állományokról, különös tekintettel a tavi planktonra (Balatoni tanulmányok alapján). Annales Instituti Biologici Tihany 27: 93–113. (Über Bestände, mit besonderer Berücksichtigung des See-Planktons (auf Grund von Balaton-Studien). In Hungarian with German summary).

Talling, J. F., 1993. Comparative seasonal changes, and interannual variability and stability in a 26-year record of total phytoplankton biomass in four English lake basins. Hydrobiologia 268: 65–98.

Tamás, G., 1955. Mennyiségi planktontanulmányok a Balatonon VI. A negyvenes évek fitoplanktonjának biomasszája. Annales Instituti Biologici Tihany 23: 95–110. (Quantitative plankton studies on Lake Balaton. VI. Biomass of the phytoplankton of the forties. In Hungarian with English summary).

Tamás, G., 1965. Horizontale Plankton-Untersuchungen im Balaton VI. über das Phytoplankton im südwestlichen Teil des Sees auf Grund von Schöpf- und Netzfilterproben vom July 1962. Annales Instituti Biologici Tihany 32: 229–245.

Tamás, G., 1967. Horizontale Plankton-Untersuchungen im Balaton V. Über das Phytoplankton des Sees auf Grund der im Jahre 1965 geschöpften und Netzfilterproben. Annales Instituti Biologici Tihany 34: 191–231.

Tamás, G., 1969. Horizontal plankton investigations in Lake Balaton VII. On the phytoplankton of Lake Balaton, based on scooped samples and filtrates taken in 1966. Annales Instituti Biologici Tihany 36: 257–292.

Tamás, G., 1972. Horizontal Phytoplankton studies in Lake Balaton based on scooped samples and filtrates taken in 1967. Annales Instituti Biologici Tihany 39: 151–188.

Tamás, G., 1974. The Biomass Changes of Phytoplankton in Lake Balaton during the1960s. Annales Instituti Biologici Tihany 41: 323–342.

Tamás, G., 1975. Horizontally occurring quantitative phytoplankton investigations in Lake Balaton, 1974. Annales Instituti Biologici Tihany 42: 219–280.

Townsend, S. A., 2005. Hydraulic phases, persistent stratification, and phytoplankton in a tropical floodplain Lake (Mary River, Northern Australia). Hydrobiologia 556: 163–179.

Utermöhl, H., 1936. Quantitative Methoden Untersuchung des Nannoplankton in Adderheldens. Handbuch der Biologische Arbeitmethoden, Abteilung IX(1).

Utermöhl, H., 1958. Zur Vervollkommnung der quantitativen Phytoplankton-Methodik. Mitteilungen der internationale Vereinigung für theoretische und angewandte Limnologie 5: 567–596.

Vörös, L., 1980. A Balaton fitoplanktonjának tömege, összetétele és diverzitása 1976-ban. Botanikai Közlemények 67: 25–33. (Masse, Zusammensetzung und Diversität der Phytoplankton im Balaton, in Jahre 1976. In Hungarian with German summary).

Willén, E., 1992. Long-term changes in the phytoplankton of large lakes in response to changes in nutrient loading. Nordic Journal of Botany 12: 577–587.

Hydrobiologia (2008) 599:239–247
DOI 10.1007/s10750-007-9202-2

ELLS 2007

Towards ecological goals for the heavily modified lakes in the IJsselmeer area, The Netherlands

Eddy Lammens · Francien van Luijn · Yolanda Wessels · Harry Bouwhuis · Ruurd Noordhuis · Rob Portielje · Diederik van der Molen

Abstract The European Water Framework Directive (WFD) obliges EU member states to define ecological goals for water bodies and, if necessary, to take measures to achieve these goals by 2015. The goals and measures for the water bodies in the IJsselmeer area of The Netherlands are elaborated in this study, following an approach described by Irmer & Pollard (2006, Alternative methodology for defining Good Ecological Potential (GEP) for Heavily Modified Water Bodies (HMWB) and Artificial Water Bodies (AWB). ECOSTAT). With the closure of the Afsluitdijk in 1932, the former Southern Sea estuary was transformed into the freshwater Lake IJsselmeer. Subsequently, a string of so-called border lakes and Lake Markermeer were created by land reclamation projects and the construction of dams. These alterations serve safety, provide drinking water supplies and created agricultural land. Owing to the change in category, the lakes are by (WFD) definition heavily modified. A natural lake rather than an estuary will be the starting point for determining ecological potential. However, damming and fixing the water table prevented the development of emergent vegetation and caused steep water-land gradients. The Maximum Ecological Potential includes the effect of these hydromorphological changes after all mitigation measures have been considered. Other pressures on the lakes are high nutrient loads, which cause phytoplankton blooms, the disappearance of aquatic macrophytes and intensive fishery, which overexploites the pikeperch and eel populations and causes indirect negative effects on water quality. Good Ecological Potential for these lakes is derived by estimating the effects of all effective hydromorphological measures that have no significant negative impact on existing functions or the wider environment, and the effects of all other measures. The suggested main measures are: construction of fish passages, adaptation of shore-lines, wind sheltered areas, reduction of nutrient load, reduction of fishery pressure and reduction of the bream stock. The effects of these measures are

Guest editors: T. Nõges, R. Eckmann, K. Kangur, P. Nõges, A. Reinart, G. Roll, H. Simola & M. Viljanen
European Large Lakes—Ecosystem changes and their ecological and socioeconomic impacts

E. Lammens (✉) · R. Noordhuis · R. Portielje · D. van der Molen
RWS RIZA, P.O. Box 17, Lelystad 8200AA, The Netherlands
e-mail: eddy.lammens@rws.nl

F. van Luijn
RWS Directie IJsselmeergebied, P.O. Box 600, Lelystad 8200AP, The Netherlands

Y. Wessels
Grontmij AquaSense, Postbus 95125, Amsterdam 1090HC, The Netherlands

H. Bouwhuis
Oranjewoud, Postbus 24, Heerenveen 8440AA, The Netherlands

calculated from multivariate analyses that establish the relationships between (a) nutrients and chlorophyll-*a*, (b) chlorophyll-*a* and water transparency, (c) water transparency, depth distribution and vegetation coverage and (d) vegetation coverage and fish community. The proposed goals will be refined using new scientific insights and further discussions with stakeholders, and will finally be reported in the River Basin Management Plan 2009. It is expected that exemptions will be used for a phased achievement of the objectives.

Keywords Water Framework Directive · Human pressures · Ecological condition · Fish · Fishery · Macrophytes · Transparency · Nutrients · Chlorophyll-*a* · Water management · Measures

Introduction

In December 2000, the European Water Framework Directive (Directive, 2000) was accepted by the member states of the EU, and these states were obliged to implement this directive. Each member state has to define ecological goals for its water bodies and take proper measures to achieve these goals by 2015. The definition of goals for artificial and heavily modified water bodies (HMWB) is given in "Overall Approach to the Classification of Ecological Status and Ecological Potential" (ECOSTAT, 2003). To overcome several limitations of this approach, Irmer & Pollard (2006) proposed an alternative approach that takes better account of current water management practice in The Netherlands. The objective of this study is to illustrate how the ecological goals for water bodies in the IJsselmeer area of The Netherlands have been worked out.

Study area

With the closure of the Afsluitdijk in 1932, this dam transformed the former Southern Sea from an estuary into the freshwater Lake IJsselmeer. The main source of freshwater is the River IJssel, a tributary of the River Rhine (Fig. 1). During the following 35 years, several polders were created for agriculture (Fig. 1). Therefore, IJsselmeer became smaller, and new lakes and a string of border lakes were created: Ketelmeer (3,195 ha), Zwarte Meer (1,691 ha), Eastern Border Lakes (5,945 ha) and Southern Border Lakes (3,961 ha). In 1975, the already-reduced IJsselmeer was divided by a dam (Houtribdijk) into two separate lakes, IJsselmeer (112,500 ha) and Markermeer (65,000 ha) (Fig. 1). Therefore, all lakes in the IJsselmeer area are more or less man-made. Damming and regulation of the water table prevented the lakes from developing in a natural way. Flood plains could not develop, as dikes created an abrupt transition from water to land. These dikes were built in the Middle Ages along the shorelines of the former

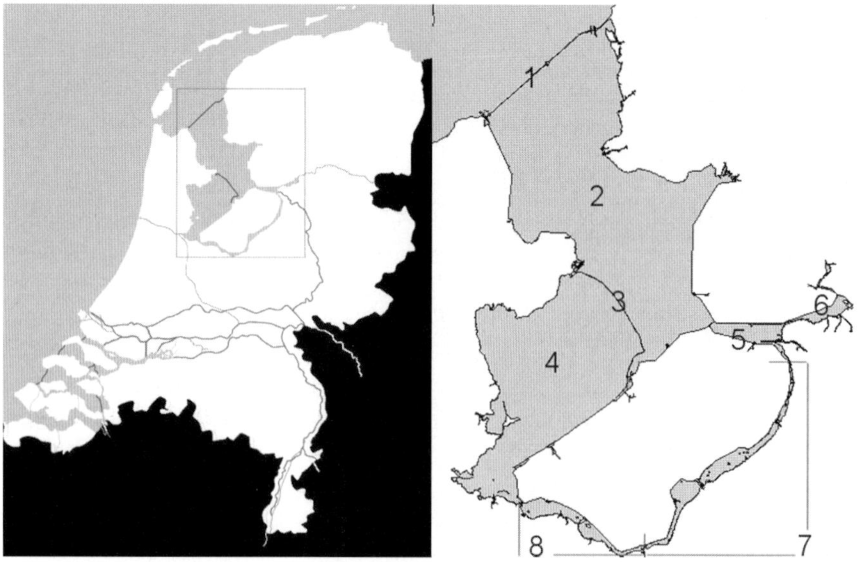

Fig. 1 IJsselmeergebied in the central part of The Netherlands. The numbers in the right figure correspond to: 1. Afsluitdijk; 2. IJsselmeer; 3. Houtribdijk; 4. Markermeer; 5. Ketelmeer; 6. Zwarte Meer; 7. Eastern Border Lakes; 8. Southern Border Lakes

Southern Sea to protect the people. Therefore, the state of the lake is comparable to the natural state in which the littoral zone is disconnected from the lake by a dike and in which the water level is regulated. The dikes and the constant water level are considered to be the two main hydromorphological pressures, which cannot be reversed without considerable risks to the people living there and to their freshwater supplies. The lakes in the IJsselmeer area are HMWB according to WFD terminology (CIS Working Group 2.2). Other pressures are high nutrient loads and overexploitation of fish populations. The P concentration in the lakes was not measured until the 1970s, when it was already high (0.2–0.3 mg l^{-1} in IJsselmeer to 0.3–1.0 mg l^{-1} in the border lakes). The nutrient problem in IJsselmeer is partly caused by local emissions, but mainly by loadings from abroad in the Rhine river basin, whereas in the border lakes, the sources are mainly water treatment plants and streams draining Dutch agricultural areas. Measures were initiated in the 1980s and resulted in a considerable reduction in P concentration. Since the creation of the lakes, there has been intensive fishery, mainly of eel, pikeperch and perch. Bream exploitation became profitable in the 1990s and bream fishery is now also intensive.

Methods

In order to identify and designate HMWB, we followed the "Guidance Document on identification and designation of Heavily Modified and Artificial Water Bodies" (Guidance, 2003). To establish ecological goals, we partly followed "Overall Approach to the Classification of Ecological Status and Ecological Potential" (ECOSTAT, 2003), but mainly used the alternative approach elaborated by Irmer & Pollard (2006). According to the former, Maximum Ecological Potential (MEP) is determined by estimating the negative effects of the irreversible pressures on biological quality elements and adding the positive effects of mitigating hydromorphological measures. In Good Ecological Potential (GEP), the biological quality elements show minor differences from MEP. According to Irmer & Pollard (2006), MEP can be defined as resulting from all possible hydromorphological measures (assuming a suitable water quality), and GEP by subtracting those measures that in combination have only a small effect.

In this study, we first prepared an inventory of the main pressures and their ecological effects. We then established a gross list of measures and subtracted those with significant negative effects on existing functions or the wider environment (remaining measures leading to MEP). Next, we shortened the list by omitting measures that together have only a small effect (remaining measures leading to GEP). Finally, we calculated the ecological effects of the measures in order to obtain quantitative estimates of GEP with respect to nutrients and to the biological quality elements—phytoplankton, macrophytes, fish and benthic invertebrates.

The GEP values for the biological quality elements are calculated in a way similar to that used to establish reference values, using the same empirical data sets for the different lake types (Van der Molen & Pot, 2006) and the 'new' hydromorphology. The indices for the biological quality elements were elaborated in 2000–2006 by expert working groups, and this is documented in Van der Molen & Pot (2006). GEP values of total P and water transparency combined with the 'new' hydromorphology were assumed to lead to the corresponding GEP status of vegetation, creating the habitat conditions for fish and thus the GEP composition of the fish.

Results

Effects of pressures

Dams, dikes and regulated water table

The regulated constant water level and steep dikes and dams prevent the development of littoral vegetation, as flood plains are absent. In Lake Markermeer, the bottom layer is mainly composed of clay and because of permanent erosion a lot of inorganic silt is suspended in the water column, causing high turbidity and preventing the growth of water plants. The closure of the Houtribdijk (Fig. 1) prevented the distribution of the silt to IJsselmeer and changed the pattern of distribution in Markermeer as well. A consequence of the Afsluitdijk (dam) is that it blocks the route of fish migration from the river to the sea and vice versa. It prevents the migration of adult salmonids from the Wadden Sea to the River IJssel, and coregonids, lampreys, sticklebacks and juveniles that must migrate to the sea to mature to adulthood.

High nutrient loads

In lakes, a high nutrient load eventually causes an increase in phytoplankton biomass and a decrease in transparency. As a consequence, submerged vegetation has less potential to develop. All lakes in the IJsselmeer area were turbid in the 1970s and 1980s, and much vegetation disappeared after the 1960s when the transparency decreased (Brouwer & Tinbergen, 1939; De Beaufort, 1954). Paleolimnological records of IJsselmeer show that in the period around 1960, the oligotrophic diatoms disappeared and only mesotrophic and eutrophic species remained. The P concentration at that time was estimated as at least 25–50 μg l^{-1} (Cremer & Bunnik, 2006), well below the value of ca. 250 μg l^{-1} that was observed during the 1980s. The transparency in the 1960s must have been ca. 1 m, but because the lake is on average 4.5 m deep, vegetation was only present in the shallow parts.

In the shallow border lakes, the increase in nutrient load caused a decrease in transparency down to 30 cm and the submerged vegetation disappeared in consequence. The fish communities associated with clear vegetation-rich water disappeared as well (Lammens et al., 2004). Perch, roach and pike almost disappeared and bream, pikeperch, ruffe and smelt became dominant.

Intensive fishery

The eel and pikeperch fisheries in IJsselmeer and Markermeer caused overexploitation of these populations. In consequence, only a few year classes are present and the populations are dominated by small individuals. This implies that prey-fish (fish <15 cm) have a good chance of survival and create good feeding conditions for piscivorous birds. On the other hand, the feeding pressure of these fish on zooplankton will be relatively high and will affect the water quality negatively by the cascading effect on the phytoplankton (Gulati et al., 1990). The bream fishery that started in the 1990s (Lammens et al., 2002) now keeps the bream population down, with positive effects on water quality, particularly in the border lakes (Breukelaar et al., 1996; Lammens et al., 2004).

Measures

Owing to safety considerations and freshwater needs, the lakes will not return to the original (brackish) status, and Good Ecological Status (GES) of a freshwater lake cannot be achieved. However, some measures may be considered to mitigate the effects of the hydromorphological alterations, such as fish passages, adapting shorelines, wind-sheltered areas and water level fluctuation. Eutrophication and commercial fisheries are considered to be independent of hydromorphological pressures and should not constrain GEP. Table 1 summarizes the measures and the motivations to include them or otherwise. Some choices are obvious (e.g. not to remove dams); others are discussed below.

Table 1 Pressures in the IJsselmeer area and their effects

	IJsselmeer	Markermeer	Borderlakes
Afsluitdijk	Change from estuary to freshwater lake		
Dikes	Steep gradient from land to water	Steep gradient from land to water	Steep gradient from land to water
Houtribdijk		Increasing the silt concentration in Markermeer	
Fixed water level	Prevents emergent vegetation to grow in potential places	Prevents emergent vegetation to grow in potential places	Prevents emergent vegetation to grow in potential places
Nutrient load (P and N)	Algal blooms, turbidity, preventing water plants to grow		
Fishery pressure	Overexploitation of eel and pikeperch	Overexploitation of eel and pikeperch	

Fish passages

A fish passage is (only) relevant to the Afsluitdijk. It is an important measure for enabling fish to migrate from the Wadden Sea to the IJsselmeer or via the IJsselmeer to the River IJssel and vice versa.

Adapt shoreline profile

In order to create a more natural shoreline, the steep land-water gradient should be altered.

However, this measure will only result in a well-developed zone of emergent vegetation if it is combined with natural water level fluctuation, which is absent and cannot be restored. Therefore, there may be only small-scale and uncertain future benefits. However, 'significant effects on the wider environment' may apply to these lakes, as the dikes may be considered a cultural heritage that should not be altered at all (Guidance, 2003).

Creating zones with wind-sheltered areas

As a mitigation measure for the dam creating the Markermeer, some wind-sheltered areas can be considered to create a better light climate and opportunities for plants to grow.

Water level fluctuation

Seasonal water level fluctuation may be beneficial for the littoral vegetation. However, in the IJsselmeer area, natural water level fluctuation conflicts with safety and fresh water needs. The margins of water level fluctuation without significant negative effects on functions are too narrow to exert significant ecological effects.

Reduction of the nutrient load

High nutrient concentrations are the most significant pressure in IJsselmeer. Phosphorus is the nutrient that limits the growth of phytoplankton and is therefore the determining factor for transparency and, accordingly, macrophytes and the biological groups depending on this habitat. Only a small part of the necessary nutrient reduction can be realized by The Netherlands itself. The remainder has to be realized by international agreements because a large part of the nutrient load comes from Germany and France. Markermeer, however, is light-limited because of a high concentration of suspended inorganic matter and the production of phytoplankton is therefore only partly dependent on phosphorus. For the border lakes, a level of 90 mg P m^{-3} is sufficient to attain a relatively stable clear water state (Portielje et al., 2005; Ibelings et al., 2007). In the eastern borderlakes this has already been achieved.

Fishery

The fishery pressure on pikeperch (*Stizostedion lucioperca*), perch (*Perca fluviatilis*) and eel (*Anguilla anguilla*) should be reduced to such an extent that a balanced population containing all year classes could develop. This implies a reduction of at least 75% of the total fishing intensity (Van Nes et al., 2002). On the other hand, fisheries may also be used as an ecological rehabilitation measure. If the chlorophyll-*a* concentration is still high, but the total P and N concentrations are within acceptable levels, a reduction of the bream stock to 25 kg ha^{-1} will help to bring the lake to clear water status (Breukelaar et al., 1996; Søndergaard et al., 2007).

Calculation of Good Ecological Potential

In the previous paragraph, we argued which hydromorphological measures are necessary to achieve MEP and which additional measures (eutrophication and fisheries) are needed to support this. These measures will lead to GEP if we subtract the effects of those with little impact (adapting shoreline, limited water fluctuation). The results are summarized in Table 1 and discussed below.

*Nutrients and chlorophyll-*a

For IJsselmeer and the border lakes we assume that the 'new' hydromorphology does not affect nutrient

Table 2 All possible measures in the IJsselmeer area to restore or improve the ecological condition

	IJsselmeer	Markermeer	Borderlakes
Removing Afsluitdijk	*Safety, *Freshwater needs		
Removing dikes	*Safety	*Safety	*Safety
Removing houtribdijk	*Safety	*Safety	
Creating fish passage in Afsluitdijk	No interference with main functions		
Adapt shoreline	Only locally feasible (ca. 1,000 ha)	Only locally feasible (ca. 1,000 ha)	Only locally feasible (ca. 1,000 ha)
Create fluctuation in water level	*Safety, *shipping, *freshwater needs	*Safety, *shipping, *freshwater need	*Safety, *shipping, *freshwater need
Creation wind sheltered areas	Not necessary	Ca. 7000 ha	Not necessary
Reduction of nutrient load	Reduction of P-load to result in 70 μg P l^{-1}		Reduction of P-load to result in 90 μg P l^{-1}
Reduction of fishery pressure	See text for explanation	See text for explanation	
Reduction of bream population	To < 25 kg ha^{-1}	To <25 kg ha^{-1}	To <25 kg ha^{-1}

* Significant negative effects on specific functions

concentration. The lack of natural retention zones causes a higher proportion of the nutrient load to be present in the water phase, but this is considered not to be significant for these lakes. In fact, the border lakes already show concentrations that resemble those of natural lakes. These natural concentrations are presented by Heinis & Evers (2007), but not yet approved because international harmonization (intercalibration) is still in progress. The natural concentration is based on a calculation with an accuracy of 90% to establish the chlorophyll-*a* concentration for top-down regulated lakes (Portielje et al., 2005). In shallow lakes (<3 m), this chlorophyll/P ratio is higher in lakes with transparencies <60 cm than in lakes with transparencies >60 cm; a transparency of 60 cm is considered to distinguish bottom-up from top-down regulated lakes. In top-down regulated lakes, the chlorophyll/P ratio varies among lake types from 200 to 500 μg chl-*a* per mg P; in bottom-up regulated lakes it varies between 500 and 1,200. This ratio is used to calculate the chl-*a* concentration given a certain concentration of total P (Portielje et al., 2005). This risk approach was taken in order to meet the WFD obligation that the supporting physical-chemical conditions must ensure GES.

For Markermeer this statistical relationship was not valid. We used the current phosphorus concentrations for GEP because the external anthropogenic nutrient loading is small and primary production is light rather than nutrient-limited.

Macrophytes

The (potential) coverage of submerged water plants is primarily determined by the light available at the lake bottom and is therefore related to light extinction and lake depth. Seeds and bottom substrate are assumed not to be limiting, as it will be usually a matter of time for the vegetation to develop. As a limit for the amount of light at the lake bottom necessary for plant growth, 3% of the amount of light available at the surface is considered necessary (Middelboe & Markager, 1997). Therefore, the extinction determines the maximal depth at which plants can grow. Another assumption is that only 65% of the zone providing all necessary conditions for plant growth would actually be covered (expert opinion). The coverage potential used is lower in very shallow areas, where wave action has more impact, and close to the maximum in areas of maximum colonized depth, where there is less light. If the extinction (or transparency) and depth zone distribution are known, the potentially covered area can be estimated, and this is considered the GEP value for submerged vegetation. For the shallow border lakes the estimated GEP is 19% coverage. For IJsselmeer and Markermeer, GEP coverage is only ca. 5% because these lakes are twice as deep as the border lakes.

Emergent vegetation such as reeds and bulrushes need dry soil to germinate and can expand vegetatively into a water depth of about 10 m, unless the shore is frequently exposed to strong winds (corresponding to roughly >1,000-m wind fetch). The area of the shore zone that becomes dry between the winter high water level and summer low water level, including the 10-m zone of vegetative expansion into the water, is considered the GEP value for emergent vegetation. The zone for emergent vegetation is limited in all lakes, seldom exceeding 1% of the surface area. Grazing by cattle or birds can prevent or reduce already-existing vegetation (Vulink & Van Eerden, 1998). The coverages for submerged and emergent vegetation for GEP are lower than the GES values for natural lake types owing to the heavily modified character of the lake (Table 1).

Fish

Data analysis showed that the presence of vegetation, the transparency of the water and the lake depth distribution are the main determinants for the fish community, since for a number of lake types the relative abundances of species (groups) have been elaborated in multi-variate analyses (Jaarsma & Klinge, 2005; Portielje et al., 2005). The following fish-indices were chosen as indicators (Van der Molen & Pot, 2006): % biomass of bream (1), % biomass of roach + perch (2), % biomass of vegetation-dependent (*limnophilous*) species (3), total number of species (4) and oxygen-tolerant species (5). The latter group is not elaborated, as the habitat for this group is not present in heavily modified conditions. For multivariate analyses the following independent variables were chosen: coverage of emergent vegetation, coverage of submerged vegetation, transparency and transparency/depth. The analysis predicted the expected contributions of these groups to the fish community and the GEP values of the indices considered. For the border lakes, GEP for % of bream was estimated as <34% and that for the % of roach + perch as >30%. The number of species was estimated as >10. All GEP values are lower than GES because the area of vegetation habitat is smaller, since the conditions are heavily modified. The % of roach + perch is an exception, but this results from the lower estimates for other species. Therefore, in this case, the value of GES is used for GEP.

For IJsselmeer and Markermeer the GES percentage of the eel and pikeperch populations exceeding legal size was estimated as 25–50%. Therefore, the GES value is again used for GEP. For IJsselmeer, the obstruction by the Afsluitdijk limits or even prevents the migration of some diadromous fish species from the sea to the freshwater and vice versa. Therefore, an indicator from an estuarine water type has been added to account for the connectivity function of the lake. The effect of fish passages is expressed as the increase in the number of diadromous fish species.

Other quality elements

The GEP has to be described for all biological quality and physico-chemical elements mentioned in Annex V of the WFD. In this article, several quality elements are elaborated. However, work is still in progress and additional objectives will be established for blooms of phytoplankton, species composition of macrophytes and benthic invertebrates. For IJsselmeer and Markermeer, several fish indicators that have been described for GES cannot yet be included in the statistical analysis because these lakes deviate too much from the other lakes in size, depth and fish community. The actual measurements for these two lakes may be compared with indicator values developed for natural lakes (Van der Molen & Pot, 2006). With respect to the physical chemistry, only nutrients and transparency have been addressed. For the other parameters, the GES values will probably be used for GEP (Table 2, 3).

Discussion

It is a WFD obligation to set ecological goals and to find the proper measures to achieve these goals. The approach described in the CIS guidance is based on comparison with natural water bodies. However, it is hard to find reference conditions for these large lakes. Most of the similar lakes in Europe are modified or are different in nature. Furthermore, suitable data are often missing. Historical data are not very useful either as the lakes used to be an estuary. Using a modelling approach and data from different sources, an attempt was made to construct reference conditions and a GES, but the values are beset by large uncertainties. Furthermore, the approach described in the CIS guidance assumes that one modification can

Table 3 Proposed values for the biological indicators representing GEP for the heavily modified lakes in the IJsselmeer area and GES for the shallow and deep hardwater lakes

	GEP IJsselmeer	GEP Markermeer	GEP border lakes	GES shallow lakes	GES deep lakes
P concentration ($\mu g\ l^{-1}$)[a]	70 (100)	100 (100)	90 (50)	90	90
Transparency (m)[a]	0.9 (0.6)	0.35 (0.25)	0.9 (1)	0.9	0.9
Phytoplankton					
Chlorophyll ($\mu g\ l^{-1}$)[a]	18 (60)	30 (50)	23 (20)	23	18
Macrophytes (% coverage)					
Submers (%)	5.3 (0.8)	4.5 (0.7)	19 (20)	25	30
Emergent (%)	0.7 (0.7)	0 (0)	0.8 (0.5)	60	60
Fish					
Bream (%)	(20)	(5)	<34 (30)	≤8	≤25
Perch + Roach (%)	(40)	(40)	>30 (40)	≥30	≥35
Limnophilous fish (%)	(0)	(0)	>1.3 (1.5)	≥40	≥3
Oxygen tolerant (%)			0	≥10	≥1
Number of species	(12)	(10)	10 (21)	≥14	≥10
Pikeperch/eel (% biomass > legal size)	>25 (<25)	>25 (<25)			≥25
Diadromous fish species	>10 (10)				

Sometimes ranges are given because international harmonization and national interpretation are not yet complete. The present situation is given in brackets

[a] Values are average for the period April–September

be mitigated by one measure. This approach turns out to be too theoretical for water bodies that suffer from multiple modifications and other pressures.

Therefore, we welcomed an alternative approach that starts from the current situation and the relevant measures (Irmer & Pollard, 2006). The goals are a result of these measures instead of the other way around. This alternative approach has been applied in the present study. All possible management measures have been considered and the effects on the lakes have been calculated. Since the same criteria are used to select measures, the ecological aim of both approaches is the same.

However, finding values for the quality elements is still difficult. We applied an empirical approach using data from the same or similar water bodies (Portielje et al., 2005) for phytoplankton, macrophytes and fish. IJsselmeer and Markermeer are very large and unique lakes that cannot be compared with the smaller and shallower border lakes. Considering size and fish community, Lake Peipsi, on the border of Estonia and Russia, is to some extent comparable to IJsselmeer and Markermeer (Haberman et al., 2000). An essential difference is that Peipsi is a natural lake with well-developed flood plains as spawning grounds for many fish species. Pike and pikeperch live together in this lake, whereas pike has no chance in IJsselmeer because of the lack of flood plains (Pihu & Kangur, 2000). Therefore, the fish community in Peipsi is not taken as a goal for IJsselmeer. However, the concentration of P (40–80 $\mu g\ l^{-1}$) in Peipsi, which is also mesotrophic, may provide an idea of what we may expect in IJsselmeer when the nutrient load in the rivers Rhine and IJssel declines.

Apart from habitat, connectivity is also important for the fish species in the lake. In former times the Southern Sea estuary was a connection with the River IJssel. The Afsluitdijk blocks the way for diadromous fish. The only entrances for these fish are sluices that are opened when the tide is low and water can flow from IJsselmeer to Waddensea. Close to high tide, when the water current is relatively low, there is a small window of opportunity for diadromous fish to enter the lake. Fish passages will increase the opportunities for these fish to enter. The prediction is that the number of diadromous fish will exceed the current number of 10 species.

Apart from the hydromorphological pressures there are also pressures related to nutrient loading and overexploitation by commercial fisheries. These pressures are the main reasons why the present state does not comply with the proposed GEP in Lake IJsselmeer. The source of the nutrients is upstream and reduction requires both national and international cooperation. Reduction of commercial fisheries may affect this sector significantly. Therefore, disproportionate costs and administrative impossibilities may lead to lowering or phasing out of goals. The eastern border lakes already comply with the proposed GEP, but the southern border lakes do not yet fulfil the demands because the nutrient loading is too high. Markermeer will comply depending on the outcome of the discussion about the creation of a wind-protected area, which will lead to a greater area of submersed water plants.

Acknowledgements We are grateful to Nico Jaarsma, Egbert van Nes, Bernard van Leeuwen, Harry Mols, Piet Jan Westendorp (Witteveen + Bos), Holger Cremer and Frans Bunnik (TNO) for cooperating in this project and we thank two anonymous reviewers for their constructive comments, which improved the text considerably.

References

Breukelaar A. W., E. H. R. R. Lammens, J. G. P. Klein-Breteler & I. Tatrai, 1996. Effects of benthivorous bream (*Abramis brama*) and carp (*Cyprinus carpio*) on sediment resuspension and the concentrations of nutrients and chlorophyll-*a*. Freshwater Biology 32: 113–121.

Brouwer G. A. & L. Tinbergen, 1939. The distribution of the mute swans, *Cygnus b. bewickii* Yarr. In the Southern Sea before and after the closure of the Afsluitdijk. Limosa 12: 1–18.

Cremer, H. & F. P. M. Bunnik, 2006. Estimating the nutrient concentration of the IJsselmeer since the origin of the lake in 1932. TNO-report 2006-U-R0133B.

De Beaufort, L. F., 1954. Changes in the flora and fauna of the Southern Sea (now IJsselmeer) after the closure in 1932. Nederlands Dierkundige Vereniging, Den Helder.

Directive, 2000. Directive 2000/60/EC of the European Parliament and of the Council of 23 October 2000 establishing a framework for Community action in the field of water policy.

ECOSTAT, 2003. Overall Approach to the Classification of Ecological Status and Ecological Potential, Working Group 2 A, Ecological Status (ECOSTAT).

Guidance, 2003. Guidance Document No. 4 Identification and Designation of Heavily Modified and Artificial Water Bodies—HMWB (2003), Final, CIS Working Group 2.2 on Heavily Modified Water Bodies, 14 January 2003.

Gulati, R. D., E. H. R. R. Lammens, M.-L. Meyer & E. van Donk, 1990. Biomanipulation, Tool for Water Management. Kluwer, Dordrecht.

Haberman, J., A. Jaani, A. Kangur, K. Kangur, R. Laugaste, A. Milius, H. Mäemets & E. Pihu, 2000. Lake Peipsi and its ecosystem. Proceedings of the Estonian Academy of Sciences. Biology/Ecology 49: 3–18.

Heinis, F. & C. H. M. Evers, 2007. Establishing GES values for nutrients in natural water bodies. RIZA Nota 2007.001.

Ibelings, B. W., R. Portielje, E. Lammens, M. L. Meijer, R. Noordhuis, M. S. van den Berg, W. Joosse & M. Scheffer, 2007. Resilience of alternative stable states during the recovery of shallow lakes from eutrophication. Ecosystems 10: 4–16.

Irmer, U. & P. Pollard, 2006. Alternative Methodology for Defining Good Ecological Potential (GEP) for Heavily Modified Water Bodies (HMWB) and Artificial Water Bodies (AWB). ECOSTAT, Ispra.

Jaarsma, N. & M. Klinge, 2005. Empirical relationships between the fish composition and water quality parameters in lakes. Rw1447-2. Witteveen + Bos.

Lammens, E. H. R. R., E. H. van Nes & W. M. Mooij, 2002. Differences in exploitation of the bream population in three shallow lake systems and their relation to water quality. Freshwater Biology 47: 2435–2442.

Lammens, E. H. R. R., E. H. van Nes, M.-L. Meijer & M. S. van den Berg, 2004. Effects of commercial fishery on the bream population and the expansion of *Chara aspera* in Lake Veluwe. Ecological Modelling 177: 233–244.

Middelboe, A. L. & S. Markager, 1997. Depth limits and minimum light requirements of freshwater macrophytes. Freshwater Biology 37: 553–568.

Portielje, R., C. Schipper & M. Schoor (eds), 2005. Effects of hydromorphological pressures on the ecological water quality elements of the WFD. Report, RIZA/2005.098X, RIKZ/2005.

Pihu, E. & A. Kangur, 2000. Main changes in the ichtyocoenosis of lake Peipsi since the 1950s. Proceedings of the Estonian Academy of Sciences. Biology/ Ecology 49: 81–91.

Søndergaard, M., E. Jeppesen, T. Lauridsen, C. Skov, E. H. Van Nes, R. Roijackers, E. H. R. R. Lammens & R. Portielje, 2007. Lake restoration: successes, failures and long-term effects. Journal of Applied Ecology 44: 1095–1105.

Van der Molen, D. T. & R. Pot (eds), 2006. Referenties en concept-maatlatten voor meren voor de Kaderrichtlijn Water. STOWA-Report 2004-42a. STOWA, Utrecht.

Van Nes E. H., E. H. R. R. Lammens & M. Scheffer, 2002. PISCATOR, an individual-based model to analyze the dynamics of lake fish communities. Ecological Modelling 152: 261–278.

Vulink, J. T. & M. R. van Eerden, 1998. Hydrological conditions and herbivory as key operators for ecosystem development in Dutch artificial wetlands. In Wallis DeVries, M. F., J. P. Bakker & S. E. Van Wieren (eds), Grazing and Conservation Management. Kluwer Academic publishers, Dordrecht: 217–252.

Hydrobiologia (2008) 599:249–257
DOI 10.1007/s10750-007-9190-2

ELLS 2007

Environmental awareness of the permanent inhabitants of towns and villages on the shores of Lake Balaton with special reference to issues related to global climate change

Anna Várkuti · Kata Kovács · Csilla Stenger-Kovács · Judit Padisák

Abstract Lake Balaton, one of the largest lakes in Europe, is an important centre for summer tourism in Hungary. The appearance of the water is, however, influenced by a number of physical and biological factors such as the frequency of storms, the hydrological balance, the water level and the growth of algae. Here we report the results of a survey designed to test the reaction of the local population to these seasonal and inter-annual variations. This survey was based on a questionnaire completed by 960 permanent inhabitants of the shoreline villages and was designed to explore their awareness of the local and regional factors responsible for the changes. About one-quarter of the respondents considered that the drying effects of global climate change had already had an effect on the Lake Balaton area. Although younger (16–26) and middle-aged (27–50) respondents benefit most from good water quality, those in the oldest age group (51–92) were more concerned about the quality of the water and broader issues such as biodiversity and the conservation of the landscape. Considering that the future of the lake depends on the youngest generation, this suggests that their environmental education could be improved to strengthen their understanding of environmental issues.

Keywords Questionnaire · Local inhabitants · Lake · Water quality · Sociological survey

Guest Editors: T. Nõges, R. Eckmann, K. Kangur, P. Nõges, A. Reinart, G. Roll, H. Simola and M. Viljanen
European Large Lakes—Ecosystem changes and their ecological and socioeconomic impacts

Electronic supplementary material The online version of this article (doi:10.1007/s10750-007-9190-2) contains supplementary material, which is available to authorized users.

A. Várkuti
Dual Language Highschool Balatonalmádi, Rákóczi Ferenc u. 38, 8220 Balatonalmádi, Hungary

K. Kovács · C. Stenger-Kovács · J. Padisák (✉)
Department of Limnology, University of Pannonia, Egyetem u. 10, 8200 Veszprém, Hungary
e-mail: padisak@almos.uni-pannon.hu

Introduction

Large lakes often fulfil a wide range of societal demands: they serve as drinking water supplies, recreation areas and conservation zones. Many, especially those located in middle latitudes, are heavily impacted by tourism and over-exploitation to the point where there is some deterioration of their original values.

Lake Balaton, in the south-west of Hungary, is the largest shallow lake in Europe. It is 77.9 km long, 7.2 km wide and has an average depth of 3.14 m. Its surface area is 596 km^2 and its catchment area is 5776 km^2. The major inflow to the lake is the Zala River. The lake was originally mesotrophic but showed signs of eutrophication in the 1960s and 1970s because of the increased nutrient load. By the 1980s some basins in the lake had reached the permanently hypertrophic stage. The implementation of more rigorous management practices for the lake then resulted in a 50–60% decrease in biologically available phosphorus and in a significant improvement of water quality (Padisák & Reynolds, 1998; Padisák et al., 2006a). Lake Balaton is one of the most popular tourist destinations in Europe. Tourism contributes 10% to the GDP of Hungary of which the share for the Balaton region is about 30%.

The sociological survey described here was designed to explore the extent to which the permanent inhabitants understood the key water quality issues in Lake Balaton and were aware of the potential effects of global climate change. We also intended to analyse the attitude and sensitivity of the population to a number of the defined environmental problems. To interpret the results of this sociological survey the following features of the lake are fundamental:

(i) A high level of inorganic turbidity is an integral characteristic of the lake. Lake Balaton is shallow, with inorganic sediments that are periodically resuspended by the wind.
(ii) Dense growths of algae were once common in the lake but are now recorded only rarely. The latest summer in which such a density of algae occurred (*Cylindrospermopsis raciborskii* bloom; Padisák and Koncsos, 2002; Padisák et al., 2006a) was in 1994, most probably because the sedimentary P-pool of the lake subsequently failed to support dense algal growth (Padisák and Istvánovics, 1997; Padisák et al., 2006b, Hajnal & Padisák, 2008).
(iii) Small areas of brown-coloured water may appear in the lake, but these are restricted to the area around the inflow of swamp-water from the Kis-Balaton reservoir.
(iv) Unpleasant odours are seldom reported along the shores of the lake but when they do occur they are connected with sporadic fish-kills in the spawning season or the concentration of the natural odours associated with swamps and emergent macrophyte. The emission of H_2S by anoxic organic sediments is in no way typical.

Our expectations concerning the results of this survey were many fold. Among them, we supposed that

- those residents who have better education and/or long-term experience of living in the lake area will have a better knowledge of water quality issues than others;
- phenomena that can be directly observed by sense organs (transparency, odour) will be judged more uniformly than those that are difficult to observe (global change, acid rain);
- the middle-aged group will be more concerned about lake water quality because their living standards depend directly on it, since they are predominantly involved in the tourism industry.

The method and the population studied

The survey was conducted between May and October 2005 and was based on voluntary written answers to a questionnaire circulated to a randomly selected sample of the permanent inhabitants of the shoreline communities above the age of 16. The title of the questionnaire was "*Questionnaire on environmental awareness related to Lake Balaton*". The original version was in Hungarian, but an English translation has been prepared and is included in the Appendix to the electronic version of this article. The survey was evaluated on the basis of a statistical analysis of the data acquired together with some supplementary interpretation.

The randomly chosen individuals whose opinions are included in the study were representative of the permanent inhabitants of the lake shoreline. The sample comprised 960 individuals (the error, in general, for the whole population is ±6.5%): 93.8% were Hungarian, 42.3% were male, 57.3% were female; 48.5% were married, 39% were unmarried, 7.7% were divorced and 3.6% were widows. All the respondents were between 16 and 92 years of age and were classified into three age groups: those between 16 and 26 (32.12%), mainly students; those between 27 and 50 (42.6%), forming the bulk of the working population and those between 51 and 92 (24.2%), mostly retired. In educational terms, the sample included 22.6% with basic schooling (up to 8 grades),

37.7% with a high school diploma and 28.4% with a university or college degree. In view of the sampling criteria (permanent residents, those who work permanently in shoreline cities and villages or have another kind of permanent, long-term observational experience), the results may represent the opinions of some 400,000–500,000 citizens.

The questionnaire was specifically designed to cover some of the key water quality issues addressed by the EC 5th Framework CLIME Project. It has two parts with two sets of questions: set I (1–7) comprises personal data (sex, age, jobs, highest schooling, marital status, citizenship) and set II (1–17) includes questions related to Lake Balaton. The questions related to global environmental changes (II/4) are combined with a series of questions that test the reactions of the respondents to other environmental problems and are designed to obtain a broad view of their opinions.

This article focuses on the analysis of variables that refer to the opinions of the respondents related to water quality (II/1–4) and to the possible costs associated with improving the environmental condition of the lake (II/15–16). Since the rate of global change is too slow to elicit an immediate reaction, we structured these questions in such a way that we could test their indirect as well as direct experience of weather-related changes in the lake environment. In Sect. II/4: "In relation to the environmental condition of the lake, what concerns permanent residents the most?", we explored the respondents' reactions to:

(i) Those characteristics that can be perceived personally and directly with the sense organs, such as the transparency of the water (II/4a), its odour (II/4b) and water level changes (II/4c); and
(ii) Those characteristics that cannot be experienced directly, i.e. the less easily quantified aspects of global climate change (II/4d) such as changes in the frequency of extreme events and the progressive increase in UV radiation.

The attributes of variables in questions II/1 and II/3 are ordinal, arranged on a decreasing scale from 1 to 4. For the other questions, we provided some set options and then gave the respondents an opportunity to add other ideas using the prompt "other, please specify…". In this way, we had both quantitative and qualitative data to analyse. Data analyses followed recommendations by Varga (2000) and Babbie (2001).

Results

Water transparency

More than half the respondents (55.4%) consider the water to be moderately non-transparent (Fig. 1a). Two contrasting opinions also appear in roughly the same ratio: 16.3% of respondents said that they had not yet experienced non-transparency, but 15.1% recognised that the water is very non-transparent. A relatively large proportion (13.2%) gave no answer. Even though moderate non-transparency is easily observable, more than 30% interpret the same objective reality in subjectively different ways. By analysing the frequencies of the opinions in relation to sex, age and level of education (data not shown), we reached the following observations:

1. The sex of the respondents had no effect on their perception of non-transparency. Most respondents (55% on average) were of the opinion that the water of Lake Balaton is moderately non-transparent and this corresponds to the objective reality.
2. Within the different age groups, the perception that the water is clear increased with the age of the respondents, whilst the perception that the water is very non-transparent decreased with age. Therefore, the younger respondents are more critical of any deterioration in water transparency.
3. The factor with the most pronounced effect on the perception of change was educational level, with the less educated group showing the biggest deviation from average.

In order to explore whether the respondents who are aware of the 'non-transparency' problem also understood the cause, we offered them a further list of options. Analysing the responses (Fig. 1b), we were surprised to find that only a small proportion of the permanent residents (23%) knew that the actual cause of the non-transparency is the mixing effect of the wind. The other respondents offered a range of

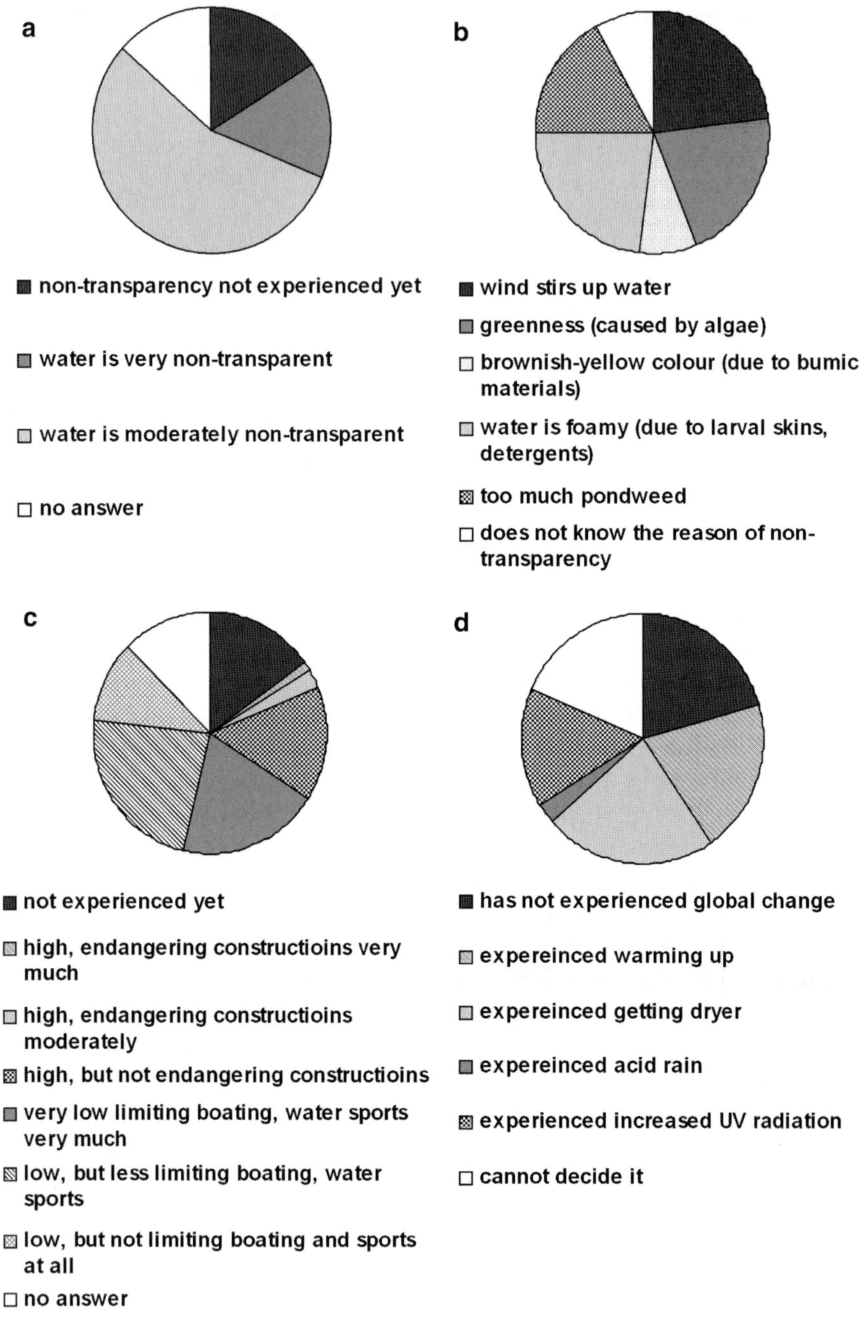

Fig. 1 The respondents' opinions on the transparency of Lake Balaton (**a**) and on why the water is not transparent (**b**). Distribution of opinions about water level changes and its possible consequences for Lake Balaton (**c**). Panel (**d**) shows the distribution of "yes" answers in categories offered in the context of whether the respondents have experienced some kind of global change

implausible explanations: 23% considered that 'foaminess' causes non-transparency even though the quality of the local water treatment is so high that detergents never reach the lake. About 20% blamed algal bloom for the non-transparency even though no serious blooms had been reported in the lake in the preceding 10 years and 17% considered that pond weed is a serious problem even though their effect is very localised (moreover, their stands block the mixing effect of wind). Some 8% thought that brownish-yellow humic compounds were the cause of non-transparency even though these effects are restricted to a very small area far from the observational range of most of the respondents.

The odour of the water

Data analysis shows that most respondents claimed that they have never experienced any odour; about one-third considered the water to be moderately smelly, and a small minority said that water is very bad-smelling. We analysed the data further to see how these opinions varied according to the sex, the age and level of education of the respondents (data not shown). These differences were quite small, but younger respondents and those who were less well educated were more critical of the water odour whilst the older and better-educated people were more likely to consider any perceived odour acceptable.

Changes in water level

In the period 2000–2003, the water budget for Lake Balaton (average depth: 3 m) was consistently negative and the water level was about 1 m lower than usual. Since this had produced a number of problems, we included some questions to explore the public response to these changes (Padisák et al., 2006b). Most respondents (Fig. 1c) were of the opinion that low water levels were a persistent problem even though this was not the case at the time of the interviews. About 34% of all respondents considered that a low water level had very little effect on boating and water sports, but about 20% believed that this was a serious problem even when the water level was relatively high. A smaller group (about 16%) believed that the water level was high, and a small group (3%) even considered that these levels of water endangered constructions. In reality, this is never the case since very high levels can be controlled very quickly and effectively through opening the Sió Canal and draining the water into the Danube. A relatively large percentage (about 12%) did not answer this question, perhaps because they were aware of the year-to-year variation in water level and a 'neither low nor high' option was provided.

Quality measures that can be related to global changes

In order to obtain the respondents' opinions on whether global climate change had influenced the characteristics of the lake, they were offered a series of multiple choices to the question "Have you experienced any way in which global environmental changes have influenced the environmental conditions of the lake?" Collating and analysing the responses (Fig. 1d), different options were given in the following percentages: 18% of the total felt that they had not yet experienced any changes that could be ascribed to global environmental change, 20% could not decide if they had experienced such changes, but the majority (62%) was convinced that such changes had already occurred. In this group, 25% said that global phenomena such as warming and drying had affected Lake Balaton, 18% expressed concern about increased UV radiation, whilst 3.4% were worried about the effects of acid rain. Since this is not an issue that could possibly affect an alkaline lake like Balaton, this response must reflect an extrapolation of concerns from other areas. When we analysed these responses in relation to the sex, age and education level of the respondents, it became clear that male and more highly educated respondents in the older age groups were more inclined to express concerns about acid rain (Fig. 2).

The importance of the lake and its reasons

In a supplementary survey designed to assess more general concerns about the quality of water in Lake Balaton (Fig. 3a), most residents (about 93%)

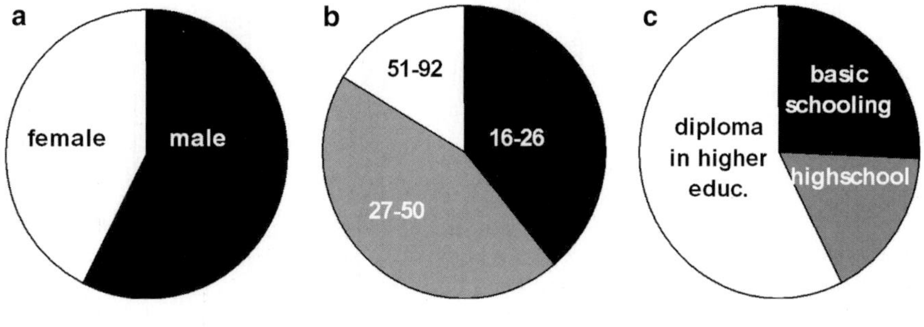

Fig. 2 Distribution of sex (**a**), age (**b**) and education level (**c**) of respondents who believed there was an acid rain effect on lake water quality

considered water quality a very important issue. To establish why this was the case, we analysed their responses to question (II/2): "From your point of view, why is good water quality important?". Of the "yes" answers, 53.9% indicated that they were concerned about biodiversity, 43.3% considered good water quality important for recreational activities, 33.1% were concerned for the landscape, but only 13.3% considered water quality important from an economic point of view. This low response was rather unexpected because a substantial number of the local inhabitants either work for tourist-related organisations or have businesses connected with tourism. Exact numbers of those involved in tourism are difficult to obtain, since indirect connections (when close relatives of a respondent are also significantly involved) were not mapped in the questionnaire. Analysis of these results in relation to the sex, age and education level of the respondents (Fig. 3) showed that sex made no significant difference (Fig 3b), but there were correlations with age (Fig. 3c) and education (Fig. 3d). The level of concern tended to increase with both age and educational level, suggesting that the older and more educated respondents were more aware of the lake's ecological importance.

Examining more closely the reasons why different ages consider water quality important, we found that the older age group is more likely to have an economic interest in the lake but the younger age group has stronger recreational interests. Respondents in the oldest age group considered good water quality very important primarily because of biodiversity and landscape. The younger (16–26) and middle-aged (27–50) respondents considered water quality less important even though they benefit from good water quality when they bathe, swim and organise water sports on the lake. The fact that the oldest age group expressed the greatest concerns about the lake is disturbing, since the future of the lake depends on the youngest generation.

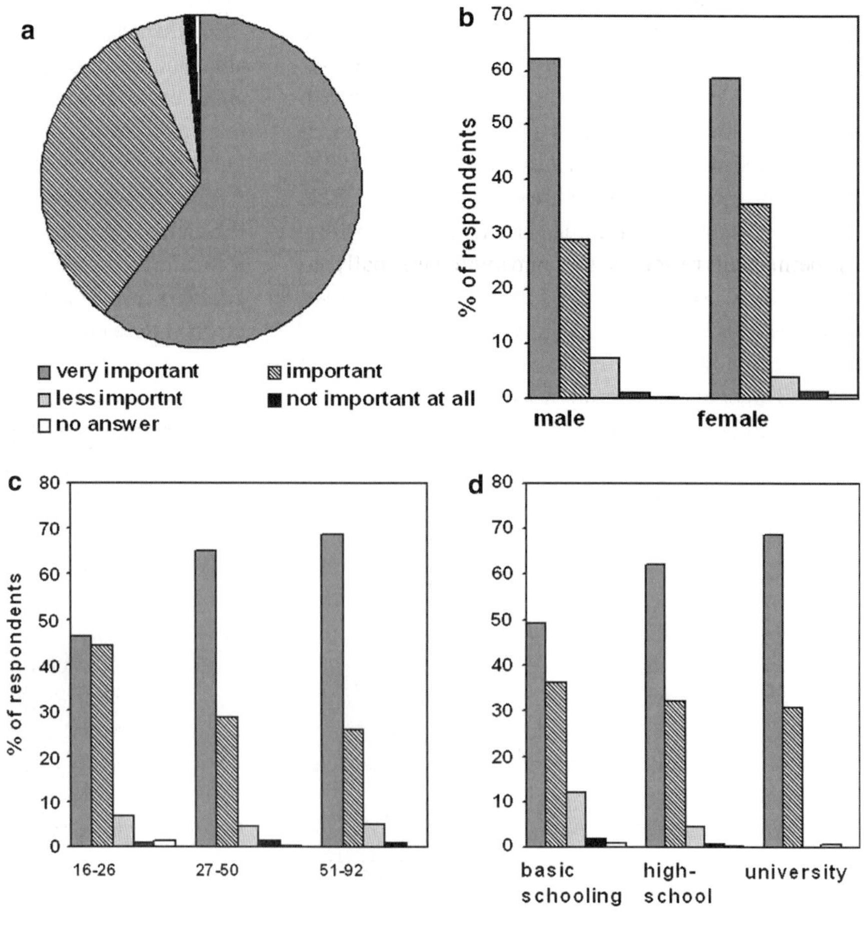

Fig. 3 Responses to the question "How important is the water quality of the lake?" (**a**) and its distribution by gender (**b**), age (**c**) and education (**d**)

Level of satisfaction with the lake water quality

Most (67%) of the respondents were satisfied with the water quality of Lake Balaton; about one-quarter were "not too satisfied", a small proportion (about 3%) were "not satisfied at all" and an even smaller proportion (about 2%) claimed that they were "very satisfied". There was no difference between the sexes in the degree of satisfaction with water quality. The proportion of satisfied respondents did, however, increase with age and with education level. We suspect that this reflects the more direct contact of the youngest age group with the water and that the correlation with age is an indirect link.

Financial resources for protection and quality improvement

Question II/15: "What financial resources should be used to cover the expenses of the protection and improvement of the conditions of Lake Balaton?" was designed to test the respondents' willingness to pay for improved water quality. The results (Fig. 4a) demonstrate an obvious preference for central support (36%) and grants from both national and international sources (29%). Others favoured support by sponsors (18%) and local authorities (16%) but only 4% were in favour of local ecotaxes. Regarding the willingness of permanent residents to contribute personally to management costs in specified ways, we obtained the responses shown in Fig. 4b. Although most respondents have real concerns about water quality, only half were prepared to pay for any improvements. Most expected these costs to be covered by the central budget, about 40% were prepared to make modest contributions, but only 10% would offer larger amounts.

Discussion

There is an increasing scientific literature on decision-making procedures in environmental policy, including issues connected to water (McDaniels et al, 1999; Alkan Olsson & Berg, 2005; Hostmann et al., 2005; Löwgren, 2005; Marttunen & Suomalainen, 2005). In line with democratic ideals, there is a growing call for greater public involvement in establishing science and technology policy (Jöborn et al., 2005; Jonsson, 2005). Various public participation procedures exist with the aim of involving and consulting the public, ranging from the public hearing to the consensus conference (Rowe and Frewer, 2000). However, there has been little systematic examination to date of the contribution of this activity towards better environmental management. Neither have there been extensive empirical studies to examine how participation processes can be made more effective (Bertram and Reynoldson, 2004). The effectiveness of any kind of communication tool depends strongly on perceptual abilities (McDaniels et al., 1997) and the observational background of the community involved as well as on their capacity to interpret simple or multi-casual phenomena.

Analysing and interpreting the answers of the 960 respondents in this study, we managed to collect information not only about what people experience or think they experience, but also what they think and how they think about the different environmental problems connected with Lake Balaton.

Respondents generally shared the same opinion about the transparency and odour of the water, probably because these properties can be experienced directly through the sense organs. They were, however, not too good at explaining the cause of moderate non-transparency. Some may have been influenced by the differences that exist between different parts of the lake, but others may simply interpret the same facts in different ways. Consequently, even an observable characteristic such as transparency can be "seen" differently, and most of the respondents cannot explain the real cause of the phenomenon. We suspect that respondents' opinions may have been influenced by environmental education and that they have 'picked up' some general knowledge that they cannot apply correctly.

Their personal experiences of recent warm, dry summers almost certainly influenced their stated concern about global warming. Only a small proportion of the respondents were worried about the increased levels of UV radiation—information that must have been gained from newspapers and television. Interestingly, a few, mainly male residents with higher education, believed that conditions in the lake had been influenced by acid rain. Since Lake Balaton is alkaline it is not affected by acidification, so this impression must have been gained from reading

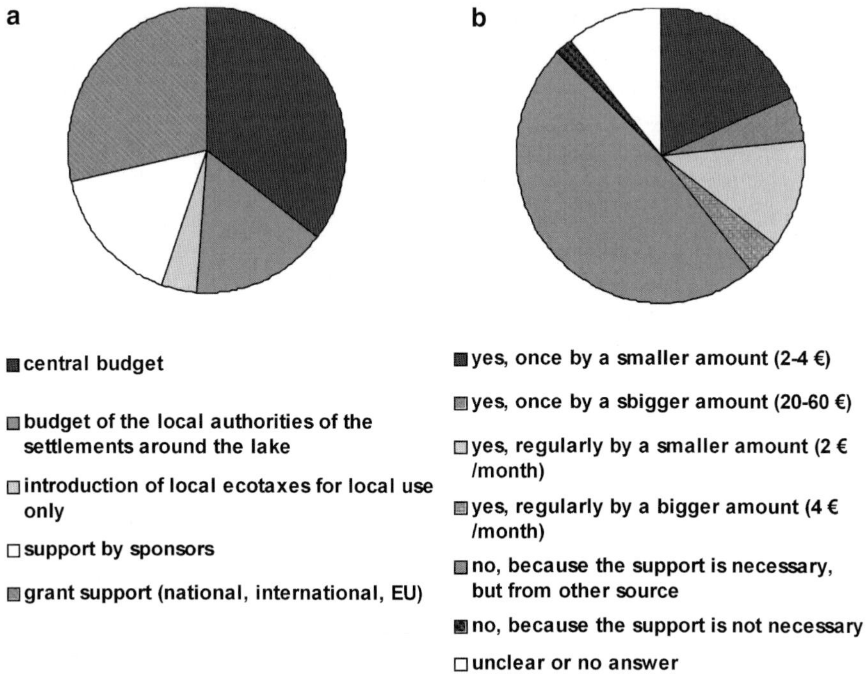

Fig. 4 The distribution of the respondents' opinions on financial resources to improve the water quality of the lake (**a**) and their willingness to contribute to the expenses to improving the environmental conditions of Lake Balaton (**b**)

general articles on the subject, e.g. in magazines dealing with fishing or aquatic sports.

Although younger and middle-aged residents benefit most from good water quality, the oldest age group is more interested in the ecological status of the lake, its biodiversity and the conservation of the surrounding landscape. This suggests that more attention should be paid to the environmental education of the younger generation by addressing their interests in more specific ways.

In the opinion of the permanent residents of Lake Balaton, financial support for the protection of the lake should first be sought from the central state budget, then the local authorities and finally grants and sponsorships. They did not favour the introduction of local ecotaxes or more direct ways of contributing personally to such expenses. It is, however, clear that the permanent residents of Lake Balaton really appreciate the benefits of the lake, its biodiversity and the surrounding landscape.

The results of this study only partly agreed with our previous expectations. In general, residents with high education interpreted observations more correctly. In many cases, however, this was not apparent in the percentage of responses but rather in deviations around the average. They were also more tolerant towards environmental issues. In this context, the surprisingly high proportion of false responses by males with high education concerning acid rain allows us to conclude that information from mass media may have priority over personal observations, which calls attention to the importance of distributing correct information instead of attracting headlines or even false information designed as "breaking news". Phenomena that can be directly observed by the sense organs (transparency, odour) were judged more uniformly than those (global change, acid rain, UV radiation) that are difficult to observe. The high diversity of opinions about easily observable phenomena (moderate non-transparency) calls attention to the importance of observational practices in environmental education. Notwithstanding our expectations, the middle-aged group did not show more concern about lake water quality. This is somewhat surprising because the living standards of many of them are directly dependent on the tourism industry. Most probably the wisdom of the older generation and the knowledge about environmental issues of the younger group counterbalanced the direct financial involvement of the active age group.

Nevertheless, the respondents showed a great deal of interest in the future of the lake and were prepared to consider different methods for preserving the lake. The responses to the survey demonstrated that a high proportion of the population were environmentally aware and recognise the important part that Lake Balaton plays in the culture and economics of Hungary.

Acknowledgements We acknowledge financial support from the European Commission's Environment and Sustainable Development Programme under contract EVK1-CT-2002-00121 (CLIME, http://www.clime.tkk.fi). We thank the pupils of the Dual Language Highschool Balatonalmádi, Hungary for their enthusiastic assistance in collecting data for this survey, and D. Glen George and Tiina Nõges for their the constructive notes on the manuscript of this article.

References

Alkan Olsson, J. & K. Berg, 2005. Local stakeholders' acceptance of model-generated data used as a communication tool in water management. Ambio 34: 507–512.

Babbie, E., 2001. The Practice of Social Research. Wadsworth/ Thomson Learning, Wadsworth.

Bertram, P. E. & T. B. Reynoldson, 2004. Developing ecosystem objectives for the Great Lakes: policy, progress and public participation. Journal of Aquatic Ecosystem Stress and Recovery 1: 89–95.

Hajnal, É. & J. Padisák, 2008. Analysis of long-term ecological status of Lake Balaton based on the ALMOBAL phytoplankton database. Hydrobiologia. doi:10.1007/s10750-007-9207-x

Hostmann, M., M. Borsuk, P. Reichert & B. Truffer, 2005. Stakeholder values in decision support for river rehabilitation. Archiv für Hydrobiologie, Supplement 155 (Large Rivers 15): 491–505.

Jöborn, A., I. Danielsson, B. Arheimer, A. Jonsson, M. H. Larsson, L. J. Lundquist, M. Löwgren & K. Tonderski, 2005. Integrated eater management for eutrophication control: public participation, pricing policy, and catchment modelling. Ambio 34: 482–488.

Jonsson, A., 2005. Public participation in water resources management: stakeholder voices on degree, scale and methods in future water management. Ambio 34: 495–500.

Löwgren, M., 2005. The water framework directive: stakeholder preferences and catchment management strategies – are they reconcilable? Ambio 34: 501–506.

Marttunen, M. & M. Suomalainen, 2005. Participatory and multiobjective development of Water Course regulation—Creation of regulation alternatives from stakeholders' preferences. Journal of Multi-Criteria Decision Analysis 13: 29–49.

McDaniels, T. L., L. J. Axelrod, N. S. Cavanagh & P. Slovic, 1997. Perception of ecological risk to water environments. Risk Analysis 17: 341–352.

McDaniels, T. L., R. S. Gregory & D. Fields, 1999. Democratizing risk management: successful public involvement in local water management decisions. Risk Analysis 19: 497–510.

Padisák, J. & V. Istvánovics, 1997. Differential response of blue-green algal groups to phosphorus load reduction in a large shallow lake: Balaton, Hungary. Verhandlungen der Internationale Vereinigung für Limnologie 26: 574–580.

Padisák, J. & C. S. Reynolds, 1998. Selection of phytoplankton associations in Lake Balaton, Hungary, in response to eutrophication and restoration measures, with special reference to cyanoprokaryotes. Hydrobiologia 384: 41–53.

Padisák, J. & L. Koncsos, 2002. Trend and noise: long-term changes of phytoplankton in the Keszthely-Basin of Lake Balaton, Hungary. Verhandlungen der Internationale Vereinigung für Limnologie 28: 194–203.

Padisák, J., G. Borics, I. Grigorszky & É. Soróczki-Pintér, 2006a. Use of phytoplankton assemblages for monitoring ecological status of lakes within the Water Framework Directive: the assemblage index. Hydrobiologia 553: 1–14.

Padisák, J., G. Molnár, É. Soróczki-Pintér, É. Hajnal & D. G. George, 2006b. Four consecutive dry years in Lake Balaton (Hungary): consequences for phytoplankton biomass and composition. Verhandlungen der Internationale Vereinigung für Limnologie 29: 1153–1159.

Rowe, G. & L. J. Frewer, 2000. Public participation methods: a framework for evaluation. Science, Technology & Human Values 25: 3–29.

Varga, A., 2000. Matematikai statisztika pszichológiai, nyelvészeti és biológiai alkalmazásokkal. [Mathematical statistics with applications in psychology, linguistics and biology]. Pólya Kiadó. Budapest (In Hungarian).

Hydrobiologia (2008) 599:259–276
DOI 10.1007/s10750-007-9233-8

Highlights of large lake research and management in Europe

Peeter Nõges · Külli Kangur · Tiina Nõges · Anu Reinart · Heikki Simola · Markku Viljanen

Abstract Lakes in Europe have a bipolar distribution by latitude with higher lake densities in the north (58–65° N) and south (38–48° N). By area, 95% of the large lakes (>100 km^2) are located at altitudes lower than 100 m above sea level (ASL) and only 1% lie higher than 1,000 m ASL. Physically large lakes exhibit several similarities to seas and oceans in their thermal structure and circulation dynamics. From the chemical point of view, lakes are important accumulation sites for substances transported from the watershed or built up in the lake itself but they may contribute positively to global greenhouse gas emission. Fauna and flora of ancient large lakes such as the Caspian Sea and Lake Ohrid include large numbers of endemic species, which become endangered if conditions change because of direct human impact, alien species invasions or climate change. Large lakes offer a wide range of ecosystem services to society, the multiple use of which creates multiple pressures on these water bodies such as nutrient load and toxic pollution, modification of hydrology and shore line structure, and shifting of the food web balance by stocking or harvesting various species. Although large lakes are among the best-studied ecosystems in the world, the application to them of environmental regulations such as the European Water Framework Directive is a challenging task and requires that several natural and management aspects specific to these water bodies are adequately considered.

Keywords Eutrophication · Toxic pollution · Climate change · Water Framework Directive · Alien species · Endemic species

Guest editors: T. Nõges, R. Eckmann, K. Kangur, P. Nõges, A. Reinart, G. Roll, H. Simola & M. Viljanen
European Large Lakes—Ecosystem changes and their ecological and socioeconomic impacts

P. Nõges · K. Kangur · T. Nõges
Centre for Limnology, Institute of Agricultural and Environmental Sciences, Estonian University of Life Sciences, Rannu, Tartu County 61101, Estonia

P. Nõges (✉)
Institute for Environment and Sustainability, European Commission, Joint Research Centre, Ispra, VA 21020, Italy
e-mail: peeter.noges@jrc.it

A. Reinart
Tartu Observatory, Tõravere, Tartu County 61601, Estonia

H. Simola · M. Viljanen
Faculty of Biosciences, Ecological Research Institute, University of Joensuu, P.O. Box 111, Joensuu 80101, Finland

Distribution and typology of large lakes in Europe

There exists no agreed definition of a large lake. Tilzer & Bossard (1992) analyzed the question of what constitutes a large lake. They did not suggest any numerical value but listed three main criteria that distinguish large lakes from smaller ones: the existence of a true and distinctive pelagic zone, the generally greater mixing depth, and the formation of areas with different physical structures causing patchiness in biological processes. Because surface area is by no means the only parameter that affects the manifestation of these large lake features, it is obvious that any attempt to create a 'natural' size classification based only on lake surface area will fail. However, for practical purposes, such classifications exist and are even widely used. For instance, the Water Framework Directive (Directive, 2000) considers a lake surface area of 10 km^2 as the boundary between medium-size and large lakes, and lakes with surface areas >100 km^2 as very large. Most countries that use the metric system traditionally divide their lakes by surface area for statistical purposes into classes of the power of 10, with the class $>10^2$ km^2 as the largest (Nixon, 1996; Lehner & Döll, 2004).

Water resources are globally unevenly distributed. Lehner & Döll (2004) investigated the latitudinal distribution of lakes and reservoirs as derived from the Global Lakes and Wetlands Database (GLWD) and found the highest concentration of lakes clearly marked throughout all size classes in the de-glaciated areas between 50° and 70° North, including Alaska, Canada, Scandinavia, and northern Russia. The latitudinal distribution of lake areas reflected the general pattern of lake numbers but had an additional and very prominent peak between 40° and 50° North, marking the dominating effect of the North American Great Lakes and the Caspian and Aral Seas.

A similar analysis based on the European Geometric Reference in Water Information System for Europe (WISE) showed a bipolar distribution of the lacustrine water bodies (lakes and reservoirs together) in Europe (Fig. 1). The northernmost peak between 65° and 68° North is mostly caused by lakes in the Kola peninsula and in northern Finland. The highest peak in the northern part (around 61° N) indicates the two large lakes in Russia, Ladoga and Onega, and the lake-rich area in southern Finland. Several large lakes in Sweden (Vänern and Vättern), Estonia (Võrtsjärv and Peipsi), and Russia (Ilmen, Rybinsk Reservoir) form a distinct peak between 58° and 59° N. A broad belt between 57° and 48° N is rather poor in lakes and reservoirs. The few large inland water bodies in this area are the Irish lakes, the heavily modified Dutch lakes IJsselmeer and Markemeer and a number of reservoirs on the Volga and Dnepr Rivers. The highest peak further south is caused mostly by the Caspian Sea but the large and deep Perialpine lakes, Lake Balaton, Neusiedler See and the Black Sea limans, also contribute to this peak. The second peak in the south around 38° N is caused by the large lakes in Turkey.

Analysis of the WISE data set revealed 192 lakes, reservoirs and coastal lagoons in Europe with surface areas exceeding 100 km^2. Seventy-five percent of the area of large lakes in Europe is accounted for by the Caspian Sea (Fig. 2); 95% is located at altitudes below 100 m above sea level (ASL), and only 1% is located above 1,000 m ASL (Fig. 2). Clustering of these water bodies on the basis of their morphometric parameters, catchment size and climatic characteristics (Nõges & de Jager, submitted) distinguished clusters that can simply be described as: (1) boreal lakes, (2) shallow lakes in maritime regions and coastal lagoons, (3) reservoirs of the Dnepr and Volga Rivers and (4) deep natural lakes in mostly high-altitude areas. Among those lake clusters, the boreal lakes and the Russian reservoirs often have a highly developed shoreline and the water bodies are divided into a number of sub-basins, which function similarly to smaller lakes. Water bodies in the shallow lake and deep lake groups, although very different from each other, exhibit most of the specific large lake features.

Large lakes as attractive study sites

Because of their high regional importance and often striking individuality, large lakes and reservoirs have become attractive study sites. Several of them such as lakes Balaton, Constance, Konnevesi, Maggiore, Neusiedl, Ohrid, Võrtsjärv and the Rybinsk Reservoir have their own limnological stations or institutes, which carry out long-term and year-round research on these water bodies. In some cases, well-equipped research vessels such as R/V Muikku on Lake Saimaa in Finland partly replace stationary lab facilities.

The focus of limnology is shifting from studies of ecosystem functioning towards studies of how that functioning will change in a changing environment. The major ongoing and foreseeable changes are associated with anthropogenic loading on the one hand – at present, mercifully offset for the most part by mitigating processes leading to diminishing strain – and climatic change on the other.

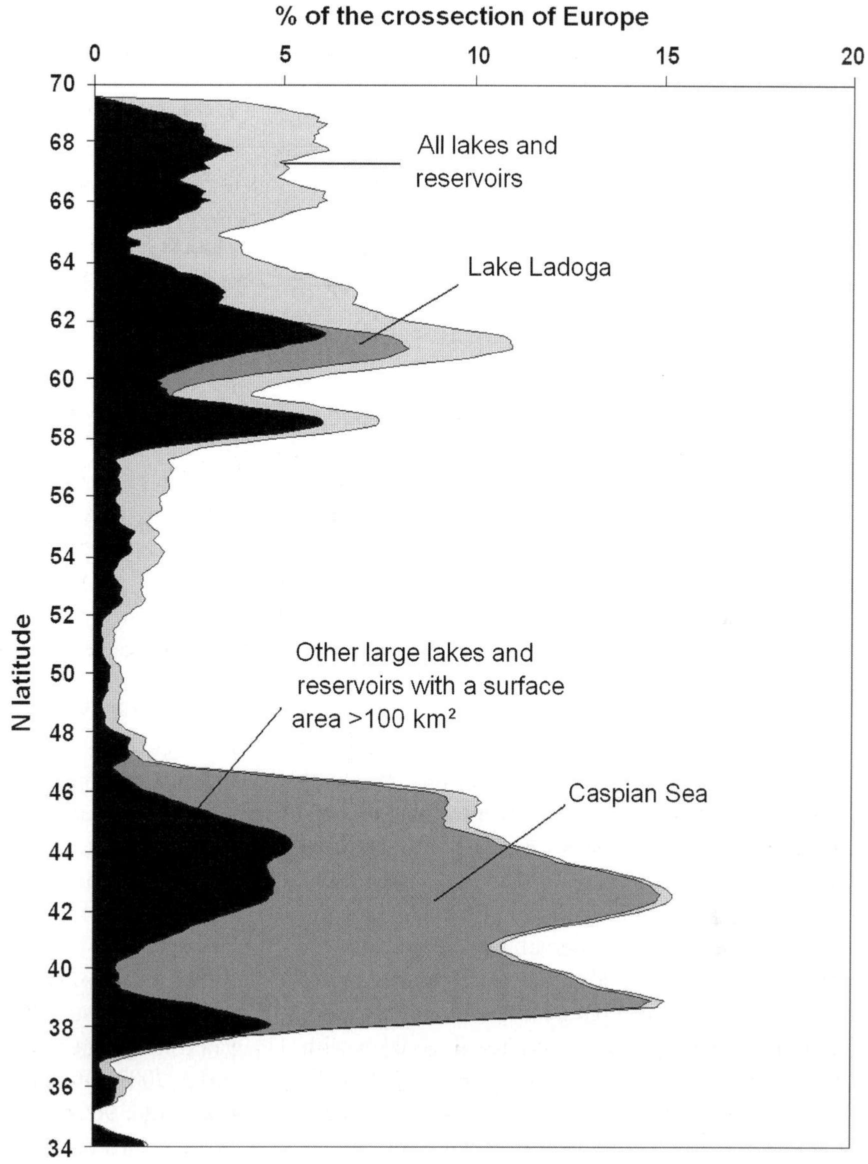

Fig. 1 Percentage of lakes within the latitudinal cross-section of Europe between −25° and 50° E. In addition, the Caspian Sea, reaching 54° E, was included in its entirety. Measurements made with a 0.1° latitudinal interval from the mosaic water bodies' layer in WISE (http://www.wise2.jrc.it) were smoothed with a moving average of 9. The mosaic is assembled from the following data sources with different spatial resolutions using the 'best available' strategy: the WFD Article 3 reports and the Geographic Information System for the European Commission (resolution variable), Shuttle Radar Topography Mission water-mask (90 m), CORINE pre-1999 landcover (25 m), Catchment characterization and modelling database (250 m), and the United Nations Landuse data (1,000 m)

For large lakes, system inertia must be particularly considered owing to their long retention times. Even quite significant trends may appear only as very subtle signals easily masked by the inevitable system noise.

Several approaches are available for tackling large lake system functioning in the medium-term, i.e., in the order of a decade, time-scales relevant for anthropogenic and climate-related changes. The approaches currently pursued can be grouped into three categories: (i) paleolimnological studies (past ecosystem development and assessment of pre-disturbance conditions); (ii) long-term monitoring of individual lake ecosystems (comprehensive data gathering of system compartments, direct observation of cause and effect processes and their variability, development of ecosystem models); and (iii) simultaneous monitoring of several lake ecosystems (observation of coherent responses of lakes to management measures or climatic forcing).

Similarity to seas and oceans

Large lakes exhibit structures and processes found in ocean basins (discrete coastal zones and shelf breaks, large scale circulation patterns, vertical stratification

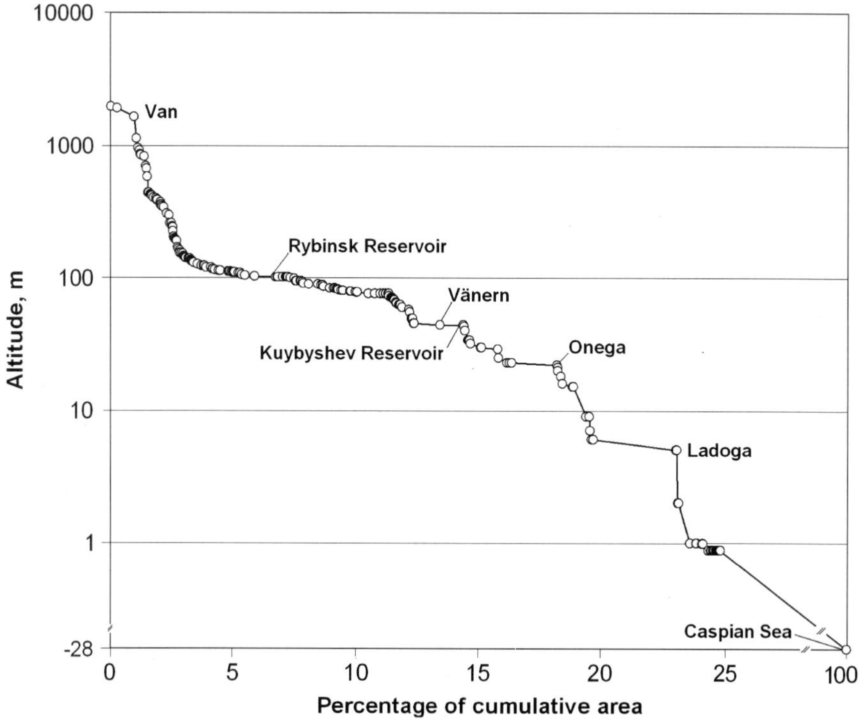

Fig. 2 Altitudinal distribution of the large lakes in Europe between −25° and 50° E. In addition, the Caspian Sea, reaching 54° E, was included in its entirety

of temperature, and biochemical processes, etc.), but they are bounded systems with constrainable inputs and outputs. Many key processes scale with lake size (Håkanson, 2004). Because of their size, large lakes exhibit complex hydrodynamic processes such as up- and down-welling, internal waves, and Kelvin waves (Paola et al., 2001). Physical oceanography has strongly influenced physical limnology (Johnson, 2003). In modelling and case studies, large lakes with their potentially well-constrained boundary conditions and high-resolution chronologies, have the potential to bridge the gap between continental-margin-scale, field-based studies and laboratory-scale experiments (Imberger, 1998; Blenckner, present issue). Application of methods such as productivity measurements in aquatic ecosystems and satellite remote sensing, commonly used in marine studies, have also become useful in large lakes research (Arst et al., present issue; Alikas & Reinart, present issue).

Dynamic processes have an impact on both the horizontal and vertical temperature distribution in large lakes and many biotic processes are largely wind-driven (Kvarnäs, 2001; Ambrosetti et al., 2003). During the stratified period, the thermocline prevents the transfer of heat and particles from the epilimnion to the lower layers and thus has implications for natural water quality. A strong thermocline acts as a "diffusion floor", suppressing vertical mixing and inhibiting the transport of mass, momentum, and heat into the hypolimnion.

The deep hypolimnion is partly isolated and participates to a lesser extent in seasonal thermal events. In deep waters, hydrostatic pressure leads to an increase in stability equivalent to a decrease in the water temperature of 0.2°C per 100 m (Wüest & Lorke, 2003). In deep lakes such as Como, Maggiore and Garda, 90% of the heat is trapped in the upper layers, which form only 20–40% of the lake volume (Ambrosetti & Barbanti, 2002).

One of the indicators of climate change is deep water warming, which is observed in large and deep European lakes and in deep tropical lakes, as well as in the Mediterranean below 2,000 m and in Greenland seas below 200 m (Dokulil et al., 2006 and many references therein). For example, the increase of deep water temperature in Lake Maggiore has accelerated from 0.015°C y^{-1} in 1963–1986 to 0.063°C y^{-1} in 1987–1998 (Ambrossetti & Barbanti, 1999); 12 deep lakes across Europe showed a coherent increase in deep water temperature of about 0.1–0.2°C per decade (Dokulil et al., 2006). Extended isolation of a potentially warmer hypolimnion from the atmosphere could drive a reduction in dissolved oxygen concentration, with potentially enormous impacts on nearly all segments of the ecosystem, particularly benthic metabolism, nutrient cycling, and fisheries.

The establishment of a so-called thermal bar is a phenomenon specific to northern large lakes such as Ladoga (Naumenko et al., 2006) or Vänern (Reinart & Reinhold, 2007) after ice break-up. The term refers to a barrier of 4°C water, which develops between the areas close to the shore and the central and more open parts of the lake. Between the thermal bar and the shore, the water warms more quickly and a thermocline is established.

Although there is a general long-term decreasing trend in the duration of ice cover around the northern hemisphere (Magnuson et al., 2000), opposite trends have also occurred during the last 50 years. For example, ice observations on Lake Ladoga since 1943 have shown an earlier freeze-up (−0.14 days/year) and no clear trend in the timing of ice break-up, resulting in a slight growth in ice duration (0.06 days/year). At the same time, the extent of areal ice cover on the lake has decreased slowly (Karetnikov & Naumenko, present issue).

Deposition sites, but not only

Sediment deposition, usually in the deepest parts of a lake basin, creates a paleolimnological record of past conditions of the lake ecosystem. Paleolimnological data are often indispensable for reconstructing changes in climate, the evolution of lakes, and the onset of human impact (Smol, 2004). The most detailed sedimentary records are provided by annually laminated or varved sediments, in which even seasonal patterns may be discernible (Lamoureux, 2001). Even in large lakes, varved sediment may deposit in deep areas, even though most large lake sediments are affected by physical or biological mixing, so that a given layer represents a more or less homogenized sample of sedimentation for some years or even decades (Håkanson & Jansson, 1983).

In addition to retrospection, quantitative paleolimnological analyses allow some prognostication, particularly in cases of ecosystem recovery from assault: pre-disturbance sediment may reveal the baseline situation towards which the system may be supposed to revert. In a paleolimnological study of the history of Lake Peipsi during the past century, Leeben et al. (present issue) found fossil pigments indicating the onset of eutrophication in the 1960s, and a continuing trend to eutrophication until the 1990s. However, a considerably earlier start for the eutrophication process was indicated by changes in the molecular size structure of pore-water-dissolved organic matter, which indicated that the contribution of autochthonous matter to the organic pool of the lake ecosystem had already started to increase around the late 1930s. The study of the diatom stratigraphy of Lake Võrtsjärv by Heinsalu et al. (present issue) demonstrates that the planktonic:periphytic diatom ratio in the sediment may be used to track water level changes; they found a good match between the ratio and instrumental water level records from 1871 up to the 1960s, before the onset of strong cultural eutrophication.

Although each lake basin is geologically a sink for organic carbon because of sediment accumulation, many lakes actually act as considerable sources of greenhouse gases owing to two major processes: heterotrophic respiration of allochthonous carbon (Jones, 2005), and methane emissions, notably from littoral macrophyte beds (Bastviken et al., 2004; Kankaala et al., 2004). The magnitude and seasonal patterns of all these processes vary widely in individual lakes. With regard to carbon sedimentation, Pajunen (2005) surveyed long-term (Holocene) carbon accumulation in Finnish lakes of different size classes, and arrived at an average value of 0.95 g C m^{-2} y^{-1} for the large Finnish lakes (>100 km^2, $n = 44$). This figure is about 30% of the corresponding estimate for a representative set of smaller lakes (1–10 km^2, $n = 30$). In Lake Vesijärvi, a large and shallow eutrophic lake (106 km^2, mean depth 6 m), Pajunen (2005) estimated the long-term C accumulation at 2 g C m^{-2} y^{-1} as an average for the whole lake bottom area, while Kankaala et al. (2004) measured annual methane emission from the littoral macrophyte zone of the same lake to be in the order of 70 g CH_4 m^{-2} y^{-1}. This illustrates the importance of the extensive reed (*Phragmites*) belts of many European lowland lakes for greenhouse gas balances. As estimated by Bastviken et al. (2004), the world's lakes contribute 6–16% of all natural methane emissions, and St. Louis et al. (2000) state that CH_4 emissions from man-made reservoirs contribute as much as 7% of the anthropogenic global warming equivalents.

Endemic, endangered and alien species in large lakes

Species diversity in general increases with lake size as larger lakes offer a greater degree of habitat diversity. This regularity has been demonstrated, for example, for bacteria (Reche et al., 2005), phytoplankton (Padisák, 2005) and birds (Sillen & Solbreck, 1977). Also, the food-chain length in 25 northern temperate lakes increases with ecosystem size (Post et al., 2000). However, lake surface area does not contribute positively to the mean richness of zooplankton (Hessen et al., 2006) and large lakes are also a harsh environment for macrophytes, since the growth of weakly rooted floating and floating-leaved plants is impeded by wind and wave action and sediment disturbance (Mäemets, 2005).

Many of the world's large lakes are famous for their endemic species – plants or animals that occur only in a single water body and nowhere else. The main premises for species endemism in large lakes are long age, unique and rather stable conditions, and long distances from other similar water bodies creating distribution barriers. Despite considerable human impact on water bodies throughout Europe, several endemic species have still survived in a number of large lakes. In the Caspian Sea on the border of Europe, 46% of the species in flora and fauna are endemic including 54 fish species, 190 zoobenthos species, and 64 zooplankton species (Zenkevich, 1963; Barannik et al., 2004). Similarly, the large number of endemic species in Lake Ohrid originates from the vast Neogenic brackish and freshwater inner seas in the southeast of Europe (Slugina, 2006). Because of their unique evolutionary history and highly specialised adaptations to the conditions in a single water body, endemic species are often the most endangered by changes in the ecosystems. The two major risks to species diversity are invasive species and eutrophication.

Many of the world's large lakes have been exposed to the introduction of exotic species (Hall & Mills, 2000). Those species have had major impacts on the ecosystems of such lakes through various processes including predation, disturbance, habitat modification and competition. The most common non-indigenous species in European large lakes are the zebra mussel *Dreissena polymorpha* Pallas (Ricciardi, 2003) and the Baikalian gammarid *Gmelinoides fasciatus* (Stebbing) (Timm & Timm, 1993). The latter is the most active invasive species in Eastern Europe (Pankova & Berezina, 2007). *G. fasciatus* was introduced in the 1960s and 1970s from the Lake Baikal basin into lakes and reservoirs in the former USSR with the purpose of enhancing fish production (Berezina, 2007). It was first found in Lake Peipsi in 1972 (Timm &Timm, 1993) and was subsequently recorded in Ladoga and Onega (Berezina, 2007); from Ladoga it was introduced to other aquatic systems.

Destruction of natural geographic barriers has resulted in expansions of several amphipod species. Since the 20th century, eight amphipod species of Ponto-Caspian, Baikalian and Atlantic origin have expanded widely within Russia (Berezina, 2007). The intentionally introduced *Pontogammarus robustoides* is the most successful amphipod invader of Lithuanian inland waters (Gumuliauskaite & Arbaciauskas, present issue). Unfortunately, as shown by Hall & Mills (2000), our understanding of the extent and impact of introductions on large lake ecosystems often remains speculative.

At the end of the 20th century, the comb jellyfish *Mnemiopsis leidyi* invaded the Caspian Sea (Ivanov et al., 2000). This organism can harvest nearly all plankton and puts the fragile ecosystem at great risk. Introductions of exotic species are a global problem that deserves global attention and understanding.

Species diversity in lakes usually increases with moderate eutrophication but drops rapidly at higher trophic states. Phytoplankton diversity peaks in the oligo-mesotrophic to moderately eutrophic portion of the trophic gradient (Padisák, 2005; Hajnal & Padisák, present issue). Macrophyte diversity also decreases with increasing nutrient loading, even though the standing stock may increase (Pokorný & Květ, 2004). In Peipsi, the expansion of reeds during the last 40 years has substantially changed the littoral communities and impoverished the floristic diversity (Mäemets, 2005). On the one hand, reed shades the light-demanding small-sized species (Mäemets & Freiberg, 2004), but on the other, the expanding reed belt provides new growth conditions for wind-sensitive shade-tolerant species (Mäemets, 2005). In contrast to Peipsi, a large-scale reed die-back has been observed in several large lakes in Central Europe during recent decades. In lake Fertö/Neusiedler See, one of the main causes of reed die-back was harvesting with heavy machines during unfavorable weather conditions, which considerably damaged reed rhizomes by rooting them out (Ágoston-Szabó et al., 2006). In Lake Constance, the extreme flood in 1999 caused serious damage and decrease of the reed stands (Schmieder et al., 2002).

Changes in fish dominants have occurred in several large lakes in Europe. In Peipsi and IJsselmeer, for example, these changes are quite similar. The fish community of Lake Peipsi has shifted from clean- and cold-water species such as vendace *Coregonus albula* (L.), whitefish *C. lavaretus* L. and burbot *Lota lota* (L.) towards more pikeperch *Sander lucioperca* and bream *Abramis brama*, which perfer productive warm and turbid waters (Kangur et al., 2007b). Warming of the aquatic environment, ongoing eutrophication and the resulting cyanocbacterial blooms have increased the mortality rate of smelt and decreased its abundance in Peipsi (Kangur et al., 2007a) as well as in IJsselmeer (Lammens, 2001).

Multiple use creating multiple pressures on large lakes

Mankind benefits from a multitude of resources and processes supplied by large lakes. These benefits, known as ecosystem services, include the use of lakes as sources of drinking water and fish, sites for recreation, waterways or places for dumping wastes. Multiple use of large lakes creates multiple pressures on the ecosystems that may lead to regime shifts, degradation of water quality and loss of some or most of the ecosystem services. Decision-making concerning the protection and management of large lakes is complicated and requires broadly based public participation. To support sustainable development, environmental education of the local population has to be on a high level (Várkuti et al., present issue).

Nutrient loading

While eutrophication has been extensively studied in Europe since the 1960s (Phillips et al., 2005; Smith et al., 2006), comparatively few studies exist on the response of large lakes to reduced nutrient loading, and most of those have focused on changes in phytoplankton (Lepistö et al., 1999; Willén, 2002, Holopainen et al., present issue), whereas other biological components have been only briefly covered (Haberman et al., 2004; Anderson et al., 2005; Jeppesen et al., 2005a, b). Most lakes in which external nutrient loading was reduced, approached a new equilibrium in 0–5 years for nitrogen and in 10–15 years for phosphorus, irrespective of hydraulic retention time (Anderson et al., 2005; Jeppesen et al., 2005a, b; Köhler et al. 2005; Søndergaard et al., 2005; Gerdeaux et al., 2006; Eckmann et al., 2007a). The effects of global change are likely to run counter to reductions in nutrient loading rather than reinforcing re-oligotrophication (Jeppesen et al., 2005a).

In the deep peri-alpine Lake Geneva, the total phosphorus concentration increased from 20 to 90 $\mu g\ l^{-1}$ during the eutrophication stage from 1960 to the mid-1970s (Anneville et al., 2007). In the early 1980s, restoration measures resulted in a progressive reduction of phosphorus concentration while the warming trend caused an increase in the stability of the water column. Contrary to expectations, phytoplankton biomass increased although P concentrations remained low. The paradoxical increase in total biomass was due to the accumulation of inedible filamentous algae.

Lake Constance underwent pronounced eutrophication until the beginning of the 1980s, mainly because of municipal phosphorus loading. Total phosphorus concentrations during winter mixing (TPmix) in Upper Lake Constance rose from 5–8 $\mu g\ l^{-1}$ in the 1950s to more than 80 $\mu g\ l^{-1}$ by the late 1970s and early 1980s (Bäuerle & Gadke, 1998; Eckmann et al., 2007a). During the following and continuing re-oligotrophication resulting from pollution control measures, TP decreased to 9 $\mu g\ l^{-1}$ by 2005, while the restoration target was set at 30 $\mu g\ l^{-1}$. While both the community structure and productivity reacted simultaneously to the increasing phase of nutrient loadings, the biotic response to decreasing P-loads was initially restricted to changes of composition; the reduction of productivity was delayed by more than a decade.

After the 1970s, the previously oligotrophic Lake Ladoga became mesotrophic, with elevated nutrient concentrations, reduced transparency and structural changes in the plankton, zoobenthos and fish communities: several species sensitive to eutrophication disappeared (Viljanen et al., 2007). Conditions in the whole lake, and especially at some of the most polluted sites, have improved since late 1990s owing to the closing of some sources of industrial pollution. The TP concentration has decreased from 25 to 15 μg l^{-1} but recovery is slow because of the long retention time (11 years). The present species composition points to mainly oligo-mesotrophic conditions in the pelagic areas but meso- or eutrophic conditions in the coastal regions.

Eutrophication of Lake Vänern, the largest lake in Sweden, caused primarily by the pulp industry and urban emissions, culminated in the late 1960s. However, the lake gradually became cleaner because of improved waste-water treatment and the closure of some of the older pulp mills (Wilander & Persson, 2001). The phytoplankton community reacted rapidly with increased taxon richness, and decreased biomasses of diatoms and cryptophytes in spring and of cyanobacteria in summer (Willén, 2001).

According to Nõges et al. (2007) and Leeben et al. (present issue), the loading of nitrogen to Lake Peipsi from the Estonian catchment decreased substantially from 1980–1991 to 1992–2004 (39%), whereas phosphorus loading decreased much less (13%). As a result, the N:P ratio decreased considerably in the loading (30%) but less in the lake water (18%). Kangur & Möls (present issue) show that in the long term, the N concentration in Peipsi is quite stable and not sensitive to year-by-year changes in riverine N load. In-lake cyanobacterial N_2 fixation compensates the decreased N-loading (Nõges et al., 2004), and the destabilization of the Lake Peipsi ecosystem results from too high a P loading (Nõges et al., 2005) and an N:P ratio below the critical value of 30 (Nõges et al., present issue).

Lake Pyhäjärvi in SW Finland, which deteriorated because of increased algal blooms, has shown some signs of recovery in recent years mainly due to intensive efforts at biomanipulation (Ventelä et al., 2007).

The impact of eutrophication on peri-alpine lake fisheries has been reported in numerous papers, and the sequence from salmonids to coregonids, from coregonids to percids, and finally from percids to cyprinids during the eutrophication since the 1970s has been documented (Gerdeaux et al., 2006; Eckmann et al., 2007a). While eutrophication continues to be a problem, re-oligotrophication has also become a problem for Central European lake fisheries (Gerdeaux et al., 2006). During the first phase of re-oligotrophication, the total fish yield remains nearly the same while coregonids become dominant, but when the TP drops below 5 μg l^{-1}, fish production and the total yield decrease rapidly (Eckmann et al., 2007a).

Climate warming and oligotrophication have largely opposing effects: species such as pikeperch, perch and many cyprinids are favored by climate change but not oligotrophication, while coregonids, salmonids, and burbot represent the opposite case.

In Lake Constance, re-oligotrophication has been reflected in an augmenting proportion of whitefish in commercial catches and an increasing age-at-capture (Eckmann & Rösch, 1998; Eckmann et al., 2007a), i.e., the fish community seems to be returning to the pre-eutrophication state.

Toxic pollution

In most European large lakes, concentrations of mercury, other heavy metals, persistent organic pollutants (POPs), adsorbable organic halogens (AOX), and DDT have substantially decreased but are still obvious in sediments and throughout the food web, and also in fish. In Vänern, alarmingly high concentrations of mercury, other heavy metals (Zn and Cd) and xenobiotic organic compounds were detected in the late 1960s, after which the pollution decreased radically (Håkanson, 1977). Despite a high mercury level in the sediments of Vänern, its levels in fish were relatively low compared to fish in other large lakes of the region that were not affected by local mercury effluents (Lindström, 2001). Currently, the situation regarding toxic pollution has greatly improved: the pollutants have largely been buried in sediments and their concentrations in the biota have decreased (Lindström, 2001).

From the 1960s until the 1990s, the concentration of DDT in fish in Lake Vättern decreased by 13%, and that of polychlorinated biphenyls (PCBs) by 5.4% (Lindell et al., 2001). The decrease was in the same order of magnitude as reported for other large lakes.

Lake Maggiore has been subject to heavy DDT contamination due to a chemical plant located near the main influent (Binelli et al., 2004). Since winter 2001–2002 a sharp increase in PCB pollution has been noticed, with values of about 3–4 μg g^{-1} lipids, probably due to the release of contaminated sediments from the numerous dams located in the watershed. Total concentrations of dioxin-like PCBs reached dangerous levels for the water community, and fish consumption may be a risk to human health especially for the resident population.

In reservoirs, commonly encountered anthropogenic organic chemicals that are poisonous, carcinogenic or suspected of serious health consequences include petroleum products, heavy fractions of mineral oil, PCBs, polycyclic aromatic hydrocarbons, phenols, various pesticide residues, and a recently discovered group of endocrine disrupters (Straškraba, 2005). In many reservoirs, there is a current risk of metal contamination, or the reservoir sediments already contain a burden of heavy metals and organic pollutants accumulated as a consequence of anthropogenic activities. In the Rybinsk Reservoir (Russia), the largest artificial water-body in Europe (4550 km^2), industrialization has resulted in contamination with PCBs (Chuiko et al., 2007). Ninety-five percent of PCBs accumulate in bottom sediments (German & Zakonnov, 2003), so after PCB discharge into the water body has been completely eliminated, their presence in the system will still be an ecological hazard for 25–30 years.

Stocking and over-fishing

Predator fishes (salmonids, pike, pikeperch) improve water quality through food chain interactions and are also the main target of sport and commercial fishery in many European large lakes. However, management of the predator–prey assemblage is still not well understood, mainly because of the complexity of the interrelated processes and the unpredictable variability in environmental factors governing the periodic recruitment of prey fish (Heikinheimo, 2000; Collares-Pereira & Cowx, 2004; Sarvala et al., 2004; Kangur et al., 2007a).

Stocking for salmonids and coregonids has been a common practice in peri-alpine deep lakes (Gerdeaux et al., 2006). Arctic char stocking is efficient and the yields of char and trout are influenced by stocking (Champigneulle & Gerdeaux, 1995), but the yield is low compared with whitefish yields. Intensive whitefish stocking is only practised in some of these lakes, such as Lake Constance and Lake Biel. The recent history of the whitefish fishery and stocking in Lake Geneva has been very positive (Gerdeaux, 2004). Although the contribution of hatchery-reared fish to the cohort 2003 was 62% in Lake Constance, it should not be concluded that stocking is necessary to maintain the current level of commercial yields, since it still remains unknown whether hatchery fish are added to the naturally recruited stock leading to stronger cohorts, or whether stocking only increases intraspecific competition without enhancing cohort size (Eckmann et al., 2007b).

Fish stockings in Finland involve practically all commercially and recreationally important fish species and all major watercourses. There has been debate about the possible role of brown trout stocking in the fluctuations of vendace and white fish stocks, and it has been recommended that brown trout stocking be restricted in vendace lakes (Helminen & Sarvala, 1994; Marttunen & Kylmälä, 1997; Vehanen et al., 1998). The gill-net fishing of whitefish is a problem in many Finnish large lakes (Heikinheimo, 2000) as many brown trout young that are still below the allowable catch size are taken as a by-catch in whitefish gill nets with small mesh sizes, which can lead to over-fishing. This decreases the profitability of brown trout stocking and weakens the potential for recreational fishing of brown trout. On the other hand, large mesh sizes in gill-net fishing may lead to the under-exploitation of whitefish, which in extreme cases causes dwarfing.

Introductions can also have negative effects on indigenous fish biodiversity (Winfield & Durie, 2004). Stocking Neusiedler See with *Ctenofaryngodon idella* led to the almost complete disappearance of submerged macrophytes, resulting in the loss of an important spawning habitat (Mikschi et al., 1996).

In Peipsi, which is inhabited by 37 fish species and is considered one of the best large fishing lakes in Europe, the recent yields of some 7,000 tons per year are about half of those 70 years ago. Over-exploitation has caused a significant decrease in the abundance of larger specimens of pikeperch both in Peipsi (Kangur et al. 2007b) and in IJsselmeer (Lammens, 2001). The natural population of eel *Anguilla anguilla* (L.) almost disappeared from the basins of Lakes Peipsi and Võrtsjärv owing to the construction of a hydropower station in the early 1950s (Kangur et al., 2002). As a result of regular stocking of elvers into Lake Võrtsjärv, eel has become the most important commercial fish there (Kangur, 1988).

The results of Huusko & Hyvärinen (2005) clearly demonstrated that in Finnish large lakes, pelagic planktivore fish (vendace) harvesting can work as a forcing agent and actually dictate the population fluctuations. Also, many top-predator fish stocks in Finnish freshwater systems have collapsed as a result of over-harvesting. Consequently, some of these communities have shifted into seemingly irreversible new states. Persson et al. (2007) have shown, for predators feeding on prey that exhibit food-dependent growth, that culling of fish prey may promote predator recovery. Fishery conflicts often prevail because of the lack of basic knowledge about the effects of long-term exploitation and policies for spatial allocation of fishing activities (Marjomäki & Huolila, 2001; Marjomäki et al., 2007).

Hydromorphological modifications

In the densely populated parts of Europe, housing and recreational development have led to extensive deterioration of the shorelines and littoral zones of most large lakes. Typical examples of lakes with very extensively altered shorelines are Balaton (Dömötörfy et al., 2003) and Constance (Ostendorp et al., 2003). Nowadays, better understanding of the pivotal importance of the littoral zone for the healthy functioning of lake ecosystems has led to stricter land use control and various littoral zone monitoring and restoration projects (Brix, 1999).

Hydrological engineering projects, with broadly different aims and variable degrees of intensity, are currently affecting practically all-large water bodies in Europe. In many cases, and especially in lowland lakes, the protection of economic interests demands control of the natural flooding of the system (van den Brink et al., 2005). Generally, the opposite case holds for hydropower: electricity production mostly calls for unnatural water level alterations and major hydromorphological modifications in the regulated basin.

Immense hydropower development projects were emblematic of the Soviet Union, and the total of more than 2000 reservoirs that were built in the former Soviet territory includes the largest reservoirs in Europe. The centralized planning system and total government control of land use made it possible to carry out the huge Rybinsk (completed in 1947) and Kuybyshevsky/Samara (1957) reservoir projects in the River Volga valley, which meant drowning numerous villages and towns and resettling over 600,000 people (Malik et al., 2000).

In the sparsely populated NW Russia, the Murmansk Oblast and Karelian Republic, several large reservoirs have been created by impounding lakes, e.g., in the River Kovda drainage area, and furthermore, several other large lakes are strongly regulated roughly within their natural bounds for hydropower production (for details, see Simola & Arvola, 2004). The typical spring draw-down of water by several meters invariably causes the thick winter ice to settle on the dried-out littoral zone, which is disastrous for winter-spawning fish and the whole littoral ecosystem.

Besides hydropower, the extensive construction projects of the Soviet waterways were also aimed at improving water traffic. Indeed, the river transport network connecting the Caspian, Black, Baltic and White Seas is a major infrastructural achievement. An unforeseen harmful consequence has been the effective spreading of numerous alien species along this traffic system in the bilge water of ships: for instance, several endemites of the brackish-water delta of the River Volga in the Caspian Sea have managed to establish themselves successfully in the Baltic (Olenin & Leppäkoski, 1999).

Various engineering operations are currently planned or pursued to protect the assets of lakes against extremes of weather or climatic change. Obvious risk areas are coastal lagoons and impoundments, such as the Dutch IJsselmeer (e.g., Breukers, 2000), for which even a small rise in oceanic level will have serious consequences. Another sensitive basin type will be the lake basins of arid climate zones, for which even subtle changes in the precipitation–evaporation balance may prove fatal. Such lakes are often additionally affected by water abstraction, e.g., for irrigation.

The huge, high-elevation Lake Van (3,700 km^2, max depth 450 m, elevation 1,719 m ASL) is a rather extreme example of an endorheic alkaline lake, for which major changes may be anticipated as a consequence of increased evaporation due to climatic change (Altunkaynak et al., 2003). The lake's paleolimnological sediment record indicates major level fluctuations connected with events during the climatic history of the Holocene (Wick et al., 2003).

Over-exploitation of the water resources of Lake Sevan caused a severe water level reduction of 19.2 m, which reduced water volume by 42.2%, decreased the hypolimnium volume by 90% and caused water quality degradation (Hovhanissian & Gabrielyan, 2000).

Climate change as an additional pressure on ecosystems

Although direct human impact on lakes still exceeds the changes attributable to climate change in most cases (Mastrantuono et al., present issue), the latter occurring on top of all other anthropogenic pressures may worsen the ecological status of lake ecosystems. According to climate change projections (Räisänen et al., 2004), the warming in Northern Europe will be greatest in winter or late autumn. By 2070–2100, Central and Southern Europe may warm locally by 6–10°C. The amount of precipitation will generally increase in Northern Europe especially in winter and will decrease in Southern and Central Europe in summer. The belt characterised by the smallest observed and projected changes in both temperature and precipitation crosses the Baltic States approximately from NW to SE. That is why several long-term changes, e.g., in Ladoga attributable to climate change are rather weak (Naumenko, present issue; Karetnikov & Naumenko, present issue).

An application of the climate scenarios combined with a physical lake model to a lake in middle Sweden predicted that by 2070, the lake will be totally ice-free 2 years out of 10 (Blenckner et al., 2002). Leppäranta & Wang (present issue) demonstrated that ice cover stability is a function of ice thickness and lake size. When the ratio of ice thickness to lake size is above $\sim 10^{-5}$, the ice cover is stable; otherwise, mechanical forcing breaks it. This transition enables a convenient distinction to be made between small and large lakes. As the climate changes, the transition size between small and large lake ice cover will change.

Other climate change effects also differ between large and small lakes. Comparing large and small lakes in Sweden, Weyhenmeyer (present issue) demonstrated that for climate change effects that act via the catchment, substantial time lags have to be expected in large lakes with a long water retention time, whereas climate change effects that act via the lake surface do not depend on lake size. It is essential to differentiate between these two types of effect in order to assess the impacts of climate change and the adaptation and vulnerability of lake ecosystems.

The effects of climate change on shallow temperate lakes will mimic the effects of eutrophication. Climate change is expected to lead to increased external loading and the critical nutrient load during eutrophication will decrease (Mooij et al., 2007). Even moderate warming has the potential to exacerbate existing eutrophication problems in the long run. Climate change will in most cases enhance the development of cyanobacterial blooms by increasing surface water temperatures and water column stability. However, the weakening of the spring turnover in deep perialpine lakes in warmer winters implies that nutrients that have accumulated in the hypolimnion during the previous stratification period will not be uniformly distributed vertically throughout the whole water column. Because of these climate-dependent differences in nutrient up-welling from year to year, phosphorus concentrations in the epilimnion of Lakes Constance (Straile et al., 2003) and Garda (Salmaso, 2005) tended to be lower after warmer winters, whereas the opposite was true for the hypolimnion.

Large lakes may have better resistance against harmful algal blooms than smaller ones. Jacquet et al. (2005) found that *Planktothrix rubescens* bloomed every year in Lac du Bourget (44 km^2) but not in Lake Geneva (582 km^2), even though the nutrient status and the phytoplankton richness were very similar in these two ecosystems. The authors suggested that this difference was caused by lower water column stability in the larger lake. Earlier and stronger water column stratification in smaller lakes, which influences light availability, nutrient inputs, sinking rates of cells, etc., has been shown to be advantageous for many bloom-forming cyanobacteria such as *Nodularia*, *Aphanizomenon* and *Planktothrix* (Kanoshina et al., 2003).

It is often difficult to detect the responses of lake ecosystem parameters to weak climate change signals and to distinguish them from those caused by large seasonal or interannual variability or other factors.

One way to overcome this problem is to study the coherence of changes over a network of water bodies. As a rule first described by George et al. (2000) and later verified in other lake studies (Dokulil & Teubner, 2002), the response of lakes to climate forcing is most coherent for physical parameters such as water temperature and ice-out dates. Changes in the chemical regimes of lakes are less coherent and depend strongly on lake type and local conditions. Because the interactions are complex, biological changes induced by climate change are less coherent.

Conventional sampling methods cannot produce enough data to describe the spatial and temporal heterogeneity in large lake systems adequately. Modern technologies using automatic measurement stations and/or satellite imagery (Alikas & Reinart, present issue) together with modelling approaches will open new horizons for studying the responses of large lakes to external forcing factors.

Magnuson et al. (2004) showed that ice-free dates have the greatest coherence because climatic variability dominates the processes involved; near-bottom water temperature in summer has lower coherence because individual morphometry (depth, fetch) influences vertical mixing; and the biological variables have the lowest coherence between lakes.

Implementation of the Water Framework Directive on large lakes

Directive 2000/60/EC of the European Parliament and Council, widely known as the Water Framework Directive (WFD), aims to protect water bodies from further degradation. As WFD requires the type-specific ecological status of water bodies to be determined (CIS, 2003a), lake size is an important characteristic of typology because large lakes react differently from small lakes to pressure factors.

There are several challenges regarding the implementation of WFD for large lakes:

1. The number of lakes of comparable size is small, and this does not allow reference sites to be used to define undisturbed conditions.
2. Historically, large lakes have usually been exploited for longer than small lakes and one has to consider carefully how far back in history to go to find reference conditions.
3. Different parts of large lakes may belong to different lake types and have different ecological qualities (CIS, 2003b). This is most relevant for the Nordic lakes with complicated shorelines, e.g., Saimaa in Finland (Pilke et al., 2002).
4. The effects of commercial fisheries on fish community structure and, through the trophic cascade, on lower trophic levels used in water quality assessment, are not explicitly addressed in the directive.
5. Large lakes may have voluntarily or accidentally introduced exotic species, e.g., zebra mussel, causing low-grade water quality according WFD, although most of the common indicators show good water quality.

As many of the large lakes have been studied seasonally for many years, they have often become valuable model objects for studying pressure–impact relationships in detail. This is needed as a basis for setting the WFD water quality class boundaries. In recent years, several large lakes have been included in European research projects.

The EC project EUROLAKES aimed to develop an integrated water resource management strategy for a number of important deep European lakes (Lac du Bourget, Bodensee, Loch Lomond, Lake Längelmävesi-Roine and Lac Léman) and their catchments. This project contributed to the estimation of ecological criteria and made recommendations for improving environmental indicators and meta-indicators in relation to the monitoring and assessment of large deep lakes and their catchments. An integrated water resource management tool for lake managers, decision makers and stakeholders was provided (Dickinson et al., 2003). In the EC project ECOFRAME, which developed ecological quality and functioning of shallow lake ecosystems with respect to the needs of WFD, the Estonian Lake Võrtsjärv was the only representative large lake (Moss et al., 2003). In the EC project CLIME, the impacts of climate on lakes in Europe were assessed and particular attention was paid to the water quality variables that are used as diagnostic elements in the WFD. The objectives of the project were to analyze the historical pattern of change observed in a network of lakes distributed throughout northern, western and central Europe and to develop models that can be used to simulate the responses of lakes to changes in the weather. Among the CLIME sites, the large lakes Peipsi, Mälaren, Balaton, Võrtsjärv, and Constance were considered (Blenckner et al., 2007; Weyhenmeyer et al., 2007). The EC project EURO-LIMPACS analyzes the impacts of climate change on the hydrological and hydrophysical conditions in deep large lakes such as Constance, Maggiore, Lomond, Hornindalsvatn and their catchments. To assess the interactions of climate change, eutrophication and restoration and their impact on WFD reference conditions, a paleoecological database from a number of European lakes has been compiled and is being used (Wade, 2006).

The WFD designates water bodies that are hydromorphologically substantially altered, e.g., by damming for hydropower, to the category of heavily modified water bodies (HMWB). According to the WFD, 'good ecological potential' should be achieved for these. Hydromorphological modifications are among most important pressures on many large lakes such as Oulujärvi and Kemijärvi in Finland or Vänern and Suorvajaure in Sweden (Marttunen et al., 2006). In the Netherlands, the former Southern Sea estuary was transformed into the freshwater Lake IJsselmeer with the closure of the Afsluitdijk in 1932. Owing to the change in category, this lake is considered an HMWB, and a natural lake rather than an estuary would be the starting point for determining its ecological potential (Lammens et al., present issue).

The WFD represents a new approach to water management in most European countries, as it stipulates a management model based on hydrological borders and watersheds as opposed to traditional political and administrative borders. River basins cut by state borders are called transboundary basins; they cover about 45% of the earth's continental land surface, excluding the Antarctic (Wolf et al., 1999). There are more than 150 large transboundary rivers and some 20 international lakes in Europe. The best-known and largest international lakes on the EU border are Lake Constance/Bodensee (Germany, Austria and Switzerland), Lake Geneva (Switzerland, France), Lake Prespa (Albania, Greece and the Former Yugoslav Republic of Macedonia), and Lake Peipsi/Chudskoye (Estonia and Russian Federation). These lakes are managed through bi- or multilateral cooperation between the member and non-member states, forming a particular exception within their own category in WFD (Grönlund & Määttä, present issue). The EC project MANTRA-East tested the applicability of WFD in EU border regions using as test area the watershed of Lake Peipsi, the largest international lake in Europe (Gooch & Stålnacke, 2006). Problems of fisheries and exotic species were addressed and various approaches (modelling, expert opinion, paleolimnology) were applied to define reference conditions and assess ecological status, which appeared to be 'moderate' (Nõges & Nõges, 2006). Premazzi et al. (2003) summarized the situation of the most important international lakes in North Italy. Lake Como, which has 10% of its drainage basin in Switzerland, has three distinct sub-basins that supposedly need to be considered as different water bodies. Lake Maggiore is the second largest Italian lake: 50% of its drainage basin is in Swiss territory while 80% of the lake surface is in Italian territory. The assessment revealed that in Lake Maggiore the ecological objectives have been already achieved; relative improvement has occurred in Lake Como, while the water quality in this lake remains below the set ecological objectives.

Lakes Ohrid and Prespa form a very unusual transboundary lake system between Albania, Macedonia and Greece (Matzinger et al., 2006). Almost the entire outflow from Lake Prespa flows into the 160 m lower Lake Ohrid through karst channels, while 65% of the transported phosphorus is retained within the aquifer, reducing the threat to Lake Ohrid, which is famous for its large number of endemic and relict species. However, a predicted potential fourfold increase in the phosphorus load from Lake Prespa could lead to a 20% increase in the phosphorus content of Lake Ohrid, which could jeopardize its fragile ecosystem. While being a potential future danger to Lake Ohrid, Lake Prespa itself is substantially endangered by water losses to irrigation, which have been shown to amplify its eutrophication.

Acknowledgments The authors are grateful to Dr. Alfred de Jager from the EC Joint Research Centre for the analysis of the spatial distribution of lakes in the WISE database and to Tuula Toivanen for technical help during the preparation of the manuscript. Financial support was provided by the Estonian Environmental Investment Centre.

References

Ágoston-Szabó, E., M. Dinka, L. Némedi & G. Horváth, 2006. Decomposition of Phragmites australis rhizome in a shallow lake. Aquatic Botany 85: 309–316.

Alikas, K. & A. Reinart, present issue. Validation of the MERIS products on large European lakes – Peipsi, Vänern and Vättern. Hydrobiologia. DOI: 10.1007/s10750-007-9212-0.

Altunkaynak, A., M. Özger & Z. Sen, 2003. Triple diagram model of level fluctuations in Lake Van Turkey. Hydrology and Earth System Sciences 7: 235–244.

Ambrosetti, W. & L. Barbanti, 1999. Deep water warming in lakes: an indicator of climate change. Journal of Limnology 58: 1–9.

Ambrosetti, W. & L. Barbanti, 2002. Physical limnology of Italian lakes. 2. Relationships between morphometric parameters, stability and Birgean work. Journal of Limnology 61: 159–167.

Ambrosetti, W., L. Barbanti & N. Salai, 2003. Residence time in lakes: science, management, education. Journal of Limnology 62(Supplement 1): 1–15.

Anderson, N. J., E. Jeppesen & M. Søndergaard, 2005. Ecological effects of reduced nutrient loading (oligotrophication) on lakes: an introduction. Freshwater Biology 50: 1589–1593.

Anneville, O., J. C. Molinero, S. Souissi, G. Balvay & D. Gerdeaux, 2007. Long-term changes in the copepod community of Lake Geneva. Journal of Plankton Research 29: 49–59.

Arst, H., T. Nõges, P. Nõges & B. Paavel, present issue. Relationships between primary phytoplankton production in situ, chlorophyll concentration and underwater irradiance in turbid lakes. Hydrobiologia. DOI: 10.1007/s10750-007-9213-z.

Barannik, V., O. Borysova & F. Stolberg, 2004. The Caspian Sea region: environmental change. Ambio 33: 45–51.

Bastviken, D., J. Cole, M. Pace & L. Tranvik, 2004. Methane emissions from lakes: dependence of lake characteristics, two regional assessments, and a global estimate. Global Biogeochemical Cycles 18: GB4009, DOI: 10.1029/2004GB002238.

Bäuerle, E. & U. Gadke (eds), 1998. Lake Constance, characterization of an ecosystem in transition. Archives Hydrobiologica Special Issues on Advanced Limnology, Vol. 53. 610 pp.

Berezina, N. A., 2007. Changes in the aquatic systems of northeastern Europe after invasion by *Gmelinoides fasciatus*. In Gerardi, F. (ed.), Biological Invaders in Inland Waters: Profiles, Distribution, and Threats. Springer, 479–493.

Binelli, A., F. Ricciardi & A. Provini, 2004. Present status of POP contamination in Lake Maggiore (Italy). Chemosphere 57: 27–34.

Blenckner, T., present issue. Models as tools for understanding past, recent and future changes in large lakes. Hydrobiologia. DOI: 10.1007/s10750-007-9210-2.

Blenckner, T., A. Omstedt & M. Rummukainen, 2002. A Swedish case study of contemporary and possible future consequences of climate change on lake function. Aquatic Sciences 64: 171–184.

Blenckner, T., R. Adrian, D. M. Livingstone, E. Jennings, G. A. Weyhenmeyer, D. G. George, T. Jankowski, M. Järvinen, C. N. Aonghusa, T. Nõges, D. Straile & K. Teubner, 2007. Large-scale climatic signatures in lakes across Europe: a meta-analysis. Global Change Biology 13: 1314–1326.

Breukers, C. M., 2000. Creating water management strategies for the northern part of Holland using a collaborative planning process. IAHS-AISH Publication 272: 51–56.

Brix, H., 1999. The European research project on reed die-back and progression (EUREED). Limnologica 29: 5–10.

Champigneulle, A. & D. Gerdeaux, 1995. Survey, management and recent rehabilitation of the Artic Char (*Salvelinus alpinus*) fishery in the French–Swiss Lake Leman. Nordic Journal of Freshwater Research 71: 173–182.

Chuiko, G. M., D. E. Tillitt, J. L. Zajicek, B. A. Flerov, V. M. Stepanova, Y. Y. Zhelnin & V. A. Podgornaya, 2007. Chemical contamination of the Rybinsk Reservoir, northwest Russia: relationship between liver polychlorinated biphenyls (PCB) content and health indicators in bream (*Abramis brama*). Chemosphere 67: 527–536.

CIS, 2003a. River and lakes – typology, reference conditions and classification systems. Common Implementation Strategy for the Water Framework Directive (2000/60/EC), Guidance document 10, European Commission, 86 pp. http://www.circa.europa.eu.

CIS, 2003b. Identification of water bodies. Common Implementation Strategy for the Water Framework Directive (2000/60/EC), Guidance document 2, European Commission, 28 pp. http://www.circa.europa.eu.

Collares-Pereira, M. J. & I. G. Cowx, 2004. The role of catchment scale environmental management in freshwater fish conservation. Fisheries management and Ecology 11: 303–312.

Dickinson, G., N. Lewis & M. O'Hare, 2003. Managing large lakes in Europe – the implications of the European Water Framework Directive for the integrated management of European deep water lakes. Department of Geography & Geomatics, University of Glasgow. http://www.geog.gla.ac.uk/olpapers/gdickinson001.pdf.

Directive, 2000. Directive 2000/60/EC of the European Parliament and of the council of 23 October 2000 establishing a framework for community action in the field of water policy. Official Journal of the European Communities L 327: 1–72.

Dokulil, M. T. & K. Teubner, 2002. The spatial coherence of alpine lakes. Verhandlungen der Internationalen Vereinigung für Limnologie 28: 1–4.

Dokulil, M. T., A. Jagsch, D. G. George, O. Anneville, T. Jankowski, B. Wahl, B. Lenhart, T. Blenckner & K. Teubner, 2006. Twenty years of spatially coherent deepwater warming in lakes across Europe related to the North Atlantic Oscillation. Limnology and Oceanography 51: 2787–2793.

Dömötörfy, Z., D. Reeder & P. Pomogyi, 2003. Changes in the macro-vegetation of the Kis-Balaton Wetlands over the last two centuries: a GIS perspective. Hydrobiologia 506–509: 671–679.

Eckmann, R. & R. Rösch, 1998. Lake Constance fisheries and fish ecology. Archiv für Hydrobiologie 53: 285–302.

Eckmann, R., D. Gerdeaux, R. Müller & R. Rösch, 2007a. Re-oligotrophication and whitefish fisheries management – a workshop summary. Advances in Limnology 60: 353–360.

Eckmann, R., M. Kugler & C. H. Ruhle, 2007b. Evaluating the success of large-scale whitefish stocking at Lake Constance. Advances in Limnology 60: 361–368.

George, D. G., J. F. Talling & E. Rigg, 2000. Factors influencing the temporal coherence of five lakes in the English Lake District. Freshwater Biology 43: 449–461.

Gerdeaux, D., 2004. The recent restoration of the whitefish fisheries in Lake Geneva: the roles of stocking, reoligotrophication, and climate change. Annales Zoologici Fennici 41: 181–189.

Gerdeaux, D., O. Anneville & D. Hefti, 2006. Fishery changes during re-oligotrophication in 11 peri-alpine Swiss and French lakes over the past 30 years. Acta Oecologica 30: 161–167.

German, A. V. & V. V. Zakonnov, 2003. Accumulation of polychlorinated biphenyls in the Sheksninskii Pool of the Rybinsk Reservoir. Water Resources 30: 524–528.

Gooch, G. D. & P. Stålnacke (eds), 2006. Integrated Transboundary Water Management in Theory and Practice: Experiences from the New EU Eastern Borders. IWA Publishing, London, UK.

Grönlund, E. & T. Määttä, present issue. Implications of flexibility in European Community environmental law: exemptions from environmental objectives in the Water Framework Directive. Hydrobiologia. DOI: 10.1007/s10750-007-9214-y.

Gumuliauskaitė, S. & K. Arbačiauskas, present issue. The impact of the invasive Ponto-Caspian amphipod *Pontogammarus robustoides* on littoral communities in Lithuanian lakes. Hydrobiologia. DOI: 10.1007/s10750-007-9209-8.

Haberman, J., E. Pihu & A. Raukas (eds), 2004. Lake Võrtsjärv. Estonian Encyclopedia Publishers, Tallinn.

Hajnal, E. & J. Padisák, present issue. Analysis of long-term ecological status of Lake Balaton based on the ALMOBAL phytoplankton database. Hydrobiologia. DOI: 10.1007/s10750-007-9207-x.

Håkanson, L., 1977. Sediments as indicators of contamination – investigations in the four largest Swedish lakes. Statens Naturvårdsverket, Solna, Rapport 839.

Håkanson, L., 2004. Lakes: Form and Function. The Blackburn Press, New Jersey: 201 pp.

Håkanson, L. & M. Jansson, 1983. Principles of Lake Sedimentology. Springer-Verlag, Berlin: 316 pp.

Hall, S. R. & E. L. Mills, 2000. Exotic species in large lakes of the world. Aquatic Ecosystems Health and Management 3: 105–135.

Heikinheimo, O., 2000. Management of coregonid fisheries: multiform and multispecies problems. University of Helsinki, Faculty of Agriculture and Forestry, Department of Limnology and Environmental Protection. Doctoral dissertation. ISBN 951-776-300-X (paperback). ISBN 951-776-302-6 (pdf).

Heinsalu, A., H. Luup, T. Alliksaar, P. Nõges & T. Nõges, present issue. Water level changes in a large shallow lake as reflected by the plankton:periphyton-ratio of sedimentary diatoms. Hydrobiologia. DOI: 10.1007/s10750-007-9206-y.

Helminen, H. & J. Sarvala, 1994. Population regulation of vendace (*Coregonus albula*) in Lake Pyhäjärvi, southwest Finland. Journal of Fish Biology 45: 387–400.

Hessen, D. O., B. A. Faafeng, V. H. Smith, V. Bakkestuen & B. Walseng, 2006. Extrinsic and intrinsic controls of zooplankton diversity in lakes. Ecology 87: 433–443.

Holopainen, A.-L., L. Lepistö, R. Niinioja & A. Rämö, present issue. Spatiotemporal and long-term variation in phytoplankton communities in the oligotrophic Lake Pyhäjärvi on the Finnish-Russian boarder. Hydrobiologia. DOI: 10.1007/s10750-007-9203-1.

Hovhanissian, R. & B. Gabrielyan, 2000 Ecological problems associated with the biological resource use of Lake Sevan, Armenia. Ecological Engineering 16: 175–180.

Huusko, A. & P. Hyvärinen, 2005. A high harvest rate induces a tendency to generation cycling in a freshwater fish population. Journal of Animal Ecology 74: 525–531.

Imberger, J., 1998. Physical Processes in Lakes and Oceans. Coastal and Estuarine Studies, Vol. 54. American Geophysical Union, Washington.

Ivanov, V. P., A. M. Kamakin, V. B. Ushivtzev, T. Shiganova, O. Zhukova, N. Aladin, S. I. Wilson, G. R. Harbison & H. J. Dumont, 2000. Invasion of the Caspian Sea by the comb jellyfish *Mnemiopsis leidyi* (Ctenophora). Biological Invasions 2: 255–258.

Jacquet, S., J.-F. Briand, C. Leboulanger, C. Avois-Jacquet, L. Oberhaus, B. Tassin, B. Vinçon-Leite, G. Paolini, J.-C. Druart, O. Anneville & J.-F. Humbert, 2005. The proliferation of the toxic cyanobacterium *Planktothrix rubescens* following restoration of the largest natural French lake (Lac du Bourget). Harmful Algae 4: 651–672.

Jeppesen, E., M. Søndergaard, J. P. Jensen, K. Havens, O. Anneville, L. Carvalho, M. F. Coveney, R. Deneke, M. Dokulil, B. Foy, D. Gerdeaux, S. E. Hampton, K. Kangur, J. Köhler, S. Körner, E. Lammens, T. L. Lauridsen, M. Manca, R. Miracle, B. Moss, P. Nõges, G. Persson, G. Phillips, R. Portielje, S. Romo, C. L. Schelske, D. Straile, I. Tatrai, E. Willén & M. Winder, 2005a. Lake responses to reduced nutrient loading–an analysis of contemporary long-term data from 35 case studies. Freshwater Biology 50: 1747–1771.

Jeppesen, E., J. P. Jensen, M. Søndergaard & T. L. Lauridsen, 2005b. Response of fish and plankton to nutrient loading reduction in eight shallow Danish lakes with special emphasis on seasonal dynamics. Freshwater Biology 50: 1616–1627.

Johnson, T. C. (ed.), 2003. Science of freshwater inland seas (SOFIS) contributions from the SOFIS working group. Large Lakes Observatory Technical Report 2003-1, University of Minnesota, Duluth.

Jones, R. I., 2005. Limnology of humic waters: special theme or universal framework? Plenary lecture. Verhandlungen der Internationalen Vereinigung für Limnologie 29: 51–60.

Kangur, A., 1988. An estimation of the eel fishery in Lake Võrtsjärv. Proceedings of the Academy of Sciences of the Estonian SSR. Biology 37: 307–311.

Kangur, K. & T. Möls, present issue. Changes in spatial distribution of phosphorus and nitrogen in the large north-temperate lowland Lake Peipsi (Estonia/Russia). Hydrobiologia. DOI: 10.1007/s10750-007-9204-0.

Kangur, A., A. Kangur & P. Kangur, 2002. The stock and yield of the European eel, *Anguilla anguilla* (L.), in large lakes of Estonia. Proceedings of the Estonian Academy of Sciences. Biology. Ecology 51: 45–61.

Kangur, A., P. Kangur, K. Kangur & T. Möls. 2007a. The role of temperature in the population dynamics of smelt

Osmerus eperlanus eperlanus m. *spirinchus* Pallas in Lake Peipsi (Estonia/Russia). Hydrobiologia 584: 433–441.

Kangur, K., Y.-S. Park, A. Kangur, P. Kangur & S. Lek, 2007b. Patterning long-term changes of fish community in large shallow Lake Peipsi. Ecological Modelling 203: 34–44.

Kankaala, P., A. Ojala & T. Käki, 2004. Temporal ans spatial variation in methane emissions from a flooded transgression shore of a boreal lake. Biogeochemistry 68: 297–311.

Kanoshina, I., U. Lips & J.-H. Leppänen, 2003. The influence of weather conditions (temperature and wind) on cyanobacterial bloom development in the Gulf of Finland (Baltic Sea). Harmful Algae 2: 29–41.

Karetnikov, S. G. & M. A. Naumenko, present issue. Recent trends in Lake Ladoga ice cover. Hydrobiologia. DOI: 10.1007/s10750-007-9211-1.

Köhler, J., S. Hilt, R. Adrian, A. Nicklisch, H. P. Kozerski & N. Walz, 2005. Long-term response of a shallow, moderately flushed lake to reduced external phosphorus and nitrogen loading. Freshwater Biology 50: 1639–1650.

Kvarnäs, H., 2001. Morphometry and hydrology of the four large lakes of Sweden. Ambio 30: 467–474.

Lammens, E. H. R. R., 2001. Consequences of biomanipulation for fish and fisheries. FAO Fisheries Circular 952.

Lammens, E., F. van Luijn, Y. Wessels, H. Bouwhuis, R. Noordhuis, R. Portielje & D. van der Molen, present issue. Towards ecological goals for the heavily modified lakes in the IJsselmeer area, the Netherlands. Hydrobiologia. DOI: 10.1007/s10750-007-9202-2.

Lamoureux, S., 2001. Varve chronology techniques. In Last, W. M. & J. P. Smol (eds), Tracking Environmental Change Using Lake Sediments. Volume 1: Basin Analysis, Coring, and Chronological Techniques. Kluwer Academic Publishers, Dordrecht: 247–260.

Leeben, A., I. Tõnno, R. Freiberg, V. Lepane, N. Bonningues, N. Makarõtševa, A. Heinsalu & T. Alliksaar, present issue. History of anthropogenically mediated eutrophication of Lake Peipsi as revealed by the stratigraphy of fossil pigments and molecular size fractions of pore-water dissolved organic matter. Hydrobiologia. DOI: 10.1007/s10750-007-9199-6.

Lehner, B. & P. Döll, 2004. Development and validation of a global database of lakes, reservoirs and wetlands. Journal of Hydrology 296: 1–22.

Lepistö, L., A. Räike & O.-P. Pietiläinen, 1999. Long-term changes of phytoplankton in a eutrophicated boreal lake during the past one hundred years (1893–1998). Algological Studies 94: 223–244.

Leppäranta, M. & K. Wang, present issue. The ice cover on small and large lakes: scaling analysis and mathematical modelling. Hydrobiologia. DOI: 10.1007/s10750-007-9201-3.

Lindell, M. J., G. Bremle, O. Broberg & P. Larsson, 2001. Monitoring of persistent organic pollutants (POPs): examples from Lake Vättern, Sweden. Ambio 30: 545–551.

Lindström, L., 2001. Mercury in sediment and fish communities of Lake Vänern, Sweden: recovery from contamination. Ambio 30: 538–544.

Mäemets, H., 2005. Relationships of the macrophyte vegetation of Estonian lakes with environmental conditions and changes in the course of anthropogenuic eutrophication. A thesis for applying for the degree of Doctor of Philosophy in Hydrobiology. Estonian Agricultural University.

Mäemets, H. & L. Freiberg, 2004. Characteristics of reeds on Lake Peipsi and the floristic consequences of their expansion. Limnologica 34: 83–89.

Magnuson, J. J., D. M. Robertson, B. J. Benson, R. H. Wynne, D. M. Livingstone, T. Arai, R. A. Assel, R. G. Barry, V. Card, E. Kuusisto, N. G. Granin, T. D. Prowse, K. M. Stewart & V. S. Vuglinski, 2000. Historical trends in lake and river ice cover in the northern hemisphere. Science 289: 1743–1746.

Magnuson, J. J., B. J. Benson & T. K. Kratz, 2004. Patterns of coherent dynamics within and between lake district at local to intercontinental scales. Boreal Environmental Research 9: 359–369.

Malik, L. K., N. I. Koronkevich, I. S. Zaitseva & E. A. Barabanova, 2000. Development of dams in the Russian federation and NIS countries, A WCD briefing paper prepared as an input to the world commission on dams, Cape Town, www.dams.org.

Marjomäki, T. J. & M. Huolila, 2001. Long-term dynamics of pelagic fish density and vendace (*Coregonus albula* (L.)) stocks in four zones of a lake differing in trawling intensity. Ecology of Freshwater Fish 10: 65–74.

Marjomäki, T. J., M. Lindroos, K. Muje, M. Sipponen & J. Karjalainen, 2007. Comparison of policies for spatial allocation of annual fishing effort between multiple stocks of vendace, *Coregonus albula* (L.). Archiv für Hydrobiologie, Advances in Limnology 60: 405–418.

Marttunen, M. & P. Kylmälä, 1997. Evaluation of the effects of fish stockings by a fish stock model in Lake Inarijärvi. Finnish Environment Institute, The Finnish Environment 117. 83 pp. (In Finnish with English abstract.).

Marttunen, M., S. Hellsten, B. Glover, A. Tarvainen, L. Klintwall, H. Olsson & T.-S. Pedersen, 2006. Heavily regulated lakes and the European Water Framework Directive – Comparisons from Finland, Norway, Sweden, Scotland and Austria. E-Water, Official Publication of the European Water Association (EWA).

Mastrantuono, L., A. G. Solimini, P. Nõges & M. Bazzanti, present issue. Plant-associated invertebrates and hydrological balance in the large volcanic Lake Bracciano (Central Italy) during two years with different water levels. Hydrobiologia. DOI: 10.1007/s10750-007-9196-9.

Matzinger, A., M. Jordanoski, E. Veljanoska-Sarafiloska, M. Sturm, B. Müller & A. Wüest, 2006. Is Lake Prespa jeopardizing the ecosystem of ancient Lake Ohrid? Hydrobiologia 553: 89–109.

Mikschi, E., G. Wolfram & A. Wais, 1996. Long-term changes in the fish community of Neusiedler See (Burgenland, Austria). In Kirchhofer, A. & D. Hefti (eds), Conservation of Endangered Freshwater Fish Species in Europe. Birkhäuser Verlag, Basel: 111–120.

Mooij, W. M., J. H. Janse, L. N. De Senerpont Domis, S. Hülsmann & B. W. Ibelings, 2007. Predicting the effect of climate change on temperate shallow lakes with the ecosystem model PCLake. Hydrobiologia 584: 443–454.

Moss, B., D. Stephen, C. Alvarez, E. Becares, W. Van de Bund, S. E. Collings, E. Van Donk, E. De Eyto, T. Feldmann, C. Fernández-Aláez, M. Fernández-Aláez,

R. J. M. Frankeng, F. García-Criado, E. Gross, M. Gyllström, L.-A. Hansson, K. Irvine, A. Järvalt, J.-P. Jenssen, E. Jeppesen, T. Kairesalo, R. Kornijów, T. Krause, H. Künnap, A. Laas, E. Lill, B. Lorens, H. Luup, M. R. Miracle, P. Nõges, T. Nõges, M. Nykänen, I. Ott, W. Peczula, E. T. H. M. Peeters, G. Phillips, S. Romo, V. Russell, J. Salujõe, M. Scheffer, K. Siewertsen, H. Smal, C. Tesch, H. Timm, L. Tuvikene, I. Tõnno, T. Virro & D. Wilson, 2003. The determination of ecological quality in shallow lakes – a tested system (ECOFRAME) for implementation of the European Water Framework Directive. Aquatic Conservation: Marine and Freshwater Ecosystems 13: 507–549.

Naumenko, M. A., present issue. Seasonality and trends in the Secchi disk transparency of Lake Ladoga. Hydrobiologia. DOI: 10.1007/s10750-007-9198-7.

Naumenko, M. A., V. V. Guzivaty & S. G. Karetnikov, 2006. Climatic trends of the water surface temperature in Lake Ladoga during ice-free periods. Doklady Earth Sciences 409: 750–753.

Nixon, S. C. (ed.), 1996. European Freshwater Monitoring Network Design. European Topic Centre on Inland Waters. EEA, Copenhagen.

Nõges, P. & T. Nõges, 2006. Indicators and criteria to assess ecological status of the large shallow temperate polymictic lakes Peipsi (Estonia/Russia) and Võrtsjärv (Estonia). Boreal Environment Research 11: 67–80.

Nõges, P. & A. de Jager, 2008. Clustering large lakes and reservoirs in Europe on the basis of their morphometry, catchment, and climatic conditions. Verhandlungen der Internationalen Vereinigung für Limnologie 30 (submitted).

Nõges, T., I. Tõnno, R. Laugaste, E. Loigu & B. Skakalski, 2004. The impact of changes in nutrient loading on cyanobacterial dominance in Lake Peipsi (Estonia/Russia). Archiv für Hydrobiologie 160: 261–279.

Nõges, T., R. Laugaste, E. Loigu, L. Nedogarko, B. Skakalski & P. Nõges, 2005. Is the destabilisation of Lake Peipsi ecosystem caused by increased phosphorus loading or decreased nitrogen loading? Water Science and Technology 51: 267–274.

Nõges, T., A. Järvet, A. Kisand, E. Loigu, B. Skakalski & P. Nõges, 2007. Reaction of large and shallow lakes Peipsi and Võrtsjärv to the changes of nutrient loading. Hydrobiologia 584: 253–264.

Nõges, T., Laugaste, R., Nõges, P. & I. Tõnno, present issue. Critical N:P ratio for cyanobacteria and N_2-fixing species in the large shallow temperate lakes Peipsi and Võrtsjärv, North-East Europe. Hydrobiologia. DOI: 10.1007/s10750-007-9195-x.

Olenin, S. & E. Leppäkoski, 1999. Non-native animals in the Baltic Sea: alteration of benthic habitats in coastal inlets and lagoons. Hydrobiologia 393: 233–243.

Ostendorp, W., M. Dienst & Schmieder, 2003. Disturbance and rehabilitation of lakeside *Phragmites* reeds following an extreme flood in Lake Constance (Germany). Hydrobiologia 506-509: 687–695.

Padisák, J., 2005. Phytoplankton. In O'Sullivan, P. E. & C. S. Reynolds (eds), The Lakes handbook, Vol. 1. Blackwell Publishing, UK: 251–308.

Pajunen, H., 2005. Lake sediments as a sink of organic carbon – long-term accumulation data from Finland. Verhandlungen der Internationalen Vereinigung für Limnologie 29: 337–339.

Pankova, E. S. & N. A. Berezina, 2007. Predation rate and size selectivity of the invasive amphipod *Gmelinoides fasciatus* preying upon the native isopod *Asellus aquaticus*. Acta Zoologica Lituanica 17: 144–150.

Paola, C., J. Mullin, C. Ellis, D. C. Mohrig, J. B. Swenson, G. Parker, T. Hickson, P. L. Heller, L. Pratson, J. Syvitski, B. Sheets & N. Strong, 2001. Experimental stratigraphy. GSA Today 11: 4–9.

Persson, L., P.-A. Amundsen, A. M. De Roos, A. Klemetsen, R. Knudsen & R. Primicerio, 2007. Culling prey promotes predator recovery – alternative states in a whole-lake experiment. Science 316: 1743–1746.

Phillips, G. L., P. E. O'Sullivan & C. S. Reynolds, 2005. Eutrophication of shallow temperate lakes. The lakes handbook. Volume 2: lake restoration and rehabilitation. Blackwell Publishing.

Pilke A., P. Heinonen, K. Karttunen, E. Koskenniemi, L. Lepisto, O-P. Pietilainen, J. Rissanen & H. Vuoristo, 2002. Finnish draft for typology of lakes and rivers. TemaNord 566: 42–43.

Pokorný, J. & J. Květ, 2004. Aquatic plants and lake ecosytems. In ÓSullivan P. E. & C. S. Reynolds (eds), The Lakes Handbook, Vol. 1. Blackwell Publishing, UK: 309–340.

Post, D. M., M. L. Pace & N. G. Hairston, 2000. Ecosystem size determines food-chain length in lakes. Nature 405: 1047–1049.

Premazzi, G., A. Dalmiglio, A. C. Cardoso & G. Chiaudani, 2003. Lake management in Italy: the implications of the Water Framework Directive. Lakes & Reservoirs: Research and Management 8: 41–59.

Räisänen, J., U. Hansson, A. Ullerstig, R. Döscher, L. P. Graham, C. Jones, H. E. M. Meier, P. Samuelsson & U. Willén, 2004. European climate in the late twenty-first century: regional simulations with two driving global models and two forcing scenarios. Climate Dynamics 22: 13–31.

Reche, I., E. Pulido-Villena, R. Morales-Baquero & E. O. Casamayor, 2005. Does the ecosystem size determine aquatic bacterial richness? Ecology 86: 1715–1722.

Reinart, A. & M. Reinhold, 2007. Mapping surface temperature in large lakes with MODIS data. Remote Sensing of Environment. doi:10.1016/j.rse.2007.05.015.

Ricciardi, A., 2003. Predicting the impacts of an introduced species from its invasion history: an empirical approach applied to zebra mussel invasions. Freshwater Biology 48: 972–981.

Salmaso, N., 2005. Effects of climatic fluctuations and vertical mixing on the interannual trophic variability of Lake Garda, Italy. Limnology and Oceanography 50: 553–565.

Sarvala, J., M. Rask & J. Karjalainen, 2004. Fish community ecology. In O'Sullivan, P. E. & C. S. Reynolds (eds), The Lakes Handbook, Vol. 1. Blackwell Publishing, UK: 538–582.

Schmieder, K., M. Dienst & W. Ostendorp, 2002. Auswirkungen des Extremhochwassers 1999 auf die Flächendynamic

and Besndstructur der Uferröhrichte des Bodensees. Limnologica 32: 131–146.
Sillen, B. & C. Solbreck, 1977. Effects of area and habitat diversity on bird species richness in lakes. Ornis Scandinavica 8: 185–192.
Simola, H. & L. Arvola 2004. Lakes of northern Europe. In Reynolds C. & P. O'Sullivan (eds), The Lakes Handbook, Vol. 2. Blackwell Publishing, UK: 117–158.
Slugina, Z. V., 2006. Endemic Bivalvia in ancient lakes. Hydrobiologia 568: 213–217.
Smith, V. H., S. B. Joye & R.W. Howarth, 2006. Eutrophication of freshwater and marine ecosystems. Limnology and Oceanography 51: 351–355.
Smol, J. P., 2004. Pollution of Lakes and Rivers: A Paleoenvironmental Perspective. Arnold Publishers, London; Oxford University Press, New York: 280 pp.
Søndergaard, M., J. P. Jensen & E. Jeppesen, 2005. Seasonal response of nutrients to reduced phosphorus loading in 12 Danish lakes. Freshwater Biology 50: 1605–1615.
St. Louis, V. L., C. A. Kelly, E. Duchemin, J. W. M. Rudd & D. M. Rosenberg, 2000. Reservoir surfaces as sources of greenhouse gases to the atmosphere: a global estimate. BioScience 50: 766–775.
Straile, D., D. M. Livingstone, G. A. Weyhenmeyer & D. G. George, 2003. The Response of freshwater ecosystems to climate variability associated with the North Atlantic Oscillation. In The North Atlantic Oscillation: Climatic Significance and Environmental Impact. Geophysical Monograph 134.
Straškraba, M., 2005. Reservoirs and other artificial water bodies. In O'Sullivan, P. E. & C. S. Reynolds (eds), The Lakes Handbook, Vol. 2. Blackwell Publishing, UK: 300–328.
Tilzer, M. M. & P. Bossard, 1992. Large lakes and their sustainable development. Aquatic Sciences 54: 91–103.
Timm, V. & T. Timm, 1993. The recent appearance of Baikalian crustacean, *Gmelinoides fasciatus* (Stebbing, 1899) (Amphipoda, Gammaridae) in Lake Peipsi. Proceedings of the Estonian Academy of Sciences. Biology. Ecology 42: 144–153.
Van den Brink, H. W., G. P. Können, J. D. Opsteegh, G. J. van Oldenborgh & G. Burgers, 2005. Estimating return periods of extreme events from ECMWF seasonal forecast ensembles. International Journal of Climatology 25: 1345–1354.
Várkuti, A., K. Kovács, C. Stenger-Kovács & J. Padisák, present issue. Environmental awareness of the permanent inhabitants of towns and villages on the shores of Lake Balaton with special reference to issues related to global climate change. Hydrobiologia. DOI: 10.1007/s10750-007-9190-2.
Vehanen, T., P. Hyvärinen & A. Huusko, 1998. Food consumption and prey orientation of piscivorous brown trout (*Salmo trutta*) and pikeperch (*Stizostedion lucioperca*) in a large regulated lake. Journal of Applied Ichthyology 14: 15–22.
Ventelä, A.-M., M. Tarvainen, H. Helminen & J. Sarvala, 2007. Long-term management of Pyhäjärvi (SW Finland): eutrophication, restoration – recovery? Lake and Reservoir Management 23 (in press).
Viljanen, M., V. Drabkova, V. Avinsky, L. Kapustina & G. Raspletina, 2007. Ecological state and monitoring of limnological and biological parameters in Lake Ladoga, Russia. Aquatic Ecosystem Health & Management 11 (in press).
Wade, A. (ed.), 2006. Monitoring and modelling the impacts of global change on European freshwater ecosystems. Science of Total Environment 365: 1–274.
Weyhenmeyer, G. A., present issue. Rates of change in physical and chemical lake variables – are they comparable between large and small lakes? Hydrobiologia (present issue).
Weyhenmeyer, G. A., E. Jeppesen, R. Adrian, L. Arvola, T. Blenckner, T. Jankowski, E. Jennings, P. Nõges, T. Nõges & D. Straile, 2007. Nitrate-depleted conditions on the increase in shallow northern European lakes. Limnology and Oceanography 52: 1346–1353.
Wick, L., G. Lemcke & M. Sturm, 2003. Evidence of Lateglacial and Holocene climatic change and human impact in eastern Anatolia: high-resolution pollen, charcoal, isotopic and geochemical records from the laminated sediments of Lake Van, Turkey. The Holocene 13: 665–675.
Wilander, A. & G. Persson, 2001. Recovery from eutrophication: experiences of reduced phosphorus input to the four largest lakes of Sweden. Ambio 30: 475–485.
Willén, E., 2001. Phytoplankton and water quality characterization: experiences from the Swedish large lakes Mälaren, Hjälmaren, Vättern and Vänern. Ambio 30: 529–537.
Willén, E., 2002. Phytoplankton in water quality assessment – an indicator concept. In Heinonen P., G. Ziglio & A. Van der Beken (eds), Hydrological and Limnological Aspects of Lake Monitoring. John Wiley & Sons Ltd, Chichester: 57–80.
Winfield, I. J. & N. C. Durie, 2004. Fish introductions and their management in the English Lake District. Fisheries Management and Ecology 11: 195–201.
Wolf, A., J. Natharius, J. Danielson, B. Ward & J. Pender, 1999. International river basins of the world. International Journal of Water Resources Development 15: 387–427.
Wüest, A & A. Lorke, 2003. Small- scale hydrodynamics in lakes. Annual Review of Fluid Mechanics 35: 373–412.
Zenkevich, L. A., 1963. Biology of seas of the USSR. Academy of Sciences of the USSR, Moscow, 739 pp. (In Russian.).